Chemistry of Atmospheres

Cover image A model simulation of ozone at 30 km altitude for 23 January 1987. A major stratospheric warming is taking place in the Northern hemisphere, and the polar vortex (with the associated low levels of ozone) has been displaced towards Europe, while tropical air (rich in ozone) is being brought across the North Pole from a region over the USSR, and moved towards Greenland and North America. (From D. J. Lary, *A photochemical general circulation model*, PhD Thesis, University of Cambridge, UK, 1990.)

Chemistry of Atmospheres

An Introduction to the Chemistry of
the Atmospheres of Earth, the Planets,
and their Satellites

SECOND EDITION

Richard P. Wayne

Dr Lee's Reader in Chemistry,
Christ Church, Oxford and
University Lecturer in Physical Chemistry,
University of Oxford

CLARENDON PRESS · OXFORD

Oxford University Press, Walton Street, Oxford OX2 6DP

Oxford New York Toronto
Delhi Bombay Calcutta Madras Karachi
Kuala Lumpur Singapore Hong Kong Tokyo
Nairobi Dar es Salaam Cape Town
Melbourne Auckland Madrid

and associated companies in
Berlin Ibadan

Oxford is a trade mark of Oxford University Press

Published in the United States by
Oxford University Press Inc., New York

First published 1985
Second edition 1991

British Library Cataloguing in Publication Data
Wayne, Richard P. (Richard Peer)
Chemistry of atmospheres.—2nd ed.
1. Solar system. Planets. Atmosphere
I. Title
551.5Q999
ISBN 0–19–855574–1
ISBN 0–19–855571–7 (pbk.)

Library of Congress Cataloging in Publication Data
Wayne, Richard P. (Richard Peer)
Chemistry of atmospheres: an introduction to the chemistry of the
atmospheres of earth, the planets, and their satellites / Richard P.
Wayne.—2nd ed.
p. cm.
Includes bibliographical references (p.) and index.
1. Atmospheric chemistry. 2. Planets—Atmospheres. I. Title.
QC879.6.W39 1991 551.5'11—dc20 90–7858
ISBN 0–19–855574–1
ISBN 0–19–855571–7 (pbk.)

Printed in Great Britain by Redwood Books, Trowbridge

Preface to the second edition

Research activity in atmospheric chemistry has continued to accelerate in the few years since the first edition of this book was published. An enormous quantity of detailed factual information has become available in areas where previously our knowledge was less well defined, and new measurements have given some of the quantitative information a more secure foundation. For a book like the present one, the greater detail and the modifications to numerical values would not, in themselves, justify the publication of a new edition. The purpose of the book is to lay down the principles of atmospheric chemistry and to provide the necessary background for study of the subject in greater depth: the older examples might be just as good as the newer in illustrating these principles. Some of the recent advances in our subject are, however, of the kind that affords new insights and interpretations, and it is to record and place in proper context these exciting, dramatic, and sometimes worrying, developments that this revised edition of the book has been prepared. In the paragraphs that follow, I have tried to highlight the most important of these developments, each of which is discussed in appropriate parts of the text.

Probably the most startling discovery about our own atmosphere was that of the 'Antarctic ozone hole' by scientists of the British Antarctic Survey. Each Antarctic spring, there now occurs substantial depletion of ozone in the stratosphere. Since we depend on the stratospheric ozone layer for protection from damaging ultraviolet solar radiation, there is obviously more than academic interest in the chemistry that leads to its thinning over the Antarctic. Although the science turns out to be fascinating, more important are the interpretations that have followed the experimental observations; the depletion could not have been predicted by our understanding of atmospheric chemistry before this discovery.

One part of the explanation for the behaviour of Antarctic ozone is associated with chemical changes occurring on the surface of the particles in polar stratospheric clouds. Chemical reactions on the surfaces of solid or liquid aerosols, or within the droplets of clouds, have long been known to play a role in a variety of atmospheric processes, although in many cases our understanding has lacked quantitative foundations. The unravelling of the mystery of the ozone hole has served to focus much attention on chemical transformations in and on particles. By chance, the period has overlapped with one in which there has been continued analysis of the atmospheric effects of large quantities of dust being injected into the atmosphere by a large volcanic eruption (El Chichón, in the spring of 1982). The heightened awareness of surface phenomena in the Earth's atmosphere has encouraged scientists to consider their influence in other atmospheres, especially those of notably dusty (Mars) or cloudy (Venus) planets. Photochemical hazes, attributable to formation of light-scattering particles, are familiar in our atmosphere, but they also seem to be present in the atmospheres of even the outermost planets of the solar system and their satellites.

There has been a growing emphasis on the way in which the composition of Earth's atmosphere is changing. The concentrations of trace gases are now known to be altering (generally increasing), not on geological time-scales, but at rates as large as one per cent per year in some cases. Noteworthy in this respect are the gases methane (CH_4), carbon monoxide (CO), and nitrous oxide (N_2O). The chemical effect of the changed burden of trace gases may extend to altering the oxidizing capacity of the atmosphere, and thus to reducing the ability of the atmosphere to attack chemically other substances released to it. The time-scales of the changes, and the period when they appear to have begun, point to man's activities as a source of the substances. Other evidence points in the same direction: for example, there is a marked geographical asymmetry in the increase of CO, which is confined largely to the highly populated and industrialized northern hemisphere. It is not only industry, however, that leads to the increases. Changes in land use and agricultural practices are also culprits, with micro-organisms acting as intermediaries in the release of many trace gases. Biomass burning has become firmly established over the last few years as another significant contributor to the total input of many trace gases.

Some compounds released to the atmosphere have no natural source, and are entirely a consequence of manufacture by man. The chlorofluorocarbons (CFCs) are typical of these species, and they have important impacts on the atmosphere. They are, for example, active in destroying stratospheric ozone; it is now fairly certain that there has been a small trend towards lower ozone concentrations, and the depletions are likely to be associated with the release of CFCs. So far as the Antarctic ozone anomaly is concerned, the causal relation between the increased atmospheric load of CFCs and the deepening and widening of the ozone hole seems definitely established.

The CFCs are also what are known as 'greenhouse gases'. That is, they trap solar infrared radiation in the lower atmosphere and so increase temperatures. Many of the other trace gases of the atmosphere whose concentrations are increasing are also greenhouse gases. Most important of all in this context is carbon dioxide; the concentration of this gas in the atmosphere has increased by about 25 per cent since the industrial revolution, mainly because of the burning of fossil fuels. Carbon dioxide and the other trace gases act synergistically in trapping infrared radiation, and there is the very real risk that the global climate may be altered significantly, with adverse socio-economic consequences. Argument continues about whether atmospheric temperatures yet show any response to the increased load of greenhouse gases, with the balance of opinion perhaps saying that they do not.

Atmospheres of bodies in the solar system other than our own planet have received their share of attention over the last five years. Voyager 2 completed its proposed odyssey through the solar system by visiting Uranus in early 1986, and then Neptune in August 1989, thus bringing resounding success to a mission that had started more than twelve years earlier. Only part of the research was concerned with atmospheric science, of course, but our understanding of the atmospheres of the outermost planets and their satellites, especially Triton, has improved immeasurably as a result of the encounters. The year 1986 also saw the 'Halley Armada' approach and investigate the best known of the comets. Some of the experiments were designed to study the gases present in the coma and the tail of the comet, and the results provide new insight into the composition and chemistry of primitive bodies in the solar system.

Alongside the new discoveries and interpretations in atmospheric science, the last five years have seen a heightened public awareness of environmental problems in which atmospheric chemistry plays a large part. Greenhouse warming and climate change, deforestation and 'slash-and-burn' agriculture, stratospheric ozone depletion, acid rain, and photochemical smog are becoming familiar terms, even if the non-scientific public feels somewhat hazy about the details. As I said in the preface to the first edition of this book, it is necessary for scientists to be well informed about these environmental issues so that they can judge both the reality of the threats and the viability of hypothetical control strategies.

In some cases, control measures have already been adopted; use of catalytic converters on automobiles has certainly had a marked effect on photochemical smog in California. But scientific advances have a way of outpacing the legislation. For example, the Montreal protocol that seeks to limit the release of CFCs to the atmosphere was based on old evidence for ozone depletion, and was concluded just as the most dramatic ozone depletions ever were being observed in the Antarctic. It seems now that a much more stringent reduction in CFC release may be needed to protect the ozone layer. Perhaps even more important than the question of ozone depletion is that of climatic change. Concentrations of trace gases, including carbon dioxide, have indubitably increased, but current evidence for a related increase in global temperatures is weak. Yet it is now that action needs to be taken if the untoward possible consequences of climate change are to be alleviated or averted. This book does not attempt to suggest answers to the problems that we now face, but it does try to lay the foundations for the study of that atmospheric chemistry on which rational decisions will have to be based.

It is once again a real pleasure to express my thanks to all those who have helped with the preparation of this book. My wife, Brenda, has continued to support me unwaveringly and to encourage me at all times. Many scientific colleagues have helped in one way or another. Some whose help was recorded in the preface to the first edition have also helped with this one. I particularly value the advice given by Karl Becker, Tony Cox, Dieter Kley, Joel Levine, Susan Solomon, Richard Stolarski, Georg Witt, and Don Wuebbles, but I hope that those not mentioned by name here will also accept my gratitude. I am also most appreciative of the support and interest of my research group in Oxford, especially for the suggestions and constructive criticism of Anne Brown and Carlos Canosa-Mas.

Oxford
April 1990   R.P.W.

Preface to the first edition

Our knowledge of the atmospheres of Earth and the planets has improved so dramatically over the last two decades that what was, in the early sixties, a rather esoteric branch of learning, has now become a 'hard' science that is a major area of interest to physicists and chemists. Two factors have brought the new-found information. Exploration of atmospheres, first our own and then those of other bodies, has been made possible by the imaginative rocket and satellite programmes of the last decades. At the same time, man has become increasingly aware that the results of many of his activities—ranging from driving motor cars to fertilizing the soil—could have damaging effects on the chemical balance of the atmosphere, and perhaps even lead to an irreversible global climatic change. Considerable effort and ingenuity have gone into separating reality or probability from speculation. The labours of laboratory kineticists and photochemists, physicists, meteorologists, mathematicians, geologists, and biologists have gone into this effort. Many uncertainties remain, and more problems have been unearthed, but we now possess a body of fact that permits of rational discussion. It seems to me that an educated scientist of the present day must have an understanding of how physics and chemistry control atmospheric behaviour. The scientist must be able to judge how a new piece of information on atmospheric behaviour fits into the procession of ideas and discoveries in the world of science. Certain of the interpretations that are now being offered are sufficiently creative and innovative to be considered as real advances in understanding our relation to the rest of the Universe. At the same time, the scientist must be concerned with the less elevated matter of man's comfort and even survival. He must certainly use his knowledge and judgement to assess possible threats to the environment. Some threats are no more than threats: others may be real, but the cure may have socio-economic consequences worse than the environmental damage; others yet again may be so serious that they must be averted at all costs. Who is to judge? Better the well-informed scientist than someone who comes to the problem with some prejudice—for *or* against action—but little knowledge. I have written this book to try to provide the connecting links between atmospheric chemistry and the traditional formal education of physics, chemistry, or biology, in the hope that the reader shall be better able to place in context new advances and new problems in atmospheric behaviour.

Meteorology has a long history; its study has been of practical importance for centuries because of the need to forecast the weather. The subject is now a fully-fledged science, and weather prediction remarkably accurate. Bit by bit the descriptive and synoptic meteorologist has become less distinguishable from his colleague, the atmospheric physicist. Most University courses in physics now include atmospheric physics as at least an option. My approach is that of a physical chemist: the idea of teaching atmospheric chemistry to chemistry undergraduates is new because our knowledge of the subject has been so recently gained (at least on a University time-scale!). For five years now, I have given a short course to third-year Oxford undergraduates on the 'Chemistry of Planetary Atmospheres'. Indeed, the content of that course, considerably expanded and rearranged, forms the basis of this book. From the point of view of a chemist, the subject actually offers an excellent vehicle for the teaching of physico-chemical concepts, quite apart from the specific interest attaching to atmospheric studies. Planetary atmospheres are giant photochemical reactors, and the atoms and 'small' molecules involved fit in well with precise descriptions of spectroscopy, reaction kinetics, photochemistry, excited states, and thermodynamics. Many of the ideas of atmospheric physics should be familiar to chemists, and teaching them can serve a useful educational function by showing a practical application of fundamental physical chemistry. What is frequently not familiar is the manner of presentation. Physicists and chemists often treat the same material in different ways, and the chemist can find unnecessary difficulties in physics, and vice versa. I have tried to write this book to be intelligible to readers approaching atmospheric chemistry from any scientific discipline.

Atmospheres in the solar system show great variety, ranging from the hydrogen and helium of the giant planets, Jupiter and Saturn, to the carbon dioxide of Venus and Mars. Along the way are the tenuous sulphur dioxide and atomic sodium atmosphere of Jupiter's satellite Io, and the dense nitrogen, argon, and methane atmosphere of Saturn's satellite Titan. Earth's atmosphere stands unique in the solar system, being oxidizing, but not completely oxidized. Life is responsible for our unusual atmosphere, which in turn is essential for the maintenance of life. Chapter 1 of this book discusses the variety of atmospheres, and the special nature of Earth's gaseous mantle. The relationship between the biosphere and atmosphere are explored, and consideration is given to the biological control of composition. Topics in atmospheric physics and meteorology that are essential to a quantitative understanding of atmospheric chemistry are presented in Chapter 2. They include the interpretation of pressure and temperature structure in the atmosphere, circulation patterns, and the formation and optical properties of clouds and aerosols. This chapter explains the conventional division of atmospheres into regions such as troposphere, stratosphere, mesosphere, thermosphere, and exosphere. Chapter 3 similarly describes topics in chemistry that are of particular importance in atmospheric processes. We deal in this chapter with the elements of photochemistry and reaction kinetics that will be used repeatedly in later parts of the book. The chapter ends with an explanation of how chemical rate processes and physical transport are incorporated into mathematical models of atmospheres that are used for diagnostic and prognostic applications.

Chapters 4 to 7 are concerned with different aspects of the chemistry of Earth's atmosphere. By and large, the energetics of chemical processes that can occur decrease with decreasing altitude, because the ultraviolet radiation that initiates photochemical change is filtered out by the atmosphere. Shorter wavelengths (higher photon energies) penetrate less deeply, so that their chemical effects are confined to high altitudes. The photons reaching the stratosphere (about 15–55 km) are energetic enough to split molecular oxygen to atoms; the atoms can then add to molecules to make ozone. This ozone is concentrated in a layer, and because it removes ultraviolet that would be hazardous for living organisms, it is an exceedingly important part of our atmosphere. Chapter 4 examines the details of stratospheric chemistry, and discusses how some of man's activities might compromise the ozone layer. Most of the mass of the atmosphere is lower down, in the troposphere (Chapter 5). In this region, the chemistry is dominated by radical reactions initiated by the hydroxyl radical. Trace gases released from the surface, mainly as a result of biological activity, undergo oxidation and conversion. Man supplements the natural emissions of hydrocarbons and the oxides of carbon, nitrogen, and sulphur, most particularly by the combustion of fossil fuels. Because of the intensity of release compared with the natural background, the delicate chemical balance of the troposphere can be upset, and 'air pollution' results. The causes of, and remedies for, common forms of air pollution are included in the discussion of Chapter 5. Ions exist throughout the atmosphere, but their concentration is small below the mesosphere. Above about 70 km altitude, ion and electron concentrations become high enough to exert an effect on radio-waves that has been recognized for sixty years or more. The region where ions are abundant is called the ionosphere, and its chemistry is presented in Chapter 6. Electronically and vibrationally excited species can emit radiation, and so contribute to the 'air-glow' that exists by day and by night. Chapter 7 reviews some of the most significant emission features of the airglow of Earth, Venus, and Mars, and examines the mechanisms by which the radiating species become excited.

Planetary exploration has led to a blossoming of interest in and knowledge about atmospheres other than our own. Chapter 8 describes the picture as it appears at the time of writing, although progress is so rapid that changes in detail are inevitable. The emphasis throughout the chapter is on explaining why a particular type of chemistry operates in a certain atmosphere. Why an atmosphere should have a particular composition, and where that atmosphere came from, are the concern of the first parts of Chapter 9, which looks at the evolution of atmospheres. The present-day composition may even hold clues about the origin of the planet or satellite to which the atmosphere belongs. Geological and geochemical evidence about the past of our own planet is sufficiently rich to enable us to construct detailed 'scenarios' for the evolution of the atmosphere, climate, and even life itself. Projections into the future are also possible. One generally accepted view, described towards the end of Chapter 9, is that man may be warming his planet up by converting back to carbon dioxide in a few centuries the fossil fuel deposits that photosynthesis created from carbon dioxide over hundreds of millions of years.

In deciding on the order of presentation just described, I tried to make the progression of topics orderly and logical, and to avoid, as far as possible, the need to 'look ahead' to later portions of the book from earlier ones. Sometimes I have chosen to delay part of a discussion, as for example the question of increasing carbon dioxide concentrations and concomitant climatic change (Chapter 9), which is an extension of the air pollution topics of Chapter 5. I have been quite liberal with cross-references—both back and forward—in order to keep the reader aware of where he has come across a subject before, or will see it again. Generally, these cross-references are to sections of a chapter, but occasionally a specific page is more appropriate. With regard to literature citations, I have been guided by my undergraduate and postgraduate students. They were unanimous in their condemnation of any form of referencing within the text, on the grounds that numbers, names, or whatever, break the flow and distract the reader's attention. At first, I was slightly uneasy about not providing chapter and verse for my statements, especially where I have chosen to present one of several disputed views. On the other hand, the book has been written as an introductory textbook rather than as a research monograph. Accordingly, there are only very few references in the text, and these are mostly to historically important work. Instead, I have provided a fairly generous set of references at the end of each chapter. I have arranged the citations more or less in the order in which topics to which they refer appear in the chapter. The references are collected in small groups preceded by a short explanation of the content of the papers. It is thus easy for the reader to find the justification for every point of substance if he has a mind to, but not if he would prefer to read on uninterrupted. Referencing in this way has the advantage that I can introduce mention of additional important topics that do not fit in to the development of the main text. Of course, I have provided attributions wherever appropriate for the tables and figures. Most references are to easily-found books or journals, although some sources (e.g. *The Stratosphere 1981*, published by the World Meteorological Organization, and some of the NASA reports) are so important that they have been included, even though they may need to be hunted down in the library. The papers and books cited themselves carry extensive literature references, so that a student can easily build up a list of original works if he has a mind to explore a subject in greater depth than is provided in this book. Where possible, I have included references to reviews or commentary-style papers.

Atmospheric chemistry is a lively and active field in which development is rapid and in which quantitative data are being modified continuously. This book presents a snapshot picture of the discipline as it appeared in 1984. Most of the underlying concepts are well established, but I am well aware of the dangers inherent in quoting 'up-to-date' numbers to illustrate the various topics treated, since refinements and new discoveries will inevitably follow soon. Speculative ideas abound in atmospheric chemistry, and I have, as far as possible, distinguished reasonably well-established fact from speculation.

A brief word about units and nomenclature may be in order. I have tried, where it is sensible, to use SI units. However, since one of the aims of this book is to enable the reader to appreciate published works in the atmospheric sciences, some exceptions are inevitable. For example, the bar (10^5 Nm^{-2}) and mbar are used almost universally as units of pressure in meteorological and atmospheric studies. Similarly, systematic chemical names are not commonly used in the literature, and I should be failing my reader if I insisted on calling, for instance, HCHO 'methanal' rather than 'formaldehyde'.

Now I come to the very pleasant task of expressing my gratitude to the many people who have helped me to make this book a reality. First, and most important, I thank my wife, Brenda, who not only provided the support and encouragement I needed to sustain me while writing, but also gave me very real practical help with her editorial skills. Then I thank my children, Carol and Andrew, who tolerated a father sometimes a little absent in spirit even if only too present in the flesh; they took a lively and informed interest in what I was writing, and made the task more enjoyable. Next, I wish to thank my research group: Tim Wallington, Martin Fowles, Peter Biggs, and Mary Davies. They accepted that, while preparing the book, their research supervisor would be more distracted than usual from their work. They also provided valuable ideas about the content and presentation, and were sympathetic listeners when I conceived an enthusiasm for, or, indeed, a passionate dislike of, some topic I was pursuing. I am grateful to the many colleagues and friends who gave me advice, information, and their time. Perhaps I might mention a few by name: Rudi Burke, Tony Cox, Bill DeMore, Hugh Ellsaesser, Garry Hunt, Don Hunten, Joel Levine, Mike McElroy, Darrel Strobel, and Georg Witt. I hope the more numerous body of un-named friends will not take this selection amiss, and will know how much I appreciated their efforts. Finally, I should like to tell my typist, Mrs Margot Long, how pleased I am that she accepted the tiresome job of typing my words. As always, she was efficient and cheerful even when faced with a much-amended manuscript that might reduce others to tears.

Oxford　　　　　　　　　　　　　　　　　　　　　　　　　　　　R.P.W.
July 1984

Contents

1 Chemical composition: a preliminary survey

1.1 Earth's atmosphere in perspective

Planet Earth possesses an atmosphere that, for hundreds of millions of years, does not seem to have been obeying the laws of physics and chemistry! Since it is the purpose of this book to interpret the composition and properties of atmospheres in terms of physico-chemical principles, an explanation of Earth's peculiar behaviour must be sought at the outset. Apparently alone amongst the planets of the solar system, Earth supports life. The composition of our atmosphere is determined by biological processes acting in concert with physical and chemical change. At the same time, our unique atmosphere seems essential for the support of life in the forms we know it. In later parts of the book we shall speculate about the origins and evolution of atmospheres, especially our own. We shall, however, start our discussion by considering the atmospheres as we find them today.

Atmospheres (Greek 'atmos' = 'vapour' + 'sphaira' = 'ball'), the gaseous envelopes around heavenly bodies, show an amazing diversity in the solar system. Venus, Earth, Mars, Jupiter, Saturn, Uranus, and Neptune all have substantial atmospheres, as does Saturn's largest satellite, Titan. Table 1.1 shows the relative abundances of various species in these atmospheres, together with information for the Sun. Data have, for the most part, been obtained directly in recent planetary missions (e.g. Viking, Pioneer Venus, Venera 11, 12, and Voyager 1, 2), although in some cases the suggested compositions depend on rational interpretation of observations made from afar. The concentrations are given as *mixing ratios* by volume; the unit is the one commonly used by atmospheric scientists, and is identical to the chemists' *mole fraction*. In the table, the mixing ratios are given as fractions, but, for minor constituents, parts per million (p.p.m., or p.p.m.v.) are frequently used. The mixing ratios quoted are really averages for the lower atmosphere of each planet; as we shall see, the relative importance of different species can vary with altitude. However, since pressure and density decrease more or less exponentially with increasing altitude (Chapter 2), most of the atmosphere is near the surface of a planet, and the lower atmosphere composition is almost the same as the average throughout the atmosphere. For the Earth, half the mass of the atmosphere lies below about 5.5 km altitude, and 99 per cent below 30 km. The total mass of our atmosphere, by the way, is about 5×10^{18} kg. A more formal division of the regions of the atmosphere, based on temperature structure, is explained in Chapter 2. It is convenient to introduce here the

Table 1.1 Solar system bodies with substantial atmospheres.

Body	Surface temperature in K	Surface pressure Earth atm	H_2	He	H_2O	CH_4	NH_3	Ne	H_2S	CO_2	N_2	O_2	CO	SO_2	Ar	N_2O
Sun	–	–	0.89	0.11	1.0(−3)	6.0(−4)	1.5(−4)	1.4(−4)	2.5(−5)	–	–	–	–	–	–	–
Venus	732	90	1(−5)[a]	2(−5)	2(−5)	6(−7)[a]	–	1.5(−5)	2(−6)[a]	0.965	0.035	2(−5)[a]	3(−5)	1.5(−4)	7(−5)	–
Earth	288	1	5.3(−7)	5.2(−6)	0 to 0.04	1.7(−6)	<1(−8)	1.8(−5)	1(−10)	3.35(−4)	0.781	0.209	4 to 20 (−8)	1.1(−10)	9.3(−3)	3.0 (−7)
Mars	223	0.006		–	3(−4)	–	–	2.8(−6)	–	0.953	0.027	1.3(−3)	7(−4)	–	1.6(−2)	–
Jupiter	170[b,c]	–	0.90	0.10	5(−6)	2.4(−3)	2(−4)	–	?	–	–	–	2(−9)	–	–	–
Saturn	130[b,c]	–	0.96	0.04	5(−6)	2.0(−3)	2(−4)	–	<4(−7)	–	–	–	–	–	–	–
Uranus	59.4[b]	–	0.85	0.15	–	<1(−7)	–	–	–	–	–	–	–	–	–	–
Neptune	59.3[b]	–	0.85	0.15	–	3(−5)	–	–	–	–	–	–	–	–	–	–
Titan	95	1.6	2(−3)	–	–	3(−2)	–	–	–	–	0.82	–	–	–	0.12	–

Notes: Numbers in parentheses are exponents: e.g. 1(−5) represents 1 × 10⁻⁵

[a] Disputed identification.

[b] Values given for altitude where pressure is that at Earth's surface.

[c] No true surface

Data from: Encrenaz, T. and Combes, M. *Icarus* **52**, 54 (1982); Holland, H. D. *The chemistry of the atmosphere and oceans*, John Wiley, Chichester, 1978: Hudson, R. (ed.-in-chief) *The Stratosphere 1981*, World Meteorological Organization, Geneva, 1981; Hunt, G. E. *Annu. Rev. Earth & Planet. Sci.* **11**, 415 (1983); Moroz, V. I. *Space Sci. Rev.* **29**, 3 (1981); Owen, T. *Planet. Space Sci.* **30**, 833 (1982); Owen, T., Biemann, K., Rushneck, D. R., Biller, J. E., Howarth, D. W., and Lafleur, A. L. *J. geophys. Res.* **82**, 4635 (1977); Pollack, J. B. Chapter 6 in *The new solar system* (eds. Beatty, J. K., O'Leary, B., and Chaikin, A.), 2nd edn. Cambridge University Press, 1981; Strobel, D. F. *Int. Rev. phys. Chem.* **3**, 145 (1983); Trafton, L. *Rev. Geophys. & Space Phys.* **19**, 43 (1981); Stone, E. C. and Miner, E. D., *Science* **233**, 39 (1986; **246**, 1417 (1989).

names of the two lowest regions, the *troposphere* and *stratosphere*, so that we can distinguish between these parts of the atmosphere. The terms reflect an important aspect of physical behaviour since 'tropos' is Greek for 'turning' and 'stratus' is Latin for 'layered'. Rather strong vertical mixing characterizes the troposphere; individual molecules can traverse the entire depth in periods between a few days in clear air and a few minutes in the updraughts of large thunderstorms. Conversely, the stratosphere is characterized by very small vertical mixing, the time-scale for transport being of the order of years. The troposphere extends through roughly the first 10–17 km of the atmosphere, and the stratosphere through the next 30–40 km. However, the altitude of the *tropopause* (the hypothetical boundary between troposphere and stratosphere) is rather variable, and depends on season and latitude. Abrupt changes in the concentrations or mixing ratios of some of the minor constituents of the atmosphere occur at the tropopause, and it is important, therefore, to specify the atmospheric region when discussing the chemistry. Although the tropospheric composition is dominant in determining total mixing ratios, chemical (and physical) processes occurring at higher altitudes can affect the atmosphere as a whole.

So far as Jupiter and Saturn are concerned, there is probably no surface marking a transition from gaseous to solid material. Instead the pressure increases with depth, and the gases of the atmosphere become more dense (perhaps liquid or even metallic). We have given quantities in Table 1.1 for these planets appropriate to that altitude in the atmosphere where the pressure is the same as that at Earth's surface. The planets missing from our table are Mercury and Pluto. Mercury has almost no atmosphere (about 10^{15} times less dense than Earth's; what there is consists of 98 per cent He and 2 per cent H). Pluto is surrounded by methane at a pressure of around 10^{-4} of an Earth atmosphere, together with some other 'heavy' gas such as neon. The volcanism of Jupiter's satellite Io indicates that gases are being released, and there is a tenuous atmosphere containing, amongst various components, neutral and ionized atomic sodium. Some other satellites of the outer planets may possess similarly tenuous atmospheres, perhaps not firmly attached to the satellite, but strung out as a torus around the parent planet. Our main concern, however, is with the massive atmospheres listed in the table.

A cursory inspection of Table 1.1 shows that the outer planets (Jupiter, Saturn, Uranus, Neptune) have atmospheres very different from those of the inner planets (Venus, Earth, Mars). Hydrogen and helium are present on the outer planets in abundances not too far removed from those on the Sun, and the atmospheres are certainly reducing. In contrast, the atmospheres of the inner planets are either oxidized or oxidizing. At this point, however, the unexpected nature of the Earth's atmosphere becomes apparent. Since Earth lies in the solar system between Venus and Mars, Earth's atmosphere might have been expected to consist primarily of the *oxidized* compound, carbon dioxide. But CO_2 is only a minor (although very important) constituent of our

Table 1.2   Actual and equilibrium concentrations of constituents of the Earth's atmosphere.

Species	Expected fractional equilibrium concentration[a]	Present fractional concentration	Output[b] kg yr^{-1}	Residence time
N_2	$<10^{-10}$	0.78	2.5×10^{11}	1.6×10^7 yr
CH_4	$<10^{-35}$	1.7×10^{-6}	7.5×10^{11}	6 yr
N_2O	$<10^{-20}$	3×10^{-7}	1×10^{11}	20 yr
NH_3	$<10^{-35}$	10^{-9}	1×10^{11}	10 days
H_2	$<10^{-35}$	5.3×10^{-7}	5×10^{10}	4 yr

Notes:

[a] Equilibrium concentrations are from Margulis and Lovelock (*Life in the universe*, ed. Billingham, J. 1981). The values are based on current oxygen concentrations and production rates.

[b] Annual outputs are estimates based on the reasoning developed towards the end of this chapter. The values depend heavily on the assumptions made in their derivation.

Representative concentrations and outputs given here are taken from: Hudson, R. ed.-in-chief) *The stratosphere 1981*, World Meteorological Organization, Geneva, 1981; Levine, J. S. and Allario, F., *Environ. Monitg & Assessm.* **1**, 263 (1982); Bolin, B. and Cook, R. B. (eds) *The major biogeochemical cycles and their interactions*, SCOPE **21**. John Wiley, Chichester, 1983.

atmosphere. It is the presence of elemental oxygen as a major constituent that poses the most serious problems. There is too much oxygen in the presence of too many gases that react with oxygen, and our atmosphere appears to be a combustible mixture. Oxygen reacts with hydrogen to form water, with nitrogen to form nitrate, with methane to form carbon dioxide and water, and so on.

The extent to which the atmosphere departs from equilibrium is shown in Table 1.2, in which expected equilibrium concentrations, calculated for the actual oxygen content of the atmosphere, are compared with the measured values. Rates of processes are sometimes slow enough to prevent achievement of thermodynamic stability, as in the conversion of diamond to graphite. Such arguments cannot be advanced to explain the discrepancies in atmospheric composition. By identifying sources of the various chemical species, as discussed in the following sections of the chapter, the annual input to the atmosphere can be estimated. Since concentrations are roughly constant over a period of a few years, there must be corresponding loss processes, and there is a *residence time* (or *lifetime*) for the species, calculable in the simplest way as the atmospheric concentration divided by the rate of supply. Some rough values of residence times are given in Table 1.2, and show clearly that there is no kinetic limitation on the establishment of chemical equilibrium. Only for the longest-lived species, nitrogen, is the lifetime comparable with geological time-scales of typically 10^7 to 10^9 years. Residence times for the other gases are much shorter, and in some cases are measured in days. The Earth's atmosphere thus maintains a steady-state disequilibrium composition with the fuel being continuously replenished.

Biological processes are dominant in the production of the oxidizable components of our atmosphere. That is, they bring about the thermodynamic disequilibrium, and effectively reduce the entropy of the atmosphere. Energy is needed to drive the entropy reduction, and it is supplied virtually entirely by the radiation from the Sun that reaches the Earth and its atmosphere. Not only does biology provide the fuel, but also the oxidant, oxygen. In the absence of life, the photochemical sources of oxygen limit the surface concentration to a value perhaps 10^{13} times smaller than the amount that the atmosphere actually contains. Consideration of the biological release rate of oxygen to the atmosphere suggests that present day concentrations are maintained with a residence time between 5000 and 10 000 years.

1.2 Land, sea, and air

Table 1.1 shows that, of all the planets, Earth has the largest relative abundance of water vapour. Although a typical concentration is 1 per cent, common experience tells us that water vapour is one of the most variable components of the atmosphere. In the tropics, water may account for up to 4 per cent (by volume) of the atmosphere, while in polar regions or in desert air the abundance may be a fraction of one per cent. Water is of peculiar interest with regard to the Earth, not only because of its importance to life, but also because it is the only atmospheric component that can exist in all three phases—solid, liquid, gas—at the temperatures found near the Earth's surface. Most of Earth's water is liquid; the term *hydrosphere* ('hydro' is Greek for water) is used to describe the condensed water environment. Only about 0.001 per cent of the total water budget is in the atmosphere at any one time, the rest being largely divided between oceans (97 per cent) and fresh water lakes or rivers (0.6 per cent). Snow and polar ice sheets account for 2.4 per cent of the water. The condensed water thus behaves as an enormous reservoir for atmospheric water vapour. Water fulfils two important non-biological functions. First, phase transformations in water are accompanied by large latent heat changes. Evaporation, subsequent transport by wind, and recondensation of water thus provide for heat transport in the atmosphere; the released heat in turn may ultimately drive the wind system. Currents in the oceans themselves provide an additional means of heat redistribution. Secondly, the oceans can hold gases in solution and provide a supply of atmospheric gases and act as a buffer against atmospheric change. Carbon dioxide is the prime example of a gas buffered by the oceans, because it undergoes reversible chemical change to bicarbonate ions on solution

$$CO_2 + H_2O \rightleftharpoons HCO_3^- + H^+, \tag{1.1}$$

and is therefore abnormally soluble. Indeed, the oceans contain around fifty times more CO_2 than the atmosphere.

Comparison of the behaviour of water vapour on Earth with that on the neighbouring planets, Mars and Venus, proves interesting. Mars is further from the Sun than is Earth, and its surface temperature is lower (typical values for Mars and Earth are 223 K and 288 K). Liquid water does not, therefore, exist on Mars, although various channel and valley forms observed on the planet's surface suggest it may have done in the past. Rather, a water-ice deposit in the north polar region of Mars is the chief water reservoir. Indeed, it is so cold in the winter polar regions of Mars that seasonal polar caps of solid carbon dioxide form there. Since CO_2 is the main constituent of the Martian atmosphere, condensation or evaporation of the solid reservoir can lead to changes in total pressure, at least locally. Venus, by contrast with Mars, has a much higher surface temperature (~ 732 K) than Earth. There are thus no oceanic reservoirs for water on Venus, and the low water-vapour content of the atmosphere (Table 1.1) represents a low planetary abundance of water. How the Venusian atmosphere came to be so hot and dry we shall explore further in Chapter 2.

The land mass—the *lithosphere* ('lithos' is Greek for 'stone')—itself exchanges gases with the atmosphere either directly or indirectly *via* the hydrosphere. Let us use carbon dioxide as our example again. On Earth, carbonate rocks such as limestone ($CaCO_3$) account for 100 000 times as much CO_2 as there is in the atmosphere. Thus the lithosphere is a potentially greater reservoir of CO_2 than are the oceans, although the rates of release may be much smaller. One indirect release process on Earth involves weathering of limestone by CO_2 dissolved in water; the process returns bicarbonate ions to the oceans:

$$CaCO_3 + H_2O + CO_2 \rightleftharpoons Ca^{++} + 2HCO_3^-. \tag{1.2}$$

A most important source of the carbonate minerals is secretion by animals and plants, and the deposition of calcite ($CaCO_3$) shells. Inorganic weathering reactions can also convert silicate rocks such as diopside ($CaMgSi_2O_6$) to carbonate:

$$CaMgSi_2O_6 + CO_2 \rightleftharpoons MgSiO_3 + CaCO_3 + SiO_2. \tag{1.3}$$

We note that on Venus the biological source of carbonate rocks is missing, that the high surface temperature favours the reactant side of equilibrium (1.3), and that there are no oceans to store CO_2. Since the total mass of CO_2 in the Venusian atmosphere ($\sim 3 \times 10^{20}$ kg) is very close to that stored on Earth as carbonate rocks, it is certainly very tempting to speculate that the total inventory of CO_2 is the same for the two planets.

1.3 Particles, aerosols, and clouds

Suspensions of particles in a gas are called *aerosols*. In principle, if the particles are liquid we refer to the aerosol as a *cloud* or *mist*, while if they are solid, the aerosol is *smoke* or *dust*. Usage of the words is, however, rather loose. More important, from our point of view, is what we mean by 'suspension' and 'particles'. Particles have a tendency to fall under the influence of gravity; this sedimentation is opposed by the kinetic energy of the surrounding molecules. For very small particles, the rate of sedimentation is so slow that macro-scale fluid motions completely dominate, and keep the particles thoroughly mixed (cf. pp. 38–39). Larger droplets reach a terminal sedimentation velocity determined by viscous drag. Some idea of the numbers of aggregated molecules, particle radii, and terminal velocities for water molecules is given in Table 1.3. Even a typical cloud drop of 10 μm diameter would take a day to fall through a cloud 1 km thick, and is, for practical purposes, permanently suspended. On the other hand, a typical raindrop falls through the cloud in under three minutes, and is not in suspension. What is it that distinguishes a 'particle' from a gas phase molecule or aggregate of molecules? The distinction is conventionally made on the basis of the light scattering properties of a body (Section 2.6). If a particle has a size comparable with, or greater than, the wavelength of light, then scattering can occur from different parts of the same particle. The consequent interference effects lead to an angular distribution of scattered intensity markedly different from that obtained with scattering from a point. Changes in scattering behaviour can be detected for particle sizes about one order of magnitude smaller than the wavelength of light, so that, for visible radiation ($\lambda \sim 500$ nm $\equiv 0.5$ μm) the corresponding detectable particle size is ~ 0.05 μm. We must emphasize that 'particle-like' properties in respect to scattering obviously depend on the wavelength of radiation used. At the lower size end of the scale, therefore, there is an imperceptible merging between what is an aerosol and what is a large molecule or cluster of molecules.

Table 1.3 Size and terminal velocities of water droplets in the atmosphere.

Radius μm	Name of drop	Approximate number of H_2O molecules per drop	Terminal fall velocity (m s^{-1})
0.1	Condensation nucleus	10^8	10^{-6}
1	Cloud	10^{11}	10^{-4}
10	Cloud	10^{14}	10^{-2}
50	Large Cloud	10^{16}	0.27
100	Drizzle	10^{17}	0.70
1000	Rain	10^{20}	6.50

Data from: Wallace, J. M. and Hobbs, P. V. *Atmospheric science*. Academic Press, New York, 1977.

Aerosols are found in most of the atmospheres we shall discuss, and they play an important part in the physics and chemistry. Water clouds cover about 50 per cent of the Earth's surface at any one time, while the surface of Venus is almost completely obscured by several cloud layers that are apparently composed of sulphuric acid droplets. Cloud cover on Mars is much more patchy, although water-ice clouds are formed in the winter polar regions. Thin carbon dioxide clouds are found at higher altitudes in many places, and thicker CO_2 clouds form lower down near the winter poles. Dust storms on Mars can grow to global proportions, and particles of a few μm in size are placed in suspension in the atmosphere. Compounds such as CH_4, NH_3, NH_4SH, and H_2O in the atmospheres of the outer planets condense out in cloud layers where the temperature and pressure are appropriate. Colours seen in the outer cloud layers of these planets are likely to be associated with solid aerosols of sulphur, phosphorus, or organic compounds. Titan has methane clouds, but in addition there is a haze brought about by aerosol particles with sizes in the range 0.1 to 0.5 μm.

Clouds obviously play a central part on Earth in the precipitation of water as rain, hail, snow, and so on. The fascinating subject of cloud physics and precipitation lies largely outside the scope of this book. We shall, however, touch upon the formation of clouds in Chapter 2. A conclusion we shall reach is that condensation of water to droplets usually starts by nucleation on foreign solid aerosol particles, known as *cloud condensation nuclei*. Hygroscopic or soluble nuclei are particularly effective, so that the presence of such aerosols in the Earth's atmosphere must be regarded as an important component of the evaporation–condensation–precipitation cycle of the Earth's water.

There is no doubt that human activities greatly increase the concentrations of solid aerosols, especially the smallest ones (radii less than about 0.5 μm, known as *Aitken nuclei*). Typical Aitken nucleus counts near the Earth's surface are 10^5 particles cm^{-3} or more over cities, 10^4 particles cm^{-3} over rural land, and 10^3 particles cm^{-3} over the sea. These particles originate primarily from deliberate combustion processes, and will be discussed further in relation to air pollution (Chapter 5). Appreciable concentrations of particles in continental and marine air indicate the existence of natural sources. So far as the global sources of all solid particles are concerned, man at present probably contributes around 20 per cent. Natural sources include evaporation of sea-spray from the ocean, wind-blown dust from surface erosion, forest fires, meteoric debris, and volcanic emissions. In addition, gas-to-particle conversion can occur in (photo)chemical processes. One of the most important examples of such a reaction is the oxidation of sulphur dioxide (natural, or released as a pollutant) to SO_3 and ultimately sulphate-containing aerosols. The colouring particles in the clouds of the outer planets, and the haze in Titan's atmosphere, must certainly come from photochemical gas-to-particle conversion.

Clouds and other aerosols may greatly modify the atmospheric balance of incoming and outgoing radiation; atmospheric and surface temperatures may be altered. Both the reflectivity and the absorptivity of the atmosphere are involved. Massive volcanic eruptions have long been thought to have a potential influence on climate. The spectacular sunsets over the entire Earth that follow a large eruption are well documented, and they are, of course, a result of increased scattering of sunlight by suspended particles. Two recent eruptions have provided an interesting comparison of climatic effects. The E1 Chichón eruption (Spring 1982) in SE Mexico spewed vast amounts of SO_2 into the atmosphere that have undergone conversion to H_2SO_4 droplets. The injection penetrated the tropopause, and the droplets are in the stratosphere from which loss is very slow. These aerosols may have led to a hemispheric cooling of the Earth of up to 0.5°C between the end of 1983 and 1985. The effect is on a par with the largest climatic perturbations recorded over the last 150 years (including the eruption of Krakatoa in 1883). By way of contrast, the Mount St Helens explosion (May 1980) is not thought to have had a major effect on temperatures, even though huge amounts of ash were injected high into the atmosphere. The difference is that the Mount St Helens explosion lacked the gases needed for the production of lasting aerosol.

As chemists, we should be alerted to the possibility of aerosols altering the

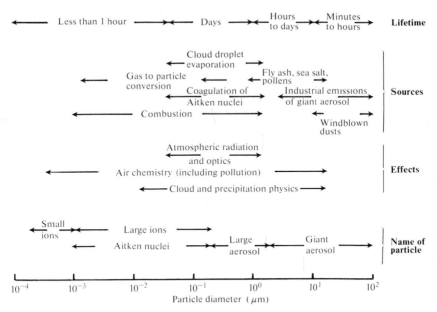

Fig. 1.1.   Names of atmospheric particles, together with effects, sources, and lifetimes. The lifetime of very small particles is short because they coagulate rapidly to form larger particles. Giant aerosol particles are shortlived because they precipitate out of the atmosphere. (Figure drawn from data of Wallace, J. M. and Hobbs, P. V. *Atmospheric science*. Academic Press, New York, 1977.)

course or rates of chemical change. Heterogeneous and catalytic processes in the atmosphere are not well identified or understood, but that does not make them unimportant! A few examples are known. For example, the rate of oxidation of SO_2 increases several-fold as the air becomes more nearly saturated with water. Sulphates can be formed by the reaction of sulphur dioxide and ammonia in cloud droplets: when the water evaporates, ammonium sulphate aerosol is left in suspension. The combined presence of soot particles and sulphur dioxide leads to enhanced oxidation rates as well as a greatly increased health hazard (see Chapter 5). The following section will describe how dependent we are on the presence of ozone in the stratosphere. An 'ozone hole' has been evident during early Spring in the Antarctic for more than a decade (see Section 4.7). It owes its origin to anomalous chemistry involving heterogeneous reactions occurring on the surfaces of polar stratospheric clouds. Water-ice particles make up these clouds, but there are indications that sulphate aerosols may also provide sites for surface reactions in the stratosphere. In that case, massive volcanic eruptions may also result in depletion of stratospheric ozone (Section 4.7.7). The involvement of particles in tropospheric chemistry is explored further in Section 5.3.9. Figure 1.1 summarizes some information about the sources, lifetimes, and effects of aerosols of different sizes.

1.4 Ozone

Ozone (trioxygen, O_3) plays a peculiarly significant part in the chemistry of the Earth's atmosphere, even though it is a 'minor' species in terms of abundance. Concentrations are rather variable, but the mixing ratio with respect to the entire atmosphere is a few tenths of a part per million. If the ozone in a column of the atmosphere were collected and compressed to 1 atm pressure, it would occupy a column about 3 mm tall. In fact, these statements about concentration conceal a very interesting feature about atmospheric ozone. Instead of being found in a constant fractional abundance, the ozone concentrations are very sharply dependent on altitude. Figure 1.2 shows a typical ozone altitude profile in two ways: as an absolute number density (concentration), and as the mixing ratio (fractional composition). The mixing ratio peaks sharply, and the ozone is pictured as being contained in an *ozone layer* about 20 km thick, and centred on an altitude of about 25–30 km. The dashed lines represent the approximate positions of the tropopause and stratopause, thus showing that the ozone layer lies within the stratosphere: indeed, as we shall see later, it is the presence of ozone that is responsible for the existence of the stratosphere. Peak fractional abundances in the ozone layer can approach 10^{-5} (10 p.p.m.).

Several factors contribute to ozone's importance, and they will be a recurrent theme in our later discussions. Perhaps the outstanding feature is the

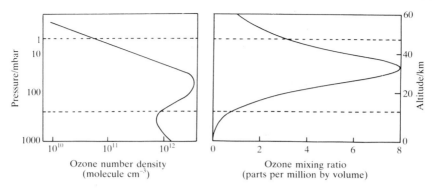

Fig. 1.2. Variation of atmospheric ozone concentration with altitude, expressed as an absolute number density and as a relative mixing ratio. (*Stratospheric Ozone 1988* UK Stratospheric Ozone Research Group, Second Report, HMSO, London, 1988.)

relationship between the absorption spectrum of ozone and the protection of living systems from the full intensity of solar ultraviolet radiation. The macromolecules, such as proteins and nucleic acids, that are characteristic of living cells, are damaged by radiation of wavelength shorter than about 290 nm. Major components of the atmosphere, especially O_2, filter out solar ultraviolet with wavelengths <230 nm; at that wavelength, only about 1 part in 10^{16} of the intensity of an overhead sun would be transmitted through the molecular oxygen. But at wavelengths longer than ~ 230 nm, the only species in the atmosphere capable of attenuating the Sun's radiation is ozone. Ozone has an unusually strong absorption just at the critical wavelengths (230–290 nm), so that it is an effective filter in spite of its relatively small concentration. For example, at $\lambda = 250$ nm, less than 1 part in 10^{30} of the incident (overhead) solar radiation penetrates the ozone layer.

Ozone is formed in the atmosphere from molecular oxygen, the necessary energy being supplied by the absorption of solar ultraviolet radiation. We have already noted that the oxygen in the Earth's contemporary atmosphere is largely biological in origin. Now we see that ozone, needed as an ultraviolet filter to protect life, is itself dependent on the atmospheric oxygen. These links further emphasize the special nature of Earth's atmosphere. Actually, the interactions are even more subtle than we have suggested. Absolute concentrations, and, indeed, the height distribution, of ozone depend on a competition between production and loss. Loss of ozone is, as we shall see in Chapter 4, regulated by chemistry involving some of the other trace gases of the atmosphere, such as the oxides of nitrogen, which are themselves at least partly of biological origin. Biological processes thus influence both the generation and destruction of ozone. We shall certainly have to examine whether human activities could interfere with the delicate balance (Section 4.5).

Energy absorbed by ozone from the solar ultraviolet radiation is ultimately degraded to heat; so, indeed, is the solar energy originally used in the formation of the ozone. The net result is a heating of the Earth's atmosphere in the region of the ozone layer that has a profound influence on atmospheric temperature structure and vertical stability (Section 2.3). Examination of the chemical routes that lead to the release of thermal energy is rather instructive. Ozone is an endothermic substance (positive enthalpy of formation from the standard state of the element):

$$\tfrac{3}{2}O_2 \rightarrow O_3; \quad \Delta H_{298}^{\ominus} = 143 \text{ kJ mol}^{-1}. \tag{1.4}$$

As a consequence, reactions involving ozone have a tendency to be exothermic and to be thermodynamically favoured. Further, rates of these reactions tend to be relatively high, so that ozone is an important chemical reaction partner. In the particular case where ozone itself absorbs a quantum of light ($\lambda \lesssim 310$ nm), the energy of the system is yet further elevated. A manifestation of this energization is that the ozone *dissociates* to form an atomic and a molecular fragment (i.e. O and O_2) *both of which are electronically excited* (cf. p. 86). The oxygen atom has an excitation energy of 190 kJ mol^{-1}, and this excess energy makes the atom highly reactive. Processes such as

$$O^* + H_2O \rightarrow OH + OH; \quad \Delta H_{298}^{\ominus} = -119 \text{ kJ mol}^{-1} \tag{1.5}$$

$$O^* + CH_4 \rightarrow OH + CH_3; \quad \Delta H_{298}^{\ominus} = -178 \text{ kJ mol}^{-1} \tag{1.6}$$

$$O^* + N_2O \rightarrow NO + NO; \quad \Delta H_{298}^{\ominus} = -340 \text{ kJ mol}^{-1} \tag{1.7}$$

(where O* represents an electronically excited oxygen atom), are all exothermic; we note that reactions (1.5) and (1.6) are endothermic with ground-state oxygen atoms for which enthalpies of reaction are 190 kJ mol^{-1} less. All reactions are fast because they have low activation energies. Water vapour, CH_4, and N_2O are important minor atmospheric constituents, while the radicals OH, CH_3, and NO are themselves highly reactive and are involved in atmospheric chemical changes of paramount significance. In each case a driving force for radical production can be the absorption of solar radiation by ozone.

1.5 Cyclic processes

Production and loss of atmospheric constituents have to be balanced if concentrations are not to vary. Over the lifetime of the Earth the atmospheric composition has almost certainly undergone considerable modification (Chapter 9), and the balance is not perfect. On short time-scales, however, a steady state obviously more or less holds for components such as nitrogen or oxygen. To a first approximation, the land, sea, air—lithosphere, hydrosphere, atmosphere—system is a 'closed' one: that is, material substance neither enters nor leaves it. In that case, the total quantity of each chemical element is fixed, although the distribution between the elemental and combined forms can alter. This conservation of elemental quantity means that if a species appears in the atmosphere (at a rate equalled by its disappearance), then the elements involved must be passing through a series of *cyclic* chemical and/or physical transformations. In the previous section we were discussing the photochemical production of O_3 from O_2. Ozone destruction in the atmosphere ultimately yields O_2, so that the number of elemental O atoms is conserved in the O_2–O_3 system by a cyclic process occurring solely within the atmosphere (and, indeed, largely within the ozone layer). Evaporation, transport, condensation, and precipitation of water are steps in a physical cycle linking the atmosphere and hydrosphere. Lithosphere, hydrosphere, and atmosphere are all involved in a cycle of which the weathering of rocks by CO_2 (described on p. 6) forms a part.

This chapter continues with a discussion of some of the major cycles occurring in the Earth's troposphere. Our concern will be to see what reservoirs there are for the species, and what determines the rate of physical and chemical interconversions. In other words, we wish to study the *budgets* involved in the cycles. Quite apart from the fundamental issue of how the Earth's atmospheric disequilibrium is sustained, an understanding of the budgets will also allow us to assess whether *anthropogenic* (man-made) sources of various species could be comparable with or larger than natural ones, and thus pose a threat of local or even global pollution (Sections 4.5 and 5.5).

sizing again the role played by life in determining the composition of our atmosphere. With the exception of the noble gases and water vapour, *all* the gases listed in Table 1.1 for Earth have a biological or microbiological source, and for species such as oxygen, methane, and ammonia, such sources may be the *only* significant ones in the contemporary atmosphere. Even apparently abiological changes such as the weathering of silicate rocks by CO_2 can be modulated by biological influences. Partial pressures of CO_2 in the soil where weathering occurs are 10–40 times higher than the atmospheric pressure, and these high partial pressures are maintained by soil bacteria. The cycles are thus *biogeochemical* in nature, and the term *biosphere* is used (by analogy with lithosphere, etc) to represent the biological component of the Earth's surroundings. The source regions of the biosphere are divided into *oxic* (containing free oxygen) and *anoxic*. Upper layers of oceans and soils are oxic, and produce fully oxidized species (e.g. CO_2) as well as partially reduced species (e.g. N_2O from bacterial processes and NH_3 from decay of animal excreta). Reduced species (e.g. H_2S, CH_4) are produced in anoxic environments such as lower soil regions or the interiors of animals. Transport and change in the biosphere itself may influence the nature and amount of gas reaching the atmosphere. Thus bacterial reduction of continental shelf sediments probably produces about 10^{12} kg yr^{-1} of H_2S. Less than one per cent reaches the atmosphere, the remainder being taken up by bacterial oxidation. Similarly, the decay of organic materials under anaerobic conditions gives rise to H_2 as the major primary product. Several groups of micro-organisms generate the intermediate (e.g. H_2S) or released (e.g. CH_4, N_2O, N_2) product.

Biospheric sinks exist for many of the atmospheric gases at the land/sea interface with the atmosphere. Respiration and other oxidative processes remove O_2, photosynthetic organisms (Section 1.5.1) remove CO_2, and certain species of plant–micro-organism systems fix nitrogen, for example. Within the troposphere itself, the chemical fates of almost all the trace gases are governed by reactions with hydroxyl radicals, which lead ultimately to oxidation. Hydroxyl radicals are themselves formed largely in reaction (1.5) between excited oxygen and water vapour; some (but not all) of the photochemical precursor of O*, ozone, is transported down from the stratospheric ozone layer. Thus one aspect of the importance of ozone photochemistry and of OH radical production begins to emerge. Methane reacts first to form methyl radicals

$$CH_4 + OH \rightarrow CH_3 + H_2O, \tag{1.8}$$

which are oxidized by O_2 in a complex set of reactions, involving formaldehyde (methanal) formation and photolysis, to carbon monoxide. Carbon monoxide from this source, as well as that released directly to the atmosphere, is also oxidized by OH in a cyclic set of reactions starting with the process

$$CO + OH \rightarrow CO_2 + H, \tag{1.9}$$

to yield CO_2 as the end product. Hydroxyl radicals are also implicated in the conversion of SO_2 to H_2SO_4, and of NO and NO_2 to HNO_3. Rain can dissolve the acids (which may perhaps serve as condensation nuclei) and *rainout* returns the sulphur- and nitrogen-containing compounds to the ground. Ammonia is very important as the only basic trace gas constituent. Although it can be dissolved and rained out of the atmosphere, it can also react with the acids to form solid aerosols of NH_4NO_3 and $(NH_4)_2SO_4$, touched on earlier (Section 1.3).

Of the gases in our atmosphere, only the 'inert' or 'noble' gases have neither a biological nor an atmospheric source. Chemical cycles are precluded because of the inertness of the noble gases. Budgets of the inert gases may therefore allow us to probe how far the lithosphere-hydrosphere-atmosphere system is truly closed. Argon is a surprisingly abundant element in the atmosphere (Table 1.1) at least so far as its isotope of atomic weight 40 is concerned. Interestingly, ^{40}Ar is *not* the natural isotope of argon (^{36}Ar with some ^{38}Ar are the natural isotopes). Radioactive decay of potassium, ^{40}K, however, does yield ^{40}Ar, and it is this radiogenic source that has provided the vast bulk of our atmospheric argon. The potassium in the Earth's crust, and deeper in the mantle, has decayed over the lifetime of the Earth to form ^{40}Ar, which has *degassed* from the solid. According to different estimates of the terrestrial potassium abundance, the present argon load (6.8×10^{16} kg) in the atmosphere corresponds to something between one-half and all of the ^{40}Ar generated within the Earth. No chemical sinks exist, and the oceans can dissolve only one per cent of the atmospheric argon. The mean annual input of ^{40}Ar to the atmosphere has thus been the present load divided by the lifetime of the Earth ($\sim 4.5 \times 10^9$ yr), and is 1.5×10^7 kg yr^{-1} globally. We shall use this result shortly. Helium is much less abundant than argon in the atmosphere: the ^{4}He isotope has a mixing ratio of 5.24 p.p.m., and ^{3}He is nearly a million times yet less abundant. Radioactive decay, this time of the ^{238}U, ^{235}U and ^{232}Th series, is again the source of the noble gas. The relative atomic ratio of ^{4}He to ^{40}Ar in many samples of natural gas lies between 0.2 and 5, so that the order of magnitude of ^{4}He/^{40}Ar in gases entering the atmosphere is unity. Why, then, is present-day ^{4}He so much less abundant— by a factor of 1782—than ^{40}Ar? The answer must be that some helium has escaped from the Earth altogether: from the top of the atmosphere to inter-planetary space. Possible mechanisms for such escape will be discussed in Section 2.3.2; other things being equal, a relatively light atom such as He will find it much easier to escape than a heavier one. Some idea of the time-scale of escape can be obtained by making an estimate of the atmospheric residence time. Assuming, as discussed above, that the volume (or atomic) rates of release of ^{4}He and ^{40}Ar are identical, then from the figures calculated earlier, the rate of release of ^{4}He is $1.5 \times 10^7 \times (4/40)$ kg yr^{-1}, where the (4/40) term reflects the differing atomic masses. The mass of helium in the atmosphere is $(6.8 \times 10^{16}/1782) \times (4/40)$ kg. For our estimate, we equate residence time with the atmospheric load divided by rate of release, and obtain 2.6×10^6 yr. This lifetime is three orders of magnitude less than the Earth's age, so that most of the helium ever released from the crust and mantle has escaped. With a different emphasis, it can also be said that 99.996 per cent of the particular helium atoms in the atmosphere when a man is born are still there when he dies!

The examination of noble gas behaviour has highlighted two new features of the fluxes of elements into and out of atmospheres. First, material can be released, for example by degassing, from the solid body of the planet. Gases can be formed, as in the cases of ^{4}He and ^{40}Ar, radiogenically, or they may have been trapped as the planet was created. Secondly, the extreme limit of the atmosphere forms an interface for the exchange of matter with 'space'. Our example was of escape, but material can also enter the Earth system at this interface. The Sun emits a continual stream of electrically charged particles, the *solar wind*, which impinges on the outer layers of a planetary atmosphere, or on the surface of a body without an atmosphere. Then meteorites, asteroids and even comets occasionally enter our gravitational field and burn up (*ablate*) or evaporate in the atmosphere.

Figure 1.3 summarizes the complex interactions that go to make up the biogeochemical cycles. The figure shows the many ways in which volatile materials can be delivered to the atmosphere or be removed from it. Solid–gas cycling occurs on a time-scale of hundreds of millions of years. Volcanic gases,

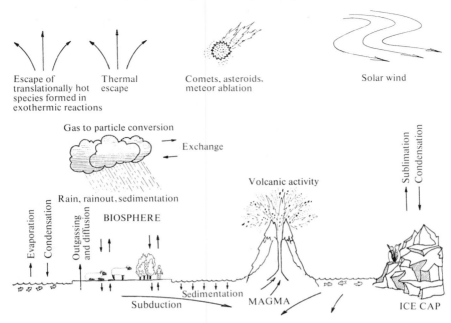

Fig. 1.3. Representation of the cyclic processes of biogeochemistry that exchange constituents between air, land, and sea. At the top of the picture, solar particles and extraplanetary objects bring matter into the atmosphere, while a certain amount escapes. Most material, however, is recycled. Even solids deposited on land and on the ocean beds can eventually be subducted to become molten and components returned to the atmosphere through volcanic activity. (Drawn by Sophie McLaughlin, with biospheric species from Carol Wayne and David Koslow.)

primarily CO_2 and H_2O, are released to the atmosphere from the Earth's crust. Carbon dioxide participates in weathering reactions that result in the deposition of carbonate sediments. Sea-floor *plates* are finally *subducted* into the mantle, volatiles trapped are released at high temperature and pressure, and recycled into the atmosphere through volcanoes along the plate interface. On a shorter time-scale, the hydrologic cycle exchanges water between the atmosphere and the condensed phases on the surface. Water vapour precipitates into the oceans, and on average the CO_2 and H_2O cycles are balanced so that hydrosphere and atmosphere are maintained at roughly constant volume. Interactions with the biosphere determine the detailed composition of the atmosphere, both with regard to major components such as O_2, and in respect of almost all the trace gases. The figure also shows that the cycles are not completely closed. Some material, *juvenile* in the sense that it has not hitherto been cycled, is released from the Earth's interior; the solar wind and debris left by stray bodies entering the atmosphere also contribute to the inward flux. Escape from the Earth's gravitational field to the interplanetary medium constitutes an outward flux.

Important biogeochemical cycles of the natural troposphere are considered in the sections that follow. The approach is concerned with budgets and lifetimes, but it should be emphasized that the actual numbers are representative only, and many are very uncertain. Some difficulty also arises over the concept of the 'natural' troposphere. It is certain that interactions with the biosphere largely determine atmospheric composition. Human life can potentially alter the composition *out of proportion* to human life's biological importance. We wish to reserve a fuller presentation of the topic for a discussion of tropospheric pollution (Section 5.5). Man's activities undoubtedly contribute to present-day budgets and reservoirs in carbon, nitrogen, and sulphur cycles. Our 'natural' troposphere ignores these contributions, and is to that extent artificial. It will, however, allow us to show up the anthropogenic sources more clearly as perturbations.

1.5.1 Carbon cycle

The major carbon species in the troposphere are carbon dioxide (~ 340 p.p.m.), methane (1.6–1.7 p.p.m.) and carbon monoxide (0.04–0.20 p.p.m.). Partial pressures of CO_2 today and in the past are of prime geochemical importance since they may influence global temperatures, the composition of marine sediments, rates of photosynthesis, and ultimately the oxidation state of the atmosphere and oceans. A pictorial representation of the complex coupling of inorganic and organic chemistry through CO_2 is given in Fig. 1.4. The boxes represent components of the atmosphere, biosphere, hydrosphere, and lithosphere. Arrows show routes for conversion between one component and another; several closed cycles can be identified on the diagram, and some cycles constitute smaller loops within larger ones. Estimates of transfer rates and reservoir capacities are given in the figure. Atmosphere–biosphere interactions completely dominate over the geochemical parts of the cycle. About 150×10^{12} kg of carbon are transferred each way each year, so that about 20 per cent of the CO_2 content of the atmosphere is converted annually. Photosynthesis by plants and micro-organisms is responsible for the intake of CO_2, while respiration and decay account for the reverse process. From the point of view of generation of organic material, the overall photosynthetic process consists of the formation of carbohydrates by the reduction of carbon dioxide,

$$n CO_2 + n H_2O \xrightarrow{\text{light}} (CH_2O)_n + n O_2, \qquad (1.10)$$

where $(CH_2O)_n$ is a shorthand for any carbohydrate. The essence of the process is the use of photochemical energy to split water and, hence, to reduce CO_2. Molecular oxygen is liberated in the reaction, although it appears at an

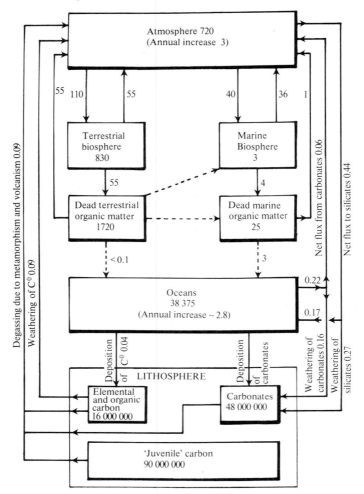

Fig. 1.4.   The carbon cycle. Arrows show the transfer of elemental carbon (C°) between the various carbon reservoirs (boxes), and indicate the closed loops that make up the cycle. Transfer rates are given in units of 10^{12} kg of C per year, and reservoir contents in 10^{12} kg of C. Burning of fossil fuels releases an additional 5×10^{12} kg yr^{-1} to yield the net increase in atmospheric burden. The figure is due originally to Holland H.D., *The chemistry of the atmospheres and oceans*, John Wiley, Chichester, 1978, and is updated with information reviewed by Clark, W. C. (ed.) *Carbon dioxide review: 1982*, Oxford University Press, Oxford, 1982; and by Bolin, B. and Cook, R. B. (eds.) *The major biogeochemical cycles and their interactions*, SCOPE 21, John Wiley, Chichester, 1983.

earlier stage in the sequence of steps than the reduction of CO_2. Light absorption is achieved by pigment systems involving various chlorophyll-cell structures. Chlorophylls are peculiarly suited to this purpose since their optical absorption is in the visible region, just where the photochemically active part of the solar radiation is highest at ground level (see Section 3.1). Furthermore, the structures of the chlorophylls make them particularly efficient photosensitizers. The biochemical reactions and cycles involved in the photosynthetic process are fascinating, but not directly relevant to our present interest. A multistep mechanism, with at least two absorption events, is needed to allow the relatively low energy photons (~ 200–300 kJ mol^{-1}) to split the O–CO (531 kJ mol^{-1}) or H–OH (498 kJ mol^{-1}) bonds. The organic chemical cycle itself is driven by energy-rich triphosphates or reduced phosphates, such as adenosine triphosphate (ATP) and reduced nicotinamide adenine dinucleotide phosphate (NADPH), so familiar in biochemistry. It is in the photochemical formation of these energy carriers that O_2 is liberated from H_2O; hydrogen is used to reduce CO_2 to carbohydrate by the energy carriers. Isotope experiments show that photosynthetically-produced O_2 does come exclusively from the H_2O and not the CO_2. So far as the atmospheric cycle is concerned, this result can be accommodated by the return of CO_2 to the atmosphere. Respiration of living organisms is the reverse of reaction (1.10),

$$(CH_2O)_n + nO_2 \rightarrow nCO_2 + nH_2O + \text{heat}, \qquad (1.11)$$

while oxidation or decay of carbohydrates likewise returns both CO_2 and H_2O out of the biosphere.

The biosphere–atmosphere cycles in Fig. 1.4 are very nearly closed, and can be treated in isolation from the rest of the carbon cycle over short times. Balances between photosynthesis and respiration/decay rates obviously alter diurnally and seasonally. As an example of diurnal change, it has been shown that, in a forest, CO_2 concentrations can rise to 400 p.p.m. at night, and drop to 305 p.p.m. at noon when photosynthetic activity is highest. Figure 1.5(a) shows how the CO_2 concentrations at a site in Hawaii have oscillated with season over many years: high CO_2 is associated with winter and spring, when there is least photosynthesis. Similar observations made near the South Pole (Fig. 1.5(b)) show very much smaller fluctuations, because there is little local photosynthetic activity, and the CO_2 remains close to the global average. We note in the figure an overall upward trend in CO_2 concentrations over the years. This increase is almost certainly a result of the burning of fossil fuels, and will be investigated further in Chapter 9.

Simple models for the distribution of carbon between reservoirs in the closed biological cycle lead to a surprising result. The way in which the rate of photosynthesis depends on CO_2 concentration is known; it is likely that the rates of transfer from the living biosphere to dead matter and of return back to the atmosphere are linear functions of the mass in the reservoir. It can then be shown that the carbon content of each reservoir is determined only

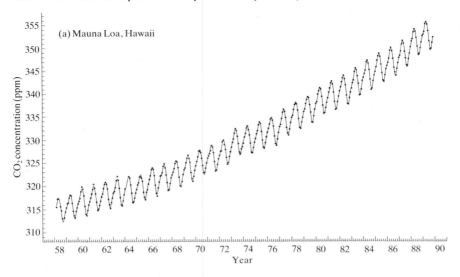

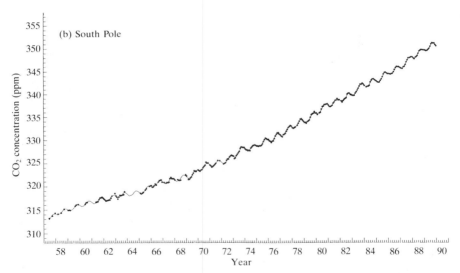

Fig. 1.5. Atmospheric carbon dioxide concentrations over more than thirty years, (a) in Hawaii, (b) near the South Pole. An increasing trend is present in the data from both stations, and is a consequence largely of man's combustion of fossil fuels. Superposed on the trend are annual oscillations caused by seasonal changes in photosynthetic activity, which consumes carbon dioxide: the amplitude of oscillation at the South Pole is smaller than that in Hawaii because there is no local vegetation, and CO_2 levels closely reflect values averaged for the globe. The smooth curve represents a fit to the data to a four harmonic annual cycle which increases linearly with time, and a spline fit of the interannual component of the

by the total carbon content of all of them. That is, the carbon content of the atmosphere is determined, and the rate of photosynthesis has no direct influence. In reality, of course, there are slow leaks to the hydrosphere and lithosphere: some dead matter (mainly marine, with a smaller terrestrial contribution) does not return to the atmosphere, and about 0.1×10^{12} kg yr^{-1} becomes buried. Losses from the rapidly cycling biosphere–atmosphere system are what determine atmospheric CO_2 pressure *and* the carbon content of the biosphere.

Chemical weathering both adds and removes CO_2 from the atmosphere. Oxidation of elemental carbon ($C°$) and organic compounds in rocks adds CO_2; decomposition of calcium and magnesium silicates [reaction (1.3)] and solution of carbonate minerals [reaction (1.2)] removes it. The weathering processes for $C°$ and carbonates are included in Fig. 1.4. Weathering of silicates consumes an additional 0.27×10^{12} kg yr^{-1} of C from CO_2, while the reverse decomposition yields 0.16×10^{12} kg yr^{-1}. Thus the total budget, in 10^{12} kg yr^{-1} to the atmosphere, is made up of: $C°$, $(0.09 - 0.12) = -0.03$; carbonates, $(0.22 - 0.16) = +0.06$; silicates, $(0.16 - 0.27) = -0.11$. Carbon leaks from the atmosphere–hydrosphere–upper lithosphere system at a rate of 0.08×10^{12} kg yr^{-1}, or roughly 0.1 per cent of the turnover by photosynthesis. This rate of removal, if continued, would deplete the atmosphere in $(720/0.08) \sim 9000$ yr, and the oceans in $(38375/0.08) \sim 480\,000$ yr. No geologic evidence exists for large fluctuations of CO_2 in the past, and it is likely that the shortfall of 0.08×10^{12} kg yr^{-1} is made up by degassing and volcanic release from the interior of the Earth of juvenile carbon or subducted carbonates. Whatever the explanation, it is clear from the calculations that the balance of CO_2 in atmospheres and oceans is a delicate one, and that a serious alteration in the leak rates could change CO_2 concentrations on a geologically short time-scale.

The lifetime of CO_2 in the atmosphere appears to be determined by the biospheric interaction, since that is the largest contributor to the annual turnover shown in Fig. 1.4. According to the numerical values used in this discussion, the residence time would be $(720/110) \sim 6.5$ yr if land photosynthesis alone were involved, or $(720/150) \sim 4.8$ yr if the marine component is

---

◁ variation. The dots indicate monthly average concentrations. Source: Keeling C. D., Bacastow R. B., Carter A. F., Piper S. C., Whorf T. P., Heimann M., Mook W. G., and Roeloffzen H., A Three Dimensional Model of Atmospheric CO_2 Transport Based on Observed Winds: Observational Data and Preliminary Analysis, Appendix A, in *Aspects of Climate Variability in the Pacific and the Western Americas*, Geophysical Monograph, American Geophysical Union, **55**, 1989 (Nov). Updated with information provided by Professor C. D. Keeling in June 1990. The results were obtained in a cooperative programme of the US National Oceanic and Atmospheric Administration (NOAA) and the Scripps Institution of Oceanography.

included. Direct evidence suggests that the actual lifetime is shorter. Nuclear weapon testing in the 1960s doubled the atmospheric level of ^{14}C, which is also produced naturally by cosmic radiation. Subsequent return to the earlier concentrations after the tests ended was much more rapid than can be explained by the rates of photosynthetic turnover and the sizes of the biospheric reservoirs. Figure 1.4 does not, however, show the exchange of CO_2 between atmosphere and ocean resulting from reversible solution, since the dissolving and evolution processes are in dynamic equilibrium. About 120×10^{12} kg yr^{-1} of C are transferred each way, so that with this term the true atmospheric lifetime of any CO_2 molecule becomes $(720/270) \sim 2.7$ yr. The oceanic reservoir of dissolved CO_2 (and HCO_3^- ions) is about 60 times bigger than the atmospheric reservoir, with the result that the partitioning of artificial ^{14}C can virtually remove it from the atmosphere in a few years.

Methane and carbon monoxide participate in cycles that are linked with each other and the main carbon cycle. Concentrations of methane are essentially independent of altitude within the troposphere but show a slight latitudinal variation: mean concentrations are (1.69 ± 0.05) p.p.m. in the Northern Hemisphere and (1.62 ± 0.03) p.p.m. in the Southern Hemisphere. Primary 'natural' CH_4 production mechanisms are enteric fermentation in animals, mostly cattle, and microbiological anaerobic decomposition of organic matter in wetlands, swamps, and paddy fields. In radiocarbon (^{14}C) dating, it is assumed that living organisms possess the same ^{14}C/^{12}C content as the atmosphere, in which the ^{14}C is continually replenished by cosmic ray bombardment of ^{14}N. Dead or fossilized material no longer incorporates new ^{14}C, so that radioactive decay reduces the ^{14}C content. The half-life of ^{14}C is only about 5600 years, so that the radiocarbon has almost completely disappeared in fossil gases, oils, and solids. The ^{14}C content of atmospheric methane is about 80 per cent that of modern wood, so that less than 20 per cent of the total CH_4 supply comes from natural gas leakage. Current estimates put the total CH_4 source strength at 0.6 to 0.9×10^{12} kg yr^{-1}. These figures imply that roughly one per cent of all photosynthetically-produced organic matter decays to produce atmospheric methane. Given the atmospheric load of 4.5×10^{12} kg, the methane could be supplied in 5 to 8 years. In fact, there is some evidence that production and loss are not quite balanced, and that the concentration of CH_4 has been increasing by about 0.01 p.p.m. yr^{-1}, starting about 150 years ago. Long-term changes could alter both the chemistry and the climate of the atmosphere, and will be discussed in Chapter 9. Loss processes for CH_4 include bacterial consumption in the soil, but by far the most important loss mechanism is the inorganic oxidation chain (p. 14 and Chapter 5) initiated by hydroxyl radicals (reaction 1.8). One net destruction cycle can be written in the simplified form

$$CH_4 + O_2 \xrightarrow[\text{light}]{\text{OH}} H_2O + CO + H_2, \tag{1.12}$$

to show that CO may be a product. A few per cent of the methane molecules produced ($\sim 0.02 \times 10^{12}$ kg yr^{-1}) cross the tropopause to enter the strato-sphere. Above the tropopause, CH_4 concentrations drop quite rapidly by a factor of about four between 15 and 40 km, suggesting that the CH_4 is rapidly destroyed. Stratospheric loss is not, in itself, important for tropospheric methane budgets, but the process does represent a chemical transport of hydrogen-containing species from troposphere to stratosphere.

Oxidation of methane is a major natural source of carbon monoxide. Non-methane organic compounds such as terpenes may also be oxidized, and contribute up to 50 per cent of the atmospheric carbon monoxide. Smaller amounts of CO are emitted by plants and micro-organisms. It used to be thought that most of the atmospheric CO was anthropogenic in origin, and that the difference in concentrations for Northern (0.07–0.20 p.p.m.) and Southern (0.04–0.06 p.p.m.) Hemispheres reflected the differences in loca-tion of industrial sources. However, the asymmetry in the natural oxidation sources could also match the concentration asymmetry. According to recent estimates, the global source of CO is about 2.8×10^{12} kg yr^{-1}, of which 60 per cent is natural. For steady-state CO concentrations globally averaged at 0.12 p.p.m., the corresponding residence time is about two months; without the natural sources, the residence time would have to be up to years. Strong hemispheric asymmetries are obviously more easily maintained if lifetimes are short. Loss of CO is, as with methane, largely (~ 90 per cent) dependent on reaction with OH in the very fast process (1.9),

$$CO + OH \rightarrow CO_2 + H. \tag{1.9}$$

The hydrogen atom can reform OH radicals in a number of ways (Chapter 5), so that the oxidation of CO to CO_2 may be regarded as catalysed by hydroxyl radicals. A microbiological sink for CO at the surface of the soil accounts for about ten per cent of the total loss.

1.5.2. Oxygen cycle

Short-term oxygen fluxes are dominated by the photosynthetic cycle. Carbon dioxide transfers between atmosphere and biosphere will correspond stoicheiometrically to oxygen transfers in the opposite direction: thus photosyn-thesis releases, while respiration and decay consume, atmospheric oxygen. Photosynthesis accounts for the annual consumption of 150×10^{12} kg of atmospheric carbon (Fig. 1.4) and release of $150 \times (32/12) \times 10^{12} = 400 \times 10^{12}$ kg of oxygen. The atmosphere contains 1.2×10^{18} kg of O_2, so that oxygen cycles through the biosphere in $(1.2 \times 10^{18}/400 \times 10^{12}) \sim 3000$ yr, a period much longer than the equivalent one of a few years for carbon because of the much greater atmospheric oxygen reservoir. Seasonal fluctuations in concentration, seen for CO_2 (pp. 19–21), are therefore damped out with oxygen.

As we discussed in Section 1.5.1, there is a small leak of carbon out of the atmosphere–biosphere system which has a strong influence on atmospheric CO_2 concentrations. This same leak is also important in the geochemistry of oxygen. Marine organic sediment deposition buries $\sim 0.12 \times 10^{12}$ kg yr^{-1} of C without decay, and so releases $\sim 0.32 \times 10^{12}$ kg yr^{-1} of O_2. That is, the leak could cause atmospheric O_2 to double in concentration in $(1.2 \times 10^{18}/0.32 \times 10^{12}) \sim 4 \times 10^6$ yr. Because O_2 is less soluble in water than CO_2, the oceans provide far less of a stabilizing reservoir for O_2. Marked variations in the atmospheric O_2 level may have occurred over geological time. Nevertheless, in the contemporary atmosphere, oxygen is consumed in oxidation of rocks. Weathering rates of elemental carbon to carbon dioxide, sulphide rocks to sulphate, and iron (II) rocks to iron (III) roughly match the leak rate. Oxidation of reduced volcanic gases (e.g. H_2 or CO) is a smaller, but not negligible, additional balancing process.

1.5.3 Nitrogen cycle

Molecular nitrogen is chemically rather inert, partly because the large $N \equiv N$ bond energy of 945 kJ mol^{-1} makes most reactions endothermic, or at least kinetically limited because of a large activation energy. Natural processes can *fix* nitrogen (bring it into combination). Lightning within the atmosphere can produce the higher oxides of nitrogen (NO, NO_2, etc.) which are converted to acids (HNO_2, HNO_3) and rained out. Biological fixation is, however, of even greater importance, at least over land. Independent micro-organisms can fix nitrogen into soils, but, for the planet as a whole, the greatest sources of naturally-fixed nitrogen are symbiotic organisms found in the root nodules of the pulses or leguminous plants (peas, beans, etc.). Assimilation of nitrogen is catalysed by the enzyme complex *nitrogenase* that brings about the reduction of N_2 to NH_4^+ ions. As in the photosynthetic conversion of CO_2 to carbohydrate, the energy-rich phosphate ATP drives the process, several molecules probably being required. The first step can be represented by the equation

$$N_2 + 8H^+ + 6e^- \xrightarrow{\text{ATP}} 2NH_4^+. \tag{1.13}$$

In the symbiotic micro-organisms, the ATP derives from the host plant, which receives up to 90 per cent of the fixed nitrogen in return. Other links in the soil microbiological chain involve *nitrifying* bacteria that oxidize NH_4^+ to NO_3^-, *denitrifying* bacteria that reduce NO_3^- to N_2, and *ammonifying* bacteria in which wastes and remains of animals and dead plants are reconverted to ammonia. The microbiological chains maintain the enormous disequilibrium between atmospheric N_2 and O_2 concentrations and those of NO_3^- ions in sea-water, while the greater solubility of common nitrate minerals compared with common carbonates is a partial cause of the dominance of N_2 over CO_2 in the Earth's atmosphere.

Estimates of global budgets and reservoirs are even more imprecise for N_2 than those for carbon, but they still indicate the major aspects of the

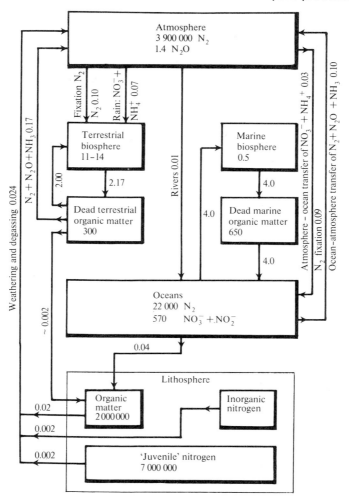

Fig. 1.6. The nitrogen cycle. See Fig. 1.4 for an explanation and for the sources of information. Reservoir contents are given in units of 10^{12} kg (of N except where N_2 or N_2O are specified), and transfer rates in units of 10^{12} kg of N per year.

nitrogen cycle. Weight ratios for C/N are about 7.9 for terrestrial plants and 5.7 for marine plants, so that the biospheric reservoirs and transfer rates in Fig. 1.4 may be converted to provide rough values for the nitrogen cycle. Figure 1.6 shows the main aspects of the cycle. In both terrestrial and marine organic systems, most of the nitrogen is recycled within the system, although a smaller fraction is transferred between biosphere and atmosphere. For the hydrosphere, some 130×10^9 kg yr^{-1} of nitrogen are transferred from the atmosphere, as against 4.0×10^{12} kg yr^{-1} cycling in the marine biosphere. Without this recycling, the lifetime of atmospheric N_2 would be about 6×10^7 yr; with the numbers used in the figure for atmospheric transfer the calculated lifetime is nearer 1.6×10^6 yr. Again, according to the figure, about 40×10^9 kg yr^{-1} of N_2 are buried as dead organic matter, mostly from the marine biosphere. Exposure, weathering, and conversion of organic and inorganic deposits, and release of juvenile nitrogen, can balance the burial rate. However, even without the restoring sources, it would take $(7.80 \times 10^6/0.040) \sim 2 \times 10^8$ yr to consume all atmospheric nitrogen. That is, the imbalance is small even on geologically-long time-scales because of the large nitrogen content of the atmosphere.

Nitrous oxide (dinitrogen oxide, N_2O) is liberated from soils as a result of incomplete microbiological nitrification or denitrification. Indications are that the biological source is dominated by nitrification. The yield of N_2O depends in a non-linear way on oxygen concentrations, ranging from 0.03 per cent (referred to overall NH_4^+ oxidation) at high oxygen levels to more than 10 per cent at low oxygen levels. Global mixing ratios seem to have increased from 0.292 p.p.m. in 1961 to 0.307 p.p.m. by 1987. Nitrous oxide is chemically significant in the atmosphere, largely as a precursor of the higher oxides of nitrogen. Reaction (1.7) with excited oxygen from ozone photolysis produces nitrogen monoxide (nitric oxide, NO) which may be further oxidized to nitrogen dioxide (NO_2). Both NO and NO_2 are involved in much of the chemistry of the troposphere and stratosphere (Chapters 5 and 4). Transport across the tropopause provides a tropospheric supply of N_2O and hence NO and NO_2 in the stratosphere.

1.5.4 Sulphur cycle

The abundances, sources, sinks, budgets, and photochemistry of atmospheric sulphur compounds are poorly understood compared to those of the carbon, oxygen, and nitrogen species considered so far. Sulphur in the atmosphere can be converted to SO_2, SO_3, and H_2SO_4. It therefore acts as an aerosol precursor, may have an effect on cloud production, and may be of climatological significance. The H_2SO_4 may also lower the pH of rain-water with deleterious consequences. In the case of the atmospheric sulphur budget, it is clear that anthropogenic sources are at least comparable with the natural ones. Volcanic activity probably produces 2×10^9 kg yr^{-1} of sulphur in the form of SO_2 (but see below). Decay of organic matter in the biosphere probably yields 58×10^9 kg yr^{-1} of sulphur over land, and 48×10^9 kg yr^{-1} over the ocean, in the reduced forms of H_2S, $(CH_3)_2S$ (dimethylsulphide), and $(CH_3)_2S_2$ (dimethyldisulphide). Sea-spray might contribute another 44×10^9 kg yr^{-1} of sulphur compounds. Atmospheric mixing ratios for the sulphur-containing compounds are rather small. Sulphur dioxide is present at a globally averaged mixing ratio of 167 p.p.m.m. (parts per million million) corresponding to a total atmospheric load of 0.9×10^9 kg S. Hydrogen sulphide is very variable, typical global mixing ratios lying between zero and 100 p.p.m.m. ($\equiv 0$ to 0.6×10^9 kg S). Lifetimes for H_2S must be only a few days even for the highest concentrations observed, and are probably more usually a few hours. Similar remarks apply to SO_2. The mechanisms for the rapid destruction of these species are uncertain, but may involve reaction with OH, or heterogeneous oxidation steps. Two more sulphur-containing gases, carbonyl sulphide (COS) and carbon disulphide (CS_2), have recently been shown to be present in the troposphere, and it has been suggested that these are the main precursors of SO_2 and sulphate aerosol. Carbonyl sulphide is thought to be the most abundant (500 p.p.m.m.) sulphur gas in the troposphere. Its distribution with latitude is uniform, consistent with a rather long lifetime of more than one year. While carbon disulphide is much less abundant, and variable (3–30 p.p.m.m.) because its atmospheric lifetime is only a few weeks, tropospheric oxidation appears to produce COS, so that CS_2 contributes to the total COS available. Most sulphur-containing gases emitted into the troposphere from natural or artificial sources are too reactive or too soluble to reach the stratosphere, but COS is an important exception. Apart from volcanic injection, tropospheric COS is the main source of stratospheric sulphur. The stratospheric sulphate layer, which influences temperature structure in the lower stratosphere and may have some effect on ozone concentrations (see Sections 1.3 and 4.7.7), thus depends on this source of sulphur. Oceans are a global source of COS contributing $0.6 \pm 0.2 \times 10^9$ kg yr^{-1}. Ash collected from the eruption of Mount St Helens (18 May 1980) gave off large amounts of COS and CS_2 at room temperature, although these gases were much less concentrated than H_2S in the gaseous part of the plume. Interestingly, SO_2 became the dominant gas only after the eruption of 15 June. Such results may be suggestive of sources, but the origins of COS and CS_2 are really unknown at present, even to the extent of whether the gases are natural or anthropogenic.

Dimethylsulphide (DMS) is an important sulphur-carrying gas that has natural sources. Large quantities are produced by algae in the oceans. For example, in spring and summer, algae along the coasts of the North Sea produce enough DMS during April and May to make a contribution after oxidation of up to 25 per cent of the H_2SO_4 burden carried in the troposphere over some parts of Europe. The mechanism of the oxidation of DMS is discussed in Section 5.3.7.

1.6 Linking biosphere and atmosphere

Figure 1.3 illustrated the cyclic processes that exchange constituents between air, land, and sea. We now present in Fig. 1.7 a diagram, devised by Professor O. Hov, that shows in more detail some major chemical systems *within* the atmosphere. Some of these systems have already been mentioned briefly, and

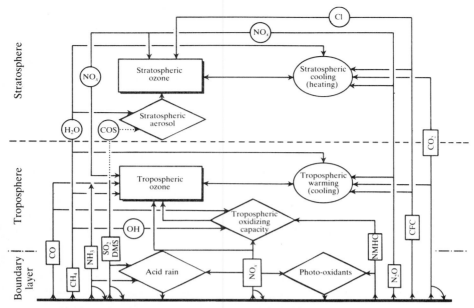

Fig. 1.7. Interactions and feedbacks between important chemical systems in the troposphere and stratosphere. The figure particularly emphasizes the factors that influence ozone concentrations in both regions as well as showing the formation of important pollutants in the boundary layer (roughly the first km of highly turbulent air next to the surface). Many of the chemical species modify atmospheric temperatures which in turn determine the rates and relative importance of the different processes. Not shown on the diagram, but also an important feedback mechanism, is the control that the chemical compounds may exert on cloud formation and thus indirectly on the biogenic source gases.

Source species are shown in the small oblong boxes, while radicals and other secondary intermediate species are shown in circles: DMS = dimethylsulphide, CFC = chlorofluorocarbons, NMHC = non-methane hydrocarbons.

Based on a diagram presented by Professor O. Hov, University of Bergen, in Varese, September 1989.

they will be the main subjects of Chapters 4 and 5 of this book. The diagram indicates clearly that not only are the systems coupled, but that there are feedback loops between them. Some of the loops are direct in the sense that they concern the chemical species present and their concentrations. Others are indirect. Carbon dioxide and several other trace gases in the atmosphere 'trap' radiation in the troposphere to warm it. This is the so-called 'greenhouse effect', and we shall explore it in greater detail in Sections 2.2 and 9.6. In the stratosphere, the same gases can lead to a cooling effect. The modification of temperature in both regions can affect the rates of chemical reactions and thus the relative importance of different chemical steps. Again, the formation of aerosols can alter the amount of photochemically active light in the stratosphere and troposphere as well as influencing the thermal radiation budget and thus temperatures. If such aerosols act as cloud condensation nuclei, then the effects can be further exaggerated. Not only may the atmospheric chemical transformations be altered, but the biological sources of the trace gases themselves are likely to respond to changes in temperature and light intensity. Many other loops of this kind can be identified.

To end this chapter, we present an idea due to Lovelock and Margulis (1974: see Bibliography), more to provoke thought than as an accepted theory. We have shown that the composition of Earth's atmosphere is displaced from equilibrium because of interactions with the biosphere. Lovelock and Margulis go further, and see the interaction between life and the atmosphere as so intense that the atmosphere can be regarded as an extension of the biosphere. The atmosphere is not living, but is a construction maintained by the biota. From this concept comes the *Gaia hypothesis* (Gaia = Earth Mother), which postulates that the climate and the chemical composition of the Earth's surface and atmosphere are kept at an optimum by and for the biosphere. The relationship between composition and the biosphere is seen as analogous to that between the circulatory blood system and the animal to which it belongs. In the case of the atmosphere, if highly improbable arrangements (equivalent to low entropy) extend beyond the boundaries of living entities so as to include also their planetary environment, then the environment and life taken together can be considered to constitute a single larger entity. It is the 'operating system' of life and its environment that is called Gaia. Chemistry, pressures, and temperatures are all regulated. Indeed, following the line of thought further, the expected efficiency of evolution would mean that every trace gas in the atmosphere had a purpose as a chemical information carrier. Control systems developed by humans have required feedback mechanisms and amplifiers. Some biogeochemical amplifiers can be identified. Chemical weathering of silicate rocks takes place where high partial pressures of CO_2 are maintained by biological oxidation. Rates of oxidation double for every 10°C temperature rise. The biota act as temperature sensors and amplifiers to control the rate of CO_2 pumping from the air. Dimethylsulphide from oceanic sources leads to substantial and amplified changes in cloud cover (see above) so that biological activity in the surface waters of the oceans may be an important link in an ecosystem regulation of temperature and illumination over the Earth. Again, we have seen the non-linear and amplified response of N_2O production to changed oxygen concentrations in the microbiological nitrification process.

Naturally, the Gaia hypothesis does not receive universal acclaim! A more cautious view (Holland, 1978: see Bibliography) does not see the close links between atmosphere, oceans, and biosphere as implying the existence of an adaptive control system. Rather, the ocean-atmosphere system has adjusted to biologic activity such as photosynthesis, and the biosphere has responded by optimizing the use of available free energy. The Gaian counter-argument asks why living organisms, which generally exhibit an economy of function, should produce, for example, trace gases such as N_2O or CH_4, unless there is some evolutionary advantage to the organism. The potential control of atmospheric composition and climate by these trace gases is then seen, in the Gaian context, as their regulatory role. In the engineering control system analogy, the difference between Gaian and non-Gaian views is concerned with whether or not there are closed feedback loops. Although resolution of the different views is not, at present, possible, the concept of closed-loop control is of more than passing interest in predicting the atmosphere's response to natural or anthropogenic disturbance.

Bibliography

Books and articles concerned with atmospheric chemistry in general.

Chemistry of the atmosphere. McEwan, M. J. and Phillips, L. F. (Edward Arnold, London, 1975.)

The chemistry of the atmosphere and oceans. Holland, H. D. (John Wiley, Chichester, 1978.)

Chemistry of the lower atmosphere. Rasool, S. I. (ed.). (Plenum Press, New York, 1973.)

Atmospheric chemistry. Heicklen, J. (Academic Press, New York, 1976.)

Energy and the atmosphere: a physical-chemical approach. Campbell, I. M. (John Wiley, London, Second edition, 1986.)

The photochemistry of atmospheres. Levine, J. S. (ed.). (Academic Press, New York, 1984.)

The planets and their atmospheres. Lewis, J. S. and Prinn, R. G. (Academic Press, Orlando, 1984.)

Air: composition and chemistry. Brimblecombe, P. (Cambridge University Press, Cambridge, 1986.)

Chemical compounds in the atmosphere. Graedel, T. E. (Academic Press, New York, 1978.)

These two books are specifically concerned with air pollution, but also contain much about atmospheric chemistry in general, especially in the earlier chapters.

Atmospheric chemistry. Finlayson-Pitts, B. J. and Pitts, J. N. Jr. (John Wiley, Chichester, 1986.)

Atmospheric chemistry and physics of air pollution. Seinfeld, J. L. (John Wiley, Chichester, 1986.)

The next book also gives a good general introduction to atmospheric chemistry, but is more specifically directed to the question of whether the biosphere and atmosphere control each other's behaviour and composition in a more than casual way (see Section 1.6).

Gaia: a new look at life on earth. Lovelock, J. E. (Oxford University Press, Oxford, 1979.)

> *This issue of* Scientific American *is devoted to a discussion of the lithosphere, hydrosphere, atmosphere, and biosphere, and the interactions between them.*

The dynamic earth. *Scient. Am.* **249**, No, 3, pp. 30–144 (Sept. 1983).

Section 1.1

The solar system. Jones, B. W. (Pergamon Press, Oxford, 1984.)

The chemistry of the solar system. Lewis, J. S. *Scient. Am.* **230**, 50 (March 1974.)

Atmospheres and evolution. Margulis, L. and Lovelock, J. E., in *Life in the universe* (ed. Billingham, J.) pp. 79–100. (M.I.T. Press, Cambridge, Mass., 1981.)

Atmospheres of the terrestrial planets. Pollack, J. B., in *The new solar system.* (eds. Beatty, J. K., O'Leary, B., and Chaikin, A.) 2nd edn. (Cambridge University Press, Cambridge, 1982.)

The atmospheres of Venus, Earth, and Mars: a critical comparison. Prinn, R. G. and Fegley, B. Jr. *Ann. Rev. Planet. Space Sci.* **15**, 171 (1987).

The atmospheres of the outer planets and satellites. Trafton, L. *Rev. Geophys. & Space Phys.* **19**, 43 (1981).

The atmospheres of the planets. Mason, B. J. *Observatory* **97**, 217 (1977).

Atmospheric composition: influence of biology. McElroy, M. B. *Planet. Space Sci.* **31**, 1065 (1983).

Section 1.2

The atmosphere and ocean: a physical introduction. Wells, N. (Taylor and Francis, 1986.)

Air–sea exchange of gases and particles. Liss, P. S. and Slinn, W. G. N. (eds.). (D. Reidel Co., Dordrecht, 1983.)

The role of the ocean in the global atmospheric cycle. Nguyen, B. C., Bonsang, B., and Gaudry, A., *J. geophys. Res.* **88**, 10903 (1983).

Role of oceans in atmospheric chemistry (Conference, Hamburg, August 1982). *J. geophys. Res.* **87**, 8769 (1982).

The dynamics of the coupled atmosphere and ocean. Charnock, H. and Philander, S. G. H. (eds.). (The Royal Society, London, 1990.)

The biogeochemistry of the air–sea interface. Lion, L. W. and Leckie, J. O. *Annu. Rev. Earth & Planet. Sci.* **9**, 449 (1981).

The carbonate–silicate geochemical cycle and its effect on atmospheric carbon dioxide over the past 100 million years. Berner, R. A., Lasaga, A. C., and Garrels, R. M. *Am. J. Sci.* **283**, 641 (1983).

Feedbacks between weathering and atmospheric CO_2 over the last 100 million years. Volks, T. *Am. J. Sci.* **287**, 763 (1987).

Coordination chemistry of weathering: kinetics of surface-controlled dissolution of oxide minerals. Stumm, W. and Wollast, R. *Rev. Geophys.* **28**, 53 (1990)

Section 1.3

The atmospheric aerosol system: an overview. Prospero, J. M., Charlson, R. J., Mohnen, V., Jaenicke, R., Delany, A. C., Moyers, J., Zoller, W., and Rahn, K. *Rev. Geophys. & Space Phys.* **21**, 1607 (1983).

Aerosols and atmospheric chemistry. Hidy, G. M. (ed.). (Academic Press, New York, 1972.)

Elemental constituents of atmospheric particulates and particle density. Sugimae, A. *Nature, Lond.* **307**, 145 (1984).

Natural organic atmospheric aerosols of terrestrial origin. Zenchelsky, S. and Youssefi, M. *Rev. Geophys. & Space Phys.* **17**, 459 (1979).

Heterogeneous interactions of the C, N, and S cycles in the atmosphere: the role of aerosols and clouds. Taylor, G. S., Baker, M. B., and Charlson, R. J., SCOPE **21**, 115 (1983).

Kinetic studies of raindrop chemistry. 1, Inorganic and organic processes. Graedel, T. E. and Goldberg, K. I. *J. geophys. Res.* **88**, 10865 (1983).

Chemistry with aqueous atmospheric aerosols and raindrops. Graedel. T. E. and Weschler, C. J. *Rev. Geophys. & Space Phys.* **19**, 505 (1981).

Review: Atmospheric deposition and plant assimilation of gases and particles. Husker, R. P., Jr., and Lindberg, S. E. *Atmos. Environ.* **16**, 889 (1982).

Volcanoes are one source of atmospheric gases and aerosols. A fascinating finding of the Voyager mission to Jupiter was the existence of intense volcanic activity on the moon Io. Major volcanic activity on Venus is strongly suspected. A recent major eruption on Earth (El Chichón) has greatly increased the aerosol burden of the atmosphere, apparently with meteorological consequences.

A comparison of volcanic eruption processes on Earth, Moon, Mars, Io and Venus. Wilson, L. and Head, J. W., III. *Nature, Lond.* **302**, 663 (1983).

The atmospheric effects of El Chichón. Rampino, M. R. and Self, S. *Scient. Am.* **250**, 34 (Jan. 1984).

El Chichón: composition of plume gases and particles. Kotra, J. P., Finnegan, D. L., Zoller, W. H., Hart, M. A., and Moyers, J. L. *Science* **222**, 1018 (1983).

El Chichón volcanic aerosols: impact of radiative, thermal, and chemical perturbations. Michelangeli, D. V., Allen, M., and Yung, Y. L. *J. geophys. Res.* **94**, 18429 (1989).

An assessment of the impact of volcanic eruptions on the Northern Hemisphere's aerosol burden during the last decade. Michalsky, J. J., Pearson, E. W., and LeBaron, B. A. *J. geophys. Res.* **95**, 5677 (1990).

Volcanic winters. Rampino, M. R., Self, S., and Stothers, R. B. *Ann. Rev. Earth planet. Sci.* **16**, 73 (1988).

Climatic effects of the eruption of El Chichón. (Many papers collected in part of a special issue.) *Geophys. Res. Lett.* **10**, 989–1060 (1983).

Increases in the stratospheric background sulfuric acid aerosol mass in the past ten years. Hofmann, D.J. *Science* **248**, 996 (1990).

COS in the stratosphere: El Chichón observations. Leifer, R., Juzdau, Z. R., and Larsen, R. *Geophys. Res. Lett.* **11**, 549 (1984).

Section 1.4

Ozone in the stratosphere is the subject of Chapter 4, and tropospheric ozone is discussed in detail in Chapter 5. References at the end of those chapters supplement the introductory articles listed here.

The chemistry of the stratosphere. Thrush, B. A. *Rev. Prog. Phys.* **51**, 1341 (1988).

Stratospheric ozone: an introduction to its study. Nicolet, M. *Rev. Geophys. & Space Phys.* **13**, 593 (1975).

Photochemical reactions initiated by and influencing ozone in unpolluted tropospheric air. Crutzen, P. J. *Tellus* **26**, 47 (1974).

Section 1.5

Cyclic processes are of paramount importance in the exchange of chemical constituents between the troposphere and the planetary surface. Several of the citations for this section are to SCOPE, which is an acronym for Scientific Committee on Problems of the Environment. The latest of the publications is

SCOPE **39**, *Evolution of the global biogeochemical sulphur cycle.* Brimblecombe, P. and Lein, A. Yu. (John Wiley, Chichester, 1989.)

Earlier SCOPE publications directly relevant to this section are:

Nitrogen, phosphorus, and sulphur: global cycles, SCOPE **7** (1975).
The global carbon cycle, SCOPE **13** (1979).
Carbon cycle modelling, SCOPE **16** (1981).
Some perspectives of the major biogeochemical cycles, SCOPE **17** (1981).
The global biogeochemical sulphur cycle, SCOPE **19** (1983).
The major biogeochemical cycles and their interactions, SCOPE **21** (1983).

The next five references provide a general introduction to biogeochemical cycles, and they are followed by two references that show the way in which atmospheric residence times of gases are estimated.

The natural and polluted troposphere. Stewart, R. W., Hameed, S., and Pinto, J., in *Man's impact on the troposphere* (eds. Levine, J. S. and Schryver, D. R.) (NASA Reference Publication, No. 1022, 1978).

Interactions of biogeochemical cycles. Bolin, B., Crutzen, P. J., Vitousek, P. M., Woodmansee, R. G., Goldberg, E. G., and Cook, R. B. SCOPE **21**, 1 (1983).

Atmospheric interactions—Homogeneous gas reactions of C, N, and S containing compounds. Crutzen, P. J. SCOPE **21**, 67 (1983).

The global troposphere: biogeochemical cycles, chemistry and remote sensing. Levine, J. S. and Allario, F. *Environ. Monitg. & Assessm.* **1**, 263 (1982).

Influence of the biosphere on the atmosphere. (Symposium covering most gases and cycles). Dütsch, H. U. (ed.). *Pure & appl. Geophys.* **116**, 452 (1978).

Residence time and variability of tropospheric trace gases. Junge, C. E. *Tellus* **26**, 477 (1974).

Residence time and spatial variability for gases in the atmosphere. Hamrud, M. *Tellus* **35B**, 295 (1983).

There is considerable evidence that concentrations of some trace gases may be increasing. There follow references to a book and two articles that address this issue specifically and that also contain much information about atmospheric concentrations and lifetimes of the compounds.

The changing atmosphere. Rowland, F. S. and Isaksen, I.S.A. (eds.). (John Wiley, Chichester, 1988.)

Trace gas trends and their possible role in climate change. Ramanathan. V., Cicerone, R. J., Singh, H. B., and Kiehl, J. T. *J. geophys. Res.* **90**, 5547 (1985).

The changing atmosphere. Graedel, T. J. and Crutzen, P. J. *Scient. Am.* **261**, 28 (Sept 1989).

References follow to detailed discussions of individual cycles or important aspects of them.

Carbon

The carbon cycle. Bolin, B., SCOPE **21**, 41 (1983).

The carbon cycle. Bolin, B. *Scient. Am.* **223**, 124 (Sept. 1970).

Changes of land biota and their importance for the carbon cycle. Bolin, B. *Science* **196**, 613 (1977).

Natural sources of atmospheric CO. McConnell, J. C., McElroy, M. B., and Wofsy, S. C. *Nature, Lond.* **233**, 187 (1971).

The cycle of atmospheric CO. Seiler, W. *Tellus* **26**, 116 (1974).

Atmospheric cycle of methane. Ehhalt, D. H. *Tellus* **26**, 58 (1974).

Elemental carbon in the atmosphere: cycle and lifetime. Ogren, J. A. and Charlson, R. J. *Tellus* **35B**, 241 (1983).

Continuing worldwide increase in tropospheric methane, 1978 to 1987. Blake, D. R. and Rowland, F. S. *Science* **239**, 1129 (1988).

Measurements of atmospheric hydrocarbons and biogenic emission fluxes in the Amazon boundary layer. Zimmerman, P. R., Greenberg, J. P., and Westberg, C. E. *J. geophys. Res.* **93**, 1407 (1988).

Oxygen

This article explains the elements of the photosynthetic process

How plants make oxygen. Govindjee and Coleman, W. J. *Sci. Amer.* **262** (2), 42 (Feb 1990).

The oxygen cycle. Walker, J. C. G. *The handbook of environmental chemistry*, **1A**, 87 (Springer-Verlag, Berlin, 1980).

Nitrogen

The nitrogen cycle. Rosswall, T., SCOPE **21**, 46 (1983).

The nitrogen cycle. Delwiche, C. C. *Scient. Am.* **223**, 136 (Sept. 1970).

Analysis of sources and sinks of atmospheric nitrous oxide (N_2O). Cicerone, R. J. *J. geophys. Res.* **94**, 18265 (1989).

Production of nitrogen oxides by lightning discharges. Tuck, A. F. *Q. J. R. Meteorol. Soc.* **102**, 749 (1976).

Nitrogen fixation by lightning. Dawson, G. A. *J. atmos. Sci.* **37**, 174 (1980).

Nitrous oxide production by lightning. Hill, R. D., Rinker, R. G., and Coucouvinos, A. *J. geophys. Res.* **89**, 1411 (1984).

Sulphur

The sulphur cycle. Freney, J. R., Ivanov, M. V., and Rohde, H. SCOPE **21**, 56 (1983).

The sulfur cycle. Kellogg, W. W., Cadle, R. D., Allen, E. R., Lazrus, A. L., and Martell, E. A. *Science* **175**, 587 (1972).

The global sulfur cycle. Friend, J. P., in *Chemistry of the lower atmosphere* (ed. Rasool, S. I.). (Plenum Press, New York, 1973.)

The homogeneous chemistry of atmospheric sulfur. Graedel, T. E. *Rev. Geophys. & Space Phys.* **15**, 421 (1977).

Human influence on the sulphur cycle. Brimblecombe, P., Hammer, C., Rohde, H., Ryaboshapko, A., and Boutron, C. F. SCOPE **39**, 77 (1989).

Photochemistry of COS, CS_2, CH_3, SCH_3, and H_2S: implications for the atmospheric sulfur cycle. Sze, N. D. and Ko, M. K. W. *Atmos. Environ.* **14**, 1223 (1980).

Dimethyl sulfide in the surface ocean and the marine atmosphere: a global view. Andreae, M. O. and Raemdonck, H. *Science* **211**, 744 (1983).

Are global cloud albedo and climate controlled by marine phytoplanckton? Schwartz, S. E. *Nature* **336**, 441 (1988).

Atmospheric dimethyl sulphide and the natural sulphur cycle. Lovelock, J. E., Maggs, R. J., and Rasmussen, R. A. *Nature* **237**, 452 (1972).

Oceanic phytoplankton, atmospheric sulfur, cloud albedo, and climate. Charlson, R. J., Lovelock, J. E., Andreae M. O., and Warren, S. J. *Nature* **326**, 655 (1987).

Phosphorus

The phosphorus cycle. Richey, J. E., SCOPE **21**, 51 (1983).

Section 1.6

On the possibility that the influence on atmospheric composition of the biota may be part of a 'deliberate' control mechanism (see also book by Lovelock above).

Atmospheric homeostasis by and for the biosphere: the Gaia hypothesis. Lovelock, J. E. and Margulis, L. *Tellus* **26**, 2 (1974).

Biological homeostasis of the global environment: the parable of Daisyworld. Watson, A. J. and Lovelock, J. E. *Tellus* **35B**, 284 (1983).

Chaos in daisyworld. Zeng, X., Pielke, R.A., and Eykholt, R. *Tellus* **42B**, 309 (1990).

The world as a living organism. Lovelock, J. E. *New Scientist* **112**, 25 (1986).

Geophysiology: the science of Gaia. Lovelock, J. E. *Rev. Geophys.* **27**, 215 (1989).

Lovelock's newer book counters some of the arguments that have been advanced against the concept of Gaia, and explores some new ideas. The article by Kirchner is representative of those that claim the Gaia hypothesis to be ill-defined, untestable, and potentially misleading.

The ages of Gaia: A biography of our living Earth. Lovelock, J. E. (Oxford University Press, Oxford, 1988.)

The Gaia hypothesis: can it be tested? Kirchner, J. W. *Rev. Geophys.* **27**, 223 (1989).

2 Atmospheric behaviour as interpreted by physics

Chemical and physical processes in atmospheres are closely interdependent. Temperatures, for example, may depend on chemical composition, while the chemical processes occurring, and their rates, may depend on temperature. It is obvious that the distinction between atmospheric chemistry and atmospheric physics is artificial. Historically, however, the atmospheric sciences have emphasized the disciplines of meteorology and climatology, the first dealing with the time-dependent behaviour of atmospheric phenomena, the second with properties averaged over the long term. Both disciplines have been treated as extensions of applied physics. Chemical change is considered within this framework mainly for the way in which it influences physical behaviour. Our approach is somewhat different, since we are concerned primarily with the chemistry of atmospheres. With this slightly altered emphasis, we ask how the physical structure of the atmosphere affects chemical processes and composition, at the same time as recognizing that the chemistry itself plays a part in determining weather, climate, and physical behaviour in general. The aim of this chapter is to explain those aspects of the physics of atmospheres that have a bearing on the chemistry. Familiar concepts of physics and physical chemistry are applied to interpret pressures, absorption of radiation, temperatures, mixing, and so on. Winds, the circulation, and cloud formation may redistribute matter and energy and they are treated briefly.

2.1 Pressures

The mean atmospheric pressure at a planet's surface (p_0) is the total atmospheric force (mass, M_A, multiplied by the acceleration due to gravity at the surface, g_0) divided by the surface area, or

$$p_0 = \frac{M_A g_0}{4\pi R_p^2} \tag{2.1}$$

for a smooth planet of radius R_p. Table 2.1 gives some physical data for those bodies with atmospheres. Pressures have now been expressed in the SI-related units of Pa (Pascal: $1 \text{ Pa} = 1 \text{ Nm}^{-1}$) rather than in multiples of Earth Atmospheres (Table 1.1). One Standard Atmosphere on Earth is defined as 101325 Pa. We may as well admit here that atmospheric scientists in general, and meteorologists in particular, have not yet adopted the units of

Table 2.1 Physical data for bodies with atmospheres

Body	Planetary radius R_p (km)	Surface pressure $10^{-5} p_0$ (Pascal) $\equiv p_0$ (bar)	Albedo	Effective temperature T_e (K)	Surface temperature T_s (K)	Surface acceleration due to gravity g_0 (m s^{-2})	Escape velocity (km s^{-1})
Venus	6050	92.1	0.77	227	732	8.60	10.3
Earth	6378	1.01325	0.29	256	288	9.78	11.2
Mars	3398	6.3×10^{-3}	0.15	217	223	3.72	5.0
Jupiter	71900[a]	[1.00[a]]	0.33	110	170[a]	22.88	59.5
Saturn	60000[a]	[1.00[a]]	0.36	80	130[a]	9.05	35.6
Uranus	26145	[1.00[a]]	~0.4	56	78[a]	7.77	21.2
Neptune	24750	[1.00[a]]	~0.4	44	72[a]	11.00	23.6
Titan	2560	1.5	0.2	85	95	1.25	2.1

[a] At the 1 bar level.

Compiled from Beatty, J. K., O'Leary, B., and Chaikin, A. (eds.) *The new solar system*, 2nd edn. Cambridge University Press, Cambridge, 1982, Kondratyev, K. Y., and Hunt, G. E. *Weather and climate on planets*, Pergamon Press, Oxford, 1982; Stone, E. C. and Miner, E. D. *Science* **212**, 159 (1981); *Science* **215**, 499 (1982); Trafton, L. *Rev. Geophys. & Space Phys.* **19**, 43 (1981).

Pa or Nm^{-2}. Instead, the millibar (mbar) is in almost universal use. The bar, and hence millibar, are exactly defined in terms of the Pascal (1 mbar $= 10^2$ Pa), but they are not part of the SI. For conformity with existing practice, we shall use the millibar throughout most of this book. We note from the definition that 1 atm is very roughly equivalent to 1 bar, or more exactly to 1013.25 mbar.

Equation (2.1) can be rearranged in order to calculate M_A from the planetary pressure and radius. For Earth, the figures in Table 2.1 give $M_A = 5.3 \times 10^{18}$ kg; the calculation does not, of course, allow for variability of pressure or of surface elevation.

Gaseous components in the atmosphere do not settle down on the planetary surface under the influence of gravitational attraction because the translational kinetic energy of the particles competes with the sedimentation forces. As a result of the competition, the density of gas falls with increasing altitude in the atmosphere. The vertical pressure profile may readily be predicted by considering the change in overhead atmospheric force, dF, for a change in atmospheric altitude, dz:

$$dF = -g\rho A \, dz, \tag{2.2}$$

in a column of gas whose density is ρ and area A. Hence

$$dp = -g\rho \, dz. \tag{2.3}$$

For an ideal gas

$$\rho = Mp/RT = mp/kT, \tag{2.4}$$

where M, m are the relative molar mass and molecular mass respectively and R, k are the gas and Boltzmann's constants. Substitution of (2.4) in (2.3) yields

$$\frac{dp}{p} = -\frac{dz}{(kT/mg)}. \tag{2.5}$$

Integration then yields

$$p = p_0 \exp\left\{-\int_0^z \frac{dz}{(kT/mg)}\right\}, \tag{2.6}$$

if p_0 is the surface pressure. Equation (2.6) is known as the *hydrostatic equation*. For a planetary atmosphere, the acceleration due to gravity, g, is nearly constant at $\sim g_0$, since the atmospheric thickness is much less than the planetary radius. In a hypothetical atmosphere of constant temperature, eqn (2.6) then reduces to

$$p = p_0 \exp(-mgz/kT). \tag{2.7}$$

Numbers of particles are proportional to pressures for a fixed temperature, so that this form is equivalent to the Boltzmann distribution, with mgz corresponding to (geopotential) energy. The quantity (kT/mg) has the units of length, and represents a characteristic distance over which the pressure drops by a factor $1/e$. It is given the symbol H_s, and is called the *scale height*.

Temperature is not independent of altitude in a real atmosphere, so that the scale height is not constant. For the Earth's lower atmosphere, the scale height, H_s, varies between 6 km at $T \sim 210$ K to 8.5 km at $T \sim 290$ K. The corresponding values near the surface of other planets are Venus: 14.9 km, Mars: 10.6 km; Jupiter (cloud tops): 25.3 km. Figure 2.1 shows pressure as a function of altitude for a 'standard' Earth atmosphere (one which represents the horizontal and time-averaged structure of the atmosphere). Since the pressure axis is logarithmic, the simplified expression (2.7) would predict a straight line, and the observed deviations reflect temperature variations in the real atmosphere.

According to the development so far, the scale height depends on the molecular or molar mass. At first sight, then, it would appear that each component of a planetary atmosphere would have its own scale height, and the pressure distribution would be specific to that species. In that case, mixing ratios even of unreactive gases would be a function of altitude. Yet, at least in the Earth's lower atmosphere, observation shows the composition, in the absence of sources and sinks, to be constant. In fact, the lower atmosphere behaves as though it is composed of a single species of relative molar mass $\sim 0.2 \times 32 + 0.8 \times 28 = 28.8$. Similarly, in the lower atmospheres of the other planets, *all* components behave as though they had a single relative molar mass determined by the relative compositional abundances: 44 for Venus and Mars, 2.2 for Jupiter. The homogeneity of lower atmospheres is a consequence of mixing due to fluid motions. Mixing on a macroscale, by convection, turbulence, or small eddies, does not discriminate according to molecular mass, so that it redistributes chemical species that gravitational attraction is trying to separate on a molecular scale by diffusion. The relative importance of the molecular and bulk motions depends on the relative distances moved between transport events in each case. For molecular motions, that distance is clearly the *mean free path*, λ_m, the average distance a particle travels between collisions. The equivalent quantity for bulk fluid motions is the *mixing length*. A simple expression for mean free path in terms of pressure, p, and molecular diameter, d, can be derived from the kinetic theory of gases.

$$\lambda_m = (1/\sqrt{2\pi d^2})(kT/p). \tag{2.8}$$

The dotted line in Fig. 2.1 shows mean free paths calculated for the Earth's atmosphere, and emphasizes the inverse pressure dependence suggested by eqn (2.8). Fluid mixing lengths tend to decrease with increasing altitude, so that there is an altitude for which molecular diffusion and bulk mixing are of comparable importance. In the Earth's atmosphere, the mean free path and mixing lengths are both of the order of 0.1–1 m at $z = 100$–120 km. At higher altitudes, where the mean free path becomes larger than the mixing length, we may expect to see mass discrimination. Observations agree with these predictions. Above 100 km, molecular nitrogen begins to exceed its ground level mixing ratio. Atomic fragments are favoured gravitationally (as well as chemically—see later) at high altitudes. Finally, at the highest levels, only the lightest species (H, H_2, He) are present.

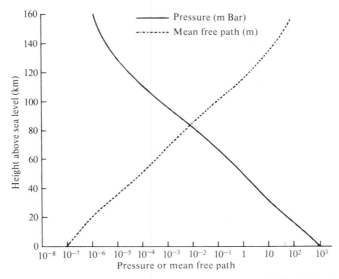

Fig. 2.1. Pressure and mean free path as a function of altitude in the Earth's atmosphere (Redrawn from Wallace, J. M. and Hobbs, P. V. *Atmospheric science*, Academic Press, New York, 1977.)

The region of transition in an atmosphere between turbulent mixing and molecular diffusion is known as the *turbopause* (or, sometimes, *homopause*). Luminous rocket trails often reveal the turbopause as a region below which the trail is violently disturbed, but above which it is not. Atmospheric structure is sometimes described in terms of composition, the well mixed region below the turbopause being called the *homosphere*, and the gravitationally separated region being called the *heterosphere*. More often, temperature structure is used to define atmospheric regions, as discussed briefly in Chapter 1, and we turn in the next sections to an examination of atmospheric temperature profiles.

2.2 Radiative heating

2.2.1 Solar radiation

Almost all the energy balance for the inner planets (Venus, Earth, Mars) is determined by solar heating.[a] Radiation that reaches the planets as heat, visible light, and near ultraviolet is emitted from the Sun's *photosphere*, which behaves nearly as a black body of temperature $\sim 5785\,\text{K}$ in these spectral regions (although there is much greater emission in the X-ray, far ultraviolet,

[a] Jupiter and Saturn have an internal heat source as well.

and radio spectral regions than a black body would allow). The total amount of energy of all wavelengths intercepted in unit time by unit surface area at the top of the Earth's atmosphere, corrected to the Earth's mean distance from the Sun, is known as the *solar constant*. The degree of variability of this 'constant', which can now be investigated by satellites in orbit outside the atmosphere, is of great potential importance in assessing climatic changes. At present, the solar flux through a surface normal to the beam is approximately 1368 W m^{-2} near the Earth. Some of the radiation is reflected by the surface and by the atmosphere; the overall reflectivity, A, of a planet is called the *albedo*, and is shown in Table 2.1. The fraction absorbed is thus $(1 - A)$. A planet also radiates thermal energy itself, and if the overall temperature of the planet is to remain constant, then the inward and outward fluxes must be the same. Let us make the simplifying assumption that the atmosphere does not itself absorb any radiation. An estimate of the *effective temperature*, T_e, of the planet can then be obtained by assuming that the planet is a black-body radiator which obeys the Stefan-Boltzmann 'fourth power' law

$$E = sT_e^4 \tag{2.9}$$

where E is the radiated flux (per unit area) and s is the Stefan-Boltzmann constant $(5.67 \times 10^{-8} \text{W m}^{-2} \text{K}^{-4})$. For a planet of radius R_p, the total emitting area is $4\pi R_p^2$, but the absorbing disc presented normal to the solar beam has an area πR_p^2, so that energy balance demands that

$$4\pi R_p^2 sT_e^4 = \pi R_p^2(1 - A)F_s, \tag{2.10}$$

where F_s is the solar flux at the edge of the planet's atmosphere. That is

$$T_e = \left\{ \frac{(1 - A)F_s}{4s} \right\}^{1/4} \tag{2.11}$$

The albedo of the Earth is ~ 0.29; using the values of F_s and s already quoted leads to a value of $T_e = 256$ K. Values of F_s for other planets can be estimated from that for Earth by using the inverse square law in conjunction with the planetary distances from the Sun. Effective temperatures calculated in this way from eqn (2.11) are listed in Table 2.1. At first sight, the agreement between the calculated effective temperatures, T_e, and the surface temperatures, T_s, might seem reasonable for Earth and Mars. However, it should be remembered that a change of 32 K at the Earth's surface, up or down, would make the planet largely uninhabitable. In any case, it is obvious that the calculation is grossly in error for Venus, where surface temperatures are 500 K higher than effective temperatures.

2.2.2 Radiation trapping: the 'greenhouse effect'

Our calculation of T_e deliberately excluded any absorption of solar or planetary radiation by atmospheric gases, and it seems reasonable to look first at atmospheric absorption as a way of reconciling effective and surface

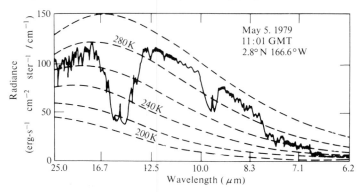

Fig. 2.2. Spectrum of infra-red emission escaping to space, as observed from outside the Earth's atmosphere by the Nimbus 4 satellite. Dashed lines represent the spectrum expected from a black body at different temperatures. Atmospheric absorbers cause the escaping radiation to come from different altitudes at different wavelengths, so that the effective temperature of the Earth's spectrum is wavelength dependent. (Figure reproduced from Dickinson, R. E., in Clark, W. C. (ed.), *Carbon dioxide review: 1982*, Oxford University Press, Oxford, 1982.)

temperatures. This idea is supported by the measured average infra-red emission temperatures of the planets as seen from *outside* their atmospheres (230, 250, 220 K for Venus, Earth, and Mars), which are quite close to the calculated values of T_e. A more detailed look at the Earth's planetary emission gives a further indication of what is happening. Figure 2.2 shows a low-resolution spectrum obtained from a satellite in a cloud-free field of view, the dashed lines indicating the expected black-body radiance at different temperatures. Over some of the spectral region, the temperature corresponds to near-surface temperatures. However, there is a huge emission temperature dip between 12 and 17 μm and smaller dips at 9.6 μm and at less than 8 μm. The three spectral regions correspond to infra-red active bands of CO_2, O_3, and H_2O respectively, and the interpretation of emission temperatures is that where the atmospheric gases have a non-absorbing 'window', the satellite views the ground or layers near it; at the wavelengths of absorption by atmospheric gases, the emission comes from higher, colder regions. Radiation from the Earth's surface has been *trapped* in the spectral regions of the absorption bands, and ultimately re-radiated to space at lower temperatures than those of the surface. Averaged over all wavelengths, the emitters are evidently about 32 K colder than the surface. As we shall see in Section 2.3, a temperature of 256 K is reached at about 6 km altitude, so that in this simplified picture one can visualize an equivalent radiating shell of the Earth lying 6 km above the surface.

Why is the radiating shell not also the absorbing shell? The answer lies in the spectral distribution of black bodies at different temperatures. Figure 2.3

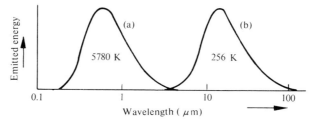

Fig. 2.3. Curves of black-body emission intensity plotted as a function of wavelength for (a) a temperature approximating to that of the Sun; and (b) the mean temperature of the Earth's atmosphere.

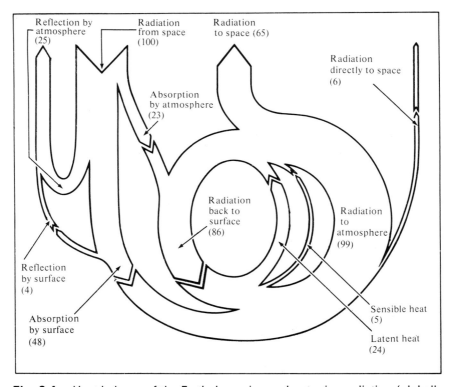

Fig. 2.4. Heat balance of the Earth. Incoming and outgoing radiation (globally averaged) are drawn in paths whose width is proportional to the contribution of each major component. (Reproduced from Clark, W. C. (ed.), *Carbon dioxide review: 1982*, Oxford University Press, Oxford, 1982, and modified for an atmospheric reflectivity of 25 rather than 26 per cent.)

shows the Planck black-body curves for 5780 K (Sun's photospheric temperature) and 256K (T_e for Earth). The bulk of the Sun's radiation lies at wavelengths where the atmospheric gases absorb only weakly. In contrast, throughout most of the wavelength region emitted by the low temperature Earth, the atmosphere is opaque. The idea is that the atmosphere lets shorter wavelength radiation in, but does not let the longer wavelength radiation out. By supposed analogy with the behaviour of panes of glass, the effect is often called the *greenhouse effect*. (In reality, greenhouses are effective more because they inhibit convection than because they trap radiation.) Absorption and emission for a particular transition occur at the same wavelengths, of course, so that one molecule can absorb radiation emitted by another molecule of the same chemical compound. Radiation is therefore passed back and forth between the infra-red active molecules, and escapes to space only from layers in the atmosphere high enough for the absorption to have become weak. The upper layers of the atmosphere suffer a net loss of energy that is relatively more than that for the lower ones, and they are therefore cooler.

The total heat budget of a planet with an atmosphere that absorbs and emits is shown pictorially in Fig. 2.4. The numerical values apply to Earth. Of 100 units of radiation reaching Earth, 47 are scattered by the cloud cover: 25 units are reflected to space, and the other 22 scattered to the ground. The average albedo of the Earth is thus a result of the 25 units reflected by clouds and 4 units reflected by the surface. Non-radiative heat losses from surface to atmosphere are due to convection of warm air (*sensible heat*) and condensation of moist air (*latent heat*).

2.2.3 Models of radiation trapping and transfer

The discussion of the last section has indicated that several chemical species can lead to radiation trapping in a planetary atmosphere. Each of these species is likely to be present with a differing mixing ratio and possibly different altitude distribution. Each will absorb or emit with an intensity, and with spectral features (vibrational, rotational, isotopic, etc.) according to its identity. Spectral lines have widths determined by collisional (pressure-dependent) and Doppler (temperature-dependent) effects. Light scattering by molecules, and especially by particles, alters the optical behaviour of the atmosphere. A proper study of radiation trapping must therefore accommodate at least these factors. The advent of high-speed computers with large memories has made it possible to investigate radiative transfer and trapping by direct *numerical modelling*. A kind of computer experiment is conducted in which all the necessary parameters can be used to determine the absorption and emission characteristics for a particular atmospheric element, and the radiative fluxes to and from that element calculated. To a large extent, computer models have removed the need to find reasonable simplifying assumptions that yield analytical algebraic expressions. At the same time, however, it must be said that an elegantly simplified theory can sometimes yield insights into the physical basis of a process that a numerical model can not. To illustrate the principles involved in radiation transfer, we develop a physical model that includes in its approximations a single absorbing species of wavelength-independent absorption, absence of scattering, and a local thermodynamic equilibrium. We shall also assume in this very crude model that all radiation is emitted or absorbed only in a vertical direction.

Absorption of radiation is governed by the interaction between photons and matter, so that if radiation of intensity I traverses thin slabs of absorber of unit area and thickness dz, the decrease in intensity is given by

$$-dI = In\sigma_a \, dz, \tag{2.12}$$

where n is the number density of absorbers, and σ_a a constant for the absorbing species and the wavelength of radiation. The constant σ_a has the units of area, and is called the absorption cross-section. Integration of eqn (2.12) for incident intensity I_0, and intensity, I_t, transmitted through a slab of thickness z yields

$$I_t = I_0 \exp\left\{-\int_0^z n\sigma_a \, dz\right\}. \tag{2.13}$$

In the case where n is independent of z, then eqn (2.13) simplifies to

$$I_t = I_0 \exp(-n\sigma_a z), \tag{2.14}$$

the familiar *Beer-Lambert* expression.

Any species that absorbs radiation also emits, and *Kirchhoff's Law* states that the emissivity and absorptivity will be identical. For short-wavelength (e.g. visible or ultraviolet) spectroscopy carried out in the laboratory, the emission intensity is almost always dominated by the absorption. In fact, eqn (2.12) should be modified to read

$$-dI = In\sigma_a \, dz - Bn\sigma_a \, dz. \tag{2.15}$$

where B is the black-body emission (per unit solid angle per unit surface area) and is a function of temperature. The modified equation can be used straightforwardly in one simplified picture of atmospheric radiative transfer. Let us first write the product $n\sigma_a z = \chi$, the *optical depth*. Reference to eqn (2.14) shows that in a homogeneous medium, an optical depth of unity corresponds to an intensity attenuation by a factor $1/e$. With χ measured from the top of the atmosphere downwards, eqn (2.15) can therefore be expressed

$$\frac{dI}{d\chi} = I - B. \tag{2.16}$$

Figure 2.5 represents a small thickness of the atmosphere through which the upward and downward radiation fluxes are represented by I_{up} and I_{down}. The net flux upwards through the layer, ϕ_r, is given by

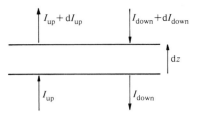

Fig. 2.5. Diagram to illustrate the upward and downward radiation fluxes discussed in the treatment of radiation transfer.

$$\phi_r = I_{up} - I_{down} \tag{2.17}$$

There is no net change in energy, since the layer does not heat up or cool down with time, so that $dI_{up}/dz = dI_{down}/dz$. Thus, $(dI_{up} - dI_{down})/dz = d\phi_r/dz = 0$, and ϕ_r is a constant independent of z (or χ). Equation (2.16) can be written explicitly for the upward and downward fluxes

$$\frac{dI_{up}}{d\chi} = I_{up} - B \tag{2.18}$$

and

$$-\frac{dI_{down}}{d\chi} = I_{down} - B. \tag{2.19}$$

Subtracting (2.18) and (2.19) gives

$$\frac{d(I_{up} + I_{down})}{d\chi} = I_{up} - I_{down} = \phi_r \tag{2.20}$$

and adding them gives

$$\frac{d\phi_r}{d\chi} = (I_{up} + I_{down}) - 2B. \tag{2.21}$$

Since ϕ_r is a constant, eqn (2.21) yields

$$I_{up} + I_{down} = 2B. \tag{2.22}$$

Substitution into eqn (2.20) and integration then gives

$$B = \tfrac{1}{2}\phi_r\chi + \text{constant}. \tag{2.23}$$

At the 'top' of the atmosphere, where $\chi = 0$, $I_{down} = 0$ and equations (2.17) and (2.22) combine to yield $I_{up} = 2B = \phi_r$ for this situation. Substitution of $\chi = 0$ and $B = \tfrac{1}{2}\phi_r$ into equation (2.23) shows that the constant is also $\tfrac{1}{2}\phi_r$, so that

$$B = \tfrac{1}{2}\phi_r(\chi + 1). \tag{2.24}$$

The result allows us to calculate the black body function B for any optical

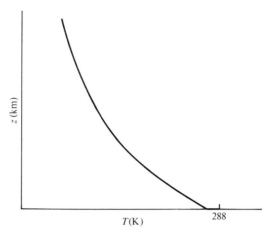

Fig. 2.6.   Radiative equilibrium temperature T as a function of altitude, z. There is a temperature discontinuity at the lower boundary. The altitude scale is arbitrary at this stage: units will appear in Fig. 2.10 (p. 56).

depth in the atmosphere, and since B is proportional to the black-body emission rate in eqn (2.9), the temperature can also be calculated as a function of χ (and, from a knowledge of n and σ_a, of z). In particular, if the temperature of the emitting edge of the atmosphere is T_e, that of a near-surface layer is T_0, and the total optical thickness of the atmosphere is χ_0, then

$$\frac{sT_0^4}{sT_e^4} = \frac{\frac{1}{2}\phi_r(\chi_0 + 1)}{\frac{1}{2}\phi_r},$$ (2.25)

or

$$T_0^4 = T_e^4(\chi_0 + 1).$$ (2.26)

Figure 2.6 shows how eqns (2.24) and (2.26) might work in the Earth's atmosphere. Since $T_0 \sim 288$ K and $T_e \sim 256$ K, χ_0 is ~ 0.6 on our model. Assuming in this very crude model that $\chi = \chi_0 \exp(-z/H_s)$, we can then generate the altitude profile of temperature predicted by the radiative equilibrium model. In comparison with the real atmosphere, the simple model predicts too steep a fall in temperature in the lowest few kilometres of the atmosphere, although it subsequently matches quite well for the next few kilometres. Convection in the lowest region reduces the *lapse rate* (rate of decrease of temperature with altitude), as we shall see in Section 2.3.

2.2.4 Trapping in real atmospheres

A real atmosphere may contain several species with infra-red active modes, and a full model will include their contributions to radiation trapping,

Table 2.2.   Contribution of atmospheric radiation absorbers to thermal trapping

Species removed	Percentage trapped radiation remaining
All	0
H_2O, CO_2, O_3	50
H_2O	64
Clouds	86
CO_2	88
O_3	97
None	100

Data of Ramanathan, V. and Coakley, J. A. Jr. *Rev. Geophys. & Space Phys.* **16**, 465 (1978).

together with the trapping produced by clouds and aerosols. Partly because the infra-red bands of the various components overlap, the contributions of the individual absorbers do not add linearly. Table 2.2 shows the percentage of trapping that would remain if particular absorbers were removed from the atmosphere. We see that the clouds only contribute 14 per cent to the trapping with all other species present, but would trap 50 per cent if the other absorbers were removed. Carbon dioxide adds 12 per cent to the trapping of the present atmosphere: that is, it is a less important trapping agent than water vapour or clouds. On the other hand, on its own CO_2 would trap three times as much as it actually does in the Earth's atmosphere. The point is of importance in seeing how far increases in the greenhouse effect could provoke climatic response to changed carbon dioxide concentrations. In this context, it is interesting to note that, since the upper layers of the atmosphere leak relatively more radiation to space than they trap, additional carbon dioxide leads to atmospheric cooling rather than warming for atmospheric layers above about 20 km. Many of the trace atmospheric gases, such as CH_4, N_2O, and NH_3, have infra-red modes active in the trapping region. These gases may therefore have a direct effect on global temperatures quite distinct from that exercised through their possible modification of concentration of major absorbers.

The influence of minor constituents is particularly marked if they absorb where there is otherwise an atmospheric window. This 'stopping up' of windows shows itself in the Venusian atmosphere. Over several years, doubt existed about whether a greenhouse effect on Venus could plausibly explain the high surface temperatures. The problem was in part to know how much solar radiation penetrated below the cloud tops, and in part to find infra-red active molecules that possessed an optical depth $\chi = (732/227)^4 - 1 = 107$.

Carbon dioxide alone cannot provide the necessary depth over the emitting spectral region. However, the Pioneer Venus and Venera 11/12 probes of 1978 have now shown that not only does enough sunlight reach the surface to fuel the effect, but that also the small H_2O and SO_2 concentrations are sufficient, together with pressure-induced transitions in CO_2, to close the spectral windows. For Mars, which also has an atmosphere predominantly composed of CO_2, the constraints are less severe, since $T_s \sim 223$ K is not so very much larger than $T_e \sim 217$ K. The requirement is for $\chi = (223/217)^4 - 1 \sim 0.12$. Radiative transfer can, in fact, account for the gross features of temperatures throughout most of the Martian atmosphere.

2.2.5 Unstable greenhouses: Venus, Earth, and Mars compared

Water vapour makes a sizeable contribution to atmospheric heating on Earth (Table 2.2). Since vapour pressures rapidly increase with increasing temperature, thus further increasing trapping, there exists a mechanism for positive feedback in the greenhouse effect. Evaporation from a planetary surface will proceed either until the atmosphere is saturated with water or until all the available water has evaporated. What happens on any particular planet will depend on the starting temperature in the absence of radiation trapping, since that will decide whether the vapour ever becomes saturated at the temperatures reached. That is, we need to compare the temperature-pressure relation with the phase diagram for the water system. Starting temperatures can be estimated in the way described for T_e in Section 2.2.1, but using throughout the albedo for cloudless Mars (0.15), instead of the real values for cloudy Venus and Earth. To illustrate the effect of water vapour, consider an initially dry atmosphere to which more and more water is added. Figure 2.7 shows the phase diagram for water, together with the planetary temperatures expected for differing water vapour pressures. On Mars and Earth, the additional heating due to liberation of water vapour is not sufficient to prevent the vapour reaching saturation as ice or liquid. However, on Venus there comes a critical vapour pressure (~ 10 mbar) when the rate of heating begins to increase dramatically: that vapour pressure is never reached at the lower temperatures on Earth or Mars. As a result, the P-T curve for the atmospheric water vapour increases more slowly than the vapour-liquid equilibrium curve. Condensation never occurs on Venus, and additional burdens of H_2O serve to increase the temperature even further. The effect is sometimes called the *Runaway Greenhouse Effect*. Certainly this positive feedback mechanism would explain why there is no surface water on Venus at the present day. Large amounts of water vapour could have been the dominant species in the early Venusian atmosphere, but photodissociation and escape of hydrogen to space (Section 2.3.2) would have removed most of the H_2O to leave the rather dry atmosphere now found (see Table 1.1).

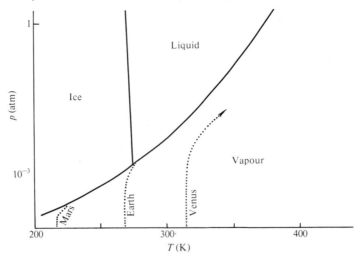

Fig. 2.7. The 'Runaway Greenhouse Effect'. The figure is a phase diagram for water onto which has been superposed as a dotted curve the 'greenhouse' temperatures produced as a function of water vapour pressure on Mars, Earth, and Venus. Starting temperatures are 'effective temperatures' calculated for the three planets using eqn (2.11) with identical values of albedo of 0.15 (appropriate to Mars today: Table 2.1). When water vapour becomes saturated with respect to ice or liquid, the increase in greenhouse effect is halted. On Venus, this saturation never occurs, and the temperature rises until all water is vaporized.

2.2.6 Diurnal and seasonal variations

On a real planet, the incident solar energy is not constant with time, since it varies during the day–night (*diurnal*) cycle, and may vary with season: both these variations are also likely to further depend on geographical location.

The rotation of a planet on its own axis is, of course, responsible for diurnal changes. Some planets rotate rather rapidly (e.g. Jupiter, in 9.8 h), some rather slowly (e.g. Venus, in 243 days). The planets also execute orbital motion around the Sun. If the equator of the planet is not in the same plane as the orbit, or if the orbit is elliptic, then the planet will experience seasonal variations of average incident radiation intensity, as illustrated in Fig. 2.8. Venus has an almost circular orbit, and very small equatorial *inclination*. It does not, therefore, experience seasons. Curiously, its slow rotation is in the sense opposite to most planets': that is, the rotation is *retrograde*, in the direction opposite to the orbital motion, and the Sun rises in the west. The contrarotation of the axial period of 243 Earth days with the orbital period of 224.7 days gives a mean Venusian day of about 117 Earth days, and illumination for half that period each cycle. Earth and Mars possess similar equatorial inclinations (23–24°), and both therefore experience seasons. In addition, the orbits are

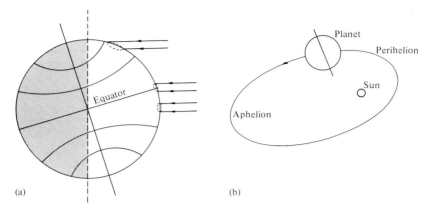

Fig. 2.8. Effects of (a) inclined; and (b) elliptic orbits on solar intensity. With an inclined orbit, the highest intensity (photons per unit area) is not necessarily at the equator, but where incident rays are normal to the surface; there are seasonal variations in the noontime angle of incidence and the length of day. An elliptic orbit causes variations as a result of changing Sun-planet distances.

slightly *eccentric* (elliptic). For example, Earth is slightly closer to the Sun during northern winter than during summer, so that the incident solar radiation is more intense in northern than in southern winter; seasonal changes due to inclination are thus reduced in the north. The opposite is true in the Southern Hemisphere, where seasonal changes due to eccentricity and inclination augment each other: other things being equal seasonal changes are thus more extreme in the Southern than in the Northern Hemisphere. Similar effects on Mars allow the polar cap in the north to persist throughout the year, while that in the south disappears during the southern summer.

Inclination has two effects on the incident solar intensity averaged over a day. Only at the equinoxes are days and nights 12 h long everywhere. At other times of year, the day period depends on date and latitude, which also determine the maximum elevation that the Sun reaches. The angle that the Sun's rays make to a normal drawn from the surface is usually referred to as the solar zenith angle, θ (Fig. 2.9). For any particular place, date, and time during day, θ is defined, the angle being minimum at any season near local noon, and largest at any time of day during local winter. Figure 2.9 shows the effect of solar zenith angle on incident intensity. A cylinder of solar radiation of area πr^2 falls on an area $\pi r(r/\cos\theta) = \pi r^2 \sec\theta$ of the planet or atmosphere. Intensity corresponds to power per unit area, so that the intensity is reduced by the factor $\sec\theta$. For $\theta \geqslant 90°$, we have night, and there is no radiation. A day's averaged intensity depends on a value of $\sec\theta$ averaged over the period for which $\theta < 90°$. The longer day (24 h) at the summer poles more than offsets the larger minimum zenith angle experienced there compared with the (12 h)

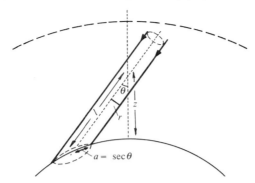

Fig. 2.9.   Intensity at the surface of a planet depends on the area illuminated. For a zenith angle θ, the intensity is diminished by a factor $\sec \theta$ compared with overhead illumination. The path, l, traversed through the atmosphere by the rays is equal to $z \sec \theta$ from any vertical altitude z.

equator. Perhaps surprisingly, the summer poles therefore receive more radiation than the equator. However, on Earth, snow, ice, and general cloudiness also tend to make the albedo near the poles greater than near the equator, so that the poles do not actually experience higher temperatures in summer.

Common experience on Earth tells us that temperatures at and near the surface change between day and night. However, for the atmosphere as a whole, the diurnal temperature variations are less than one per cent, because the heat capacity of the atmosphere is sufficient to take up the solar energy absorbed during the day with little temperature change. Venus has such a long diurnal period (~ 117 days: see p. 50) that significant temperature differences might be expected between day and night. In the event, the high atmospheric pressures (~ 90 atm—see Table 2.1), and hence mass, buffer the atmosphere against temperature changes. Entry probes from the Pioneer Venus mission showed very small day-night variation of temperature in the lower atmosphere: at nominal zero altitude the atmospheric temperatures were 732.4 K from the day probe and 733.8 K from the night probe. There are, however, marked altitude fluctuations of day-night temperature differences. It may be that wind, eddy, and wave transport modify the effects of high thermal inertia. The Martian day, at 24.6 h, is comparable with the Earth's, but several factors encourage relatively large diurnal atmospheric temperature changes. First, the atmospheric pressure is very low—less than one per cent of Earth's surface pressure (Table 2.1)—so that the atmospheric thermal inertia is much smaller than is ours. Secondly, the surface is arid and dry, with small thermal conductivity or capacity. Like in Earth's desert regions, ground surface temperatures show large day-night variations, but, unlike Earth, there are no huge oceanic areas where temperatures stay nearly constant. Finally, the major component of the Martian atmosphere is the 'greenhouse' gas CO_2, so that the atmosphere is sensitive to changing infra-red flux from the variable temperature surface. The two Viking landers descended through the Martian atmosphere at different local times of day (as well as at somewhat different latitude and season). Differences in lower atmospheric temperatures measured by the two landers seem consistent with the picture of diurnal as well as seasonal temperature variability in the Martian atmosphere.

Solar zenith angle has been viewed in our treatment as affecting the intensity of the solar radiation only through the influence on illuminated area. There is, however, a second influence of zenith angle in an atmosphere that absorbs or scatters solar radiation. Absorption of light depends on the total number of absorbing atoms or molecules that are encountered by a light beam, the form of relationship being shown in eqn (2.13). For any depth of atmosphere, d_a, that an overhead beam traverses, an oblique ray passes through a longer layer of path $d_a \sec \theta$. In general, then, the larger the zenith angle, the more will be the absorption in upper layers of the atmosphere, and the less radiation will penetrate to the lower levels. The translation of these ideas into quantitative terms requires a knowledge of altitude-concentration profiles and absorption cross-sections of all species capable of absorbing or scattering at the wavelength of interest. Qualitatively, experience tells us that more sunlight is scattered at sunrise or sunset, and the intensity of near ultraviolet radiation at the surface (at, say, $\lambda > 310$ nm) falls with increasing zenith angle much faster than does visible radiation because the ultraviolet passes obliquely through the ozone layer. To illustrate the magnitude of the effect, let us consider an unreal ozone layer that is 20 km thick containing 3×10^{12} molecules cm^{-3} of ozone homogeneously mixed throughout so that the simplified eqn (2.14) can be used. At $\lambda = 310$ nm, σ_a is $\sim 10^{-19}$ cm^2 for ozone, so that the transmitted intensity is a fraction $\exp(-3 \times 10^{12} \times 10^{-19} \times 2 \times 10^6) \sim 0.55$ of the incident intensity for overhead illumination ($\theta = 0$). Now if the zenith angle is say $75.5°$ ($\sec \theta = 4$), the optical path becomes 80 km rather than 20 km, and the transmitted light is a fraction 0.09 of the incident radiation. The dramatic fall is partly a result of the exponential in eqn (2.13) or (2.14), but $\sec \theta$ itself becomes very sensitive to θ at large angles, becoming 5.75 at $\theta = 80°$ and 11.5 at $\theta = 85°$. Thus for a zenith angle of $85°$ (near sunset), our hypothetical transmitted intensity at $\lambda = 310$ nm becomes 10^{-3} of the incident intensity. A final point we should remember about radiation absorption in atmospheres is that upper layers may be illuminated long after the ground and lower atmospheric layers are in shadow.

2.3 Temperature profiles

2.3.1 Troposphere, stratosphere, and mesosphere

Radiative transfer in the atmosphere tends to produce the highest temperatures at the lowest altitudes, as illustrated by the model calculation represented in Fig. 2.6. At first sight, it might seem inevitable that convection would arise in this situation, since the hot, lighter, air initially lies under the cold, heavier air. In an atmosphere, the behaviour is somewhat more complex because gases are compressible, and pressures decrease with increasing altitude. A rising air parcel therefore expands, does '$p\,dV$' work against the surrounding atmosphere, and is somewhat cooled. It can be that the temperature drop resulting from expansion would exceed the decrease in temperature of the surrounding atmosphere: in that case convection will not occur.

Simple thermodynamic arguments allow us to calculate the temperature decrease with altitude that is expected from the work of expansion. We imagine a packet of dry[‡] gas that is in pressure equilibrium with its surroundings, but thermally isolated from them. This packet of gas is imagined to be moveable up or down in the atmosphere, and the rate of change of temperature with altitude (lapse rate) can be calculated. Since no heat flows (adiabatic conditions) and the gas is dry, the temperature profile calculated is the *dry adiabatic lapse rate*. The First Law of Thermodynamics can be expressed as

$$dU = dq + dw, \tag{2.27}$$

where dU is change in internal energy, dq the heat supplied to the system and dw the work done on it. Here $dq = 0$ (adiabatic) and $dw = -p\,dV$. From the definition of enthalpy, H,

$$dH = dU + p\,dV + V\,dp. \tag{2.28}$$

In our calculation, therefore,

$$dH = V\,dp. \tag{2.29}$$

The heat capacity of the gas at constant pressure, C_p, is defined as $(dH/dT)_p$, so that

$$C_p\,dT = V\,dp. \tag{2.30}$$

We already have an expression for dp in the atmosphere from the differential form of the hydrostatic equation (2.3), and on substitution into (2.30) we obtain

$$C_p\,dT = -V\rho g\,dz. \tag{2.31}$$

For unit mass of gas, for which $V = 1/\rho$, we replace C_p by the value appropriate to unit mass (c_p), with the result

$$-\frac{dT}{dz} = \frac{g}{c_p} = \Gamma_d. \tag{2.32}$$

The dry adiabatic lapse rate, Γ_d, thus depends only on the acceleration due to gravity on a planet, and the average heat capacity per unit mass of the

[‡] 'Dry' here means containing no condensable material, so that the discussion can be applicable to planets other than Earth.

atmospheric gases. For Venus, Earth, Mars, and Jupiter, the calculated values of Γ_d are 10.7, 9.8, 4.5, and 20.2 K km^{-1}. This information now allows us to determine whether any particular dry atmosphere is stable or unstable with respect to convection. If the actual temperature gradient in the atmosphere, $-(dT/dz)_{atm}$, is less than Γ_d, then any attempt of an air packet to rise is counteracted by $p\,dV$ cooling that makes it colder and more dense than its surroundings. The atmosphere is stable. Conversely, if there were a tendency for $-(dT/dz)_{atm}$ to be greater than Γ_d, convection would be set up. Such convection will restore the temperature gradient until the atmosphere is stable again, so that actual atmospheric lapse rates rarely exceed Γ_d by more than a very small amount.

Presence of condensable vapours in the atmospheric gases complicates matters. Condensation to liquid or solid releases latent heat to our hypothetical air parcel. For a saturated vapour, every decrease in temperature is accompanied by additional condensation. Qualitatively, it is obvious that the *saturated adiabatic lapse rate*, Γ_s, must be smaller than Γ_d. The derivation of an expression, similar to (2.32), for Γ_s proceeds essentially as before, but with an additional term for the latent heat of condensation as the vapour load changes with temperature. Γ_s is not constant, but depends on temperature and pressure. In the Earth's atmosphere, where the important condensable vapour is water, Γ_s varies from about 4 K km^{-1} near the ground at 25°C to 6–7 K km^{-1} at around 6 km and -5°C. The stability conditions in saturated atmospheres are the same as in unsaturated ones, but with Γ_s replacing Γ_d. Partial saturation is treated in two stages, with Γ_d being applicable until temperatures are low enough to cause condensation, and Γ_s thereafter. In these circumstances, a situation known as *conditional instability* can arise. If the atmospheric lapse rate is less than Γ_d but more than Γ_s, the atmosphere is stable *unless* some kind of forced lifting (e.g. by winds) raises gas to an altitude where condensation occurs. From this point on, the atmosphere is unstable.

We now have enough information to interpret, in broad terms, some further features of atmospheric structure. Figure 2.6 shows the temperatures in the Earth's atmosphere predicted by the very simple radiative transfer model. More sophisticated models show similar trends. At low altitudes, the negative slope of the temperature–altitude relationship greatly exceeds even the dry adiabatic lapse rate. Convective instability therefore mixes the atmosphere until the adiabatic lapse rate is nearly reached. In fact, the average lapse rate on Earth (resulting from an appropriate combination of Γ_d and Γ_s, as well as global circulation) is 6.5 K km^{-1}. If we draw a line of this slope together with the radiative transfer predictions on the same diagram (Fig. 2.10), we see that the two lines intersect at ~ 11 km. Below that altitude, convection keeps even the largest atmospheric lapse rate only marginally more than the adiabatic lapse rate. Above, however, the atmosphere is stable, and the atmospheric temperatures are, to a first approximation, determined by radiative transfer for that part of the atmosphere where trapping rather than escape of radiation

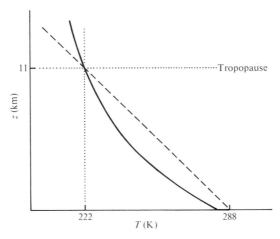

Fig. 2.10. A dry adiabatic lapṡe line (~ 6 K km^{-1}) has been added to the radiative transfer model results of Fig. 2.6. The atmosphere is convectively unstable below ~ 11 km.

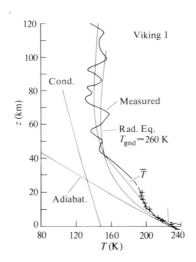

Fig. 2.11. Temperature structure of the Martian atmosphere as measured by instruments on board the Viking 1 lander. [Seiff, A. and Kirk, D. B. *J. geophys. Res.* **82**, 4364 (1977).]

occurs. The critical level corresponds to the tropopause, while below is the turning, turbulent troposphere and above the layered, stable stratosphere, as described without explanation in Chapter 1.

A Martian altitude-temperature profile, obtained in the Viking 1 experiments, is shown in Fig. 2.11. The mean lapse rate up to 40 km is about

1.6 K km^{-1}, which is markedly subadiabatic, as indicated by the adiabatic line. It is also clear that the CO_2 condensation boundary lies well below the temperatures measured in these low latitudes (22.3 °N) at mid-afternoon (landed 4:13 p.m. local time). It follows that clouds and hazes seen in the Mars atmosphere at similar places and seasons are probably water ice rather than CO_2. Contrary to some earlier predictions there is no sharply defined tropopause, and radiative transfer can explain the temperature profiles at least down to 1.5 km from the surface. However, there is a clear change in slope of the profile a few kilometres above the surface, which suggests a turbulent region, produced by winds rather than by natural convection. The subadiabatic profile itself may be 'stabilized' by direct absorption of sunlight by the widely-distributed atmospheric dust on Mars. Thermal 'tides', caused by the diurnal fluctuations discussed earlier, are thought to be responsible for the oscillatory excursions of the temperature profile above ~40 km. Venusian atmospheric temperatures determined by the Pioneer-Venus sounder are displayed in Fig. 2.12. Pronounced changes in lapse rate below 60 km are

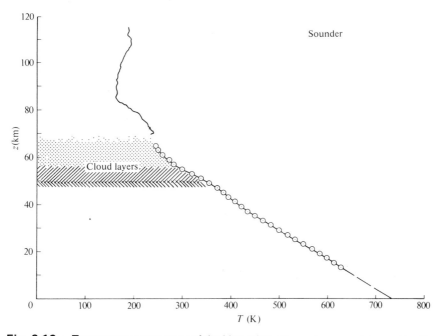

Fig. 2.12.   Temperature structure of the Venusian atmosphere as observed by the Pioneer-Venus 'sounder' probe. Data below ~65 km, shown by the circles, were obtained by direct sensing of temperature and pressure. Data at higher altitudes were derived from measurements of atmospheric densities defined by probe deceleration during high-speed entry into the atmosphere. [Seiff, A., Kirk, D. B., Young, R. E., Blanchard, R. C., Findlay, J. T., Kelly, G. M., and Sommer, S. C. *J. geophys. Res.* **85**, 7903 (1980).]

evident in higher resolution graphs, and several changes can be correlated with the clouds and cloud boundaries. In the middle cloud, for example, the lapse rate is close to adiabatic, but at the upper boundary makes a sudden change to stability as the upper cloud is entered. The main conclusions are that a major stable layer 25 km deep exists just below the clouds, below which there is evidence for convective overturning. Just above the clouds, the lapse rate becomes stable, and a 'stratosphere' begins which extends upwards to 110 km, becoming nearly isothermal above 85 km. From $z = 58$ to 59 km, at $T = 273$ K, the lapse rate drops from ~ 10 K km^{-1} to ~ 4 K km^{-1}, suggesting a phase change process. Water is the obvious candidate for phase change at these temperatures, but since other evidence suggests that the atmosphere is very dry, H_2SO_4 freezing may be responsible (80 per cent H_2SO_4 freezes at 270 K).

We have discussed the detailed structure of temperature profiles on Mars and Venus before considering Earth, because there is an additional and interesting feature in the Earth's atmosphere. Figure 2.13 for Earth presents roughly the same information as Figs 2.11 and 2.12 did for Mars and Venus. The troposphere and tropopause are evident enough, but beyond the tropopause the atmospheric temperatures only follow the radiative lapse rate in a very small region, and then they start to *increase* again in the stratosphere. Regions of negative lapse rate are said to constitute *inversions*, with hotter air on top of cooler. Obviously, inversion layers are peculiarly stable against vertical motions. Inversions can sometimes arise near the ground because of particular meteorological and geographical conditions. Such inversions can trap pollutants and prevent their dissipation, as we shall see in Chapters 4 and 5. Inversions have also been detected in the atmospheres of Jupiter and Saturn (at pressures and temperatures of about 135 mbar, 110 K; 70 mbar, 82 K respectively). On Earth, the stratospheric inversion is a result of heating by absorption of solar ultraviolet radiation in the ozone layer. Ozone is formed photochemically from O_2 and the 'layer' structure owes its existence to a peak in absorption and in reaction rates (Section 4.3.2). Too low in the atmosphere there are insufficient short-wavelength photons left to dissociate much O_2, while too high there are insufficient O_2 molecules to absorb much light and to associate with O atoms to make O_3. A series of chemical reactions concerned in the formation and destruction of ozone ultimately releases the chemical energy of O_2 dissociation, while the solar ultraviolet absorbed by ozone itself is also liberated as heat. As a consequence, the heating in the stratosphere is related to the ozone concentration profile (although modified according to the exact mechanism of conversion of ultraviolet to heat energy). On planets such as Mars and Venus, with little oxygen, there is virtually no ozone layer and no stratospheric heating. It is noteworthy that the ozone layer absorbs wavelengths from the Sun that do not reach the Earth's surface, and that the heating is achieved *in situ* rather than by absorption of re-radiated infra-red. Figure 2.13 shows that at altitudes above about 50 km, the heating

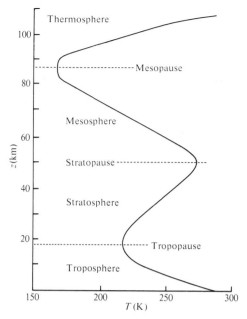

Fig. 2.13. Temperature-altitude profile for the Earth. The curve shown represents the mean structure for latitude 40 °N during June. (Data from Houghton, J. T. *The physics of atmospheres*. Cambridge University Press, 1977).

effect is too weak to compete with the cooling processes, and temperatures decrease again. Conventional nomenclature ascribes the name *mesosphere* to this next atmospheric region, and *stratopause* to the upper boundary of the stratosphere. The lapse rate is subadiabatic in the mesosphere (about 3.75 K km^{-1} in the mid-profile of our figure), so that the mesosphere is stable with respect to convective motions. The lowest temperatures in the entire atmosphere of the Earth are found in the mesosphere, and are lowest over the *summer* polar regions. Finally, the temperature stops falling at the *mesopause*.

2.3.2 Thermosphere, exosphere, and escape

All atmospheres eventually become so thin that collisions between gaseous species become very infrequent. One consequence may be that energy is not equilibrated between the available degrees of freedom. In particular, translational temperatures may be much higher than vibrational or rotational temperatures, and atomic species with high kinetic energy may not be sharing that energy with molecular species. Since the loss of energy to space from atmospheric constituents has been interpreted as radiation from infrared-active molecular vibrations, it can immediately be seen that inefficient re-equilibration between energy modes will lead to a bottle-neck for energy loss and excess translational temperatures. That region of a planetary atmosphere that shows, from these causes, increasing temperature with altitude is called the *thermosphere*. For Earth, the residual cooling mechanisms are probably infra-red emission from transitions between the angular momentum levels of atomic oxygen, together with some emission from nitric oxide. The cooling efficiency is low, and thermospheric temperatures are correspondingly high. Thermal inertia is very small, so that there is a big diurnal temperature variation. At $z = 250$ km translational temperatures are typically 850 K at night and 1100 K during the day; but at these altitudes the energetic particles that constitute the solar wind interact with the atmosphere, so that temperatures also depend on whether the Sun is 'quiet' or 'active'. By way of contrast, the thermospheric temperatures of both Mars and Venus are much lower than Earth's because of the preponderance of CO_2 that can itself radiate. Venusian thermospheric temperatures are only ~ 300 K during the day and ~ 100 K at night; Martian thermospheric temperatures are ~ 300 K, with smaller diurnal fluctuations.

It is worth emphasizing that the high temperatures in the Earth's thermosphere do not reflect a large energy source, but rather the extreme thinness of the atmosphere and its inefficiency at disposing of energy by radiative transfer. The solar radiation responsible for the heating lies in the extreme ultraviolet, at $\lambda \leqslant 100$ nm, and the primary effect is photoionization. Only about three parts in 10^6 of the total solar energy are radiated in this wavelength region, so that the source really is weak, although the energy of individual photons is high. By way of comparison, about one part in 10^4 of the solar energy lies at wavelengths between 100 and 200 nm (far or vacuum ultraviolet) and is absorbed (by O_2) at altitudes from 50–110 km. Much more ultraviolet solar energy (~ 2 per cent) is absorbed by the ozone layer, while ~ 98 per cent of the energy lies at 'near' ultraviolet, visible, and longer wavelengths ($\lambda > 310$ nm) and is the energy responsible for surface and near-surface heating *via* the greenhouse mechanism.

Sufficiently fast-moving species may *escape* from the gravitational influence of the planet altogether, or perhaps go into orbit around it. Two conditions must be met for escape to occur. First of all, the escaping constituent must not suffer any collisions as it leaves the planet. A *critical level*, z_c, can be defined for which a proportion e^{-1} of particles capable of escape experience no collisions as they pass out of the atmosphere. Intuitively, it would seem that this altitude is the one at which the mean free path, λ_m, [cf. eqn (2.8)] becomes equal to the scale height for the particle in question, and, indeed, this condition can easily be shown to fit the definition of z_c. For the Earth's atmosphere, z_c is 400–500 km, and beyond that altitude the region of the atmosphere is the *exosphere*. The second condition to be fulfilled for escape is that the particle has sufficient translational energy (i.e. starting velocity) to escape completely the gravitational attraction of the parent planet. An analogy can be seen with the ionization potential of an atom, where we are interested in removing an electron from the Coulombic attraction of the nucleus. In the gravitational case, the potential energy of a mass m at a distance a from the centre of a planet, mass M_p, is GmM_p/a, where G is the gravitational constant. If the few hundred kilometres contribution made by z_c to a are ignored, a can be set at the planetary radius R_p, and $GM_p = gR_p^2$. For a velocity v of the particle, the kinetic energy is $\frac{1}{2}mv^2$, so that the condition for escape is

$$\tfrac{1}{2}mv^2 > mgR_p, \tag{2.33}$$

or

$$v > (2gR_p)^{1/2}. \tag{2.34}$$

The minimum velocities meeting this condition are the *escape velocities* given in Table 2.1. Although it is only a critical (mass-independent) *velocity* that is needed, the *energy* is still mass-dependent. For Earth and Venus, the velocities are 10–11 km s^{-1}, but, put in energy terms, that corresponds to ~ 0.6 eV or ~ 58 kJ mol^{-1} per atomic mass unit. From Mars, the escape energies are one-fifth of those from Earth, and from Titan only one twenty-eighth.

We now have to consider escape mechanisms. Sir James Jeans formulated the details of the process that bears his name. Escape is seen to involve the high velocity particles in the tail of the Maxwell velocity distribution. Atomic hydrogen has a most probable velocity of ~ 3 km s^{-1} at 600 K, and the fraction of atoms with $v > 11.2$ km s^{-1} is just greater than 10^{-6}. Atomic oxygen, on the other hand, has a most probable velocity of ~ 0.8 km s^{-1}, and the fraction with a velocity greater than the escape velocity is 10^{-84}! Escape of hydrogen is reasonably probable, but little oxygen can have escaped from the Earth over its lifetime by the thermal mechanism. Several non-thermal mechanisms have been proposed for escape. Many of these employ the energy of a chemical (particularly photochemical or ionic) reaction to impart high translational energy to product fragments. A typical example is *dissociative recombination* between ions and electrons (see Chapter 6). The process is essentially the reverse of ionization, and the ionization energy has to be carried off as translation of fragments, as in the process

$$O_2^+ + e \rightarrow O^t + O^t, \tag{2.35}$$

where the superscript represents translational excitation. Even though some energy also goes into electronic excitation of atomic oxygen, 2.5 eV are released as translational energy to each O atom. The equivalent velocity is ~ 5.5 km s^{-1}, sufficient to escape from Mars, for example. Ions are particularly important both because of their excess energy over the neutral parents, and also because they can be accelerated in planetary electric fields. A few tens of volts acceleration can dominate over any available photon or chemical energies. In fact, the *charge-exchange* process

$$H^{+t} + H \rightarrow H^+ + H^t, \tag{2.36}$$

with accelerated H^+ ions, is thought to be a more important source of 'hot' H^t atom escape from the Earth's atmosphere than the Jeans mechanism.

There are three major stages in the escape process: transport of the constituent through the atmosphere, conversion to the escaping form (usually atom or ion), and the actual escape. Normally one of these will be the slowest, and rate-determining. On Earth, for example, hydrogen loss is limited by the upward diffusion flux and not by any process involved in the conversion to escaping H atoms. With this knowledge, the flux of escaping hydrogen can be estimated from the diffusion rate through the stratosphere (and hence through the mesosphere and to the exosphere). Mixing ratios for H_2O, H_2, and CH_4 yield an escape flux of $\sim 2.7 \times 10^8$ H atoms $cm^{-2} s^{-1}$ from the atmosphere, *regardless* of the escape mechanism.

Preferential escape of a lighter isotope is possible, since for a given available energy the lighter isotope will have a larger velocity. Isotopic fractionation can then occur, and give valuable clues about the history of an atmosphere. The changes in isotopic composition depend, for example, on whether all the original inventory of an element was in the early atmosphere, or whether much was in surface or interior reservoirs. We shall meet several examples in later chapters, and defer further discussion until then.

2.3.3 Vertical transport

Vertical mixing and redistribution of atmospheric constituents is assured in the troposphere because of the turbulence that characterizes the region. Above the tropopause, however, where lapse rates are subadiabatic (or even negative, as in some regions of the Earth's atmosphere), how, and how fast, are chemical species transported in the vertical plane?

In the last section, we suggested that the escape of hydrogen from our atmosphere is kinetically limited by the rate of diffusive motion through the upper atmosphere. Diffusion in one form or another does indeed provide a mechanism for transport through a non-turbulent atmosphere. The type of diffusion most familiar to chemists is *molecular diffusion*, in which molecules (or atoms) move in response to a concentration gradient, and in such a direction as to try to remove the gradient. At the level of individual particles, random molecular motion is destroying a special arrangement (higher concentration, and thus more molecules, in one volume element than in another). Thermodynamically, the system is behaving so as to maximize the entropy in accordance with the Second Law. Similar remarks apply to the velocities of particles as to concentrations. A system in which molecules with high velocity are separated from those with low velocity possesses a temperature gradient, and the statistical rerandomization corresponds to a heat flow, or thermal conduction, in response to that gradient. If the 'hot' molecules are also *chemically* different from the surroundings, as they might well be in an atmosphere, then thermal conduction also corresponds to an identifiable redistribution of matter, and the process is called *thermal diffusion*.

Calculation of the rates of transport in an atmosphere by diffusion is only slightly different from the procedure familiar in laboratory chemical problems. We shall illustrate the problem for diffusion in one dimension. The starting point for molecular diffusion, which we review here for one dimension, is *Fick's First Law*,

$$\Phi = -D(\partial n/\partial z), \qquad (2.37)$$

which states that the flux, Φ, of molecules across unit area in unit time is proportional to the concentration gradient in the z direction, $\partial n/\partial z$, that drives the process. The constant of proportionality, D, is the *diffusion coefficient*; for gases, elementary kinetic theory gives

$$D = \tfrac{1}{3}\bar{c}\lambda_{\mathrm{m}}, \qquad (2.38)$$

where $\bar{c}$ is the mean molecular velocity and λ_{m} the mean free path. Equation (2.8) for λ_{m} thus shows that D is inversely proportional to pressure. The flux of molecules leads to redistribution, and the rate of change in an element ∂z is the difference between the flux entering and the flux leaving. Hence

$$\frac{\partial n}{\partial t} = -\frac{\partial \Phi}{\partial z} = -\frac{\partial}{\partial z}\left(-D\frac{\partial n}{\partial z}\right). \qquad (2.39)$$

If D is independent of z, then the second equality on the right of (2.39) can be written $D(\partial^2 n/\partial z^2)$, and we have *Fick's Second Law* of diffusion.

In an atmosphere consisting of a compressible fluid, the modification to the simple diffusion equations in the vertical direction starts from a recognition that $\partial n/\partial z$ is not zero in a completely mixed atmosphere because pressure and number density of all components decrease with z according to the hydrostatic equation (2.5) or (2.6). That is, the driving force for diffusive separation of a component i is not the actual concentration gradient $\partial n_i/\partial z$ (where the subscripts refer to the particular component), but rather the difference between the actual concentration and the equilibrium (perfectly mixed) gradient, $(\partial n_i/\partial z)_{\mathrm{e}}$. Equation (2.37) is then rewritten

$$\Phi_i = -D\left\{\frac{\partial n_i}{\partial z} - \left(\frac{\partial n_i}{\partial z}\right)_{\mathrm{e}}\right\}. \qquad (2.40)$$

Let us write, for convenience, $n_i = f_i N$, where f_i is the mixing ratio of the species and N the total atmospheric number density. Remembering that $(\partial f_i/\partial z)_{\mathrm{e}} = 0$ for perfect mixing, we can rewrite (2.40) in the form

$$\Phi_i = -D\left(N\frac{\partial f_i}{\partial z}\right), \qquad (2.41)$$

and eqn (2.39) for the rate of change of concentration becomes, with D independent of z,

$$\frac{\partial n_i}{\partial t} = N \frac{\partial f_i}{\partial t} = -\frac{\partial \Phi_i}{\partial z} = D \left(\frac{\partial N}{\partial z} \cdot \frac{\partial f_i}{\partial z} + N \frac{\partial^2 f_i}{\partial z^2} \right). \tag{2.42}$$

The differential $(\partial N/\partial z)$ is equivalent to the differential form of the hydrostatic equation (2.5) in terms of number density and scale height $H_s = kT/mg$,

$$\frac{\partial N}{\partial z} = -\frac{N}{H_s} - \frac{N}{T} \frac{\partial T}{\partial z}, \tag{2.43}$$

with the term in $(\partial T/\partial z)$ coming from the temperature-dependent relationship between number density and pressure. Substitution of eqn (2.43) into (2.42), leads, after cancelling N on both sides, to

$$\frac{\partial f_i}{\partial t} = D \left(\frac{\partial^2 f_i}{\partial z^2} - \frac{1}{H_s} \frac{\partial f_i}{\partial z} - \frac{1}{T} \frac{\partial T}{\partial z} \cdot \frac{\partial f_i}{\partial z} \right). \tag{2.44}$$

The differential equation thus shows how an atmospheric mixing ratio will evolve with time under the influence of diffusion, and is the quantitative expression of the rate of diffusive transport in the vertical direction.

Molecular diffusion, and where appropriate thermal diffusion, are able to account for the rates of vertical transport in the upper mesosphere and thermosphere of the Earth's atmosphere. A number of indications suggest, however, that vertical mixing in the stratosphere is much more rapid than can be accounted for by simple diffusion mechanisms, even though the rates are also much less than those operating in the troposphere. Information about vertical transport in the stratosphere includes observations of tracer species injected by volcanic eruptions or by nuclear weapon testing. More indirectly, distributions of various minor constituents (e.g. CH_4 or N_2O) can be modelled for different hypothetical transport rates to obtain agreement with observation. To account for the relatively rapid mixing, the concept of *eddy diffusion* is introduced. Instead of individual molecules moving independently, small packets of gas are envisaged as executing eddy motions that can still be treated by the methods of the kinetic theory of gases. An eddy diffusion equation can then be derived, identical in all respects to eqn (2.44), with the exception that the molecular diffusion coefficient, D, is replaced by an eddy diffusion coefficient, K_z. An idea of the relative magnitudes of the rates of diffusion by molecular and eddy mechanisms is given by comparing the values of D and K_z for a mid-stratospheric altitude. At say 35 km, $D \sim 2 \times 10^{-3} \, m^2 \, s^{-1}$, while values of K_z typically adopted are about $10^2 \, m^2 \, s^{-1}$. A given 'tagged' molecule might then be expected to move 1 km vertically in a period measured in hours for the eddy mechanism, but in years for the molecular mechanism.

Transfer of chemical constituents in both directions across the tropopause is clearly of great importance. Many of the minor reactants in the stratosphere are initially released, naturally or anthropogenically, to the troposphere. Downward transport may be an important stratospheric loss mechanism for those species or their products, as well as a source of ozone to drive tropospheric chemistry. Exactly how the transfer takes place remains unclear. For nearly forty years, it has been supposed that the stratosphere contains very little water vapour (a few parts per million) because gases entered the stratosphere through a region cold enough to trap out the water. The requisite cold trap temperature is about 189 K, and the proposal is that troposphere to stratosphere exchange takes place in the tropics, which is where the lowest tropopause temperatures are experienced. However, recent high-flying aircraft experiments over Panama have found minimum H_2O concentrations a few kilometres higher than the levels of temperature minimum, and the smallest mixing ratios (3.5 p.p.m.) were significantly lower than the saturation values (6.1 p.p.m.) at the observed tropopause temperature. The implication is that the air enters the stratosphere in a region where the tropopause is colder than that at Panama, and the tropical western Pacific is one possible region. These tropical regions are, as we shall see in Section 2.4, associated with rapidly rising moist tropospheric air masses. Trace constituents, and water vapour in particular, could be transferred either by a steady rising motion or in violent cumulonimbus cloud convection that sometimes overshoots the tropopause by several kilometres. There is also known to be an exchange of stratospheric and tropospheric air at middle and high latitudes as a result of *tropopause folding*. Thin (~ 1 km) laminar intrusions of stratospheric air enter the troposphere for perhaps 1000 km parallel to the tropospheric jet stream (Section 2.4), and then become mixed with the turbulent tropospheric air. Radioactive materials from stratospheric nuclear weapon testing have been found to leave the stratosphere in regions largely limited to mid-latitudes in spring, and to be associated with the jet stream.

2.4 Winds

Horizontal motions of the atmosphere are central to the interests of meteorology. From the viewpoint of atmospheric chemistry, winds are responsible for the transport and mixing of chemical constituents. We shall examine in this section the barest outline of wind systems.

At and near ground level, winds are very variable and turbulent, gusting and changing direction frequently. Away from surface features and friction, the motions are more regular, and averaged over many days or weeks a clear and reproducible pattern emerges. The term *general circulation* is used to describe the global average winds that we need to discuss. Regular wind regimes have been recognized since the early days of sailing ships, with the tropical 'trade' winds in the Atlantic and Pacific (north-east in the Northern Hemisphere, south-east in the Southern) being very important. Even today, aircraft flying the transatlantic routes at mid-latitudes take longer in the east-west direction than on the return journey because of the prevailing high-speed westerly (from the west) *jet stream* winds at aircraft altitudes near the top of the troposphere.

Winds may be regarded as the air flow which is a response to pressure differentials between different locations on a planet. Temperature differences may be the cause of the pressure variations, as we shall see shortly. Thus the winds, by transporting heat (in both sensible and latent forms), tend to remove those temperature differences, in accordance with thermodynamic expectations. Equator-to-pole temperature differences on Earth are certainly smaller than would exist in the absence of atmospheric motions, although the oceans also play a large role. The Martian atmosphere is so thin that heat transport is ineffective in diminishing latitudinal temperature variations, and the winter polar regions become so cold ($\sim 150\,K$) that CO_2 can condense out. In turn, this seasonal condensation leads to a horizontal transport phenomenon unique to Mars, that of *condensation flow*, which is still a response to temperature differentials. Venus has an atmosphere so massive that it can maintain surface temperatures identical, within a few degrees, between poles and equator.

Rotation of a planet has a very dramatic effect on the atmospheric circulation, but we shall first see how temperature differentials can lead to pressure-driven winds. Our experience on Earth tells us that *surface* pressures do not vary dramatically from one place to another. The same is not true aloft in the atmosphere, and the various forms of the hydrostatic equations (2.5)–(2.7) all indicate that pressures fall off more rapidly with altitude when the temperatures are low. Put slightly differently, the scale height, H_s ($= kT/mg$), is proportional to temperature. If one location experiences a higher surface temperature than another, then for similar lapse rates, the temperatures higher in the troposphere will bear the same relationship, and the pressure aloft over the warmer area will be higher. The consequent force will accelerate the air mass until kinetic energy losses (e.g. by friction) are matched. Behaviour of this kind on a small scale is very familiar in coastal areas, where the land mass temperature fluctuates diurnally much faster than the sea mass. During the afternoon the land is at a higher temperature than the sea, and air aloft is at higher pressure over the land. Flow of air from land to sea therefore occurs, with the return air at low altitudes flowing from sea to land: the familiar *sea breeze*. At night, the rapid cooling of the land mass leads to the reverse motions, and a *land breeze* is experienced. On a global scale, exactly the same kind of wind motions occur, with hot air rising near the equator and being forced towards the cooler poles by the higher pressures in the higher temperature regions. Each hemisphere has its own circulation cell—named, after its discoverer, a *Hadley cell*—and they converge near the equator in the *Intertropical Convergence Zone* (ITCZ). Until recently, it was thought that the tropical trade winds were separated by a region of calm winds called the 'doldrums', but it now seems that the transition occurs a few degrees north of the equator, in the very narrow ITCZ belt, and is characterized by very strong upward motion and heavy rainfall. The returning surface air becomes saturated with water as it passes over the oceans. As it rises in the ITCZ, water condenses and heavy precipitation occurs. Release of latent heat by condensation increases the convective instability, reduces the lapse rate, and hence increases the driving pressure differential. It can readily be seen now why the ITCZ is a prime candidate region for the convective overshoot from troposphere to stratosphere suggested in Section 2.3.3.

Hadley circulation might well consist of two cells each encompassing a hemisphere from equator to pole, and with the flows directly north-south. For Earth, the real Hadley circulation only extends within the tropics (up to the 'horse latitudes'), and it has a strong westerly component aloft, and a corresponding easterly component on the return surface flow. Both the westerly directional component and the limited span of the Hadley cell are a consequence of planetary rotation. Atmospheric gases possess mass, and if they rotate more or less with the planet they therefore possess angular momentum. North-south motions imply a change in radius of rotation, decreasing to near zero at the poles. Yet angular momentum must be conserved, and the atmosphere achieves this conservation by developing zonal motion (that is in the direction of rotation of the Earth). For example, an air parcel at rest with respect to the equator must develop a zonal (west to east) velocity by the time it reaches $30\,°N$ of $134\,m\,s^{-1}$. The hypothetical force producing this motion perpendicular to the initial direction of transport is called the *Coriolis force*. The horizontal component of the Coriolis force is directed perpendicular to the horizontal velocity vector: to the right in the Northern Hemisphere and to the left in the Southern Hemisphere. It can readily be shown that for an air parcel of horizontal velocity V_h at latitude ϕ on a planet with angular velocity Ω, the horizontal component of the Coriolis force, F_c, has a magnitude

$$F_c = 2\Omega V_h \sin \phi. \tag{2.45}$$

That is, the force is minimum at the equator and maximum at the poles. Trade winds on Earth clearly fit the requirements of angular momentum conservation. The winds aloft, from equator to poles, gain a component from west to east in both hemispheres, and the returning surface trade winds are north-easterly (from the north-east) and south-easterly, north and south of the equator. At mid and high latitudes, the Coriolis force increases [cf. the $\sin \phi$ term in eqn (2.45)] and winds flow mostly in an east-west direction. Fluid dynamic instabilities actually break up the Hadley cell. Each hemisphere has, according to one interpretation, three cells, of which the middle one circulates in a thermodynamically indirect sense, with air rising at the warm end and sinking at the cold. Quite small perturbations result in a drastic circulatory change. Figure 2.14 indicates both the surface winds and atmospheric cells envisaged by the three-cell pattern.

Fluid dynamic instabilities of a different kind produce a wave-like progression of low- and high-pressure regions (*cyclones and anticyclones*) around middle latitude areas of the Earth. Disturbances of this kind can be

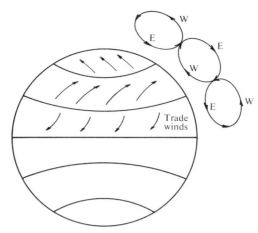

Fig. 2.14. Organization of circulation systems on Earth into three cells in each hemisphere.

both reproduced in the laboratory in experiments with heated rotating fluids, and predicted by numerical models. The instabilities arise when latitudinal temperature differentials become too large, and they are called *baroclinic instabilities*. Eddy motions (*baroclinic waves*), resulting from the instability, transport heat both poleward and vertically upward (thus reducing the tropospheric lapse rate). They also transport momentum to the upper troposphere and thus maintain the high velocity jet stream. Four to six pairs of waves typically encircle the Earth at any one time.

Large amplitude planetary scale waves in the stratosphere can play an important role in the transport and variability of inert and photochemically active species. A dynamical phenomenon of considerable interest in the Earth's upper atmosphere is the *Sudden Stratospheric Warming* (SSW) that occurs about once every one to three years in the Northern Hemisphere winter. A large growth in wave amplitude occurs over a two-week period, and temperatures at 40 km altitude can increase by as much as 80 K at high latitudes. This heating is sufficient to reverse the usual latitudinal temperature gradients and winds, the normal westerlies temporarily changing to easterlies. These planetary scale eddies then transport heat, and presumably trace gases, polewards during the warming.

Having established that the east-west component of the Earth's winds is intimately bound up with the effects of planetary rotation, it is interesting to compare the circulation patterns with those of other planets. Rapid rotations of Jupiter and Saturn (periods 9.8 h, 10.23 h) seem to be responsible for the alternating cloud bands of different colours, seen so beautifully in the Voyager pictures. The cloud bands are separated by jet streams, and the Great Red Spot and other such features seem to have a fluid-dynamic origin. There is little pole-to-equator energy transfer at the level of the clouds. Venus rotates very slowly (period 243 days), so that Coriolis forces and associated instabilities are weak. Nevertheless, the winds in its lower and middle atmosphere blow primarily in an east-west direction in the sense of the planet's rotation. Close to the cloud tops, the wind velocities are a phenomenal $100\,\mathrm{m\,s^{-1}}$ over the entire surface (but only about $1\,\mathrm{m\,s^{-1}}$ near the surface). Transport of momentum by a combination of a hemispheric Hadley circulation and eddies is thought to generate the high east-west velocities. The centrifugal force due to the winds themselves may be responsible for the east-west directionality; the north-south component is characterized by much more modest wind speeds (5–$10\,\mathrm{m\,s^{-1}}$) at the cloud tops.

Winds on the planets seem, then, to be explicable largely in terms of a thermally driven circulation modified by Coriolis forces and baroclinic instabilities. Additional features, such as stationary eddies (resulting from topographical or temperature contrast) and thermal tides (especially important on Mars—see Section 2.2.6) also influence transport of heat and momentum (e.g. to the stratosphere) to produce the observed general circulation at all altitudes.

2.5 Condensation and nucleation

Cloud formation is intimately coupled, *via* optical and latent heat effects, into the radiation balance, atmospheric temperature structure, and circulation patterns that we have been discussing in the preceding sections. Just in terms of albedo, clouds have a remarkable influence: a cloudless Earth, with the albedo of Mars (0.15), would have an average ground-level temperature of ~ 304 K, some 16 degrees higher than the present-day value, with consequent melting of polar ice caps.

In this section, we wish to examine rather more closely how condensation can occur. The principles apply to formation of liquid droplets or solid particles, and so are relevant to cloud or aerosol formation in any atmosphere. We shall, however, illustrate the problem initially with respect to water clouds. Section 1.3 contains a hint that condensation requires the presence of small particles to act as condensation nuclei. The problem concerns the vapour pressure produced over a curved surface. As we shall show shortly, the equilibrium vapour pressure at any temperature is larger for a drop than the saturated vapour pressure over a plane surface, and the excess vapour pressure increases as the radius of the drop becomes smaller. The atmosphere must therefore be supersaturated with respect to the thermodynamic (plane surface) vapour pressure if droplets are to condense. A water droplet of radius 0.01 μm, for example, requires a surrounding atmosphere supersaturated by about 12.5 per cent[a] if it is to survive. However, a droplet this size contains about 10^5 water molecules, a number exceedingly unlikely to come together at the same time by chance. A smaller initial droplet would contain fewer molecules, but require an even larger supersaturation. Atmospheres *are* sometimes supersaturated, but 12.5 per cent is about the maximum ever observed. One way out of the apparent dilemma is provided by condensation on nuclei. It is certainly suggestive that rainfall is greater over industrial areas where particle counts are highest, and that more rain falls on weekdays than at weekends. Before exploring the action of condensation nuclei, however, we should see why vapour pressure is influenced by curvature.

All liquids (and solids) possess an energy at the interface with their vapour that we call the *surface energy*. The energy is a function of surface area, σ, and can conveniently be expressed as $dG = \gamma\,d\sigma$ where dG is the surface contribution to Gibbs free energy and the constant γ is called the *surface tension*. It is in order to minimize the surface energy that liquids try to form spherical drops (the shape with minimum surface to volume ratio). The surface energy augments the chemical potential of the liquid phase, so that the equilibrium pressure of the vapour must be correspondingly larger. Some quantitative manipulation may be carried out to determine the free energy change associated with condensation, ΔG_{cond}, in producing a droplet of radius r. For a plane surface, we would write

$$G_{liq} = G_{liq}^{\ominus} \tag{2.46}$$

$$G_{vap} = G_{vap}^{\ominus} + nRT\ln p, \tag{2.47}$$

where n is the number of moles in the system. Thus,

$$\Delta G_{cond} = \Delta G_{cond}^{\ominus} - nRT\ln p. \tag{2.48}$$

At equilibrium $\Delta G_{cond} = 0$ and $p = p_{sat}$, so that

$$\Delta G_{cond}^{\ominus} = -nRT\ln p_{sat}. \tag{2.49}$$

For the curved-surface droplet, of surface area σ, eqn (2.46) must be modified to

$$G_{liq} = G_{liq}^{\ominus} + \gamma\sigma, \tag{2.50}$$

and

$$\Delta G_{cond} = \Delta G_{cond}^{\ominus} + \gamma\sigma - nRT\ln p \tag{2.51}$$

$$= \gamma\sigma - nRT\ln(p/p_{sat}). \tag{2.52}$$

---

[a] If the saturated vapour pressure over a plane surface is p_{sat} and the vapour pressure actually present is p, then a supersaturation of 12.5 per cent means $p/p_{sat} = 1.125$. The term *relative humidity* is given to the ratio expressed as a percentage ($100p/p_{sat}$), so that percentage supersaturation is relative humidity less 100.

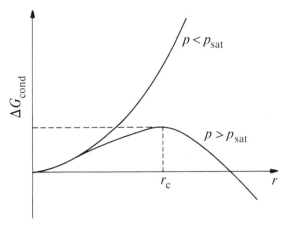

Fig. 2.15. Change in Gibbs Free Energy, ΔG_{cond}, for condensation as a function of droplet size. If the vapour pressure is greater than the saturation vapour pressure with respect to a plane surface, droplet formation can be stable once a critical radius, r_c, has been exceeded.

For a droplet of radius r, $\sigma = 4\pi r^2$, and the number of moles, n, is $(4/3)\pi r^3 \rho/M$ where ρ is the liquid density and M the relative molar mass. Hence

$$\Delta G_{cond} = 4\pi r^2 \gamma - \tfrac{4}{3}\pi r^3 \frac{\rho}{M} RT \ln\left(\frac{p}{p_{sat}}\right). \tag{2.53}$$

Figure 2.15 shows ΔG_{cond} calculated from this expression for a droplet in an undersaturated ($p < p_{sat}$) and a supersaturated ($p > p_{sat}$) vapour. In the first case, ΔG_{cond} becomes more positive with increasing drop radius, so there is no tendency for a drop to form. On the other hand, for the supersaturated vapour, ΔG_{cond} passes through a maximum at a droplet radius r_c, and then becomes smaller for increasing droplet radius. The maximum thus corresponds to a position of *unstable equilibrium*. If an embryonic droplet has radius $r < r_c$ it tends to evaporate, but if $r > r_c$ it tends to grow spontaneously at the expense of the vapour. Differentiation of eqn (2.53) with respect to r to obtain the position of maximum ΔG_{cond} allows calculation of the critical radius for any degree of supersaturation. The result is

$$r_c = \frac{2\gamma}{(\rho/M)RT\ln(p/p_{sat})}, \tag{2.54}$$

and is known as the *Kelvin equation*. It can be rewritten in terms of the vapour pressure, p_{drop}, needed to sustain any drop of radius r in (unstable) equilibrium

$$p_{drop} = p_{sat}\exp(2\gamma M/r\rho kT). \tag{2.55}$$

The positive exponential in eqn (2.55) now shows explicitly that $p_{drop} > p_{sat}$,

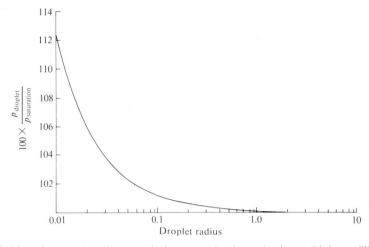

Fig. 2.16. Supersaturation needed to sustain drops in (unstable) equilibrium. Data are calculated from eqn (2.55) for pure water at 5°C and expressed as a percentage relative humidity with respect to equilibrium vapour pressure over a plane surface. (From Wallace, J. M. and Hobbs, P. V. *Atmospheric science.* Academic Press, New York, 1977.)

and that p_{drop} increases with decreasing r. Figure 2.16 shows supersaturation as a function of droplet radius calculated from eqn (2.55).

We are now in a position to see why condensation nuclei can promote condensation. Suppose water is condensed on a completely wettable nucleus of diameter $0.2\,\mu$m. The water film would be in unstable equilibrium with air supersaturated by as little as 0.6 per cent, a very reasonable atmospheric value. Slight increases in drop size would now lead to additional growth of the droplet as the free energy decreases yet further.

A most interesting situation arises if a condensation nucleus is soluble in water. Saturated vapour pressures are lower over solutions than over pure solvent, as a result of the lowered chemical potential of the solution. Solution droplets can therefore exist in equilibrium with their surroundings even when the relative humidity is less than 100 per cent (unsaturated air). The thermo-dynamic advantage gained by dissolving explains why salts can *deliquesce* in moist, but not necessarily saturated, air. Growth of the droplet may be limited, because at some stage the solution becomes sufficiently dilute that the droplet vapour pressure equals that of the surroundings. Additional condensation would tend to increase the droplet vapour pressure beyond that of the sur-roundings and evaporation would follow. Conversely, evaporation from the equilibrium droplet would make the solution more concentrated and favour more condensation. Droplet size is therefore stabilized, and the droplet is in stable equilibrium with its surroundings. Haze droplets are often of this stabilized kind. Only if the stable equilibrium diameter is also near the critical value, r_c, for the unstable equilibrium of surface-tension-controlled growth can the particle go on condensing yet more material.

Heterogeneous nucleation on pre-existing particles, although one of the most important mechanisms, is not the only way in which clouds and aerosols can be formed. Ion nucleation may be an important source of aerosol, especially in the stratosphere. Because of charge-induced dipole Coulombic forces, ions have a greater tendency to attract molecules around them than do neutral species, between which the intermolecular forces are much weaker. Large 'cluster' ions are well known in the laboratory, and will be discussed in Chapter 6. In the stratosphere, balloon-borne mass spectrometers find the most abundant ions to be of the type $HSO_4^-(H_2SO_4)_n(H_2O)_m$ with n and m up to 4 or 5. For $n > 3$, these ions already develop properties similar to $H_2SO_4 - H_2O$ solution droplets. Ultimately, the ions can become neutralized to yield sulphuric acid aerosol. The essential feature of ionic processes such as this is the reduction of free energy due to the electrostatic term that allows particle growth to start at lower supersaturations or smaller radii than would otherwise be possible.

Once a liquid or solid particle has passed its critical size for condensation, it may grow by several mechanisms. Continued condensation is one possibility, larger drops (lower saturated vapour pressure) growing at the expense of smaller ones (higher pressure). Vapour pressure differences are even more important in the growth of ice and snow crystals in water supercooled with respect to freezing. At say $- 10\,^\circ C$, saturated vapour pressures over water are about 10 per cent larger than over ice, so that ice grows at the expense of liquid. *Agglomeration* or *coalescence* represents another very important mechanism for growth. A large particle moving through a field of smaller ones can sweep them up. Raindrops coalesce in clouds in this way, and if updraughts compete with gravitational settling to keep drops in the cloud, there may be time for considerable growth. A large raindrop (~ 2.5 mm radius at the beginning of a shower) will need to have collected up the equivalent of $(2500/2.5)^3 = 10^9$ droplets of initial size $2.5\ \mu m$! At least it is clear why heavy rain is associated with clouds of considerable thickness (e.g. cumulus) and drizzle with thinner ones (e.g., stratus).

2.6 Light scattering

Electromagnetic radiation may be scattered by a medium even when it is not absorbed, and, as suggested in Section 1.3, the scattering behaviour depends on the size of the scattering body relative to the wavelength of radiation. Electrons in any object are driven into oscillation by incident light, and in due course those electrons re-emit radiation. For a perfectly homogeneous non-absorbing medium, all secondary re-emitted waves interfere destructively except in the original direction of the incident beam. To an observer, therefore, the medium appears to have had no effect. If, however, there are inhomogeneities in the medium, scattered radiation in other directions does not cancel exactly, so that light can be observed away from the original direction of propagation. At the same time, the intensity of the forward beam is decreased by an amount equivalent to the light scattered. Analogous equations to those for absorption can be written, (2.12) being replaced by

$$-\mathrm{d}I = In\sigma_s\,\mathrm{d}z, \qquad (2.56)$$

where σ_s, the *scattering cross-section*, replaces σ_a, the absorption cross-section. Integrated forms are exactly equivalent to (2.13) and (2.14). A quantity referred to as the *scattering area coefficient*, κ, compares (as a ratio) the scattering cross-section with the geometrical cross-sectional area of the scattering particles.

Particles which are small compared with the wavelength of the incident light have their entire electronic distribution distorted by the electromagnetic radiation. Scattering in these circumstances is called *Rayleigh scattering*. Conversely, several different waves can be scattered from large particles, and interference of the various waves produces a different scattering behaviour, both in magnitude and angular distribution. The comparison between wavelength λ and particle radius r can be expressed most conveniently in terms of the dimensionless size parameter $\alpha = 2\pi r/\lambda$. For $\alpha \ll 1$, we are in the Rayleigh scattering regime. Molecular diameters for gases are typically one thousand times smaller than visible and near ultraviolet wavelengths, so that non-aggregated species in the atmosphere cause Rayleigh scattering at these wavelengths. Similarly, weather radar operates at 'microwave' wavelengths of a few centimetres, so that scattering by raindrops ($r < 2.5\,\mathrm{mm}$) is also Rayleigh in nature.

The scattering parameter, κ, depends not only on α, but on the refractive index of the particles. For the Rayleigh scattering case of $\alpha \ll 1$ it can be shown that κ is proportional to α^4: that is, the scattering intensity is proportional to $1/\lambda^4$ for a fixed radius particle. That result explains why light scattered from an aerosol-free sky is blue, while that transmitted through long pathlengths at sunrise and sunset is red. Fourth-power proportionality emphasizes the wavelength dependences. Red light ($\lambda \sim 650\,\mathrm{nm}$) and blue light ($\lambda \sim 450\,\mathrm{nm}$) are scattered in the relative ratio $1:(650/450)^4 = 1:4.4$. Rayleigh scattering theory predicts that the scattered intensity at any angle θ to the incident beam is proportional to $(1 + \cos^2\theta)$ for unpolarized radiation. Maximum intensity therefore occurs in the forward ($\theta = 0$) and backward ($\theta = 180°$) directions. As the inequality $\alpha \ll 1$ ceases to hold, so the angular dependence of scattering ceases to follow the simple Rayleigh formula, and κ also deviates from its α^4 dependence. For $\alpha \gtrsim 50$, κ becomes ~ 2, and the angular distribution of scattered radiation can be described by conventional geometrical optics. Visible radiation ($\lambda \sim 500\,\mathrm{nm}$) thus undergoes geometrical scattering by

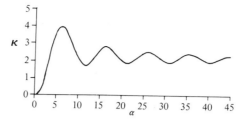

Fig. 2.17. Scattering area coefficient, κ, as a function of size parameter, α, for a refractive index of 1.33 (From Wallace, J. M. and Hobbs, P. V. *Atmospheric science*. Academic Press, New York, 1977.)

particles of radius greater than about 4 μm. This size range includes 'typical cloud drops' ($\sim 10\,\mu$m) and larger particles. Distinctive optical phenomena such as glories, halos, and rainbows are produced in this regime.

We have so far avoided discussion of the region for which α is neither much less than nor much greater than unity. For visible radiation, the range of sizes involved is from a few nm to a few μm: that is, it spans exactly that range of sizes conventionally associated with Aitken particles, condensation nuclei, and aerosols in general. A theory of scattering applicable to all values of α was worked out by Gustav Mie as long ago as 1908; the term *Mie scattering* is used more restrictively for α typically in the range 0.05 to 50. The angular distribution of scattered radiation is very complicated, and varies rapidly with α in this range. Scattering area coefficients (κ) show an oscillatory dependence on α, as illustrated in Fig. 2.17 for refractive index 1.33 (water). Note that for $\alpha \sim 5$, corresponding to $r \sim 0.4\,\mu$m in visible light, the scattering is particularly intense. Colours associated with Mie scattering of sunlight by haze, smoke, dust, etc., depend first on whether the particle sizes cover several oscillations of the $\kappa - \alpha$ relationship: if they do, the scattering is fairly neutral. For uniform particles, however, the colour cast will depend on whether κ increases or decreases with α (and, thus, in the reverse sense, with λ).

Bibliography

Books on general atmospheric physics and meteorology.

The *atmosphere*. Lutgens, F. K. and Tarbuck, E. J. (Prentice-Hall, Englewood Cliffs, New Jersey, 1982.)

Atmospheres. Goody, R. M. and Walker, J. C. G. (Prentice-Hall, Englewood Cliffs, New Jersey, 1972.)

Atmospheric science. Wallace, J. M. and Hobbs, P. V. (Academic Press, New York, 1977.)

The physics of atmospheres. Houghton, J. T. 2nd edn. (Cambridge University Press, Cambridge, 1986.)

Aeronomy (Parts A & B). Banks, P. M. and Kockarts, G. (Academic Press, New York, 1973.)

Theory of planetary atmospheres. Chamberlain, J. W. and Hunten, D. M. 2nd edition (Academic Press, New York, 1987.)

Atmospheric physics. Iribarne, J. V. and Cho, H.-R. (D. Reidel, Dordrecht, 1980.)

Weather and climate on planets. Kondratyev, K. Y. and Hunt, G. E. (Pergamon Press, Oxford, 1982.)

> *The book listed below describes the major techniques available for determining temperatures, densities, and composition in the atmosphere. The special issue of* Phil. Trans. *discusses in situ methods as well as remote sounding.*

Remote sounding of atmospheres. Houghton, J. T., Taylor, F. W., and Rodgers, C. D. (Cambridge University Press, Cambridge, 1984.)

The middle atmosphere as observed from balloons, rockets and satellites. *Phil. Trans. R. Soc.* **A296**, 1–268 (1980).

Section 2.2

Atmospheric Radiation Goody, R. M. and Yung, Y. L. (Oxford University Press, Oxford, 1989.)

Models for infrared atmospheric radiation. Tiwasi, S. N. *Adv. Geophys.* **20**, 1 (1978).

A comparison of the contribution of various gases to the greenhouse effect. Rohde, H. *Science* **248**, 1217 (1990).

The effect of clouds on the Earth's solar and infrared radiation budget. Herman, G. F., Man-Li, Wu. C., and Johnson, W. T. *J. atmos. Sci.* **37**, 1251 (1980).

The runaway greenhouse: a history of water on Venus. Ingersoll, A. P. *J. atmos. Sci.* **26**, 1191 (1969).

Response of Earth's atmosphere to increases in solar flux and implications for loss of water from Venus. Kasting, J. F., Pollack, J. B., and Ackerman, T. P. *Icarus* **57**, 335 (1984).

Section 2.3.2

> *Escape of gases from the atmosphere.*

Escape of atmospheres and loss of water. Hunten, D. M., Donahue, T. M., Walker, J. C. G., and Kasting, J. F. in *Origin and evolution of planetary and satellite atmospheres.* Atreya, S. K., Pollack, J. B., and Matthews, M. S. (eds.). (University of Arizona Press, Tucson, 1989.)

Hydrogen and deuterium loss from the terrestrial atmosphere: a quantitative assessment of non-thermal escape fluxes. Yung, Y. L., Wen. J-S., Moses, J.I., Landry, B. M., Allen, M., and Hsu, K.-J. *J. geophys. Res.* **94**, 14971 (1989).

Thermal and non-thermal escape mechanisms for terrestrial bodies. Hunten, D. M. *Planet. Space Sci.* **30**, 773 (1982).

Escape of atmospheres, ancient and modern. Hunten, D.M., *Icarus* **85**, 1 (1990).

Section 2.3.3

> *Vertical transport mechanisms. The first paper describes mathematical models for vertical transport (and is thus also relevant to the models discussed in Section 3.6). Exchange of minor components between troposphere and stratosphere is discussed*

in the next papers, and the possibility that water vapour is removed at a 'cold trap' near the tropopause is considered.

Dynamic effects on atomic and molecular oxygen distributions in the upper atmosphere: a numerical solution to equations of motion and continuity. Shimazaki, T. *J. atmos. terr. Phys.* **29**, 723 (1967).

The transport of minor atmospheric constituents between troposphere and stratosphere. Robinson, G. D. *Q. J. R. Meteorol. Soc.* **106**, 227 (1980).

Stratospheric-tropospheric exchange processes. Reiter, E. R. *Rev. Geophys. & Space Phys.* **13**, 459 (1975).

Why is the stratosphere so dry? Newell, R. E. *Nature, Lond.* **300**, 686 (1982).

The distribution of water vapour in the stratosphere. Harries, J. E. *Rev. Geophys. & Space Phys.* **14**, 565 (1976).

Stratospheric H_2O. Ellsaesser, H. W., Harries, J. E., Kley, D., and Penndorf, R. *Planet. Space Sci.* **28**, 827 (1980).

Mesospheric water vapor. Deguchi, S. *J. geophys. Res.* **87**, 1343 (1982).

Section 2.4

Winds, circulations, and models for them.

Symmetric circulations of planetary atmospheres. Koschmieder, E. L. *Adv. Geophys.* **20**, 131 (1978).

The range and unity of planetary circulations. Williams, G. P. and Holloway, J. L., Jr. *Nature, Lond.* **297**, 295 (1982).

On the sensitivity of a general circulation model to changes in cloud structure and radiative properties. Hunt, G. E. *Tellus* **34**, 29 (1982).

Sections 2.5 and 2.6

The formation of aerosols and clouds, and the optical properties of molecules and particles in the atmosphere. See also the references for Section 1.3.

Particulate matter in the atmosphere: primary and secondary particles. Chapter 12 in *Atmospheric chemistry.* Finlayson-Pitts, B. J. and Pitts, J. N. Jr. (John Wiley, Chichester, 1986.)

Growth laws for the formation of secondary ambient aerosols: implications for chemical conversion mechanisms. McMurry, P. H. and Wilson, J. C. *Atmos. Environ.* **16**, 121 (1982).

Stratospheric aerosols: observation and theory. Turco, R. P., Whitten, R. C., and Toon, O. B. *Rev. Geophys. & Space Phys.* **20**, 233 (1982).

Stratospheric aerosol particles and their optical properties. Cadle, R. D. and Grams, G. W. *Rev. Geophys. & Space Phys.* **13**, 475 (1975).

Light scattering in planetary atmospheres. Hansen, J. E. and Travis, L. D. *Space Sci. Rev.* **16**, 527 (1974).

The scattering of light and other electromagnetic radiation. Kerker, M. (Academic Press, New York, 1969.)

Optical properties of aerosols—comparison of measurements with model calculations. Sloane, C. S. *Atmos. Environ.* **17**, 409 (1983).

Rayleigh scattering by air. Bates, D. R. *Planet. Space Sci.* **32**, 785 (1984).

3 Photochemistry and kinetics applied to atmospheres

Solar radiation not only heats planetary atmospheres, in the ways described in Chapter 2, but it also drives much of the disequilibrium chemistry of those atmospheres through photochemically initiated processes. The consequences of primary photochemical change are often decided by the relative rates of competing chemical reactions: that is to say, kinetically. It is the purpose of this chapter to summarize the principles of photochemistry and kinetics as they are applied to atmospheric chemistry. Attention is paid to features that would be thought of as esoteric detail in laboratory chemistry, but that are of major significance in atmospheres.

3.1 Photochemical change

An atmosphere is a giant photochemical reactor, in which the light source is the Sun. Radiation, generally in the visible and ultraviolet regions, either fragments the atmospheric constituents to produce atoms, radicals, and ions, or excites the constituents, without chemical change, to alter their reactivity.

To understand why visible and ultraviolet radiation are so important, we must first examine the magnitude of one quantum of electromagnetic radiation, the *photon*. According to *Planck's Law*, the energy of one photon of frequency v is hv, where h is Planck's constant. Photochemists usually write 'hv' in chemical equations as a shorthand for the photon that is a reactant, e.g.

$$O_3 + hv \rightarrow O_2 + O. \tag{3.1}$$

Planck's constant is known, so that the energy per photon can be calculated immediately. Chemists often find it convenient to consider energies (and other thermodynamic quantities) for one mole (i.e. $\sim 6.022 \times 10^{23}$ molecules or atoms) of material. Scaled in this way, the energy per 'mole' of photons, E, is given by

$$E = Lhv, \tag{3.2}$$

where L is the Avogadro constant; in terms of wavelength of radiation, λ, and velocity of light, c, eqn (3.2) becomes

$$E = Lhc/\lambda. \tag{3.3}$$

Substituting values for L, h, and c, we find that with λ expressed in nanometres (nm)

$$E = 119625/\lambda \, \text{kJ mol}^{-1} \qquad (3.4)$$

$$= 1239.8/\lambda \, \text{eV}.^{\ddagger} \qquad (3.5)$$

Thus the red extreme of the visible spectrum (~ 800 nm) corresponds to about $150 \, \text{kJ mol}^{-1}$, and the violet extreme (~ 400 nm) to twice that energy. At shorter wavelengths lies the ultraviolet, conventionally subdivided into 'near' ($\lambda \sim 400$–200 nm), 'vacuum' (VUV, $\lambda \sim 200$–100 nm) and 'extreme' (EUV, $\lambda \sim 100$–10 nm) regions, with successively higher energies. X-rays, gamma rays, and galactic cosmic radiation (GCR) constitute the shortest ($\lambda < 10$ nm) wavelengths of the electromagnetic spectrum. The very high photon energies are associated with an ability to *penetrate* as well as to ionize atmospheric gases. We shall have to consider such radiation (especially GCR) as a source of ionization (Chapter 6), but the interactions are not usually thought of as photochemical. The photon energy of red light is comparable with the bond energies of rather loosely bound chemical species. Of common gaseous inorganic species, ozone is, in fact, the only compound with such a small bond energy (the O—O_2 energy is $\sim 105 \, \text{kJ mol}^{-1}$); nitrogen dioxide, with an O—NO bond energy of $\sim 300 \, \text{kJ mol}^{-1}$ ($\cong 399$ nm) is more typical. Ionization becomes possible at the shorter wavelengths (e.g. ionization of NO at ~ 135 nm). The point is that the visible region contains the lowest energy photons of the entire electromagnetic spectrum that are capable of promoting chemical change in single quantum events. Many *more* photons arrive each second at a planet at longer wavelengths, but they can only heat the atmosphere up.

It so happens that the wavelengths at which chemical change becomes possible also correspond roughly to the energies at which *electronic transitions* are excited in atoms and molecules. Longer wavelengths tend to excite molecular vibrations or rotations. Although high vibrational levels are involved in some photochemical processes, electronic excitation is the spectroscopic step most frequently associated with photochemical change. Much of the discussion of the following sections will therefore centre around the fates of electronically-excited species. We should note here that electronic excitation not only carries with it the energy absorbed, but also a new electronic structure of the chemical species. A profound influence on the chemistry and reactivity can result. Electronic transitions are governed by the ordinary rules of spectroscopy. We suggest here, as a reminder, the types of consideration that apply. First of all, absorption of radiation only occurs if an upper energy level of the species exists which is separated from the lower level by an energy equal to that of the incident photon ('*resonance condition*'). Intensities of transitions are determined in part by the *electronic transition moment*, which itself depends

---

$\ddagger$ The electron volt, eV, is a unit of energy commonly used in physics and chemical physics, especially where ionic species are involved. It is the work done in moving L electrons through a potential difference of 1 volt, and is numerically equivalent to $96.485 \, \text{kJ mol}^{-1}$.

on the wavefunctions of lower and upper states. *Selection rules* aim to avoid the calculation of the transition moment by stating conditions which need to be satisfied if the transition moment is not to vanish. These rules are often formulated for *electric dipole* interactions, which are the most intense. Familiar examples of the rules are that $\Delta S = 0$, or that, in atoms, $\Delta L = \pm 1$. *Forbidden transitions* can be observed either because the quantum number for which the rule is stated does not rigorously describe the system (e.g. **L** and **S** are not 'good' quantum numbers in heavy atoms because of spin-orbit coupling), or because the transition occurs by an interaction (e.g. magnetic dipole) other than that envisaged. On top of the electronic contribution to intensity, the vibrational and rotational components must be considered for molecular species. Overall intensities of absorption depend on the populations of the lower (and upper) states of the species. Even very weak, highly forbidden, transitions can contribute to atmospheric absorption because of the large optical paths involved.

Small, light chemical species generally show intense electronic absorption at shorter wavelengths than more complex compounds. Molecular oxygen absorbs strongly for $\lambda < 200\,\text{nm}$, H_2O for $\lambda < 180\,\text{nm}$, CO_2 for $\lambda < 165\,\text{nm}$, while N_2 and H_2 absorb significantly only for $\lambda < 100\,\text{nm}$. It is the limitation on laboratory experiments in air, resulting from the O_2 absorption, that has led to the term 'vacuum' ultraviolet for $\lambda < 200\,\text{nm}$. Atmospheres tend to act as filters cutting out short-wavelength radiation, since the absorptions of their major constituents are generally strong at wavelengths shorter than the threshold value. As a result, photochemically active radiation that penetrates deeper into an atmosphere is of longer wavelength, and the chemistry characterized by lower energies, than that absorbed higher up. The principle is well exemplified by the chemistry of Earth's atmosphere. Tropospheric photochemistry is dominated by species such as O_3, NO_2, SO_2, and HCHO which absorb in the near ultraviolet. At progressively higher altitudes, photodissociation of O_2 and photoionization phenomena become the most important processes. At ground level, only radiation with $\lambda \gtrsim 300\,\text{nm}$ ($\cong 400\,\text{kJ mol}^{-1}$) remains, and the peak intensity is at $\lambda \sim 500\,\text{nm}$ ($\cong 240\,\text{kJ mol}^{-1}$). The particular significance of energy storage in the chlorophyll–photosynthetic system, alluded to in Section 1.5.1, now becomes clear, since direct, single photon, decomposition of H_2O or CO_2 is energetically impossible near the Earth's surface.

3.2 Photochemical primary processes

Absorption of a photon of photochemically active radiation leads to electronic excitation, a process we will represent symbolically as

$$AB + h\nu \rightarrow AB^*. \tag{3.6}$$

Many fates of the excited AB* molecule are recognized, and several of them occur in atmospheres. Figure 3.1 summarizes the processes most frequently encountered. Routes (i) and (ii) lead to fragmentation of one kind or another. Process (iii) is the re-emission of radiation: if the optical transition is allowed the luminescence is called *fluorescence*, and otherwise *phosphorescence*. Routes (iv) and (v) involve population of excited species other than those first produced by excitation. Intramolecular energy transfer (iv) generates a new electronic state of the same molecule by a radiationless transition, while intermolecular transfer (v) excites a different molecule, often chemically distinct from the absorbing species. Quenching (vi) is a special case of intermolecular energy transfer, where electronic excitation is degraded to vibrational, rotational, and translational modes. Pathway (vii), chemical reaction, includes all processes where reaction is possible for, or rates are enhanced with, electronically excited reactants.

The *quantum yield* of a photochemical reaction is defined in general as the number of reactant molecules decomposed for each quantum of radiation

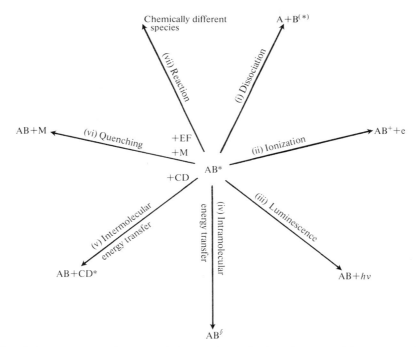

Fig. 3.1. Pathways for loss of electronic excitation that are of importance in atmospheric chemistry. The use of the symbols * and § illustrates the presence of electronic excitation: the products of any of the processes *may* be excited. With the exception of pathways (i) and (iv), excited atoms can participate as well as excited molecules.

absorbed, although it is often more helpful to define the quantum yield for a particular reaction pathway. One reactant species is electronically excited for each quantum of radiation absorbed. If the *primary* photochemical step is seen as the excitation step (3.6) taken together with the subsequent immediate fate of the excited species, then the sum of primary quantum yields for all the steps (i) to (vii) in Fig. 3.1 is obviously unity. The primary quantum yield for loss of AB has a maximum value of unity, since, for example, quenching or fluorescence, which restore AB, may compete with dissociation.

We turn now to a more detailed examination of the processes leading to chemical change.

3.2.1 Photodissociation and photoionization

Fragmentation of a chemical species following absorption of light is one of the most important photochemical processes in atmospheric chemistry. Photodissociation may come about when the energy of the absorbed photon exceeds the binding energy of the chemical bond under consideration. That is, the species AB* excited initially in the absorption event (3.6) can lie energetically above the dissociation threshold in the molecule, and the bond then rupture in some way. Two main mechanisms are recognized for this photochemical rupture, and they are referred to as *optical dissociation* and *pre-dissociation*. We shall now consider the essential features of the two mechanisms, and illustrate the principles with reference to the molecules O_2 and O_3 that are of such central significance to the chemistry of Earth's atmosphere. Potential energy curves for some electronic states of O_2 are shown in Fig. 3.2. The labelling of states X, A, B, etc. is explained in the caption.

Optical dissociation is characterized by dissociation from the electronic state to which absorption first occurred. The spectrum of the absorption leading to dissociation is a continuum, since the fragments may bear translational energy away from the dissociating molecule, and such energy is essentially continuous. At some longer wavelength, the spectrum may be banded as a result of absorption to the discrete vibrational levels in a spectral region where dissociation does not follow absorption. The bands get closer together as the restoring force for the vibration gets weaker. Ultimately, a continuum is reached: the energy corresponding to the onset of the continuum ('convergence limit') is the dissociation energy to the products formed. Absorption from the 'X' to the 'B' state in O_2 (the so-called Schumann-Runge system) is of this kind, and Fig. 3.3 shows how the sharp banded spectrum that starts at $\lambda \sim 200$ nm converges to a limit at $\lambda \sim 175$ nm.

Examination of the potential energy curve of Fig. 3.2 shows that one of the two atomic fragments produced by extending the O—O bond length in the B state is, in fact, itself excited. The observed convergence limit at $\lambda \sim 175$ nm thus corresponds to the formation of one ground state (^3P) and one excited (^1D) atom. When two fragments lie on a smooth extension of the parent's

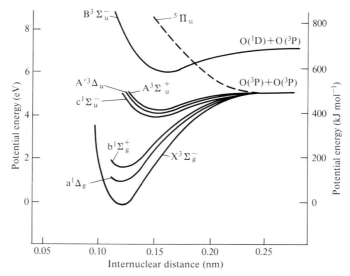

Fig. 3.2. Potential energy curves for some of the states of molecular oxygen. The ground state is labelled X, and successive higher levels of the same spin multiplicity are identified by the letters A, B, C, etc. Levels of different spin multiplicity are identified by the lower case letters a, b, c, etc. Spectroscopic term symbols for atoms and simple molecules provide more specific information about the electronic states. The symbols are constructed according to the form ^{2S+1}L (atoms) or $^{2S+1}\Lambda$ (linear molecules) where $2S + 1$ is the spin multiplicity and the angular momentum quantum number L (or Λ) is replaced by S, P, D, F ... (or $\Sigma, \Pi, \Delta, \Phi$ ...) according to its value 0, 1, 2, 3, ... Additional elements to the molecular term symbol may appear to describe symmetry properties of the wavefunction. In general, the actual term symbols are of little consequence for our purposes, but they still provide the most convenient way of identifying particular electronic states.

The dashed curve is representative of several repulsive states (e.g. $^5\Pi_u$, $^1\Pi_u$, $^3\Pi_u$) that correlate with $O(^3P) + O(^3P)$ and that pre-dissociate the $B^3\Sigma_u^-$ state. [After Krupenie, P.H. *J. Phys. Chem. Ref. Data* **1**, 423 (1972)].

potential energy curve (or surface), the fragments and parents are said to *correlate* with each other (see Section 3.3). For allowed optical absorption in diatomic molecules, the upper molecular state usually correlates with atoms of which at least one is excited, as in the O_2 case just discussed. Weak, 'forbidden', transitions can populate a state that correlates with ground state fragments. For example, the A state in O_2 correlates with two $O(^3P)$ atoms (Fig. 3.2). The A ← X absorption ('Herzberg I' system) is weak because it is forbidden (for an electric dipole interaction). It is, however, important in the atmosphere, since optical dissociation can occur for photon energies corresponding to the formation of two ground state oxygen atoms at wavelengths

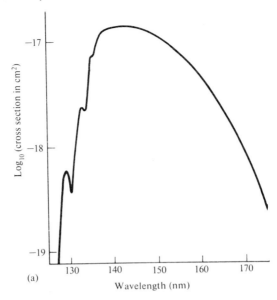

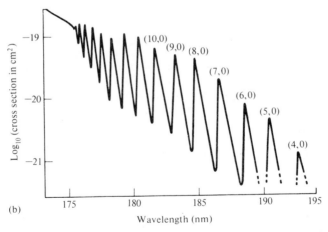

Fig. 3.3. Absorption spectrum of molecular oxygen in the $B^3\Sigma_u^- \leftarrow X^3\Sigma_g^-$ Schumann-Runge system: (a) the continuum, and (b) the discrete banded region. (From McEwan, M. J. and Phillips, L. F. *Chemistry of the atmosphere*. Edward Arnold, London, 1975.)

(~ 185 nm to 242 nm) where solar intensities are relatively higher than at the wavelengths needed for Schumann-Runge dissociation.

We explained the continuum in the absorption spectrum seen in optical dissociation as a consequence of the nearly continuous—properly speaking, very closely spaced—translational energy levels. For wavelengths shorter than the convergence limit for dissociation, the excess photon energy must be dissipated as translational motion apart of the daughter species. Translationally 'hot' products are of potential interest in their own right, because they may be more reactive than species that are in thermal equilibrium with their surroundings. The upper states discussed (B, A) for oxygen are bound, and the continuum follows a discrete banded region of the spectrum. In other molecules, absorption may be to a repulsive, unbound, upper state. The absorption spectrum will then be continuous over all wavelengths of the transition, and be accompanied by fragmentation.

Our second mechanism for photodissociation, *pre-dissociation*, is distinguished from optical dissociation by the involvement of an electronic state of the molecule different from that initially populated. The word was adopted to describe the spectroscopic appearance of an absorption system. Absorption spectra of the simpler molecules show considerable sharp rotational structure, but in some cases this structure becomes blurred, leading to a diffuseness of the bands at a wavelength longer than that corresponding to the optical dissociation limit: hence 'pre'-dissociation. The spectroscopic diffuseness may be accompanied by chemical fragment formation, which is the process of interest to us. Photodissociation is thus occurring for wavelengths longer than (i.e. at energies less than) the convergence limit for optical dissociation in the absorption system. For example, some fragmentation of O_2 arises at $\lambda > 175$ nm for absorption in the B ← X (Schumann-Runge) system.

Pre-dissociation is now understood to arise from the 'crossing' of electronic states, and the occurrence of radiationless energy transfer between them. The dashed curve in Fig. 3.2 shows a pre-dissociating state in oxygen. A radiationless transition (i.e. intramolecular energy transfer, process (iv) of Fig. 3.1) near the crossing point of the two potential energy curves can lead to population of the repulsive state, which is unstable with respect to formation of *ground state* fragment atoms. Pre-dissociation thus provides a route for product formation in an allowed transition, but without the need for wavelengths short enough to yield excited fragments.

We have seen that one or more of the products of photodissociation may be electronically excited. For a triatomic, or larger, molecule, more than one set of chemically distinct products can be formed. Only laboratory experiment can unequivocally identify what chemical species are formed, in what electronic states, and with what quantum efficiencies. Theory can, however, offer guidance in two ways. First, the incident photons must have enough energy to bring about any proposed change, so that thermochemical information can be used to show whether a particular photodissociative channel is energetically *possible*. Secondly, considerations such as the need to conserve quantum mechanical spin and orbital angular momentum can indicate whether the channel is *probable*.

We shall discuss this latter point more fully in Section 3.3, but, for the time being, we illustrate it in terms of the photolysis of ozone, reaction (3.1)

$$O_3 + h\nu \rightarrow O_2 + O. \tag{3.1}$$

Ozone absorbs strongly in the ultraviolet region, as shown in Fig. 3.4. A much weaker absorption exists in the red region, with a peak at $\lambda \sim 600$ nm. Depending on the photon energy, the atomic and molecular oxygen fragments of reaction (3.1) can themselves be electronically excited. Table 3.1 shows the wavelength thresholds required thermodynamically for the production of various combinations of states of O and O_2. The table does not indicate whether or not the particular fragments really are formed. However, the spin conservation theory suggests that the products of photolysis must *both* be triplets or *both* be singlets. The lowest energy singlet pair is $O(^1D) + O_2(^1\Delta_g)$, and hence from Table 3.1, the expected threshold wavelength for spin-allowed

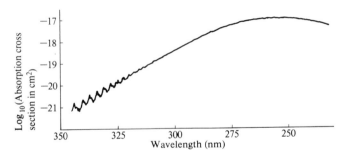

Fig. 3.4. Absorption spectrum of ozone in the ultraviolet region at 300 K. (Data of Bass, A. M. and Paur, R. J., represented at the International Workshop on Atmospheric Spectroscopy, Rutherford Appleton Laboratory, Chilton, July 1983.)

Table 3.1 Wavelength thresholds (in nm) for ozone photodissociation channels.

Molecule Atom	$O_2(^3\Sigma_g^-)$	$O_2(^1\Delta_g)$	$O_2(^1\Sigma_g^+)$	$O_2(^3\Sigma_u^+)$	$O_2(^3\Sigma_u^-)$
$O(^3P)$	1180	612	463	230	173
$O(^1D)$	411	310	267	168	136
$O(^1S)$	237	199	181	129	109

Data derived from: Baulch, D. L., Cox, R. A., Hampson, R. F., Jr., Kerr, J. A., Troe, J., and Watson, R. T., *J. Phys. Chem. Ref. Data* **9**, 295 (1980); and Moore, C. E. *Atomic energy levels*, Vol. 1, NSRDS-NBS 35, Washington, DC, 1971.

$O(^1D)$ formation is $\sim 310\,nm$. Figure 3.5 shows experimentally measured quantum yields for $O(^1D)$ production as a function of wavelength. The quantum yield does, indeed, increase from near zero to its limiting value (~ 0.9) over a narrow wavelength range centred on $\lambda = 310\,nm$. Direct determinations also show that $O_2(^1\Delta_g)$ is actually formed with the same efficiency as $O(^1D)$ at $\lambda < 300\,nm$, in exact agreement with the spin conservation arguments. A knowledge of the efficiency of $O(^1D)$ production is of particular importance in atmospheric chemistry, because $O(^1D)$ can initiate many other reactions (see Section 1.4 for a preliminary discussion).

The quantum yield–wavelength curve of Fig. 3.5 leads us to the next topic. It might be thought that, since the energy for a particular dissociative channel is well defined, the onset of dissociation ought to be abrupt. That is, figures such as 3.5 ought to be step functions, with the step rising at the wavelength corresponding to the critical energy. Although reasonably steep, the curve for $O(^1D)$ formation from O_3 is certainly not a step function. The shape of the curve can be explained if, in polyatomic molecules, energy contained in internal vibrations and rotations can assist the photon energy in causing dissociation. Not all modes are equally 'active' in this way, and the efficiency of energy utilization may be dependent on the internal energy. Nevertheless, photon energies somewhat smaller than the bond dissociation threshold can be effective in rupturing the bond. Internal energy depends on temperature, so that the exact shape of the quantum yield–wavelength curve will also be temperature-dependent.

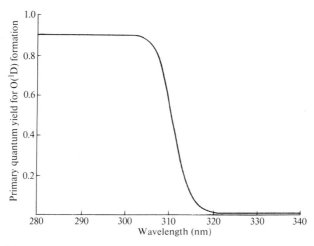

Fig. 3.5. Primary quantum yield for $O(^1D)$ formation in the photolysis of ozone displayed as a function of photolysis wavelength. (Curve drawn from data presented in Hudson R. (ed.-in-chief) *The stratosphere,* 1981, World Meteorological Organization, Geneva, 1981.)

Considerable emphasis has been placed on the exact magnitude of quantum yields for primary dissociative processes and on the wavelength and temperature dependences. The reason is that calculated photolytic production rates of species such as $O(^1D)$ in atmospheres are particularly sensitive to these factors. A change in the nature of products, or, indeed, the onset of photolysis, is very often associated with a rapid change in absorption cross-section, for good spectroscopic reasons. Comparison of Figs. 3.4 and 3.5 for ozone will illustrate the effect clearly: for $\lambda = 304$ to 319 nm, the absorption cross-section drops by a factor of 10, while the quantum yield for $O(^1D)$ formation falls from 0.9 to ~ 0. The rate of $O(^1D)$ formation depends on the product of quantum efficiency, absorption cross-section, and incident flux. The product of the first two terms changes extremely rapidly with wavelength, so that the rate of $O(^1D)$ production will be critically dependent on how the incident intensity varies with wavelength. As an example, near the ground the incident intensity increases by a factor of more than ten between $\lambda = 300$ and $\lambda = 320$ nm. A small error in the wavelength scale for Fig. 3.5 could thus have an enormous effect on the predicted rates of $O(^1D)$ formation in the troposphere.

Photoionization may be regarded as a special case of photodissociation, but one in which the products are a positively charged ion and an electron. The processes

$$O + h\nu \rightarrow O^+ + e, \tag{3.7}$$

$$O_2 + h\nu \rightarrow O_2^+ + e, \tag{3.8}$$

are typical of photoionization. In general, the energies (*ionization potentials*) required to remove an electron from an atom or molecule are larger than those needed to split a molecule into chemical fragments. Some typical ionization potentials are shown in Table 3.2, together with the corresponding wavelengths. Even the molecules with the lowest ionization potentials (NO, NO_2) require vacuum ultraviolet radiation for ionization. Photoionization is thus only a significant source of ions in the *upper* atmosphere of Earth. Electronically excited atoms or molecules can often use their excitation energy as a contribution towards the ionization potential. Thus the photoionization wavelength threshold for $O_2(^1\Delta_g)$, which has 0.98 eV excitation energy, is raised to 111.8 nm. As with dissociative processes, the effect of small wavelength shifts can be large in atmospheric chemistry, because of the rapid variation of incident intensity with wavelength. In the ionizing ultraviolet region, the most abundant gases such as O_2, N_2, and CO_2 absorb strongly, so that their absorption is imposed on top of the spectral distribution of the solar radiation, even at high atmospheric altitudes. Since the molecular absorptions generally show some structure, there are spectral 'windows' through which radiation is transmitted to greater depth in the atmosphere, and these windows may be the major contributors to ionization. One of the most curious coincidences in Earth's atmosphere is the almost perfect spectral match between a sharp dip in the absorption of O_2 and the Lyman-α (resonance) line of atomic hydrogen at $\lambda = 121.59$ nm. At wavelengths below

Table 3.2 Ionization potentials for some atmospherically important species, together with equivalent threshold wavelengths.

Species	Ionization energy		Equivalent wavelength (nm)
	eV	kJ mol^{-1}	
Na	5.1	496	241.2
Mg	7.7	738	162.1
NO	9.3	886	135.0
NO_2	9.8	940	127.2
O_2	12.1	1165	102.7
H_2O	12.6	1214	98.7
O_3	12.8	1233	97.0
N_2O	12.9	1246	96.0
SO_2	13.1	1264	94.6
H	13.6	1312	91.2
O	13.6	1314	91.1
CO_2	13.8	1331	89.9
CO	14.0	1352	88.5
N	14.5	1403	85.2
H_2	15.4	1488	80.4
N_2	15.6	1503	79.9

Data obtained from:

(monatomic species), Moore, C. E., *Ionization potentials and ionization limits derived from the analysis of optical spectta*, NSRDS-NBS 34, Washington, DC (1970);

(diatomic species), Huber, K. P. and Herzberg, G. *Constants of diatomic molecules*. Van Nostrand-Reinhold, New York, 1979;

(triatomic species), Herzberg, G. *Electronic spectra and structure of polyatomic molecules*. Van Nostrand-Reinhold, New York, 1966.

~ 190 nm, the solar emission is almost entirely in the form of atomic lines, and the strongest individual feature is the Lyman-α line. Because of the chance existence of the O_2 window, the strongest solar feature is also least attenuated in Earth's atmosphere. Table 3.2 shows that NO (as well as NO_2 and some metal atoms) can be ionized by Lyman-α radiation.

Mechanisms of photoionization are analogous to those for dissociation, both direct ionization and *pre-ionization* (*auto-ionization*) being recognized. Excited electronic (and, in molecules, vibrational and rotational) states of the ions may be generated. Any excess of energy between the ionizing photon and the resultant ion is carried off as translation by the electron, a fact utilized in photoelectron spectroscopy.

3.2.2 Reactions of electronically excited species

Electronically excited atoms and molecules can be formed photochemically either directly by optical pumping (absorption) or as products of a photofragmentation process (see preceding section). Electron impact can also lead to electronic excitation, and is often important in planetary atmospheres. A typical example is the process

$$N_2 + e \rightarrow N_2^* + e, \tag{3.9}$$

where N_2^* can be $N_2(^3\Sigma_u^+)$. We are now interested in the reactivity of excited states, regardless of the excitation mechanism.

There are two ways in which the reactivity of a species can be influenced by its electronic state. First, the energetics of a reaction are altered, with thermodynamic and kinetic consequences. A reaction favours products only if $\Delta G^\ominus$ for the reaction is negative. If a process is endothermic ($\Delta H^\ominus$ positive) then the condition for $\Delta G^\ominus$ will be met only if the enthalpy term is dominated by a large increase in entropy (i.e. $T\Delta S^\ominus > \Delta H^\ominus$). In general, this thermodynamic condition is likely to be met only for slightly endothermic reactions or at high temperatures. However, if a reactant is excited, and if the excitation energy can contribute to the reaction energetics, then $\Delta H^\ominus$ may become negative for the excited state reaction where it was positive for the ground state. Kinetic limitations may be even more important than thermodynamic ones. As we shall see in Section 3.4.1, rates of reaction are often proportional to a factor $\exp(-E_a/RT)$, where E_a is the *activation energy* for the reaction. For an endothermic reaction, it can be shown that E_a is $\gtrsim \Delta H^\ominus$ (see Section 3.4.1) and the exponential term can make the rate of reaction insignificant. Activation energies are usually positive even for exothermic reaction. In most cases, then, excess energy in the reactants will increase the rate of reaction so long as that energy can be used in overcoming the activation barrier.

Let us now consider some examples. We have already mentioned several times the production of hydroxyl radicals by excited oxygen atoms. The process

$$O(^3P) + H_2O \rightarrow OH + OH, \tag{3.10}$$

is about $70\,kJ\,mol^{-1}$ endothermic, while the reaction with $O(^1D)$,

$$O(^1D) + H_2O \rightarrow OH + OH, \tag{3.11}$$

is exothermic by $120\,kJ\,mol^{-1}$ because of the $190\,kJ\,mol^{-1}$ excitation energy of $O(^1D)$. Similarly, the reaction of $O(^1D)$ with CH_4 is exothermic, but would be endothermic by $11\,kJ\,mol^{-1}$ for ground state $O(^3P)$. Both reactions

$$O(^3P) + O_3 \rightarrow 2O_2 \tag{3.12a}$$

$$O(^1D) + O_3 \rightarrow 2O_2 \text{ and } 2O + O_2 \tag{3.12b}$$

are exothermic. However, at a temperature of $220\,K$ (typical of the lower stratosphere), the second reaction proceeds more than 3×10^5 times as fast as the first for equal reactant concentrations. Most of the difference (but not all: see later in this section) results from the activation energies of the reactions: $17.1\,kJ\,mol^{-1}$ in (3.12a), ~ 0 in (3.12b). Obviously, the $17.1\,kJ\,mol^{-1}$ can be supplied by the $190\,\text{kJ mol}^{-1}$ excitation energy of $O(^1D)$. Another interesting example is the reaction of excited N_2 in the $^3\Sigma_u^+$ state (produced by optical pumping, or by electron impact in reaction 3.9) with O_2:

$$N_2^* + O_2 \rightarrow N_2O + O. \tag{3.13}$$

It has been suggested that the reaction may produce N_2O about once in every two collisions between the reactants. For ground-state N_2, the reaction is endothermic and negligibly slow. Process (3.13) is a possible source of N_2O at altitudes above $20\,\text{km}$, and could appreciably augment the biogenic sources from the Earth's surface.

We turn now from the influence of excess energy to that of electronic structure on the reactivity of electronically excited states. Chemists are certainly inclined to believe that the properties and reactions of a substance are influenced by the number and arrangement of electrons, especially in the outer shell. Since electronic excitation implies some alteration in structure, species might be expected to show a particular reactivity for any electronic state. Nothing at this stage suggests that an excited state will be more (or, indeed, less) reactive than the ground state. It is, of course, necessary to disentangle the two influences of energy and of structure, since any excited state is bound to be energy rich. The easiest way to determine the specific influence of structure is to extrapolate measured reaction rates to infinite temperature, which removes the effect of activation energy since $\exp(-E_a/RT)$ is now unity. Carrying out the extrapolation with rate data for reactions (3.12a) and (3.12b) leads to the conclusion that $O(^1D)$ is *intrinsically* about twenty times more reactive than $O(^3P)$ in the reaction with ozone. Conversely, the intrinsic reactivity of nitrogen atoms with ground-state oxygen

$$N(^4S) + O_2(^3\Sigma_g^-) \rightarrow NO + O, \tag{3.14}$$

is several orders of magnitude *greater* than the intrinsic reactivity with the first excited $^1\Delta_g$ state of O_2. Very often the effects of electronic arrangement on reactivity are a reflection of the ability to pass smoothly from reactants to products without a change of, for example, spin angular momentum. We have already encountered another facet of the same phenomenon when discussing the spin states of the products of ozone photolysis. It is appropriate to consider now the underlying principles of momentum-conserved reactions.

3.3 Adiabatic processes and the correlation rules

Potential energy curves, such as those shown in Fig. 3.2, represent the molecular potential energy as a function of internuclear distance between the constituent atoms of a diatomic molecule. We have seen in Section 3.2.1 that a molecular state is said to *correlate* with particular atomic fragments if a continuous curve connects atoms and the diatomic molecule. Thus, in the oxygen case to which Fig. 3.2 applies, the B state correlates, for example, with $O(^1D) + O(^3P)$, while the ground X state correlates with two ground-state, ^3P, oxygen atoms. Bringing together a ^1D and a ^3P atom cannot, therefore, generate O_2 in the ground state in a process involving a continuous potential energy curve. This argument can be extended to the reaction

$$O(^1D) + O(^3P) \rightarrow O(^3P) + O(^3P), \tag{3.15}$$

which is clearly exothermic. To get from reactants to products, it is necessary to start on the B curve, and then cross to the pre-dissociating repulsive state. In the context of reaction mechanisms, the term *adiabatic* has the special meaning of 'remaining on the same curve', so that reaction (3.15) can only occur *non-adiabatically*. Non-adiabatic reactions, which involve curve crossing, are nearly always less efficient than adiabatic ones.

At the end of this section, we shall discuss how the potential energy of a polyatomic system can be described by a *potential energy surface* rather than by the potential energy *curve* used in connection with diatomic systems. For the time being, we need note only that with reactions involving more than two atoms in total, an adiabatic reaction will be one that is conducted on a single potential energy surface. That is, reactants and products are connected by a continuous surface. Where such connection does not exist, non-adiabatic surface crossing must occur, which may be very inefficient.

Why is it that crossing from one curve or surface to another is sometimes so inefficient? The answer lies in the probabilities of radiationless transitions, for which selection rules may be formulated just as for optical (radiative) transitions. In the example with which we started this section, reaction (3.15) cannot occur adiabatically because of the selection rule that spin must remain unchanged ($\Delta S = 0$) in a radiationless transition. Reference again to Fig. 3.2 will show that the dashed curve is of different spin multiplicity (quintet: i.e. $S = 2$) from the B state (triplet: i.e. $S = 1$), so that the radiationless transition is formally forbidden and acts as a bottle-neck in the non-adiabatic process. In this very simple case, the electronic states and term symbols for all the atomic and molecular species can be identified. For polyatomic reaction systems, such detailed and specific information may not exist. On the other hand, it may still be possible to say if a reaction *cannot* proceed adiabatically on the basis of the information available.

Correlation or *conservation rules* are formulated in terms of the angular momentum or symmetry properties that do exist for reactants or products. Often, only spin angular momentum, S, can be specified and our discussion will be illustrated in terms of this quantum number. However, it must always be borne in mind that orbital momentum, parity, and so on, must all be conserved where they have a meaning.

Consider a hypothetical reaction between two species A and BC

$$A + BC \rightarrow [ABC] \rightarrow AB + C. \tag{3.16}$$

At some stage, A, B, and C are all interacting as a transient reaction complex [ABC]. The basis of the conservation rules is that, for an adiabatic reaction, [ABC] can be reached without change of potential energy surface, either from the reactants or from the products. In that case, the transient must have its total spin, S_{ABC}, produced either by a combination of the reactant spins, S_A and S_{BC}, or by a combination of the product spins, S_{AB} and S_C. The only potential difficulty concerns the addition of spins. Angular momenta sum vectorially, but with quantized momenta the resultant must also be quantized. Spins S_A and S_{BC} therefore produce resultants $|S_A + S_{BC}|, |S_A + S_{BC} - 1|, \ldots$ $|S_A - S_{BC}|$, while S_{AB} and S_C produce $|S_{AB} + S_C|, |S_{AB} + S_C - 1|, \ldots |S_{AB} - S_C|$. The question now is whether or not the first list and the second have one or more values in common. If not, then adiabatic reaction *cannot* occur. If there are common values then adiabatic reaction *may* be possible. Let us consider a concrete example. What combinations of the singlet and triplet spin states can the O_2 products of the reactions

$$O(^3P, {}^1D) + O_3(^1A) \rightarrow O_2 + O_2 \qquad (3.17)$$

possess? For $O(^3P)$, the spins of the reactants are 1 and 0, which combine to give a total spin of 1. Two triplet products ($S_{O_2} = 1, S_{O_2} = 1$) can give a total spin of 2, 1, or 0, so that the reaction to two triplets *could* occur adiabatically on a triplet surface ($S = 1$). One triplet and one singlet ($S_{O_2} = 1, S_{O_2} = 0$) give uniquely a total spin of 1, so that again adiabatic reaction is possible on a triplet surface. Two singlet products ($S_{O_2} = 0, S_{O_2} = 0$) can only combine to $S = 0$, so that adiabatic reaction is *not* possible. With $O(^1D)$, the surface correlating with the reactants is a singlet, so that the product O_2 molecules in this case must be *both* singlets or *both* triplets for adiabatic reaction to occur (on the singlet surface). All other reaction possibilities involve spin forbidden crossings between singlet and triplet reaction surfaces, and these non-adiabatic processes are likely to be inefficient.

Let us now see how the spin conservation rule could be applied to a *photochemical* process. In particular, we shall examine what are the 'spin-allowed' fragments of ozone photolysis in reaction (3.1), since the outcome has already been presented without proof in Section 3.2.1. The argument goes in the following way. Ground state ozone is a singlet ($S = 0$), and the strong ultraviolet absorption is likely to be spin allowed ($\Delta S = 0$), so that the transient excited state of ozone yielding $O + O_2$ is also a singlet. Products lying on the same potential energy curve (or surface) as a precursor have to have spins that can add to give the spin of the precursor. A value of $S = 0$ can be obtained from products both possessing $S = 0$ or both possessing $S = 1$, but *not* from one triplet and one singlet. In the ozone case, then, $O(^1D)$ production, if spin allowed, must be accompanied by a singlet state O_2 product. Alternatively, both atom and molecule can be triplets.

A certain amount of caution is needed in applying the spin and other conservation rules. First, they are valid only so far as the quantum number, such as **S**, to which they pertain is a good description of the system. Where elements of large atomic number are concerned, **S** is a notoriously poor quantum number. In a reactive interaction ('collision': see later) **S** may be a poor representation even where only 'light' elements are concerned. The other point that should be made is that the rules show *excluded* possibilities rather than permitted ones. For example, that a triplet surface can connect reactants with products does not necessarily mean that reaction occurs over the triplet surface. Let us revert to our original two-atom reaction (3.15). According to the spin rules developed in connection with reaction (3.17), singlet + triplet reactants *can* give two triplet products (on a triplet surface). We know, however, from the potential curves (Fig. 3.2) that reactants and products are *not*, in fact, connected by a continuous curve, triplet or otherwise. The reaction as written could occur non-adiabatically, via the pre-dissociating (dashed) curve. The curve actually drawn is a quintet state, so that the curve crossing is forbidden and likely to be slow.

This section ends with a short digression about potential energy surfaces for polyatomic systems, since we have already alluded to such surfaces and will need to refer to the subject in more detail in the following section. It is easy to represent, on a two-dimensional curve, interactions and reactions between atoms, since we need to display the energy as a function of only one interatomic distance. Potential energies for polyatomic molecules are not so easily represented. Even with a triatomic system, say ABC [Fig. 3.6(a)], we need either three distances (r_{AB}, r_{BC}, r_{AC}) or two distances and an angle (e.g. r_{AB}, r_{BC}, $\angle ABC$) to define the relative positions of A, B, and C. Four dimensions are then required to show potential energy as a function of atomic coordinate. An improvement for the triatomic molecule can be achieved if one variable, such as the bond angle, is fixed. In that case, the potential energy can be described by a three-dimensional *potential energy surface*, using one coordinate for energy, and the other two for the two interatomic distances. The potential energy surface can be displayed in two dimensions by a contour map. Without the restriction on bond angle, or with more than three atoms in the molecule, then the three-dimensional surface becomes instead a multi-dimensional potential energy *hypersurface*. There is no real difficulty in concept here, since we wish only to know what the potential energy is for the atoms in given positions. The difficulty is in making drawings in two- or three-dimensional space!

Our remarks about potential energy surfaces apply both to surfaces for bound polyatomic molecules and to polyatomic systems created transiently during the course of a chemical reaction. For example, in the reaction represented by eqn (3.16), a transient triatomic molecule ABC is formed as the reactant A approaches BC. In this case, however, the energy of the ABC system increases as the AB distance decreases, rather than reaching the minimum characteristic of the stable configuration of a bound triatomic molecule. The contour map of Fig. 3.6(b) is of this repulsive (reactive) type. At large r_{AB}

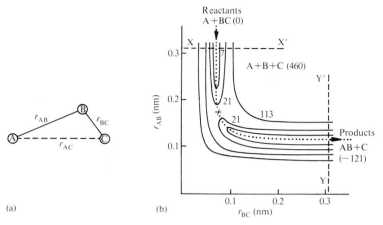

Fig. 3.6. (a) Distances in the A ... B ... C system. (b) Typical potential energy surface (contour map) for a particular ABC angle. The lowest energy path from reactants (A + BC) to products (AB + C) is shown by the dotted line, which represents the 'reaction coordinate'. On this path, the highest energy is reached at the point marked '$\neq$', which is the 'transition state' or 'activated complex'. The potential energy surface is actually one computed *ab initio* for the reaction $F + H_2 \rightarrow H + HF$, and numerical values on the contour lines are energies in kJ mol^{-1} referred to the reactants. [After Muckerman, J. T. *J. Chem. Phys.* **54**, 115 (1971).]

(reactants far apart) a cross-section through the surface along the line XX' would be the potential energy curve for the free diatomic reactant molecule BC. When r_{BC} is large, the products have separated, and the cross-section along YY' is the potential energy curve for the product AB.

The rate of passage over the potential energy surface from reactants to products is the subject of reaction kinetics, and it is to this topic that we now turn.

3.4 Chemical kinetics

Very many of the elementary reaction steps which make up the chemistry of planetary atmospheres involve atoms, radicals, excited states, and ions as reactants or products. The species involved are highly reactive, and laboratory investigations of such reactions have, therefore, to look for special methods of generating the reactants, and of measuring their concentrations. For kinetic experiments, time resolution must be provided, and secondary reactions avoided as far as possible. Over the last few decades, enormous progress has been made in experimental kinetics, with methods such as flash photolysis and flow techniques having provided much information applicable to atmospheric studies. Kinetic data for use in atmospheric chemistry have been critically evaluated by bodies such as NASA, the National Bureau of Standards (NBS) and the Committee for Data Analysis (CODATA), and compilations from these sources are periodically updated.

There are, however, limitations on the temperature, pressure, concentration, and rates that can be achieved or studied. Quite often, the laboratory data cannot be obtained under conditions identical to those present in an atmosphere. Extrapolation of the laboratory results is therefore necessary in order that the kinetic data may be used in interpreting atmospheric chemistry. Dangers abound in long-range extrapolation of data obtained over a limited span of experimental conditions. At least the extrapolation should be based on a believable theory for the process concerned. To highlight the areas of importance, this section reviews some aspects of chemical kinetics.

A reaction,

$$A + B \ldots \to products, \tag{3.18}$$

proceeds with a rate proportional to the concentrations raised to some power

$$\text{Rate} = -\frac{dn_A}{dt} = -\frac{dn_B}{dt} = kn_A^\alpha n_B^\beta. \tag{3.19}$$

The powers α and β are the *order* of reaction with respect to A and B, and $\alpha + \beta$ is the overall order; the constant of proportionality, k, is the *rate coefficient* (rate 'constant'). The *molecularity* of a reaction is the number of reactant molecules written in the stoicheiometric equation. Thus order is an experimental quantity, molecularity an arbitrary theoretical one. An *elementary* reaction step is conceived as one that cannot be split into any chemically simpler processes. For truly elementary steps, order and molecularity are in general identical. Thus, if reaction (3.18) is elementary, and the only reactants are A and B, it is both bimolecular and overall second-order: first-order in each of the components A and B. Our review will start with a discussion of elementary bimolecular processes; unimolecular and termolecular reactions are then considered later.

3.4.1 Bimolecular reactions

As two reactant molecules approach each other closely enough, the energy of the reaction system rises, as indicated by the potential energy surface (Fig. 3.6(b)) for the A + BC reaction (3.16). The contours of the surface show that there is a 'valley' which provides the lowest energy approach of the reactants, and the dotted line in the figure is that path. There comes a point, marked '$\neq$', beyond which the energy starts to decrease again, and so product formation is now energetically favourable. In our 'mountain' analogy, the point $\neq$ is a 'col' or 'pass'. Figure 3.7 shows the energy of the ABC system as a function of distance travelled along the low path for an exothermic reaction. The

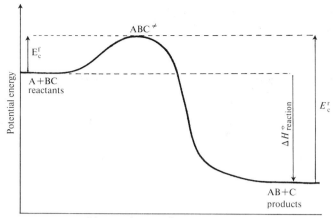

Fig. 3.7. The relationship between barrier energies for forward (E_c^f) and reverse (E_c^r) reaction and the enthalpy of reaction ($\Delta H_{reaction}^{\ominus}$). The diagram is essentially a cross-section of a potential energy surface (e.g. Fig. 3.6(b)) along the reaction coordinate. If the potential energies are expressed an enthalpies, then

$$\Delta H_{reaction}^{\ominus} = H_{products} - H_{reactants}$$

$$E_c^f = H_{ABC}^{\neq} - H_{reactants}$$

$$E_c^r = H_{ABC}^{\neq} - H_{products}$$

so that

$$E_c^f - E_c^r = \Delta H_{reaction}^{\ominus}$$

abscissa labelled 'reaction coordinate' is essentially the dotted line of Fig. 3.6(b) pulled out straight. The diagram shows the relationship between the critical energies (and presumably the measured activation energies: see p. 90 and p. 98) for forwards (E_c^f) and reverse (E_c^r) directions of a reaction. For an exothermic reaction as shown (left to right), ΔH is negative, and $E_c^r = E_c^f - \Delta H_{reaction}$. Furthermore, an endothermic reaction (right to left) must be accompanied by an activation energy (E_c^r) at least as large as the endothermicity, as stated in Section 3.2.2.

The reaction pathway shown is only one of an infinite number of possibilities. In principle, if the potential energy surface is known, it is possible to calculate, using the laws of mechanics, the path followed for any initial 'starting' distance and direction of approach of the reactants A and BC. For any given speed, and internal excitation (vibration, rotation) of the reactants, the fraction of 'trials' leading to product formation can then be assessed, and is related to the probability of reaction. The ordinary macroscopic rate coefficients, k, of eqn (3.19) can then be determined from a sum over the distributions of translational velocity, and vibrational and rotational excitation, appropriate to any temperature T: for thermal equilibrium, the Maxwell-Boltzmann distributions will be used. Such methods of *molecular dynamics* would be ideal for predicting rate coefficients *if only* it were possible to calculate from first principles the potential energy surfaces for all reactions. Unfortunately, it is not yet feasible to perform these 'a priori' calculations of surfaces, except for the very simplest reactions. Modern experimental kinetics can show the probability of passing from one set of reactant states (translational, vibrational, rotational, etc.) to one set of product states: so-called 'state-to-state' kinetics. The results of such experiments can then be used to test hypothetical potential energy surfaces. In this case, however, the molecular dynamic calculations have lost their predictive value.

Two simplifications are commonly adopted to overcome the lack of knowledge of complete potential energy surfaces for reaction. The first of these is the *collision theory* (CT). Reactant molecules are assumed to be hard spheres (radii r_A and r_{BC} in our example), and reaction is taken to be possible only if two conditions are met: a collision must occur, and the energy of collision along the line of centres must equal or exceed the energy required, E_c, to reach a critical configuration ($ABC^{\neq}$ in Fig. 3.7). The rate of reaction according to this theory is readily shown to be given by

$$-\frac{dn_A}{dt} = -\frac{dn_{BC}}{dt} = n_A n_B \sigma_c \bar{c} \exp(-E_c/RT), \qquad (3.20)$$

where σ_c is the cross-sectional area for collision (*collision cross-section*) given by

$$\sigma_c = \pi(r_A + r_B)^2, \qquad (3.21)$$

and $\bar{c}$ is the mean relative velocity of molecules for temperature T

$$\bar{c} = \left(\frac{8kT}{\pi\mu}\right)^{1/2}; \quad \mu = \frac{m_A m_{BC}}{m_A + m_{BC}}. \qquad (3.22)$$

Equation (3.20) certainly has the correct concentration dependence for an elementary bimolecular reaction, so that the rate coefficient can be written

$$k = \sigma_c \bar{c} \exp(-E_c/RT). \qquad (3.23)$$

Experimentally, many second-order rate constants are found to follow a temperature law embodied in the *Arrhenius expression*

$$k = A \exp(-E_a/RT), \qquad (3.24)$$

where E_a is an experimental *activation energy* and A is the 'pre-exponential' factor. At this stage in the development, it is usual to identify E_c with E_a, and thus to ask if $\sigma_c \bar{c}$ is to be compared with A. However, it should not be forgotten that $\bar{c}$ is dependent on $T^{1/2}$ [cf. eqn (3.22)], while A, in the simplest formulation, is not temperature dependent. Over limited temperature ranges, any $T^{1/2}$

dependence of experimentally-determined A factors may be undetectable. That does not mean, though, that eqn (3.24) will be suitable for extrapolation over extended temperature ranges. A more telling difficulty concerns the absolute magnitudes of A and $\sigma_c \bar{c}$. For typical atmospheric reactants, with collision radii ~ 400 pm and relative molecular masses ~ 30, $\sigma_c \bar{c}$ is $\sim 3 \times 10^{-10}$ cm^3 molecule^{-1} s^{-1} at 300 K. The product $\sigma_c \bar{c}$ is called the *collision frequency factor*. Except for the very simplest of reactants, experimental A factors are usually less than, and often much less than, the collision frequency factor. An explanation for the lack of agreement is sought in terms of molecular complexity, with the existence of special geometric arrangements needed during the collision to bring reactive parts of the molecules together (*steric requirements*) and of special needs for the distribution of internal energy. That explanation takes us well away from the idea of hard-sphere reactants.

The alternative simplification adopted in the interpretation of bimolecular reactions is that of the *transition state theory* (TST) or *activated complex theory* (ACT), a theory sometimes rather hopefully also called 'absolute rate theory'. Quasi-equilibrium is assumed between reactants and the critically-configured ABC$^{\neq}$ molecule, in order to calculate concentrations of ABC$^{\neq}$ (the *transition state*). Rates of reaction can then be obtained from the rate at which ABC$^{\neq}$ passes to products (as a result of translational or vibrational motions along the reaction coordinate). Equilibrium constants can be expressed in statistical thermodynamic terms, and if the formulation is also valid for the quasi-equilibrium, where the system is at a (free) energy maximum rather than minimum, then the resultant rate coefficient, k, is given by

$$k = \frac{\mathbf{k}T}{h} \frac{q''_{ABC^{\neq}}}{q'_A q'_{BC}} \exp(-E_c/RT). \tag{3.25}$$

Total (volume-independent) partition functions are written as q'_A, q'_{BC}, $q''_{ABC^{\neq}}$ for reactants and transition state. The double prime on $q''_{ABC^{\neq}}$ indicates that the motion along the reaction coordinate has been factorized out (and a numerical constant introduced). In TST, then, the internal motions neglected in CT are expressly taken into account through the use of the partition functions. For the special case where both reactants are monatomic, and thus hard spheres, both CT and TST give identical algebraic expressions even though the underlying concepts are quite different. TST concentrates only on that region of the potential energy surface around the transition state (for the calculation of $q''_{ABC^{\neq}}$), while CT is interested only in the height of the energy barrier at the transition state. It is the calculation of $q''_{ABC^{\neq}}$ that offers most difficulties in the practical implementation of TST. Spectroscopic parameters for the reactant molecules are usually available, so that q'_A, q'_{BC} are readily estimated. However, knowledge of the shape of, and the forces acting at, the transition state implies that the potential energy surface is itself known, at least in the region of ABC$^{\neq}$. Usual practice is to make an 'informed guess' at the magnitude of $q''_{ABC^{\neq}}$, based on a hypothetical interaction mechanism and a corresponding model for the transition state. Considerable differences in predicted pre-exponential factors naturally follow from models of the transition state that are, for example, linear, bent, or cyclic. In a more limited way, TST can suggest a sensible order of magnitude for the pre-exponential factor. The three total partition functions in eqn (3.25) are each the product of translational, rotational, and vibrational partition functions. The translational parts can all be calculated, and orders of magnitude for rotational and vibrational parts employed in accordance with the number of each of these modes that exist in A, BC, and $ABC^{\neq}$. More important, from our point of view, is that the temperature dependence for every partition function can be evaluated as a power law. That is, eqn (3.25) can be rewritten as

$$k = A'T^n \exp(-E_c/RT), \tag{3.26}$$

where A' is the temperature-independent part of the pre-exponential function, and n some exponent chosen from the nature of the reactants (monatomic, diatomic, etc.) and a model of the transition state. For the hard sphere (CT) case, $n = 0.5$, from eqns (3.20) and (3.22). In the more general case, n can be positive or negative. The most sensible procedure in temperature extrapolation thus seems to be first to predict n from a model of the reaction, and then to fit the experimental data to equation (3.26) with that value of n.

Regardless of whether the CT or TST simplifications are made, the rate of reaction is in part controlled by the energy of a critical, transition state, configuration, an energy which has as its counterpart the activation energy of experimental kinetics. The energy barrier arises because the reactant molecules are forced close together (closer than the sum of their radii in the hard sphere collision approximation), and reactant bonds have to be broken while product bonds are made. Certainly, the energy required is less than that required first to break reactant bonds and then to form product molecules in separate steps. There is, however, no suggestion that the energy of the reactant system *decreases* at any stage in the passage from separated reactants to the transition state. Such a decrease in energy would correspond to long-range attractive forces, and might lead to an increased collision frequency, and to an A factor that exceeded $\sigma_c \bar{c}$. Many examples of this type of behaviour are in fact known, even with neutral reactants. The effects are, however, strongest and most common with charged reactants. In *ion-molecule* reactions, such as

$$O^+ + CO_2 \rightarrow O_2^+ + CO, \tag{3.27}$$

the ion can induce a dipole in the neutral reactant, and the resultant attractive force can both balance the ordinary chemical activation barrier as well as make the real encounter rate greater than the gas kinetic collision frequency factor for neutral molecules. Near-zero activation energies are thus often found in this type of reaction, and the pre-exponential factors are typically $\sim 10^{-9}$ cm^3 molecule^{-1} s^{-1} (i.e. ~ 3–4 times larger than the values for neutral reactants). Because the long-range attractive forces dominate the potential energy, high velocities of approach are counterproductive in promoting reaction, and some *negative temperature coefficient* of rate constant may be observed. The stronger (or longer range) the interaction, the larger the rate coefficient. For ion reactions with neutral molecules possessing permanent (rather than induced) dipoles, pre-exponential factors are increased by another two or three times. Thus charge transfer from O^+ to the dipolar molecule H_2O,

$$O^+ + H_2O \rightarrow H_2O^+ + O, \qquad (3.28)$$

has a rate coefficient of $2.3 \times 10^{-9} \, cm^3 \, molecule^{-1} \, s^{-1}$ at 298K, and nc activation energy. The long-range interactions are yet larger, of course, for two reactants both of which are charged. Positive ion-negative ion, or positive ion-electron reactions are characterized by rate coefficients 3–4 orders of magnitude faster than typical gas kinetic collision frequency factors. For example, the rate coefficient (298 K) for neutralization of NO^+ by an electron

$$NO^+ + e \rightarrow N + O \qquad (3.29)$$

is $4.5 \times 10^{-7} \, cm^3 \, molecule^{-1} \, s^{-1}$.

The possibility was raised in the last paragraph that high approach velocities, associated with high temperatures or energies, might reduce reactivity rather than increase it, particularly if the activation barrier is rather small. Both experimental measurements, and reaction dynamic calculations using suitable potential energy surfaces, have shown that this effect can also occur with neutral reactants. Furthermore, vibrational or rotational energy in the reactants can affect reactivity in unexpected ways. The relative importance of translational and internal energy in overcoming an activation barrier depends on the form of the potential energy surface for reaction. We seem to be returning here to the 'state-to-state' approach to kinetics, or at least to kinetics from particular reactant states to all product states. In atmospheric chemistry the question is important, since pressures and thus non-reactive collision rates may not be sufficient to keep reactants in thermal equilibrium. Much interest surrounds the possibility of species being created energy-rich and showing an enhanced reactivity. Our treatment of electronically excited species (Section 3.2.2) had as one of its premises that excess energy might be capable of overcoming an activation barrier. Translational and vibrational reactant excitation have both been proposed as possible ways in which reactivity could be enhanced. For example, 'hot' $O(^3P)$ could be created in the quenching of $O(^1D)$ by atmospheric N_2, or directly in ozone photolysis at wavelengths longer than the $O(^1D)$ production limit (310 nm: see Section 3.2.1). Translational excitation might then help overcome the endothermicity and the barrier to reaction of a process such as reaction (3.10) of $O(^3P)$ with H_2O to form OH. Similarly, absorption of solar radiation by H_2O in overtone and combination bands has been considered as a mechanism for enhancing the rate of reaction (3.10), some energy being carried in this case by the internal vibrational excitation of the water molecule. Another typical example of this reasoning concerns the production of NO in the Earth's atmosphere. Reaction of ground state $N(^4S)$ with oxygen

$$N(^4S) + O_2(^3\Sigma_g^-) \to NO + O \tag{3.14}$$

has a substantial activation barrier. Nitrogen monoxide (nitric oxide, NO) is a key compound in both neutral and ionic atmospheric chemistry, and suggestions have abounded about possible enhancement of the rate of (3.14) at various levels in the atmosphere as a result of electronic excitation in N or in O_2. It now appears that photolysis of N_2,

$$N_2 + h\nu \to N(^4S) + N(^2D), \tag{3.30}$$

gives not only an electronically excited $N(^2D)$ atom, but also a hot $N(^4S)$ atom with sufficient translational excitation to produce additional NO *via* reaction (3.14) in the Earth's lower thermosphere. Assessment of the reactivity of vibrationally or translationally excited reactants almost always assumes that all the excitation energy can be used to reduce an activation energy. The more complicated behaviour revealed by experimental and theoretical state-to-state kinetics for energies beyond the critical threshold suggests that the atmospheric calculations of reactivity should be approached with some diffidence.

3.4.2 Unimolecular and termolecular reactions

If chemical reaction requires collision, or at least close proximity, between the reactants, then it might seem that all thermal processes ought to be kinetically of second order. Unimolecular, first-order, elementary processes appear to lack the necessary approach of reactants, and termolecular, third-order, steps suffer from the impossibility of a simultaneous collision between three hard-sphere reactants. The explanation for first- and third-order thermal kinetics shares common ground, and a simple introduction is provided here.

No obstacle exists to understanding how single-step unimolecular and first-order decomposition occurs in a molecule AB that *already* has more than enough energy in it to break one of its bonds. We have seen examples in photodissociation of diatomic and triatomic molecules, where optical- or pre-dissociation populates vibrational levels of AB sufficient to cause fragmentation. The rate of fragmentation may depend on the rate at which energy can accumulate in the bond to be broken, but the reaction will be kinetically of first order. *Chemical activation* offers another route to high vibrational excitation. For example, the reaction of ClO with NO can produce a highly excited $ClONO^\dagger$ molecule (the dagger representing vibrational excitation):

$$ClO + NO \to ClONO^\dagger. \tag{3.31}$$

This excited $ClONO^\dagger$ can then either split up to the reactants again, or form Cl and NO_2:

$$ClONO^\dagger \rightarrow Cl + NO_2. \tag{3.32}$$

Reaction of the excited $ClONO^\dagger$ is a unimolecular, first-order, elementary reaction.

It is in interpreting *thermal* unimolecular reactions that some difficulty arises, since the formation of an excited $AB^\dagger$ molecule involves collisions between the AB species, and might therefore be expected to show *second*-order kinetics. A basic understanding was provided by Lindemann, who suggested that thermal first-order reactions were not true elementary steps, but rather involved at least three elementary processes

$$AB + AB \xrightarrow{\ k_a\ } AB^\dagger + AB \text{ collisional activation} \tag{3.33}$$

$$AB^\dagger + AB \xrightarrow{\ k_d\ } AB + AB \text{ deactivation} \tag{3.34}$$

$$AB^\dagger \xrightarrow{\ k_r\ } A + B \qquad \text{reaction.} \tag{3.35}$$

If reaction (3.34) dominates as a loss process for $AB^\dagger$ over (3.35) then the concentration of $AB^\dagger$ is almost at its thermal equilibrium value, while the rate-determining step for reaction is the first-order process (3.35). Overall first-order kinetics follow. It is obvious, however, that at sufficiently low concentrations of AB, there becomes a point at which (3.34) is rate limiting, and the kinetic behaviour will be second order. Transition from first to second order behaviour is, indeed, seen at low enough pressures in this kind of thermal unimolecular reaction. Quantitative expression of these ideas can be obtained by a *steady state* treatment for the concentration of $AB^\dagger$ (see Section 3.5). The result for the rate of loss of AB is

$$-\frac{dn_{AB}}{dt} = k_1 n_{AB} = \frac{k_a k_r n_{AB}^2}{k_d n_{AB} + k_r}, \tag{3.36}$$

where k_1 is the experimentally defined pseudo-first-order rate coefficient. So long as $k_d n_{AB} \gg k_r$, the reaction is first order, but if n_{AB} is reduced to the point at which the reverse inequality holds, then the reaction becomes second order. At high concentration, the limiting value of k_1 (referred to as k_1^∞) is equal to $(k_a k_r / k_d)$ and is thus truly first order, being independent of n_{AB}. The low pressure limit, k_1^0, is equal to $k_a n_{AB}$ and is itself first order in pressure, or second order overall.

Considerations about high and low pressure extrapolations of rate data are most frequently met in atmospheric chemistry in connection with termolecular reactions. As with unimolecular reactions, termolecular processes have orders variable with pressure, being third order at 'low' pressure and second order at 'high' pressure. Such reactions are extremely important in combination processes, and we can see why by first looking at the reaction of two atoms to form a diatomic molecule. A typical case is the combination of two $O(^3P)$ atoms, for which some potential curves were given in Fig. 3.2. Even if the combining atoms have no relative translational energy, the newly-formed O_2 molecule has the $O + O$ combination energy stored in it: that energy is the $O—O$ bond energy, and the O_2 is chemically activated $O_2^\dagger$ at its dissociation limit. *Unless* some energy is removed within one vibrational period, the molecule will fall apart again as the internuclear distance increases on the first oscillation. Energy can be removed in collisions, and we usually represent the species that dissipates energy by the symbol M. In atmospheric chemistry, the most important M may be the most abundant 'bath' gas (e.g. N_2 or O_2 on Earth). The overall reaction is now written

$$O + O + M \rightarrow O_2 + M, \tag{3.37}$$

which is a termolecular step. The redissociation that has been prevented is the unimolecular dissociation of $O_2^\dagger$ equivalent to step (3.35), and the process deactivating $O_2^\dagger$ is the equivalent of (3.34). The same general considerations about flow of energy thus apply to both unimolecular and termolecular reactions. If the newly formed molecule is larger than diatomic, then there are several vibrational modes into which the bond combination energy can flow. The lifetime of the newly formed molecule can thus correspond to many vibrational periods before the energy flows back to the critical bond. With a large enough polyatomic molecule, the lifetime can be so great that collisional removal of excess energy (*stabilization*) is no longer rate determining, and the combination process then exhibits second-order kinetics. The analogue of expression (3.36) can be derived from the single excitation level kinetic scheme

$$A + B \xrightarrow{k_c} AB^\dagger \qquad \text{combination} \tag{3.38}$$

$$AB^\dagger + M \xrightarrow{k_s} AB + M \quad \text{stabilization} \tag{3.39}$$

$$AB^\dagger \xrightarrow{k_r} A + B \qquad \text{reaction.} \tag{3.35}$$

The result is

$$\frac{dn_{AB}}{dt} = k_{II}n_A n_B = \frac{k_c k_s n_A n_B n_M}{k_s n_M + k_r}, \tag{3.40}$$

where k_{II} is the experimentally-defined pseudo-second-order rate coefficient corresponding to k_I in eqn (3.36). We see straightaway that if $k_r \gg k_s n_M$, the reaction is third order, with $k_{II}^0 = (k_c k_s / k_r) n_M$. If, however, $k_r \ll k_s n_M$, then the reaction is second order, with $k_{II}^\infty = k_c$. Increased complexity in the molecule AB reduces the value of k_r, because the combination energy is distributed amongst more vibrational modes. The concentration, or pressure, of third-body M at which third-order behaviour turns over to second-order kinetics is thus lower the more complex the molecule produced. 'Complex' is only a relative term here: combination of two hydrogen atoms to form H_2 is third order up to 10^4 atm, while combination of two CH_3 radicals to form C_2H_6 is second order at all but the lowest pressures. However, it so happens that the reactants in several combination reactions of great atmospheric importance, such as

$$OH + NO_2 + M \rightarrow HNO_3 + M, \tag{3.41}$$

or

$$O + O_2 + M \rightarrow O_3 + M, \tag{3.42}$$

are of just that molecular size that complex intermediate-order kinetics are displayed at some point in the atmospheric pressure range.

Expressions (3.36) and (3.40) represent the variations of experimentally determined rates of reaction with pressure. The pseudo-first- or second-order rate coefficients k_I or k_{II} can be conveniently expressed in terms of the high and low pressure limiting values k_I^∞, k_I^0, or k_{II}^∞, k_{II}^0. For example, k_{II} in eqn (3.40) can be expressed as

$$k_{II} = \frac{k_{II}^0 k_{II}^\infty}{k_{II}^0 + k_{II}^\infty}. \tag{3.43}$$

Remembering that k_{II}^0 is itself first order in pressure, it can be seen that eqn (3.43) represents in outline the variation of k_{II} with pressure that is found experimentally. Unfortunately, however, the equation does not match experimental data in detail, so that it cannot be applied directly to the calculation of rates at intermediate pressures. The reasons for the failure are known. The reactions and the rate coefficients k_a or k_c, k_d or k_s, and k_r should have been defined for each individual quantized vibrational level of $AB^\dagger$, and the individual rates summed to give the total rate. It is, perhaps, easy to see that the more energy available (beyond the critical amount needed to break a particular bond), the more rapid will be the fragmentation (i.e. the larger will be k_r). Related to this point is the implication that energy stored in any vibrational mode can be made available to the critical bond. Experimental evidence largely favours the flow of energy between modes as being fairly free, and the distribution as being near-statistical. An additional complication involves the inter-conversions of vibrations and rotations in the fragmenting molecule. The theory has been extended, modified, and manipulated over the years by Rice, Ramsperger, Kassel, and Marcus, and the familiar initials RRKM are used to designate their formulation. With sufficient sophistication of the input information, very good agreement can be obtained between theory and experiment. Correspondingly, one could have confidence in the extrapolation of data obtained in an intermediate concentration regime to either high-pressure (first-order) or low-pressure (second-order) limits. However, application of RRKM theory to real processes of atmospheric importance is in practice rather difficult, and an alternative, much simpler, approach is now almost universally adopted. This approach has its origins in work by Troe (see Bibliography) on the theoretical prediction of unimolecular reaction rate parameters. However, with k_{II}^0 and k_{II}^∞ known, Troe has shown that a simplification of his theory allows the right-hand side of eqn (3.43) to be multiplied by a *broadening factor*, F, that is a function of (k_{II}^0/k_{II}^∞). For many atmospherically important termolecular reactions F may be calculated from a simple mathematical expression.

Third-order reactions often show decreasing rate with increasing temperature: they have a negative temperature coefficient. The reason is that the larger the thermal kinetic energy possessed by the reactants A and B in process (3.38), the more internal vibrational energy will be stored in the $AB^\dagger$ molecule produced. As pointed out earlier, the chance of the critical bond energy finding its way back to a breakable bond is thus increased, and k_r is larger. Since k_c and k_s are only slightly affected by temperature, it follows from eqn (3.40) that the rate of reaction will decrease with increasing temperature. Thermal energy in effect assists the newly-formed molecule to split up again, thus slowing the rate of combination. In the third-order limit, k_{II}^0 is inversely proportional to k_r (see above). Theory suggests that the temperature variations of k_r should be better expressed in terms of a power, T^n, rather than as a conventional activation energy. Hence experimental measurements of k_{II}^0 as a function of temperature should be fitted against a T^{-n} law to allow rational interpolation or extrapolation to atmospheric temperatures. Typical measured values of n are 2.5 to 3.1 for reaction (3.41) and 1.7 for reaction (3.42). Models of the transition state for bond association reactions also suggest that at the high-pressure limit, k_{II} should possess a negative exponent of temperature.

Two other combination mechanisms of atmospheric interest can be included in this section, although they are not termolecular. Both mechanisms highlight the need to dissipate the energy of the combination process. Energy can be carried off as photons, so that if the combination can populate, directly or indirectly, an excited state (usually electronically excited) which can radiate, then the nascent molecule can be stabilized. An example of this process is the *two*-body combination of O and NO

$$O + NO \rightarrow NO_2 + h\nu. \tag{3.44}$$

The process is probably responsible for some continuum emission in the atmosphere. However, three-body combination of O and NO is much more efficient than two-body radiative combination at all but very low pressures. In ion-neutralization processes such as reaction (3.29), the initial step is the formation of a neutral molecule which is highly excited by an amount equivalent to the ionization potential. Unless the energy can be dissipated, the diatomic neutral molecule will re-ionize. We are not interested here in chemical stabilization, but rather in stabilizing the removal of charged species. Energy can be conserved in the neutralization by fragmentation of the excited NO, with translational motion carrying off the excess of ionization energy (9.2 eV) over bond dissociation energy (6.4 eV). Ionized *atoms* obviously cannot combine with electrons in this way. Radiative stabilization thus becomes the only (non-collisional) neutralization mechanism. However, the radiative transitions needed are usually forbidden for the neutral atoms formed, so that the rates of neutralization are much slower, by about five orders of magnitude, for atomic than for molecular ions.

3.5 Multistep reaction schemes

Chemistry in planetary atmospheres is made up of complex interactions of elementary reactions. *Consecutive* and *parallel* steps involve reactive *intermediates* in competitive processes. In this section we discuss the kinetics of such processes as an introduction to the discussion of models of atmospheric chemistry in Section 3.6. Reaction intermediates of particular interest include atoms, radicals, ions, and excited species. Most of these intermediates are highly reactive, and, with one or two exceptions, cannot be 'stored' in a laboratory for long periods because they are lost on the walls of the containing vessel, or react with each other. Such intermediates are not necessarily *unstable*, and chemical lifetimes of isolated atoms or radicals in the absence of surfaces can be virtually infinite. Many excited state species *are* unstable, since they may possess enough internal energy to fragment, and they may also be able to lose their energy by emission of radiation. An excited species that cannot undergo loss by an allowed radiative transition is said to be *metastable*.

Multi-step reaction schemes are interpreted kinetically by writing down the differential equations, such as (3.19), for all the species of interest, including the intermediates. Solution of these equations then allows prediction of the concentration–time variation of each of the species. Unfortunately, analytical solution of the many simultaneous differential equations is rarely possible. Numerical solution has become a widely-used alternative since the advent of high-speed computers and good techniques for dealing with differential equations. Such methods do not, however, afford much insight into the underlying chemistry of the system. For some highly reactive intermediates, the *Stationary State Hypothesis* (SSH) provides a simplification that will permit algebraic solutions to the kinetic equations. Consider an intermediate X that is created in a process whose rate is constant, and whose loss rate increases with increased [X]. After the reaction is started, [X] will increase until the rate of loss is equal to the rate of formation. A steady state for [X] has been reached, and $d[X]/dt = 0$. To illustrate the steady state method let us calculate the concentration of $O(^1D)$, assumed to be formed at some altitude in the Earth's atmosphere solely by oxygen photolysis and lost by quenching with nitrogen and oxygen:

$$O_2 + hv \rightarrow O(^1D) + O(^3P); \quad \text{Rate} = I_{abs} \tag{3.45}$$

$$O(^1D) + M \xrightarrow{k_q} O(^3P) + M \tag{3.46}$$

$$M = N_2 \text{ or } O_2.$$

The kinetic equation that describes these processes is

$$\frac{d[O(^1D)]}{dt} = I_{abs} - k_q[O(^1D)][M]. \tag{3.47}$$

If $O(^1D)$ is in a stationary state, then we set the differential equal to zero, and

$$[O(^1D)]_{ss} = I_{abs}/k_q[M], \tag{3.48}$$

where the subscript indicates a steady state concentration. Such a value might now be used to estimate the rate of a minor process, involving $O(^1D)$, such as (3.11) which generates OH. The problem is to know if the concentration of $O(^1D)$ calculated using the SSH bears any relationship to actual concentrations. Our two-reaction example has been chosen because it can also be solved analytically. So long as I_{abs} and M are independent of time, eqn (3.47) can be integrated to yield

$$[O(^1D)] = (I_{abs}/k_q[M])(1 - \exp(-k_q[M]t)), \tag{3.49}$$

where t is the time for which the system has been illuminated. This expression for $[O(^1D)]$ approaches the steady state expression so long as $k_q[M]t \gg 1$, the error in applying the SSH being less than one per cent for $k_q[M]t \geqslant 4.6$. At an atmospheric altitude ($\sim 80\,\text{km}$) where $[M] \sim 3 \times 10^{14}$ molecule cm^{-3}, and the composite k_q for N_2 and O_2 is $\sim 3 \times 10^{-11}$ cm^3 molecule^{-1} s^{-1}, the non-steady and steady state concentrations of $O(^1D)$ are therefore nearly identical for illumination times exceeding $4.6/(3 \times 10^{14} \times 3 \times 10^{-11}) \sim 5 \times 10^{-4}$ s. The SS hypothesis can thus be applied so long as I_{abs} and $[M]$ remain constant over, say, 10^{-3} s. In our example, $[M]$ (and $[O_2]$, which is the $O(^1D)$ precursor) are both time invariant, and solar intensities (I_0) change diurnally over periods of hours rather than milliseconds, so that $[O(^1D)]_{ss}$ is a good approximation for $[O(^1D)]$. As solar intensity varies during the day, so the steady state concentration of $O(^1D)$ adjusts (on the time-scale of milliseconds) to a new value. Now let us consider what happens with an intermediate that is much less reactive than $O(^1D)$. A good example is the ground state $O(^3P)$ formed photolytically in (3.45); for simplicity, we ignore the $O(^3P)$ atom from reaction (3.46). Three-body combination of O with O_2,

$$O + O_2 + M \rightarrow O_3 + M, \tag{3.42}$$

is the major loss process for $O(^3P)$, so that eqns (3.48) and (3.49) for $O(^1D)$ are replaced by

$$[O(^3P)]_{ss} = I_{abs}/k_t[O_2][M] \tag{3.50}$$

and

$$[O(^3P)] = (I_{abs}/k_t[O_2][M])(1 - \exp(-k_t[O_2][M]t)), \tag{3.51}$$

where k_t is the third-order rate coefficient for reaction (3.42). For temperatures $\sim 200\,\text{K}$, appropriate to our hypothetical reaction altitude (80 km),

$k_t \sim 1.4 \times 10^{-33} \, \text{cm}^6 \, \text{molecule}^{-2} \text{s}^{-1}$. The exponential in (3.51) is thus < 0.01 for $t \geqslant 4.6/[1.4 \times 10^{-33} \times 0.2 \times (3 \times 10^{14})^2] \sim 1.8 \times 10^5 \, \text{s} \sim 50 \, \text{h}$. Steady-state conditions take much longer to achieve than the time-scale over which solar intensities vary; a steady state for $O(^3P)$ is *not* set up, and the SSH is inappropriate. The difference in behaviour of $O(^1D)$ and $O(^3P)$ arises because of the relative values of the terms $k_q[M]$ and $k_t[O_2][M]$, which are the *pseudo-first-order* loss rates for the two intermediates respectively. For the highly reactive excited atom, the loss rate is large, and a steady state is achieved rapidly; for the less reactive ground state atom, the steady state is not reached on the time-scale over which reaction conditions remain constant.

3.6 Models of atmospheric chemistry

Interpretations of atmospheric behaviour can be tested by comparing the predictions of a *model* of the atmosphere with results from measurements on the atmosphere itself. Models of atmospheric physics have been implicit in much of our earlier discussion (e.g. of radiative heating, Section 2.2). We now indicate how chemical changes can be accommodated in the models. *Diagnostic models* are used to assess hypotheses about the physics and chemistry of atmospheres from a knowledge of present-day physical and chemical structure. *Prognostic models* are of interest in evaluating the future behaviour and evolution of an atmosphere subject, for example, to changes in natural and artificial inputs of trace species. Such models can also be used to determine whether an assumed set of 'starting' conditions could lead to observed present-day atmospheres.

Consider a volume element in an atmosphere small enough to be uniform witk respect to all variables such as temperature, density, composition, and so on. If the rate of flux to and from that element of heat, radiation, matter, etc., from all other atmospheric elements is calculated, then the rate of change of the various physical parameters can be established. Chemical change within the volume element requires only slight modification for the local alterations in composition, and possibly in physical conditions. The *continuity equations* involved in the models are typified by the one for matter, which states that the net flow of mass into unit volume per unit time is equal to the local rate of change of density. Solution of the continuity equations for every physical and chemical parameter of interest, and for every volume element in the atmosphere, should then lead to a self-consistent model of atmospheric behaviour that mimics in all respects the temporal and spatial changes in the real atmosphere. According to this view, with sufficient input information, and a fine enough grid size for the volume elements, all meteorological as well as chemical phenomena could be simulated by the model. Such *three-dimensional models* (3-D) would be ideal for studying atmospheric chemistry. Computer-numerical solutions are naturally used in models, but the 3-D models are so demanding of computer time and memory that, at present, the chemistry has to be simplified drastically. Details of the chemistry are thus lost. Simpler models may then provide more useful chemical information, and we now turn to a discussion of some possible approximations.

Box models assume uniform mixing of individual constituents of an atmosphere. The chemistry is thus treated exactly as in a well-mixed laboratory system. The time evolution of each species is controlled by the chemical interactions alone. Rates of reaction are calculated as described in the previous section, with the steady state hypothesis applied if necessary and appropriate. For long-lived species [e.g. the $O(^3P)$ atom for the conditions specified in the example of Section 3.5], diurnal (but probably not seasonal) averaging may be reasonable. Such models are most useful in the analysis of global budgets of long-lived species.

One-dimensional (1-D) models are designed to simulate the vertical distribution of atmospheric species, but not any horizontal variations. The models thus represent horizontal averages, but do include atmospheric transport, attenuation and scattering of solar radiation, and detailed chemistry. Where vertical winds are weak or absent, as in the Earth's stratosphere, vertical transport is assumed to occur *via* the eddy diffusion mechanism (see Section 2.3.3). Equation (2.44), which is appropriate to molecular diffusion of a particular species, i, in the absence of chemical reaction, has to be modified. Allowance is made for the rate of production, P_i, and of loss, L_i, of the species in all possible chemical reactions, and the molecular diffusion coefficient is replaced by the (vertical) eddy diffusion coefficient, K_z:

$$\frac{\partial f_i}{\partial t} = K_z \left(\frac{\partial^2 f_i}{\partial z^2} - \frac{1}{H_s} \frac{\partial f_i}{\partial z} - \frac{1}{T} \frac{\partial T}{\partial z} \frac{\partial f_i}{\partial z} \right) + (P_i - L_i)/N. \tag{3.52}$$

As pointed out in Section 2.3.3, K_z is a phenomenologically-constructed parameter based on observations of a minor constituent whose chemistry is presumably well known. Chemical tracers (e.g. CH_4, N_2O) and radionucleides from past atmospheric nuclear tests (^{14}C, ^{90}Sr, ^{95}Zr, ^{185}W) have been used to provide source data. Typical K_z values are $10-500 \, m^2 \, s^{-1}$ between 50 and 80 km in the Earth's atmosphere, and $10 \, m^2 \, s^{-1}$ in the dynamically active troposphere. The stability of the lower stratosphere leads to values of 0.1 to $1 \, m^2 \, s^{-1}$.

Averaging of the 1-D model results must be performed to achieve a link with physical reality. A sensible altitude spacing must be adopted, and results calculated over averaged time steps. An extreme case puts the chemistry in a stationary state competition with vertical transport (i.e. $\partial f_i/\partial t = 0$ in eqn (3.52) for one or more constituents). Diurnal averaging procedures, on the other hand, permit considerable savings of computer time when solution is required of time-dependent problems over long durations. Fully time-dependent solutions are more useful diagnostically and prognostically, and must be used for many short-lived species. Such solutions are, however, complicated, and expensive computationally.

Two-dimensional (2-D) models provide spatial resolution in vertical and meridional directions (i.e. latitudinal, but not longitudinal, resolution). Mean horizontal motions must now be incorporated into the two-dimensional analogue of eqn (3.52), and the eddy diffusion coefficient has to be treated as a tensor (2×2 matrix) quantity. Averaged meridional velocities can be obtained, from direct observation, for the Earth's troposphere and stratosphere. Some problems exist about interpretations of eddy motions, which include, in two dimensions, large-scale internal waves. Nevertheless, 2-D models are far more realistic than 1-D models. The motivation for constructing 2-D models is that meteorological quantities on a rotating planet are not expected to show variations with longitude if averaged over several days. In practice, latitudinal gradients in the Earth's atmosphere are generally larger than those in the longitudinal direction. The chemical reactions are handled in *Eulerian models* by allowing them to occur in a hypothetical box fixed in space through which material is transported. In *Lagrangian models*, on the other hand, a definite parcel of air is followed on its trajectory through the atmosphere, and the chemical changes allowed to take place within this moving parcel. Coupling of transport and heating effects is an important part of the models. Feedback between chemistry and other aspects of the model should ideally be accommodated because of the potential interactions. For example, the reactions determining the concentration of a 'greenhouse' gas (Section 2.2.2) such as ozone may be highly temperature sensitive, and the concentration of the gas itself will influence atmospheric temperatures, radiation balance, and dynamics (cf. Sections 4.5.9 and 4.5.10).

Because atmospheres are three-dimensional fluids, it is clear that complete simulation of the radiative-chemical-dynamical behaviour of an atmosphere requires a 3-D model. A 3-D general circulation model (GCM) can, in principle, include feedback mechanisms, such as those from changes in composition, to the transport process. As pointed out earlier, full 3-D models with chemistry place enormous demands on computer power. 3-D models are, however, indispensable in examining the coupling effects (e.g. radiative-transport) that are incompletely treated in simpler models. They are also of great value in assessing the accuracy of spatial and time averaging techniques needed in the 1-D and 2-D models. Progress in 3-D modelling is currently rapid both because of increasing programming efficiency and the ever-increasing power of computers.

Even for 1-D and 2-D approximations, the chemistry may impose an excessive computational load. For example, for the Earth's stratosphere, about fifty chemical species are involved in nearly 200 reactions. There are thus fifty (simultaneous) continuity equations to be solved at each grid point. Direct numerical solution of the differential equations may require inordinate computer memory and run times. The problem is obviously particularly pressing when a model is used to test prognostically the effects on an atmosphere of altering concentrations of trace species because the calculation must be repeated many times. An approach to reduce the computational needs consists of identifying closely coupled chemical species that can be summed into *families*. For the Earth's atmosphere, 'odd oxygen' (i.e. O, O_3 but not O_2: 'odd' means 'not even' number of oxygen atoms) is one such family. O_3 is rapidly converted photolytically to O (reactions 3.1 + 3.46), while in the lower stratosphere and below, reaction (3.42) converts O back to O_3. Other important families, as we shall see, include NO_x ($x = 1, 2$), HO_x ($x = 0, 1, 2$) and ClO_x ($x = 0, 1, 2$). A reduced set of equations is solved for the families, and then each family is partitioned into its components. This latter stage requires some (inspired) guesswork, but 1-D models using 'family' and direct solutions compare well.

Bibliography

Sections 3.1 and 3.2

Books on general photochemistry

Photochemistry. Calvert, J. G. and Pitts, J. N., Jr. (John Wiley, New York, 1966.)
Photochemistry of small molecules. Okabe, H. (John Wiley, New York, 1978.)
Principles and applications of photochemistry. Wayne, R. P. (Oxford University Press, Oxford, 1988.)
Photochemistry and spectroscopy. Simons, J. P. (John Wiley-Interscience, London, 1971.)

> *Many of the quantitative data in this classic and seminal work by Leighton are, of course, superseded. However, the book still provides a sound exposition of photochemical change in the atmosphere and of the principles determining solar irradiances and ultraviolet penetration.*

Photochemistry of air pollution. Leighton, P. A. (Academic Press, New York, 1961.)

Articles on photochemical change in the atmosphere

The atmosphere as a photochemical system. Chapter 5 in McEwan, M. J. and Phillips, L. F. *Chemistry of the atmosphere.* (Edward Arnold, London, 1975.)
Photodissociation effects of solar uv radiation. Simon, P. C. and Brasseur, G. *Planet. Space Sci.* **31**, 987 (1983).

> *An example of the use of satellites to monitor the intensity and spectral distribution of ultraviolet radiation incident on the atmosphere.*

18 months of UV irradiance observations from the Solar Mesosphere Explorer. London, J., Bjarnason, G. G., and Rottman, G. J. *Geophys. Res. Lett.* **11**, 54 (1984).

Section 3.2.1

> *These articles discuss in detail the photochemistry of ozone, a molecule central to much of the chemistry of the Earth's atmosphere.*

The photochemistry of ozone. Wayne, R. P. in *The handbook of environmental chemistry*; *volume 2, part E* Hutzinger, O. (ed.). (Springer-Verlag, Berlin, 1989).
The photochemistry of ozone. Wayne, R. P. *Atmos. Env.* **21**, 1683 (1987).

Section 3.2.2

> *Consideration of enhanced reactivity generally emphasizes the effects of electronic excitation. In this paper, the possibility is explored that translational excitation might contribute to increased reaction rates.*

The possible effects of translationally excited nitrogen atoms on lower thermospheric odd nitrogen. Solomon, S. *Planet. Space Sci.* **31**, 135 (1983).

Section 3.3

> *Although the title of this chapter in a book seems rather restricted, the chapter provides an excellent introduction to the subject of the conservation rules and correlation diagrams. Further explanations are to be found in the texts given as references in connection with Section 3.4.*

On the rapidity of ion-molecule reactions. Talrose, V. L., Vinogradov, P. S., and Larin, I. K., in *Gas phase ion chemistry* (ed. Bowers, M. T.) Vol. 1, pp. 305–47. (Academic Press, New York, 1979.)

Section 3.4

> *Books on the theories of chemical kinetics. The first book contains chapters covering many topics of central importance to atmospheric chemistry, including the influence of temperature on bimolecular, unimolecular, and termolecular reactions, the reactions of electronically- and vibrationally-excited species, and on ion-molecule reactions and the influence of translational and internal energy on them.*

Reactions of small transient species. Fontijn, A. and Clyne, M. A. A. (eds.). (Academic Press, New York, 1984.)
The theory of the kinetics of elementary gas phase reactions. Wayne, R. P. in *The theory of kinetics* (ed. Bamford, C. H. and Tipper, C. F. H.) *Comprehensive chemical kinetics*, Vol. 2. (Elsevier, Amsterdam, 1969.)
Kinetics and dynamics of elementary gas reactions. Smith, I. W. M. (Butterworth London, 1980.)
Molecular reaction dynamics and chemical reactivity. Levine, R. D. and Bernstein, R. B. (Oxford University Press, Oxford, 1987.)

> *Theoretical and observational aspects of the reactions of ionic species in the gas phase. Descriptions of some experimental techniques are to be found in references given at the end of Chapter 6.*

Classical ion-molecule collision theory. Su, T. and Bowers, M. T., in *Gas phase ion chemistry* (ed. Bowers, M. T.) Vol. 1, p. 84. (Academic Press, New York, 1979.)

Temperature and pressure effects in the kinetics of ion molecule reactions. Meot-Ner, M., in *Gas phase ion chemistry* (ed. Bowers, M. T.) Vol. 1, p. 198. (Academic Press, New York, 1979.)

Chemical reactions of anions in the gas phase. DePuy, C. H., Grabowski, J. J., and Bierbaum, V. M. *Science* **218**, 955 (1982).

Isotope exchange in ion-molecule reactions. Smith, D. and Adams, N. G., in *Ionic processes in the gas phase* (ed. Almoster Ferreira, M. A.). (D. Reidel, Dordrecht, 1984.)

Elementary plasma reactions of environmental interest. Smith, D. and Adams, N. G. *Topics curr. Chem.* **89**, 1 (1980).

Section 3.4.2

The last three references for this section describe the development of Troe's simplified treatment for the prediction of unimolecular reaction rates; the final one provides an accessible synthesis of the ideas.

Current aspects of unimolecular reactions. Holbrook, K. A. *Chem. Soc. Rev.* **12**, 163 (1983).

Current aspects of unimolecular reactions. Quack, M. and Troe, J. *Int. Rev. phys. Chem.* **1**, 97 (1981).

Predictive possibilities of unimolecular rate theory. Troe, J. *J. phys. Chem.* **83**, 114 (1979).

Specific rate constants $k(E, J)$ for unimolecular bond fissions. Troe, J. *J. chem. Phys.* **79**, 6017 (1983).

The first of these three references is a review that discusses the care needed in applying laboratory kinetic data to atmospheric systems and that presents critically the types of data available. The second two references are to typical data compilations that are used by atmospheric modellers.

Kinetics of thermal gas reactions with application to stratospheric chemistry. Kaufman, F. *Annu. Rev. phys. Chem.* **30**, 411 (1979).

Evaluated kinetic and photochemical data for atmospheric chemistry. Baulch, D. L., Cox, R. A., Crutzen, P. J., Hampson, R. F., Jr., Kerr, J. A., Troe, J., and Watson R. T. *J. Phys. Chem. Ref. Data* **11**, 327 (1982).

Chemical kinetics and photochemical data for use in stratospheric modeling: evaluation number 9. DeMore, W. B., Molina, M. J., Sander, S. P., Golden, D. M., Hampson, R. F., Kurylo, M. J., Howard, C. J., and Ravishankara, A. R. (NASA, Washington, DC, 1990.)

These reviews discuss the reactions of some specific radicals and atoms of special significance to atmospheric chemistry.

Gas phase reactions of hydroxyl radicals. Baulch, D. L. and Campbell, I. M. *Gas kinetics and energy transfer* Specialist Periodical Reports Chem. Soc., **4**, 137 (1981).

Kinetics and mechanisms of the reactions of the hydroxyl radical with organic compounds in the gas phase. Atkinson, R., Darnall, K. R., Lloyd, A. C., Winer, A. M., and Pitts, J. N., Jr. *Adv. Photochem.* **11**, 375 (1979).

Kinetics of gaseous hydroperoxyl radical reactions. Kaufman, M. and Sherwell, T. *Prog. Reaction Kin.* **12**, 1 (1983).

Elementary reactions of atoms and small radicals of interest in pyrolysis and combustion. Baggott, J. E. and Pilling, M. J. *Annu. Rep. Pro. Chem.* **C79**, 199 (1983).

Atom and radical recombination reactions. Troe, J. *Annu. Rev. phys. Chem.* **29**, 223 (1978).

The kinetics of radical-radical processes in the gas phase. Howard, M. J. and Smith, I. W. M. *Prog. Reaction Kin.* **12**, 55 (1983).

Section 3.6

> *The first reference provides a straightforward introduction to the practice of atmospheric modelling, and the others discuss particular aspects of constructing models. Many further references to models will be found at the ends of later chapters, especially Chapter 4 and Chapter 5.*

Modelling chemical processes in the stratosphere. Chang, J. S. and Duewer, W. H. *Annu. Rev. phys. Chem.* **30**, 443 (1979).

A comparison of one-, two- and three-dimensional representations of stratospheric gases. Tuck, A. F. *Phil. Trans. R. Soc.* **A290**, 477 (1979).

A modified diabatic circulation model for stratospheric tracer transport. Pyle, J. A. and Roger, C. F. *Nature, Lond.* **287**, 711 (1980).

4 Ozone in Earth's stratosphere

4.1 Introduction

Ozone plays several extremely important roles in the Earth's atmosphere, as we have already seen in Chapters 1 and 2. First, ozone absorbs virtually all solar ultraviolet radiation between wavelengths of about 240 and 290 nm which would otherwise be transmitted to the Earth's surface. Such radiation is lethal to simple unicellular organisms, and to the surface cells of higher plants and animals. Ultraviolet radiation in the wavelength range 290 to 320 nm (so-called UV–B) is also biologically active, and prolonged exposure to it may cause skin cancer in susceptible individuals (see Section 4.5.1). Secondly, upper atmospheric meteorology is greatly influenced by the heating that follows absorption by ozone of UV, visible, and thermal IR (9.6 μm) radiation. Stratospheric air is statically stable because of the increase in temperature with altitude. Ozone absorption also provides a significant energy source for driving the circulation of the mesosphere, and for forcing tides in the upper mesosphere and thermosphere. Although the importance of atmospheric ozone has been recognized for more than sixty years, research has intensified dramatically over the last twenty. Interest has been stimulated by concern that a variety of human influences might lead to detectable changes in the abundance of stratospheric ozone. An increasing awareness of the part played by minor atmospheric constituents in the determination of natural ozone concentrations coincided with an understanding that man could inject into the stratosphere substances that, because of the stratospheric stability, could build up to levels where significant depletion of ozone might occur. The truly global problem posed by anthropogenic pollution of the stratosphere captured the imagination of the scientific community and led to the identification, much more clearly than before, of the factors that control ozone concentrations in the unpolluted atmosphere. It is now apparent that stratospheric ozone concentrations have been declining over the past two decades (Section 4.6), and dramatic depletions of ozone over the Antarctic (Section 4.7) each year—the so-called 'Antarctic ozone hole'—can only be explained in terms of chemistry perturbed by man's release of halogen-containing compounds. In this chapter we discuss the 'natural' stratosphere first, and then consider how the ozone layer might be modified by man. We must, however, remember that the present understanding of the natural stratosphere has grown out of the active programme of research motivated by the perceived threat of man's activities. Facing that programme was the formidable task of bringing together the several different aspects of the problem: transport and dynamics, the complex chemistry and photochemistry of the stratosphere, sources and sinks of trace constituents, and the intensity and variability of solar radiation, to name a few. An elaborate series of field measurements has been carried out to verify and consolidate the understanding gained. One clear result is that it is not enough to examine the stratosphere in isolation, since its chemistry and its dynamics are linked with those of the troposphere.

4.2 Observations

The aeronomy and meteorology of atmospheric ozone have been studied for nearly a century and a half, since the suggestion[a] by Schönbein in 1840 of an atmospheric constituent having a peculiar odour (Greek for 'to smell' is *ozein*). Ozone's existence in the troposphere was established in 1858 by chemical means.[b] Subsequent spectroscopic studies in the visible and ultraviolet regions showed, as early as 1881[c], that ozone is present at a higher mixing ratio in the 'upper' atmosphere than near the ground. In the early part of this century, quantitative analysis, particularly by Fabry and by Dobson, had shown the existence of the ozone layer (although its estimated altitude was misplaced).

We now know that ozone is found in trace amounts throughout the atmosphere, with the largest concentrations in a well-defined layer at altitudes between about 15 and 30 km (cf. Sections 1.4, 4.3.2, and Fig. 1.2 (p. 11)). The ozone layer is, however, a highly variable phenomenon. Total ozone densities in an atmospheric column near the pole can increase by 50 per cent in ten days in spring, and daily variations of up to 25 per cent have been recorded. Concentrations at particular altitudes show even greater variability, eight-fold increases in a few days being not unusual in the lower stratosphere (say 13–19 km altitude).

Average column densities of ozone vary with season and latitude. Figure 4.1 (a) shows some typical column density measurements[d] in contour map

---

[a] Schönbein, C. F., Recherches sur la nature de l'odeur qui se manifeste dans certaines actions chimiques. *C. R. Acad. Sci. Paris* **10**, 706 (1840).

[b] Houzeau, A., Preuve de la présence dans l'atmosphère d'un nouveau principe gazeux, l'oxygène naissant. *C. R. Acad. Sci. Paris* **46**, 89 (1858).

[c] Hartley, W. N., On the absorption of solar rays by atmospheric ozone. *J. Chem. Soc.* **39**, 111 (1881).

[d] The geophysical literature often uses 'Dobson Units' (DU) to describe atmospheric column ozone abundances in order to honour Dobson's pioneering work on stratospheric ozone. One DU is the thickness in the units of hundredths of a millimetre that the ozone column would occupy at standard temperature and pressure (STP: $0°C$ and one atmosphere). As stated at the beginning of Section 1.4, the ozone column compressed in this way would typically be 3 mm thick, so that the atmospheric abundance is 300 DU. In this book, we keep to molecular units more familiar to chemists and physicists. The conversion factor is roughly $1\ DU = 2.69 \times 10^{16}$ molecule cm^{-2}.

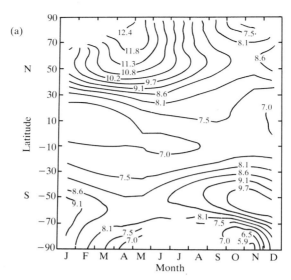

Fig. 4.1(a). Observed atmospheric ozone abundance as a function of altitude and time of year. The contour lines are total overhead column abundances in units of 10^{18} molecule per cm^2 area. The data are averages for the period 1979–1987 obtained by the Total Ozone Mapping Spectrometer (TOMS) on the Nimbus 7 satellite. Redrawn from *Scientific assessment of stratospheric ozone: 1989*. World Meteorological Organization, Geneva, 1990.

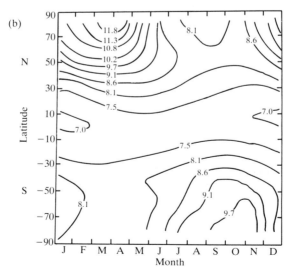

Fig. 4.1(b). Atmospheric ozone abundance as a function of altitude and time of year, calculated using the LLNL 2-D model. The contour lines are to be compared with the observed values that appear in Fig. 4.1(a). Redrawn from *Scientific assessment of stratospheric ozone: 1989*. World Meteorological Organization, Geneva, 1990.

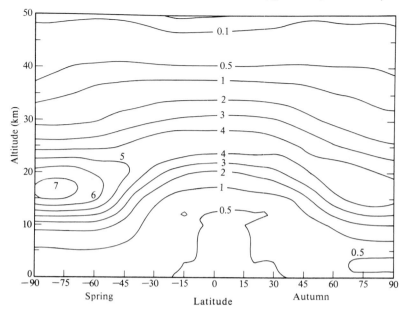

Fig. 4.2. Ozone concentrations (zonally averaged) as a function of altitude for spring (22 March). Units are 10^{12} molecule cm^{-3}. [Redrawn from: Johnston, H. S. *Rev. Geophys. Space Phys.* **13**, 637 (1975).]

form. The largest amounts of ozone in the Northern Hemisphere are found at polar latitudes in spring. Figure 4.1 (b) shows some calculated abundances for comparison, and it will be discussed in Section 4.4.5. In the Southern Hemisphere, the spring maximum occurs at mid-latitudes rather than at the pole. An altitude distribution of ozone for Northern Hemisphere spring (22 March) is indicated in Fig. 4.2. We now have to ask if the absolute measured concentrations of ozone can be interpreted in terms of photochemical kinetics.

4.3 Oxygen-only chemistry

4.3.1 Reaction scheme

The first approach to a theoretical explanation of the ozone layer was that of Chapman[a], who, in 1930, proposed a static pure-oxygen photochemical steady state model. The reactions in Chapman's scheme were

---

[a] Chapman, S., A theory of upper-atmosphere ozone. *Mem. Roy. Meteorol. Soc.* **3**, 103 (1930).

$$\begin{array}{lcr} & & \text{Change in odd oxygen} \\ O_2 + h\nu \rightarrow O + O & +2 & (4.1) \\ O + O_2 + M \rightarrow O_3 + M & 0 & (4.2) \\ O_3 + h\nu \rightarrow O + O_2 & 0 & (4.3) \\ O + O_3 \rightarrow 2O_2 & -2 & (4.4) \\ [O + O + M \rightarrow O_2 + M & -2 & (4.5)] \end{array}$$

Reaction (4.5) is now known to be too slow for it to play a part in stratospheric chemistry. Both photolytic reactions (4.1) and (4.3) can yield excited fragments, but collisional deactivation to $O(^3P)$ is the (almost) exclusive fate of any $O(^1D)$ formed. For the purposes of discussing oxygen-only chemistry, no account need be taken of excitation.

Reactions (4.2) and (4.3) interconvert O and O_3. Even at the top of the stratosphere, where pressures are lowest, reaction (4.2) has a half-life of as little as ~ 100 s. Ozone likewise has a very short photolytic lifetime in reaction (4.3) during the day. It is this rapid interconversion of O and O_3 that provides the rationale for the concept of 'odd' oxygen (O and O_3, see p. 112). If the odd oxygen concentration is defined as $[O] + [O_3]$, then odd oxygen is produced only in reaction (4.1) and lost only in reactions (4.4) (and (4.5)). Reactions (4.2)

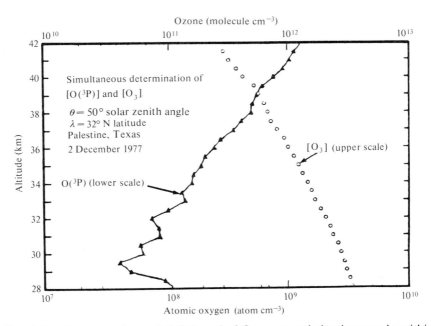

Fig. 4.3. Concentrations of $O(^3P)$ and of O_3 measured simultaneously within the same element of the stratosphere. (From Hudson, R., ed.-in-chief, *The strato-sphere 1981*, World Meteorological Organization, Geneva, 1981.)

and (4.3) 'do nothing' so far as odd oxygen is concerned, but merely determine the ratio $[O]/[O_3]$. Changes in the number of odd-oxygen species are shown alongside the chemical reactions in eqns (4.1) to (4.5). Reaction (4.2) becomes slower with increasing altitude, while reaction (4.3) becomes faster. Atomic oxygen is thus favoured at high altitudes and ozone at lower ones. Simultaneous measurements of $[O(^3P)]$ and $[O_3]$, plotted in Fig. 4.3, show that these expectations are borne out experimentally. Ozone is the dominant form of odd oxygen below ~ 60 km, and in the stratosphere (< 45 km) is responsible for more than 99 per cent of odd oxygen. The rate of production of ozone in the stratosphere may thus be equated with the rate of the O atom formation, which is twice the rate of O_2 photolysis in reaction (4.1). Steady-state analysis for odd oxygen shows that the concentration is equal to $(2P/k_1)^{1/2}$, where P is the rate of photolysis of O_2 in reaction (4.1) and k_1 is the rate coefficient for loss of odd oxygen in reaction (4.4). Given that most odd oxygen is in the form of ozone, it follows that stratospheric ozone concentrations are proportional to the square root of the oxygen photolysis rate, according to the steady state model.

After sunset, atomic oxygen concentrations fall very rapidly at relatively low altitudes ($\lesssim 40$ km) where the sink reactions (4.2) and (4.4) remain, but the sources (4.1) and (4.3) are cut off. With little atomic oxygen present, ozone is no longer destroyed in reaction (4.4) even though none is formed. Diurnal variations in stratospheric O_3 are therefore expected—and found—to be small. Above about 42 km, diurnal changes in concentration become increasingly pronounced, as daytime O_3 photolysis becomes faster, and conversion of O back to O_3 becomes slower.

4.3.2 Chapman layers

The existence of a layer-like structure is expected for any species such as O_3 whose concentration depends on the photochemical production rate in an atmosphere of varying optical density. In Section 2.3.1 we saw qualitatively that at high altitudes the solar intensity is also high, but the concentration of O_2 is too low to give large rates of O atom formation by photolysis. At low altitudes, there is plenty of O_2, but little radiation that can dissociate it. Somewhere in between, there is a compromise that maximizes the rate of O atom production. Chapman first discussed the formation of layers in this way, and the mathematical function that describes the shape of such a layer is called a *Chapman function*.

Let us show a simplified version of the derivation of the Chapman function. Atmospheric number density, n, can be obtained for this purpose, from eqn (2.7), p. 38, for constant scale height $H = kT/mg$, and with n substituted for p,

$$n = n_0 \exp(-z/H). \qquad (4.6)$$

For an *increase* in altitude dz, the atmospheric path that the Sun's rays have

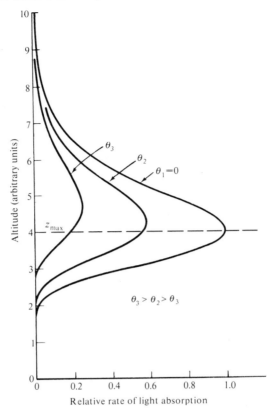

Fig. 4.4.   Photochemical energy deposition in Chapman layers. The figure shows the rate of light absorption as a function of altitude for three different zenith angles, θ, as calculated from the simplified Chapman function, eqn (4.10). For $\theta = 0$, z_{max} is the altitude at which the rate is a maximum.

to traverse is *decreased* by $dz \sec\theta$ for a zenith angle θ (Fig. 2.9, p. 52), so that eqn (2.12) for optical absorption becomes

$$dI = In\sigma_a(dz \sec\theta). \tag{4.7}$$

Combining eqns (4.6) and (4.7) yields

$$dI/I = d(\ln I) = n_0\sigma_a \sec\theta \exp(-z/H)\,dz. \tag{4.8}$$

On integrating this expression with the boundary condition $I = I_\infty$ at $z = \infty$ (limit of the atmosphere), we obtain

$$I = I_\infty \exp(-n_0\sigma_a H \sec\theta \exp(-z/H)). \tag{4.9}$$

The rate, P, at which energy is removed from the incident radiation is the decrease in intensity per unit path traversed. That is,

$$P = dI/(dz \sec\theta) = (dI/dz)\cos\theta$$
$$= I_\infty n_0 \sigma_a \cos\theta \exp(-z/H - n_0\sigma_a H \sec\theta \exp(-z/H)). \qquad (4.10)$$

Figure 4.4 is a sketch of the form of this function for three zenith angles. A clear maximum exists in the rate of light absorption. Since H and n_0 are properties of a particular atmosphere, the altitude of maximum light absorption depends only on the absorption cross-section, σ_a, and zenith angle, θ. For absorption by molecular oxygen in the Earth's atmosphere at $\lambda \sim 220$ nm, that altitude is about 35 km with an overhead Sun ($\theta = 0$). The ozone layer is thus predicted by this theory to be centred on 35 km. Increasing zenith angles lead (Fig. 4.4) to a decrease in the magnitude of the maximum and an increase in its altitude.

4.3.3 Comparison of experiment and theory

The simplified treatment of oxygen photolysis, given in the last section, can be improved to allow for latitudinal variations of solar intensity and zenith angle, real atmospheric temperatures, curvature of the Earth, and so on. Each wavelength makes its own contribution to the photolysis rate, so that the ozone production rate has to be summed over all photochemically-active wavelengths. Analytical expressions such as (4.10) are obviously not available, and numerical methods are used instead. Figure 4.5 gives rates of ozone formation *via* reactions (4.1) and (4.2) calculated in this way. The contour lines are zonal averages (over longitude at a given latitude) for the Northern Hemisphere spring equinox (22 March). The production rates are therefore relevant to the conditions for which ozone *concentrations* are shown in Fig. 4.2.

Comparison of Figs. 4.2 and 4.5 shows straight away that the highest oxygen photolysis rates do not correspond with the highest ozone concentrations. Photolysis rates are highest near the equator, while ozone concentrations are at a maximum in northern regions. At the equator, the ozone layer is centred on about 25 km where the production rate is insignificant, while the production of atomic oxygen reaches a maximum at ~ 40 km. Near the poles, the maximum production rate is displaced to higher altitudes, but the largest ozone concentrations are found at lower altitudes, and there is a marked north–south asymmetry. The steady state oxygen-only photochemical model predicts that ozone concentrations should be proportional to $P^{1/2}$ (p. 121), so that the lack of correspondence of $[O_3]$ and P indicates an inadequacy of the model. Vertical and horizontal transport are clearly of great importance in redistributing stratospheric ozone. Indeed, the comparison of Figs. 4.2 and 4.5 can be said to reveal the occurrence and direction of air motions in the stratosphere. Observations confirm this view. For example, the rapid build-up of ozone at the spring pole (Section 4.2) is

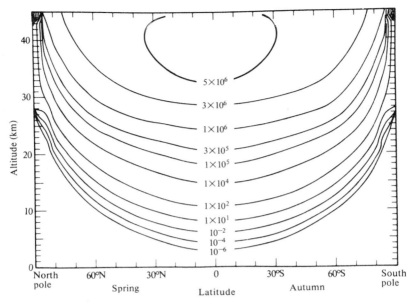

Fig. 4.5.   Rate (zonally averaged) of ozone formation from the photolysis of O_2 in units of molecule $cm^{-3} s^{-1}$. (Source as for Fig. 4.2.)

strongly correlated with northward and downward air transport that could bring ozone in from high-altitude equatorial regions where the production rate is highest.

If a large amount of horizontal north–south transport is to be responsible for redistribution of ozone, then the (photo)chemical lifetime of the ozone must clearly be greater than the time taken for the transport to occur. Since seasonal changes, and north–south reversals, are seen in the ozone concentrations, it follows that the transport time-scale of interest must be 3–4 months. The photochemical lifetime can be estimated by equating it with the time taken to replace the ozone in any air parcel: that is, the concentration of O_3 divided by the local rate of its formation. Reference to Figs 4.2 and 4.5 shows that near the equator this lifetime ranges from $10^{12}/10^4 = 10^8$ s $\sim$ 3 yr at an altitude of 15 km to $10^{12}/5 \times 10^6 = 2 \times 10^5$ s $\sim$ 2 days at an altitude of 40 km. The photochemical lifetime reaches the critical limit of 3–4 months lowest ($\sim$20 km) in the atmosphere near the equator, and highest ($\gtrsim$40 km) at high latitudes in winter. At altitudes higher than these, most of the ozone must be produced locally, but at lower altitudes it can survive long-range transport.

We now have to ask whether or not the oxygen-only mechanism can account for the absolute ozone concentrations observed, if due allowance is made for horizontal and vertical transport. Answers to this question can be obtained in several ways. Models can be set up to predict vertical concentration–altitude profiles for comparison with experiments. Such models generally match the *shapes* of the observed profiles (such as those in Fig. 1.2), with the altitude of ozone maximum being more or less correctly predicted. Invariably, however, the absolute concentrations calculated from the oxygen-only scheme are too high. Total overhead column abundances of ozone are also found experimentally to be lower than those calculated. Typical measured column abundances are 8.8×10^{18} molecule cm^{-2}, while 1-D models including vertical transport give about twice this value with oxygen-only chemistry. Because the rate of reaction (4.4) is approximately proportional to $[O_3]^2$, the result suggests that the reaction accounts for only a quarter of the actual ozone loss. Similar results are obtained by examining the predicted global rates of production and destruction of ozone in order to remove errors arising from horizontal transport. For the spring equinox data, the integrated rate of ozone production is 4.86×10^{31} molecule s^{-1} of which 0.06×10^{31} molecule s^{-1} are transported to the troposphere. The calculated rate of loss of odd oxygen in reaction (4.4) is 0.89×10^{31} molecules s^{-1}, leaving 3.91×10^{31} molecule s^{-1} unaccounted for. It this figure really represented an unbalanced rate of ozone formation, the present global quantity of ozone would double in two weeks! Ozone is being produced five times faster than it is being destroyed by the Chapman mechanism. Since ozone concentrations are not rapidly increasing, we conclude that something other than the Chapman reactions is very important in destroying ozone in the natural stratosphere.

4.4 Influence of trace constituents

4.4.1 Catalytic cycles

Until about 1964, it was thought that the Chapman, oxygen-only, reactions could explain atmospheric ozone abundances. However, improved laboratory measurements of rate coefficients, especially for reaction (4.4), revealed the discrepancy discussed in the last section. Reaction (4.4) is too slow to destroy ozone at the rate it is produced globally. An additional, or faster, loss process for odd oxygen is needed.

At first sight, no trace component in the stratosphere could be responsible for loss of odd oxygen, since the species involved would be rapidly consumed. This objection disappears if the trace constituent participates in a *catalytic* process that removes O or O_3. The idea of catalytic atmospheric loss processes had originated with Bates and Nicolet[a] in 1950. Serious consideration of the influence of such processes on ozone concentrations followed the discovery that the laboratory rate measurements could also be affected by trace impurities reacting in a cyclic manner.

---

[a] Bates, D. R. and Nicolet, M., The photochemistry of atmospheric water vapour. *J. geophys. Res.* **55**, 301 (1950).

The essence of catalytic schemes for loss of odd oxygen is the provision of a more efficient route than the direct one for reactions (4.4) and (4.5). A chain mechanism that achieves the same result as reaction (4.4) can be represented simplistically by the pair of reactions

$$X + O_3 \rightarrow XO + O_2 \tag{4.11}$$

$$XO + O \rightarrow X + O_2 \tag{4.12}$$

Net $$O + O_3 \rightarrow O_2 + O_2.$$

The reactive species X is regenerated in the second reaction of the pair, so that its participation in odd-oxygen removal does not lead to a change in its abundance. Since the overall reaction consists of two separate steps, each one must be exothermic to be efficient (see Section 3.4.1). That requirement places a constraint on the X-O bond energy such that $107 \, kJ \, mol^{-1} < D(X\text{-}O) < 498 \, kJ \, mol^{-1}$. Rates of the catalytic destruction of ozone may then exceed those of the elementary step (4.4) either because [X] exceeds [O], or because the rate coefficient for reaction (4.11) exceeds that for (4.6). The rather large activation energy for reaction (4.4) of $17.1 \, kJ \, mol^{-1}$ does, as we shall see shortly, often favour the indirect route at low stratospheric temperatures.

Several species have been suggested for the catalytic 'X' in the atmosphere. The most important of these for the natural stratosphere are X = H and OH, X = NO, and X = Cl; the catalytic cycles are then said to involve HO_x, NO_x, and ClO_x species[a]. Reactions (4.11) and (4.12) are replaced by:

Cycle 1:

$$H + O_3 \rightarrow OH + O_2 \tag{4.13}$$

$$OH + O \rightarrow H + O_2 \tag{4.14}$$

Net $$O + O_3 \rightarrow O_2 + O_2$$

Cycle 2:

$$OH + O_3 \rightarrow HO_2 + O_2 \tag{4.15}$$

$$HO_2 + O \rightarrow OH + O_2 \tag{4.16}$$

Net $$O + O_3 \rightarrow O_2 + O_2$$

[a] The use of the representation NO_x refers to NO and NO_2, and if a concentration of NO_x is specified it means the sum of the concentrations of these two oxides. Another representation, NO_y, is often also found in the literature; it means 'odd nitrogen', and is defined as the sum of NO_x and all oxidized nitrogen species that represent sources or sinks of NO_x through processes that occur on relatively short time scales. Important compounds, besides NO and NO_2, that contribute to NO_y include N_2O_5, HNO_3, HNO_4, $ClONO_2$, and the compound PAN that will be discussed in Section 5.3.4. Although N_2O is an NO_x source in the upper atmosphere, it is *not* normally considered a component of the NO_y compounds.

Cycle 3:

$$NO + O_3 \rightarrow NO_2 + O_2 \qquad (4.17)$$

$$NO_2 + O \rightarrow NO + O_2 \qquad (4.18)$$

Net $\qquad O + O_3 \rightarrow O_2 + O_2$

Cycle 4:

$$Cl + O_3 \rightarrow ClO + O_2 \qquad (4.19)$$

$$ClO + O \rightarrow Cl + O_2 \qquad (4.20)$$

Net $\qquad O + O_3 \rightarrow O_2 + O_2.$

Rate parameters for these reactions are given in Table 4.1, and show how the catalytic cycles can be faster than the direct reaction (4.4). The first row of the table ('X = O_2') shows that the activation energy for reaction (4.4) with ozone is far higher than that for any of the other reactions of atomic oxygen. At a temperature of 220 K, typical of the stratosphere, the rate coefficient for the direct reaction is thus orders of magnitude smaller than the corresponding value for the catalytic reactions. Whether or not the catalytic reactions will be actually *faster* then depends on the relative concentrations of XO and of O_3. Figure 4.6 shows the relative contribution to ozone destruction of the HO_x, NO_x, and ClO_x cycles as well as of the oxygen-only 'Chapman' reactions. It is apparent that the catalytic cycles, especially the one involving NO_x, make a major contribution to destruction of odd oxygen. Cycle 3 (NO_x) dominates in the lower stratosphere, with cycles 1 and 2 (HO_x) becoming relatively more important higher up.

Table 4.1 Rate parameters for some reactions in catalytic cycles.

Reaction	$X + O_3$			$XO + O$		
X	A	E_a	k_{220}	A	E_a	k_{220}
'O_2'		(i.e. reaction $O_3 + O$):		8×10^{-12}	17.1	6.8×10^{-16}
H	1.4×10^{-10}	3.9	1.7×10^{-11}	2.3×10^{-11}	~ 0	2.3×10^{-11}
OH	1.6×10^{-12}	7.8	2.2×10^{-14}	2.2×10^{-11}	-0.1	3.7×10^{-11}
NO	1.8×10^{-12}	11.4	3.5×10^{-15}	9.3×10^{-12}	~ 0	9.3×10^{-12}
Cl	2.8×10^{-11}	2.1	8.7×10^{-12}	4.7×10^{-11}	0.4	3.7×10^{-11}

A and E_a are the Arrhenius parameters, and k_{220} is the rate coefficient calculated for a temperature of 220K. Units for A and k_{220} are cm^3 molecule^{-1} s^{-1}, and those for E_a are kJ mol^{-1}.

Data taken from compilation of *Atmospheric ozone 1985*. World Meteorological Organization, Geneva, 1986.

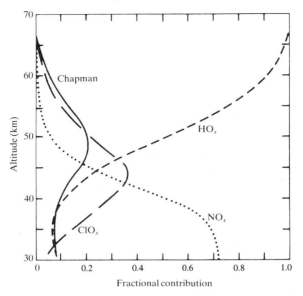

Fig. 4.6.   Fraction of the odd oxygen loss rate due to the 'Chapman', HO_x, NO_x, and ClO_x mechanisms (based on a mid-latitude diurnal average calculation). From *Atmospheric ozone 1985*. World Meteorological Organization, Geneva, 1986.

The set of catalytic cycles based on the $X \rightarrow XO \rightarrow X$ processes (4.11) and (4.12) gives a first indication of how the natural ozone balance is maintained in the atmosphere, although the simple ideas have to be modified before a fully quantitative result emerges. First of all, even within a 'family' (e.g. HO_x), catalytic cycles can be identified which do not fit into the category so far discussed. For example, the sequence of

Cycle 5:

$$OH + O \rightarrow H + O_2 \qquad (4.14)$$

$$H + O_2 + M \rightarrow HO_2 + M \qquad (4.21)$$

$$HO_2 + O \rightarrow OH + O_2 \qquad (4.16)$$

Net $\qquad O + O + M \rightarrow O_2 + M$

has the same overall effect as reaction (4.5), which destroys two odd-oxygen species. Direct reaction was disregarded as being of no consequence in the atmosphere, but the catalytic cycle 5 is of major importance above 40 km. Similarly, at lower altitudes (< 30 km) the cycle

Cycle 6:

$$OH + O_3 \rightarrow HO_2 + O_2 \tag{4.15}$$

$$HO_2 + O_3 \rightarrow OH + O_2 + O_2 \tag{4.22}$$

Net $\qquad O_3 + O_3 \rightarrow O_2 + O_2 + O_2$

becomes important because it does not utilize atomic oxygen, whose concentration is very low at those altitudes. Partitioning of the overall odd hydrogen budget between the reactive species H, OH, and HO_2 thus determines which sets of reactions really lead to destruction of odd oxygen. The identification of particular cycles and of where they are atmospherically important is somewhat artificial but does give an insight into the chemistry.

Investigation of the contribution made to ozone loss by each family of species is obviously the next step. Table 4.2 gives the result in terms of the total ozone columns calculated from a 1-D model in which various combinations of the catalytic species are present. The altitude of the peak ozone concentration is also shown for each calculation. Since measured column abundances are ~ 8–9×10^{18} molecule cm^{-2}, it is apparent that the catalytic cycles *can* explain atmospheric ozone concentrations if the trace constituents are present at levels consistent with atmospheric observations. The table also shows, though, that the interpretation requires further consideration. The effects of the various catalytic families are definitely not additive. Indeed, adding NO_x to the calculation leads to *less* destruction of O_3 than is obtained with ClO_x, or, especially, $HO_x + ClO_x$ alone. The explanation is that the

Table 4.2   Ozone columns calculated for various families of chemical species present in the atmosphere.

Family present	Ozone column molecule cm^{-2}	Altitude of O_3 maximum (km)
None (i.e. Champman chemistry)	16.1×10^{18}	23
NO_x	8.3×10^{18}	24
HO_x	9.8×10^{18}	26
ClO_x	7.5×10^{18}	25
$NO_x + HO_x$	9.4×10^{18}	24
$NO_x + ClO_x$	8.3×10^{18}	24
$HO_x + ClO_x$	6.3×10^{18}	28
$NO_x + HO_x + ClO_x$	9.4×10^{18}	24

The model is one-dimensional, and incorporates mixing ratios of 18×10^{-9} NO_x, 1×10^{-9} HO_x, and 1.5×10^{-9} ClO_x at an altitude of 35 km.

Data taken from compilation of Hudson, R. (ed.-in-chief) *The stratosphere 1981.* World Meteorological Organization, Geneva, 1981.

various families are not isolated; rather, members of one family can react with members of another. Two reactions have shown themselves to be of particular importance in this context, and, as it turns out, to the overall balance of odd oxygen in the stratosphere. They are

$$HO_2 + NO \rightarrow OH + NO_2, \tag{4.23}$$

and

$$ClO + NO \rightarrow Cl + NO_2. \tag{4.24}$$

Reaction (4.23) is so central to stratospheric chemistry that each change in rate coefficient reported from different laboratory experiments has necessitated drastic revision of stratospheric models. The reactions do not appear to have any unusual characteristics, and we shall have to examine the chemistry of the catalytic cycles in more detail to discover why the two processes are so important.

4.4.2 Null cycles, holding cycles, and reservoirs

In competition with every catalytic cycle that destroys odd oxygen, cycles can be written that interconvert the species X and XO without odd-oxygen removal. Such processes are *null cycles* and can be illustrated by writing catalytic and null cycles alongside each other for NO_x:

Cycle 3 (catalytic)

$$NO + O_3 \rightarrow NO_2 + O_2 \quad (4.17)$$
$$NO_2 + O \rightarrow NO + O_2 \quad (4.18)$$

Net $\quad O + O_3 \rightarrow O_2 + O_2$

Cycle 7 (null)

$$NO + O_3 \rightarrow NO_2 + O_2 \quad (4.17)$$
$$NO_2 + h\nu \rightarrow NO + O \quad (4.25)$$

$$O_3 + h\nu \rightarrow O_2 + O.$$

Reaction (4.17) converts NO to NO_2 in both cases, but photolysis of NO_2 in cycle 7 leads to atomic oxygen formation. Cycle 7 has the overall effect of ozone photolysis, reaction (4.3), formally photosensitized by NO_2: it is a 'do-nothing' cycle so far as odd oxygen is concerned. However, that fraction of the NO_x that is tied up in the null cycle is ineffective as a catalyst. During the day, when reaction (4.25) can occur, a given atmospheric NO_x concentration then destroys less odd oxygen than originally anticipated because of the null cycle.

Competitive processes abound in stratospheric chemistry. Reactions (4.18) and (4.25) are not, for example, the only reactions open to NO_2. The nitrate radical, NO_3, can be formed by reaction of NO_2 with ozone:

$$NO_2 + O_3 \rightarrow NO_3 + O_2. \tag{4.26}$$

Most atmospheric NO_3 is removed by photolysis during daytime. The products of photodecomposition are $O + NO_2$ (in which case there is no net loss of odd oxygen), and $O_2 + NO$ (in which case odd oxygen is consumed): the competition between these processes depends on the quantum yields for the different channels at the wavelengths available for photolysis, a matter still not fully resolved. In addition, some NO_3 reacts in the three-body process

$$NO_3 + NO_2 + M \rightarrow N_2O_5 + M. \tag{4.27}$$

Now, the importance of N_2O_5 is that it is rather unreactive in the stratosphere. Ultimately, it thermally or photochemically decomposes back to $NO_2 + NO_3$, so that its formation does not constitute a permanent loss of odd nitrogen. On the other hand, it does behave as an unreactive *reservoir* of NO_x, containing typically 5 to 10 per cent of the total NO_x budget. The cycle involving its formation and destruction is then a *holding cycle*.

Holding cycles involving members of two families turn out to be even more important. Three-body formation of nitric acid ($HONO_2$) is an important storage step in the HO_x–NO_x system

$$OH + NO_2 + M \rightarrow HONO_2 + M. \tag{4.28}$$

Nitric acid is photolysed to regenerate $OH + NO_2$, but the process is relatively slow, and about half the stratospheric load of NO_x is stored in the nitric acid reservoir. For the ClO_x family, hydrochloric acid (HCl) is the main reservoir, reaction of Cl with stratospheric methane being the major source. The holding cycle for Cl involving HCl as reservoir can then be written

Cycle 8:

$$Cl + CH_4 \rightarrow CH_3 + HCl \tag{4.29}$$

$$OH + HCl \rightarrow H_2O + Cl \tag{4.30}$$

Net $$CH_4 + OH \rightarrow CH_3 + H_2O.$$

About 70 per cent of stratospheric ClO_x is thought to be present as HCl.

As the understanding of stratospheric ozone chemistry has become more sophisticated, so yet more exotic molecules have been called into service as reservoir species. The list includes HOCl (hypochlorous acid), HO_2NO_2 (HNO_4, pernitric acid) and $ClONO_2$ (chlorine nitrate): the 'trivial' names of the compounds are almost always used in the published literature, and are given here for identification. Suggested routes to the compounds mentioned once again emphasize the interaction between families:

$$ClO + HO_2 \rightarrow HOCl + O_2, \tag{4.31}$$

$$HO_2 + NO_2 + M \rightarrow HO_2NO_2 + M, \tag{4.32}$$

$$ClO + NO_2 + M \rightarrow ClONO_2 + M. \tag{4.33}$$

Of these three reservoir species, $ClONO_2$ is the best characterized in the stratosphere, measured mixing ratios reaching about 1×10^{-9} at an altitude of 25–35 km. Chlorine nitrate plays several important roles in stratospheric chemistry. Formation of $ClONO_2$ strongly couples the chlorine and nitrogen cycles, and at high concentrations of stratospheric chlorine could lead to non-linear responses of ozone to chlorine perturbations (see Section 4.7). The observed diurnal variations in $[ClO]$ (Section 4.4.5) appear to be driven by $ClONO_2$ chemistry. For HNO_4, there is a tentative identification of the molecule in the stratosphere, and $HOCl$ has not been identified positively, although its detection is desirable for testing current understanding of stratospheric chlorine chemistry.

The reservoir compounds are of very great importance in stratospheric chemistry. They act to divert potentially catalytic species from active to inactive forms, but the compounds remain available to liberate active catalysts. The assumed rates of production and destruction have a large influence on the predictions of stratospheric models. We shall see in Section 4.7 that unexpected release of active catalytic species from reservoir compounds is probably responsible for the phenomenon of the Antarctic ozone hole.

We now return to the importance of reactions (4.23) and (4.24) in stratospheric chemistry. In chemistry involving O_x and NO_x reactions alone, there are null cycles (e.g. cycle 7) and holding cycles. For HO_x or ClO_x on their own, there are no corresponding null cycles. Reactions (4.23) and (4.24) are critical in providing effective, but otherwise absent, null cycles for the HO_x and ClO_x families. Cycles 9 and 10 show these null cycles:

Cycle 9:

$$OH + O_3 \rightarrow HO_2 + O_2 \tag{4.15}$$

$$HO_2 + NO \rightarrow OH + NO_2 \tag{4.23}$$

$$NO_2 + h\nu \rightarrow NO + O \tag{4.25}$$

Net $\quad O_3 + h\nu \rightarrow O_2 + O$

Cycle 10:

$$Cl + O_3 \rightarrow ClO + O_2 \tag{4.19}$$

$$ClO + NO \rightarrow Cl + NO_2 \tag{4.24}$$

$$NO_2 + h\nu \rightarrow NO + O \tag{4.25}$$

Net $\quad O_3 + h\nu \rightarrow O_2 + O.$

Not only are these cycles effective null paths for HO_x or ClO_x, but they also provide an additional null cycle for NO_x. It is the photolytic step (4.25) that completes the null cycle in each case. In the real atmosphere, photolysis of NO_2 is, of course, in competition with reaction (4.18) which would lead to an overall loss of two odd oxygens.

Reaction (4.23) can also participate in rather unusual cycles that generate odd oxygen. Oxidation of CO to CO_2 usually destroys odd oxygen, but the sequence

Cycle 11:

$$OH + CO \rightarrow H + CO_2 \qquad (4.34)$$

$$H + O_2 + M \rightarrow HO_2 + M \qquad (4.21)$$

$$HO_2 + NO \rightarrow OH + NO_2 \qquad (4.23)$$

$$NO_2 + hv \rightarrow NO + O \qquad (4.25)$$

Net $\qquad CO + O_2 + hv \rightarrow CO_2 + O$

creates an oxygen atom. Similarly, the combination of OH and HO_2 as a direct step is a loss process for HO_x:

$$OH + HO_2 \rightarrow H_2O + O_2. \qquad (4.35)$$

However, catalysed by NO_x, and with the participation of reaction (4.23), a cyclic process with the effect

$$OH + HO_2 + 2hv \rightarrow H_2O + O + O, \qquad (4.36)$$

can be written. The importance of a knowledge of the rate coefficient for reaction (4.23) can now be appreciated. Sequences of steps that would destroy odd oxygen if the rate coefficient were small become diverted into null or even generating cycles for larger values. We shall see in Sections 4.5.2 and 4.5.4 how predicted effects on ozone of anthropogenic pollution have depended on the rate coefficients used in the models.

4.4.3 Natural sources and sinks of catalytic species

All three catalytic families HO_x, NO_x, and ClO_x appear to be present in the 'natural' atmosphere unpolluted by man's activities. In the contemporary atmosphere, however, the background concentrations, especially of ClO_x, have already been supplemented by anthropogenic sources. We shall consider possible perturbations of the stratosphere in Section 4.5; measurements on, and models of, the atmosphere are concerned with actual total loadings of trace species, regardless of their origin.

Slow changes, over periods of tens of years or more, may be affecting concentrations of trace atmospheric constituents, but since the lifetimes (chemical or physical) of *most* of the species in the stratosphere are shorter, sources and sinks must be more or less in balance. Precursors of the atoms and radicals that participate in the catalytic cycles themselves have no source above the Earth's surface, and have, therefore, to be transported through the troposphere to the stratosphere. The probable exception to this generalization is a small contribution to NO_x, since N_2 and O_2 are the major atmospheric components.

Most of the stratospheric NO_x originates with tropospheric N_2O, whose source is largely biological (see Chapter 1). Excited atomic oxygen, $O(^1D)$, derived largely from ozone photolysis, reaction (4.3), and to a lesser extent from oxygen photolysis (reaction 4.1) at higher altitudes, can react with N_2O to yield NO,

$$O(^1D) + N_2O \rightarrow NO + NO, \qquad (4.37)$$

and thus initiate the NO_x chemistry. The photolytic production of $O(^1D)$ and the reasons for its reactivity were reviewed in Chapter 3 and will not be discussed further here. A great deal of ingenuity has gone into finding upper atmospheric sources of N_2O. Suggestions have included the reaction of $N_2(A^3\Sigma_u^+)$ with O_2 (p. 91), or of excited O_3, produced in the $O + O_2$ association reaction (4.2), with N_2:

$$O_3^* + N_2 \rightarrow N_2O + O_2. \qquad (4.38)$$

No *direct* evidence shows that such processes are of stratospheric significance.

Several other sources of NO_x exist, but they are of lesser importance than the N_2O source. Cosmic rays penetrate the atmosphere and contribute to NO production in the altitude range 10–30 km. Ionization reactions involving N_2 lead, after several steps, to NO. The source is probably the major one during the polar night, although it is not important on a global scale. Short-wavelength ultraviolet, and solar particle, ionization in the mesosphere and thermosphere also lead to NO production. However, rapid photodissociation at levels above the stratopause seems to prevent this high-altitude source from reaching the stratosphere. It is also unlikely that tropospheric NO or NO_2 (e.g. that produced by lightning) can survive long enough to be transported in significant quantities to the stratosphere.

Reactions of $O(^1D)$ with water vapour and methane,

$$O(^1D) + H_2O \rightarrow OH + OH \qquad (4.39)$$

$$O(^1D) + CH_4 \rightarrow OH + CH_3, \qquad (4.40)$$

are the main sources of OH radicals. As discussed in Chapter 2, the stratosphere is very dry , probably because tropospheric water vapour has to pass through a 'cold trap' at the tropopause. Much stratospheric H_2O is in fact a product of CH_4 oxidation. Nevertheless, CH_4 constitutes more than a third of the total $[CH_4] + [H_2O]$ in the lower stratosphere, so that reaction (4.40) is important. In fact, the reaction contributes more than one HO_x species, since subsequent oxidation of CH_3 to CO yields two or three more odd-hydrogen species. We shall discuss this oxidation in more detail in connection with tropospheric chemistry (Chapter 5).

The only precursor of ClO_x that is established to be natural is methyl chloride (CH_3Cl). Tropospheric mixing ratios show little difference between Northern and Southern Hemisphere, suggesting a non-industrial source. The main contributor appears to be the world's oceans. Burning of vegetation also produces CH_3Cl, and forest fires must be regarded as natural events. Primitive agriculture employs 'slash and burn' techniques that lead to elevated methyl chloride levels near the practising communities, but, properly speaking, that is an anthropogenic source. Volcanic emissions can also show high CH_3Cl levels. Tropospheric lifetimes for methyl chloride are estimated as ~ 1 year, and some CH_3Cl can be transported across the tropopause. Stratospheric concentrations show a decrease with altitude, as expected if CH_3Cl is reactive, but extrapolate well at lower altitudes to the tropospheric mixing ratios. Once in the stratosphere, CH_3Cl is primarily removed by reaction with OH,

$$CH_3Cl + OH \rightarrow CH_2Cl + H_2O, \qquad (4.41)$$

although above $\sim 30\,km$ about one third is photolysed. In either case, the Cl atom becomes able to enter the ClO_x cycle.

Alternative natural sources of ClO_x have been investigated. These include volcanic release of HCl, either slowly emitted to the troposphere followed by transport to the stratosphere, or, in major volcanic eruptions, directly to the stratosphere. Acidification (e.g. by H_2SO_4) of marine aerosols containing NaCl is another potential source of HCl. None of these processes seems capable, however, of providing enough ClO_x to have a marked influence on atmospheric ozone concentrations.

Sink processes for the catalytic families largely involve slow transport of the reservoir species across the tropopause. In the troposphere, the species then dissolve in water, and are subsequently 'rained out'. Reservoir compounds such as $HONO_2$, HOCl, HCl, and probably $ClONO_2$ and HO_2NO_2, are all highly soluble, and are thus specially good candidates for rainout. Radical–radical reactions such as process (4.35) are important in determining the total concentrations of active reaction intermediates. A role for metals has also been suggested as a possibility in chlorine chemistry. Metals such as sodium and calcium are deposited by ablating meteors (p. 292 and p. 323) near the mesopause. A steady downward flux of metal atoms must be present throughout the mesosphere and stratosphere. Compounds such as NaO, NaOH, CaO, and CaOH might be formed in the stratosphere. Reaction with HCl, Cl, and ClO might then lead to the sequestering of ClO_x as NaCl. The initial impetus for these ideas came from an erroneous identification of a stratospheric ion as $NaOH^+$. However, the metals *ought* to be present in the stratosphere, and current interest centres on whether species like NaOH and NaCl are photo-dissociated. For the case of NaCl, the overall effect would be conversion of HCl to Cl, so that, far from diminishing the impact of ClO_x on stratospheric ozone, the meteoric metals would augment it! Recent experiments have shown that particles of NaCl can react, at least at 298 K, with the reservoir compounds N_2O_5 and $ClONO_2$ to release chlorine in photochemically active forms (as $ClNO_2$ and Cl_2 for the two reactions respectively). The processes are analogous to particle reactions that are thought to be important in producing the Antarctic ozone hole (cf. Section 4.7.4), but demonstrate here the existence of mechanisms for converting metal halides into catalytically active gaseous species.

4.4.4 Summary of chemistry

The information presented in the previous sections can most conveniently be summarized in flow charts showing source, radical, and sink species, and their reactive interconnections: such diagrams are sometimes referred to as *Nicolet diagrams* to honour a pioneer in atmospheric chemistry. Figures 4.7–4.9 are the flow charts for the HO_x, NO_x, and ClO_x species respectively. For ClO_x the important anthropogenic sources, to be discussed later, are included. The reaction partner that effects a transformation is given within the arrows, photochemical change being represented by *hv*. Cycles are readily identifiable as closed loops of arrows, and the reservoirs are generally the non-radical species in the centre of the diagrams that are formed and destroyed by radical and photochemical processes.

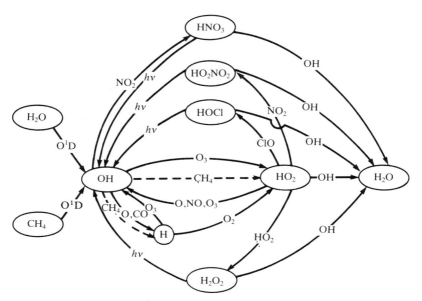

Fig. 4.7.   Chemical cycles for HO_x trace species. Source as for Fig. 4.6.

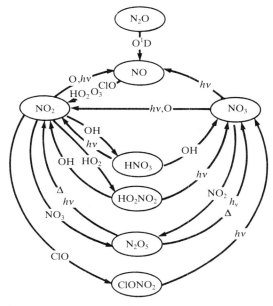

Fig. 4.8. Chemical cycles for NO$_x$ trace species. Source as for Fig. 4.6.

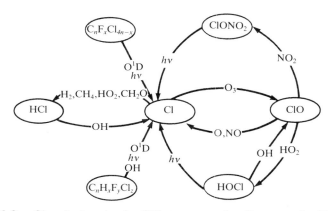

Fig. 4.9. Chemical cycles for ClO$_x$ trace species. Source as for Fig. 4.6.

Table 4.3 summarizes the 24-hour average contributions of the major different catalytic reactions to ozone loss at different altitudes. The slight shortfall of the totals from 100 per cent reflects the occurrence of minor processes not listed.

Table 4.3   Relative contributions of the major rate-limiting chemical reactions to the removal of ozone.

Altitude (km)	Reaction				
	$O + O_3$ (percentage contribution)	$O + NO_2$	$O + ClO$	$O + HO_2$	$HO_2 + O_3$
50	25	7	4	52	–
45	29	24	10	31	–
40	18	53	16	10	–
35	11	68	13	4	1
30	10	69	8	2	3
25	2	78	5	1	8
20	1	70	1	1	26

Data taken from compilation of Hudson, R. (ed.-in-chief) *The stratosphere 1981.* World Metereological Organization, Geneva, 1981.

4.4.5 Comparison of experiment and theory

Catalytic cycles for stratospheric odd-oxygen chemistry are thought to explain a loss of odd oxygen beyond that provided by the Chapman, oxygen only, reactions. Since these cycles are invoked to reconcile production and destruction rates of ozone (Section 4.4.1), a first test of the extended chemistry must be to see if experimental ozone concentrations are correctly predicted. The link between experiment and theory is provided by a model that incorporates trace constituents and their chemical reactions. Results of modelling include, therefore, not only the ozone data, but also information about the concentrations of the trace constituents and reactive intermediates in the chemical scheme. Confidence in the completeness of a chemical mechanism, and in the correctness of the rate parameters used, is obviously increased the larger the number of observable variables that can be matched with prediction. Conversely, good agreement between experiment and theory tends to lend credibility to the modelling procedures used, and to the approximations inevitably adopted. Depending on the detail available from the model and in the observations, comparisons may be made at various levels of refinement, ranging from globally-averaged column abundances to values for a particular geographic location, altitude, and time.

A more sophisticated test of the concepts of stratospheric ozone chemistry is provided by measurements of the concentrations of the minor trace species that play a part in the theories as sources, sinks, or, especially, reservoirs and intermediates. Critical tests of photochemical theories require the simultaneous measurement of several coupled species within the same air mass. Observations of concentration–altitude profiles, and of the seasonal and even diurnal variations of abundance, provide excellent material for comparison with the predictions of models. Above all, the degree of match between experiment and prediction for the atomic and radical intermediates of the postulated reactions provides an indication of how complete and realistic the theory is.

A wide variety of experimental techniques now exists for the measurement of atmospheric trace species. They include absorption and emission spectroscopy in spectral regions ranging from the microwave to the ultraviolet. In addition, various grab-sampling and matrix-isolation experiments have been described. Ground-based measurements have been supplemented by experiments carried on balloons, aircraft, and rockets. But perhaps the most important advance of the 1980s has been the harnessing of satellite and space-shuttle platforms to provide long-term and global measurements of trace gases in the atmosphere. The acronyms for these instruments include the shuttle-borne ATMOS (a high-resolution infra-red spectrometer), and LIMS (a limb infra-red monitor), SAMS (a stratospheric and mesospheric sounder), SBUV and TOMS (ozone vertical and column distribution) on Nimbus 7. Other experiments include the Stratospheric Aerosol and Gas Experiments (SAGE and SAGE 2) and the Solar Mesosphere Explorer (SME).

One technique for the *in situ* detection of atomic and radical intermediates which has proved particularly valuable in atmospheric studies is that of *resonance fluorescence*. The basis of the technique is that electronic excitation of a species by absorption of light may be followed by fluorescent emission [pathway (iii), Fig. 3.1] at the same wavelength: that is, the fluorescence is 'resonant' in energy with the exciting radiation. The wavelength(s) of absorption and emission are sharply defined for atoms and small radicals or molecules. For example, the ground-state hydrogen atom, $H(^2S)$, can absorb radiation at $\lambda = 121.6$ nm (the Lyman-α line of atomic hydrogen) to reach the first excited, ^2P, electronic state. One fate of $H(^2P)$ is radiation of the resonant Lyman-α line

$$H(^2S) + h\nu \rightarrow H^*(^2P) \quad \text{absorption} \tag{4.42}$$

$$H^*(^2P) \rightarrow H(^2S) + h\nu \quad \text{resonance fluorescence.} \tag{4.43}$$

Since the radiation is isotropic, it can be detected off-axis from the exciting beam. Observation of Lyman-α fluorescence from a system illuminated by $\lambda = 121.6$ nm radiation (obtained from, say, a hydrogen discharge lamp) thus demonstrates the presence of H atoms. The intensity of fluorescence is proportional to the concentration of hydrogen under suitable conditions. In the atmospheric experiments, a lamp and a shaped tube that is to act as a fluorescence cell are borne aloft on a balloon (or rocket) and parachuted down. Air passes through the tube, and the intensity of resonance fluorescence provides a measure of the concentration of the species for which the lamp provides specific excitation. Many of the atmospheric experiments have been carried out by Dr Jim Anderson, of Harvard University, in whose skilled hands atoms and radicals can be detected at concentrations of 10^6 particles cm^{-3} (1.7×10^{-15} M, for chemists used to thinking in molarities!). A clever trick enables detection of radicals such as HO_2 and ClO, which do not fluoresce. By injection of NO, carried with the payload, into the test cell, OH or Cl can be produced stoicheiometrically. Both OH and Cl are fluorescent. The conversion involves the two reactions that are also so important in the chemistry of the atmosphere itself:

$$HO_2 + NO \rightarrow OH + NO_2 \qquad (4.23)$$

$$ClO + NO \rightarrow Cl + NO_2. \qquad (4.24)$$

A modification[a] to the balloon experiments allows the fluorescence probe to be lowered, like a Yo-Yo, more than ten kilometres from the balloon by a winch. The tethered instrument can thus make repetitive measurements of altitude profiles by reeling down or up, and the technique obviously represents a substantial advance over a conventional 'one-off' balloon experiment.

We now look at a few specific examples to give an indication of the state-of-the-art in the early 1990s in both modelling and atmospheric observations. For ozone, the observational data base is sufficiently large that sophisticated tests of two-dimensional models can be made. The largest and most well-established variation of ozone occurs on a seasonal time-scale. An important test of 2-D models is therefore their ability to simulate seasonal variations of ozone. Historically, the most extensive part of the ozone data base is the record of columnar ozone obtained from a network of ground stations observing ultraviolet absorption. Satellite measurements of back-scattered ultraviolet over a few years suggest that the principal features of these long term records are realistic. Figure. 4.1 (a) has already presented seasonal and latitudinal variation of column ozone abundances obtained from ground observations. Results from a 2-D model, with an advanced chemical reaction scheme, are given in Fig. 4.1 (b). All the general features of Fig. 4.1 (a) are reproduced, and the model exhibits excellent qualitative agreement with measurements. Spring–autumn asymmetries show up well in the model results. North–south asymmetry is not so well reproduced. Data for circulation are more readily available for the Northern Hemisphere, and it is probable that predictions for that hemisphere are more realistic. Nevertheless, there seems to be notably less ozone at high southern latitudes in September to November than the model would suggest. We shall see later (Section 4.7) that this lack of agreement probably reveals a shortcoming of the model; specifically, the model does not include heterogeneous chemistry in its formulation. Altitude profiles are also quite well predicted by models, at least for northern latitudes, as Fig. 4.10 indicates for some mid-spring comparisons. Such small

[a] See, for example, Stratospheric ClO in situ detection with a new approach. Brune, W. H., Weinstock, E. M., Schwab, J. J., Stimpfle, R. M., and Anderson, J. G. *Geophys. Res. Lett.*, **12**, 441 (1985).

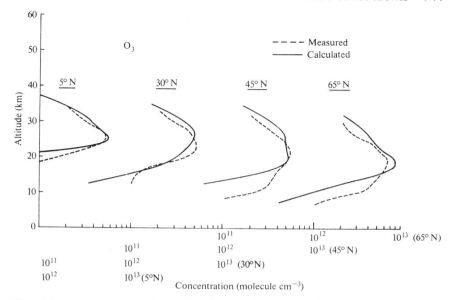

Fig. 4.10.   Observed and calculated vertical concentration profiles of ozone as a function of latitude in the Northern Hemisphere for mid-spring [From: Miller, C., Filkin, D. L., Owens, A. J., Steed, J. M., and Jesson, J. P. *J. geophys. Res.* **86**, 12039 (1981).]

discrepancies as there are between model and observation are usually ascribed to errors in transport contributions.

Instruments on the Solar Mesosphere Explorer (SME) satellite, launched in October 1981, simultaneously measure ozone density, temperature, and solar ultraviolet flux in the upper stratosphere and mesosphere (50–80 km altitude). The altitude region is one which is ideally suited for testing the photochemical theory without the additional variable of atmospheric transport (cf. p. 123). Results from global observations taken over the six-month period January to June, 1982, show that ozone concentrations vary greatly in time and place. Ozone density changes from day to day, and with the seasons, and the principal cause of these changes is the variation in atmospheric temperature. An inverse relationship exists between ozone concentration and temperature, this dependence being observable not only in seasonal patterns, but also in orbit-to-orbit variations during dramatic atmospheric temperature fluctuations such as stratospheric warmings (p. 68). This behaviour matches that expected, with most of the temperature dependence coming from the oxygen-only Chapman reactions

$$O + O_2 + M \rightarrow O_3 + M \tag{4.2}$$

$$O + O_3 \rightarrow 2O_2. \tag{4.4}$$

In the atmospheric regions near the stratopause, to which the results refer, ozone concentrations are controlled largely by the HO_x catalytic cycle and the Chapman reactions (cf. Table 4.3 and Fig. 4.6). Reaction (4.2) increases in rate with decreasing temperature, and reaction (4.4) decreases in rate, while the HO_x-catalysed processes themselves show little temperature dependence. Ozone concentrations are therefore expected to increase with decreasing temperature, as observed by the SME instruments. Absolute ozone abundances are somewhat larger than 2-D model predictions (possibly because the rate of O_2 photolysis was overestimated in the calculations), but the model and observations show similar trends with respect to altitude and latitude.

A critical test of models and the chemistry they incorporate is the ratio of the atomic oxygen concentration to the ozone concentration. Fortunately, there exists a simultaneous measurement of $[O(^3P)]$ and $[O_3]$, the results of which have been given in Fig. 4.3. The ratio $[O(^3P)]/[O_3]$ is plotted out in Fig. 4.11. Since the experiments refer to a particular place and time, a 1-D model can be used to generate the theoretical ratio. The solid line in the figure shows how good the agreement is between the calculated and measured ratios.

Hydroxyl radicals are not only central to the HO_x catalytic chain, but they are also responsible for coupling between the NO_x and ClO_x families. Thus

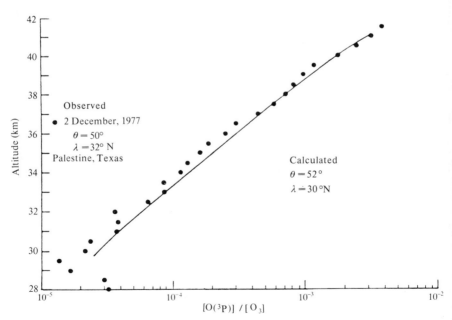

Fig. 4.11.   Ratio $[O(^3P)]/[O_3]$ derived from the measurements of Fig. 4.3, and as calculated by Logan, J. A., Prather, M., Wofsy, S. C., and McElroy, M. B. *Phil. Trans. R. Soc.* **A290**, 187 (1978).

OH is the most important member of the HO_x family, and a key species in atmospheric chemistry. Several techniques exist for determining stratospheric [OH], including the *in situ* resonance fluorescence method. However, most measurements are confined to altitudes greater than about 28 km. Profiles of [OH] can also be inferred by assuming a steady state between sources and sinks of OH, and using models to calculate [OH] that are constrained by observations of other trace chemical constituents. Where the data are available, [OH] measurements are consistent with model predictions, mixing ratios ranging from roughly 10^{-11} at 29 km to 10^{-9} at 42 km. The hydroperoxyl (HO_2) radical is another important species. Mixing ratios measured by balloon-borne FTIR spectroscopy range from about 1.7×10^{-11} at 23 km to 4.5×10^{-10} at 49 km by day; abundances are much lower at night. Below 40 km, the measured concentrations agree well with predictions. Measurements made previously by other methods were much higher, and were particularly hard to reconcile with the relatively low values of [OH] observed at the same altitudes, as well as leading to concentrations of H_2O_2 and HNO_4 (HO_2NO_2) well above the experimental upper limits.

Odd-nitrogen species are represented in the observations mainly by NO, NO_2, and HNO_3, for which latitudinal, altitudinal, seasonal, and diurnal variations have been reported. The total column abundance of the nitrate radical, NO_3, shows a maximum in spring (40°N), and decreases by a factor of three at other seasons. Polar measurements have been used to elucidate the effects of stratospheric temperature on the abundance of the radical as well as to demonstrate the removal of NO_3 by photolysis as the sun rises. Night-time profiles have been obtained by absorption measurements from a balloon platform using the setting star Arcturus or the rising planet Venus as light sources. Mixing ratios at 44°N range from as little as 2×10^{-12} at 20 km to 4×10^{-10} at 40 km. Quantitative measurements of the important molecules N_2O_5 and $ClONO_2$ are still disputed, but for the latter molecule measured altitude profiles (mixing ratios of 0.5 to just over 1×10^{-9} in the altitude range 19–30 km) seem to fit model predictions quite well.

In general, the gross features of the variations of NO_x ($NO + NO_2$) and NO_y (all reactive odd-nitrogen species) are well simulated by models, although there are some quantitative disagreements. For example, the observed columns of NO and NO_2 show sharp variations at high latitudes in winter that are not reproduced in models that consider only homogeneous gas-phase chemistry (cf. Section 4.7.6).

Chlorine-containing species measured in the stratosphere include halo-carbons, HCl, HOCl, and $ClONO_2$, as well as the important intermediates Cl and ClO. Halocarbon (p. 161–2) and $ClONO_2$ (above) concentrations are discussed elsewhere in this chapter. Hydrogen chloride has been deter-mined as a function of altitude, and in a more limited way as a function of latitude and season. The mixing ratios of HCl at mid-latitudes range from 4×10^{-10} at 20 km to 2×10^{-9} at 40–50 km. Model calculations predict

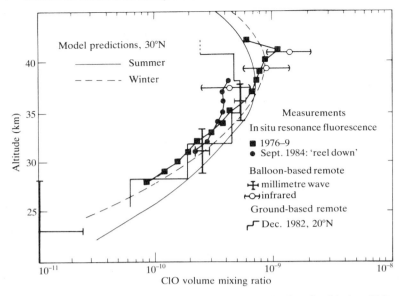

Fig. 4.12.   Comparison between measured and calculated mid-day ClO concentrations. Source as for Fig. 4.6.

similar concentrations at most altitudes. Balloon-borne FTIR measurements
of HOCl give mixing ratios from 3.5×10^{-11} (25 km) to 1.6×10^{-10} (35 km);
below 31 km, the mixing ratios are lower at night. There is excellent agreement
between these experimental values of [HOCl] and the predictions of models.

Measurements also exist of stratospheric abundances of the intermediate
species, Cl and ClO. The data on Cl obtained by direct methods (resonance
fluorescence) are sparse because of the extreme difficulty of the experiments,
but the concentrations match the predictions of models within a factor of two.
For ClO, the data base is much more extensive. Figure 4.12 shows some data
for vertical profiles of mixing ratios obtained generally in northern mid-
latitudes. The predictions of model calculations for a representative latitude
in summer and winter are also indicated, and they show a good match with
the atmospheric observations. Diurnal variations of [ClO] have also been
investigated, and Fig. 4.13 shows the results of some ground-based experiments. The diurnal changes in radical abundance provide a critical test of the
coupling between stratospheric Cl–N–H chemistry, and their interpretation
is free of uncertainties concerning long-term transport processes. Photolysis of
ClONO$_2$ during the day, and its formation from ClO and NO$_2$ during the
night, can explain the broad features of the observations, as the results of the
two representative models show.

Figure 4.14 provides a summary of measurements of some of the concentrations of trace constituents obtained by the shuttle-launched ATMOS

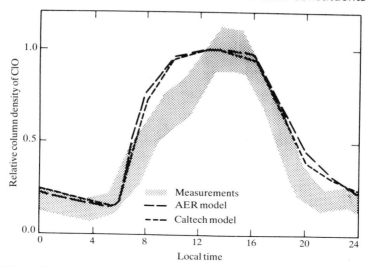

Fig. 4.13. Diurnal variation of [ClO] as measured by ground-based millimetre-wavelength spectroscopy (shaded area represents error limits about the mean) and as predicted by two models. Source as for Fig. 4.6.

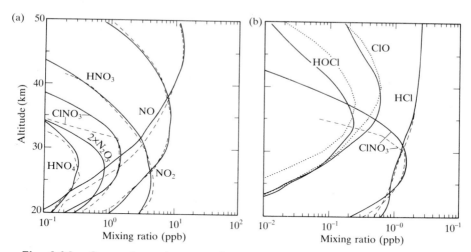

Fig. 4.14. Comparison of measured mixing ratio profiles for active trace gases with predictions of a stratospheric model: (a) nitrogen species; (b) halogen species [McElroy, M. B., and Salawitch, R. J., *Science* **243**, 763 (1989).]

experiments for active nitrogen (a) and chlorine (b) compounds. Note how $ClONO_2$ belongs to both classes, emphasizing its importance as an agent of coupling between them. The broken and dotted lines show predictions of a model. These comparisons, and the discussion of the preceding paragraphs, suggest that the general pattern of stratospheric chemistry is now quite well understood. There remain some quantitative difficulties, and quite exacting demands will be made on the precision of atmospheric measurements if the problems are to be resolved by determining trace-constituent concentrations in the unperturbed stratosphere.

4.4.6 Response of stratospheric ozone to perturbations

The last section was concerned with the validation of stratospheric chemistry from measurements of concentrations of the chemical species that participate in the expected reactions. An alternative, and additional, source of information can come from a study of the response of the atmosphere, and especially its ozone content, to perturbations. Man cannot ethically, of course, embark deliberately on potentially damaging experimental perturbation of the atmosphere, and, in any case, the desired perturbations may not be feasible because of the scale of the exercise. However, atmospheric nuclear tests in the past might have inadvertently perturbed the stratosphere, and they are important for study, because ozone data records exist for comparison with the models. A variety of natural phenomena can also affect the stratosphere. Included in these are impulsive phenomena such as solar particle storms and volcanic eruptions, as well as more gradual variations of influences such as solar UV output or galactic cosmic ray intensity. We shall now consider some of these perturbations and compare the expected with the observed effect on stratospheric ozone.

During 1961 and 1962, the US and USSR detonated about 340 megatons of nuclear explosives in the atmosphere. The shock wave of the explosion heats the air initially to temperatures of several thousand degrees K, and the concentration of NO in equilibrium with atmospheric N_2 and O_2 becomes significant. Calculations for the 1961–2 series of tests indicate that 2–4×10^{34} molecules of NO were deposited in the stratosphere. Since the ambient stratospheric loading of NO_x is 4–15×10^{34} molecules, a significant depletion of stratospheric ozone is conceivable. In fact, the ozone record shows a decrease for 1957–61, followed by a substantial *increase* in 1962–70. Nimbus 4 satellite observations of ozone in the area above the French and Chinese thermonuclear tests of 1970 also do not give clear-cut results. Several explanations have been put forward to explain why no large decreases in ozone were observed after the tests. One of the most interesting reveals how the understanding of stratospheric chemistry has evolved since 1974, when calculations were predicting 8 to 10 per cent ozone reduction for a 58 megaton explosion. Most of the NO injections occurred between 20 and 30 km.

Reaction (4.23), as we now understand (cf. p. 133 and p. 158), leads to *increases* in stratospheric ozone for NO_x injection at altitudes below ~ 20 km. According to current models, a maximum ozone reduction of ~ 2 per cent is expected if the 58 megaton bomb cloud rises to 33 km, but an *increase* of ~ 1 per cent occurs if the cloud rises only to 19 km. A further confusion is introduced by measurements made by an instrumented aircraft that was flown into the nuclear cloud seven days after a 4 megaton atmospheric test explosion in China in late 1976. No measurable change in either NO or O_3 concentration was detected as the aircraft passed through the cloud edges. Seven days after the explosion, [NO] is expected to be from 11 to 80 times that in the unperturbed stratosphere, and [O_3] should be reduced by up to 20 per cent. A plausible 'explanation'—for which there is no direct experimental evidence—is that H_2O is totally dissociated by the fireball, and that the greatly enhanced HO_x concentrations remove the NO_x species in reactions such as (4.28) and (4.32). The salutary point of all the nuclear explosion results and their explanations really concerns the readiness of some interpreters to find a special reason why an effect, predicted by their favoured model, was not, in fact, observed.

Charged particles from the Sun are continually entering the Earth's atmosphere, generally at high latitudes because the moving charges are guided by the Earth's magnetic field lines. Bombardment of the atmosphere by the particles is a source of odd hydrogen and nitrogen. Every few years, a burst of solar activity leads to great enhancements of the particle flux known as a *solar proton event* (SPE). During an SPE, NO_x and HO_x are produced very rapidly, and only above about 60° latitude. Any resulting changes in ozone concentration are relatively easy to identify because of this SPE 'signature', and SPEs thus act as natural 'experiments' to check elements of the stratospheric chemistry problem. Between 4 and 9 August 1972, an SPE of unusually large magnitude occurred, producing large amounts of NO in the stratosphere at high latitudes. It was later discovered that large depletions of ozone had been recorded by the backscatter ultraviolet instrument on the Nimbus 4 satellite. At an altitude of 42 km, ozone concentrations were reduced by 15 per cent, and the depletion persisted for almost thirty days. The SPE thus demonstrated qualitatively the validity of the theory of catalytic destruction by NO_x of ozone. Quantitative comparisons of observed and predicted ozone depletions are also in good agreement. Figure 4.15 compares the ozone depletions measured two weeks after the August 1972 SPE with some model predictions. The two curves are displaced by only one kilometre in the vertical, giving considerable confidence in the modelling assumptions. Chlorine and HO_x chemistry have to be included in the reaction scheme, and the models have shown the importance of including feedback mechanisms. Depletion of ozone will lead to a decrease in stratospheric temperature, since solar heating by ozone is a major contributor to the radiative balance. According to the model, cooling of 6–10 K in the upper stratosphere would follow the 1972 SPE ozone depletions. Rocketsonde temperature measurements indeed show a

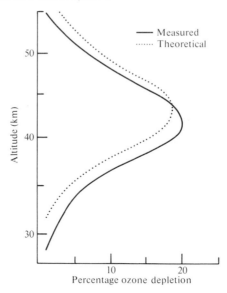

Fig. 4.15. Measured and theoretical depletion of ozone two weeks after the Solar Proton Event (SPE) of August 1972. The data are corrected for the seasonal variation. [From McPeters, R. D., Jackman, C. H., and Stassinopoulos, E. G. *J. geophys. Res.* **86**, 12071 (1981).]

cooling of ~ 6 K after the SPE. Lower temperatures reduce the rate of odd-oxygen destruction, and there is thus a tendency for ozone depletion to be damped out. That is, there is a *negative feedback* mechanism which opposes changes in ozone concentrations.

Analysis of other SPE events (e.g. January and September 1971, and the very intense event of 13 July 1982) has shown ozone depletion only *during* the event, with rapid (~ 36 h or less) recovery. Very large ozone depletions were also observed at the time of increased proton flux in the 1972 events. These short-lived changes are associated with HO_x production by the incoming charged particles. Chemical lifetimes of HO_x are small (a few hours) compared with those for NO_x (months), so that excess HO_x destruction of ozone is short-lived. Calculations for the 1972 event suggest that 1.5 NO molecules, and 2 odd hydrogens, are produced for each ionization. With these figures, and the measured proton flux, [NO] is shown to increase from $\sim 1 \times 10^9$ to $\sim 4 \times 10^9$ molecule cm^{-3} at 45 km, while $[HO_x]$ increases from $\sim 4 \times 10^7$ to a maximum of 1.1×10^8 molecule cm^{-3}. The corresponding half-lives at 45 km are >1 month and ~ 4 hours. In the 1982 SPE, many low-energy protons were deposited but they penetrated only to the mesosphere, where the SME satellite (p. 141) detected ozone depletions of up to 60 per cent at ~ 75 km and high latitudes.

The solar ultraviolet irradiance determines the overall level of stratospheric ozone, principally through the rate of odd-oxygen production in photo-dissociation processes, but also by affecting atmospheric temperature and dynamics. Some time variations exist in the solar UV irradiance as a result of solar rotation, changes in the Sun–Earth distance between summer and winter, and variations in solar activity (coupled to the sunspot cycle). Natural changes in ozone concentrations due to solar UV variability must be taken into account in any attempt to isolate long-term anthropogenic effects on the ozone layer. Until recently, such investigations have been hampered by the uneven geographical distribution of ground-based ozone monitoring stations. Satellite instruments, however, now provide a potential for following global trends in ozone concentration, especially since the results depend on the response of one instrument rather than the many (cross-calibrated) devices of the ground network. Analysis of the infra-red and ultraviolet data from the Nimbus satellites shows sizeable annual, semi-annual, and quasibiennial vari-ations in ozone abundances. These variations can be filtered from the data to seek a possible relationship between global ozone and solar activity (see Section 4.6). A strong correlation is indeed found, with ozone exhibiting an 11-year cycle of 2–3 per cent changes. Model calculations show that the ozone variations require the solar flux at wavelengths between 175 and 210 nm to increase by 15 per cent between solar minimum and solar maximum. Recent studies suggest that solar UV does, in reality, vary by comparable amounts. Solar measurements aboard the Atmospheric Explorer E satellite, for example, give evidence of a 21 per cent variation at wavelengths near 180 nm. The model shows that total ozone abundances respond to changes in solar intensity within one or two months, in accord with observation showing an in-phase relationship between ozone and solar flux.

Stratospheric aerosols of natural and artificial origin can affect the chemistry and climatology of the Earth (see Bibliography). One source of the aerosols is the volcanic eruptions of significant size that occur every few years. The explosive eruptions of Mount St. Helens and especially of El Chichón are examples of recent events that have injected large quantities of material into the stratosphere. Appreciable effects on stratospheric ozone might be expected, but quantitative modelling is difficult because of the many potential ways in which O_3 might be influenced. Water and hydrochloric acid injected above the local tropopause add to the HO_x and Cl_x inventories. Sulphur compounds (SO_2, OCS, CS_2) are likely to increase the loading of stratospheric aerosol and thus alter, by scattering and absorption, the ultraviolet flux available for odd-oxygen production as well as changing stratospheric temperature and dynamics. More important, it is now apparent that aerosol can have a direct chemical effect on stratospheric chemistry (Section 4.7): evidence presented in Section 4.7.7 suggests that perturbed stratospheric chemistry did follow the El Chichón eruptions. So far as homogeneous chemistry is concerned, the HCl injected might also be expected to provide enough chlorine to add to the background depletion of ozone in the ClO_x cycle. In the six months following the El Chichón eruptions, column HCl under the cloud of volcanic debris was measured to be enhanced by 40 per cent. Ozone depletions of as much as 10 per cent were found at altitudes between 18 and 25 km that could be attributed through their timing and properties to the El Chichón event. However, the period involved in 1982–83 was also a period of tropospheric climatic anomalies that makes it difficult to ascribe the stratospheric changes to the volcanic injections alone. Further analysis of the observations may make possible elucidation of the impact of changes in stratospheric aerosol and chlorine compounds on ozone abundances following the El Chichón eruption, and volcanic eruptions in the future may offer scope for investigating the response of the stratosphere to sudden disturbances that will inevitably occur. We turn now, however, to man's own contribution to the chemical inventory of the stratosphere.

4.5 Pollution of the stratosphere

Human impact on the global environment was stressed by the Study of Critical Environmental Problems (SCEP) in 1970. SCEP noted the vulnerability of the stratosphere, where the air is both thin and stable against vertical mixing. Pollutants introduced into the stratosphere by man would have a lifetime for physical removal by transport of several years, and might therefore build up to globally damaging levels. The situation is quite different from pollution of the troposphere or of the boundary layer, where lifetimes are small and effects are *generally* localized (see Chapter 5). SCEP considered particularly the role of supersonic stratospheric transport (SST) aircraft that could emit, into the stratosphere, H_2O, CO, and NO_x. The special feature of these aircraft is not that they are supersonic, but that for efficiency they fly in the stratosphere. Most of the immediate interest centred on the injection of water vapour into the stratosphere, and SCEP felt that the quantity of artificial NO_x was so much less than that of stratospheric O_3 that only very small reduction of O_3 could result. However, in the same year, Paul Crutzen, then working in Oxford, had drawn attention[a] to the *catalytic* role of NO_x in the natural stratosphere, and Harold Johnston in California pointed out[b] that artificial injection of NO_x could bring about a disproportionately large reduction in ozone that would not be calculable from a simple reaction stoicheiometry. So started the 'SST controversy'! The US set up a Climatic Impact Assessment Program (CIAP) of interdisciplinary research, and the UK and France followed with COMESA (Committee on the Meteorological Effects of

[a] Crutzen, P. J., The influence of nitrogen oxides on the atmospheric ozone content. *Q. J. Roy. Meteorol. Soc.* **96**, 320 (1970).

[b] Johnston, H. S., Reduction of stratospheric ozone by nitrogen oxide catalysts from supersonic transport exhaust. *Science*, **173**, 517 (1971).

Stratospheric Aircraft) and COVOS (Conséquences des Vols Strato-sphériques). An enormous, and extraordinarily valuable, scientific effort was put in motion, and it is hard to overestimate the advance in understanding of our atmosphere that has followed. At the same time, a curious 'political' current underlay many of the early discussions, perhaps because the US Congress had refused to fund the Boeing SST prototype programme in March 1971, while a decision had to be reached about landing rights in the US for the Anglo–French Concorde. Environmental opposition to the SSTs had focused on noise pollution in the vicinity of airports, and on sonic booms. The ozone depletion aspect of the SST debate appears to have been an 'eleventh hour' consideration. Even at the end of the CIAP programme, there was controversy concerning the timing of the release, as well as the wording, of the executive summary.

One consequence of the SST studies was the identification of a series of potential modifiers of the ozone layer, some of which are anthropogenic. The agents considered include a BrO_x cycle, as well as changes in the NO_x, HO_x, and ClO_x catalytic cycles that we have already considered for the natural stratosphere. Other influences on ozone are also seen to result from the release of infra-red active gases that can modify stratospheric temperatures, and of species such as CO that can indirectly modify (via reaction with OH) strato-spheric composition. Certain of the agents can be injected directly into the stratosphere, while others may be of tropospheric origin, but of sufficient tropospheric stability to be transported to the stratosphere. In the subsequent parts of Section 4.5, we look at some of the individual mechanisms by which stratospheric ozone can be influenced, and we also attempt to identify the effects of coupled perturbations by several agents. Section 4.6 will examine the evidence that stratospheric ozone is already showing a response to increased loads of catalytically active species. The most spectacular response of all, the development of ozone 'holes', warrants a section of its own (Section 4.7). First, however, we must see why possible changes to atmospheric ozone have aroused so much concern.

4.5.1 Consequences of ozone perturbation

A reduction in stratospheric ozone could have biological consequences because of increased levels of UV-B (290–320 nm) at the ground (see Section 4.1). Changes in ozone concentration could also have climatological effects because of altered stratospheric heating.

Environmental worries about increased UV-B radiation are concerned with deleterious effects both on human beings and on all other plant or animal species. Fear of cancer has made human skin cancer the most publicized potential effect of ozone depletion. It should, however, be pointed out that the kinds of cancer that can definitely be attributed to sunlight (basal cell and squamous cell cancers) are not terribly dangerous, since, if caught in time, they may be successfully treated. There are much more dangerous cancers of the skin, the *melanomas*, but these are relatively rare, and are usually found on parts of the body that are not exposed to sunlight.

A generally adopted, but not necessarily justifiable, measure of the action of ultraviolet radiation on human skin is the spectral efficiency curve for producing *erythema*. Erythema is a reddening resulting from dilation of small blood vessels just below the epidermis. Skin cancer, on the other hand, is a manifestation of abnormal growth in the living cells of the epidermis, and is not closely related to the dilation of underlying vessels. Dead cells on the surface of the skin absorb and scatter radiation to modify the erythemal action spectrum. Absorption and scattering by particles of the pigment melanin, which gives dark colour to skin, further complicates matters. Just what wavelength range is cancer-producing is important in calculating the effects of ozone depletions. Most investigators have used a rule of thumb that a one per cent depletion in ozone leads to a two per cent increase in UV-B. Two-dimensional models show that only in middle latitudes in winter does the UV-increase to ozone-depletion ratio (the *radiation amplification factor*, RAF) approach 2, being less than unity for much of the globe. However, for the spectrally-weighted erythemal curve, the RAF is 1.9 to 2.2. For DNA destruction the RAF is rather higher, being from 2.5 to 2.8.

Measurements of human skin cancer incidence and of UV radiation dose have now been carried out together in several geographical areas, including Australia, the US, New Guinea, and Ireland. For the US at least, it seems certain that more than 90 per cent of skin cancer *other than melanoma* is associated with sunlight exposure, and that the damaging wavelengths are in the UV-B region most affected by changes in ozone concentration. Skin cancer incidence in mid-latitudes doubles for regions increasingly near the equator for each 8° to 11° of latitude, or about 1000 km. The situation with respect to the very serious skin cancer melanoma is confused. Incidence of melanoma does increase with decreasing latitude, suggesting that UV radiation is a contributing factor, but this contribution is compounded by occupational exposure and other factors. Figure 4.16 shows data on melanoma mortality rates as a function of latitude: there appears to be a statistical correlation, but not necessarily a causal one. In the US, a two per cent increase in UV-B is epidemiologically connected with a two to five per cent increase in basal skin cancer, and a four to ten per cent increase in (the more serious) squamous cell cancer. Estimates such as these are highly dependent on location, sex, skin-type, life-style, and so on. During 1990, there were in the US about 500 000 cases of basal cell cancer, 100 000 cases of squamous cell cancer, and 27 600 cases of melanoma.

The epidemiological data are usually used to show how many additional cases of skin cancer would be caused by a certain reduction of stratospheric ozone. According to the figures given in the last paragraph, a depletion of 7 per cent ozone, for example, would be expected to lead to an additional 20 000

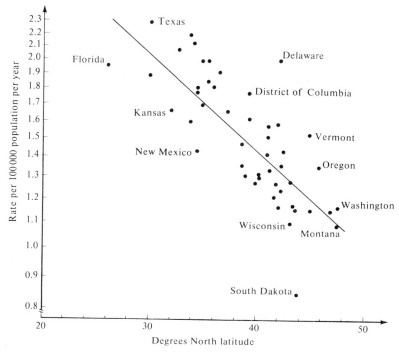

Fig. 4.16. Variation with latitude of human death due to skin melanoma among white males in the United States excluding Alaska and Hawaii. (From Rowland, F. S., Chapter 4.6 in Coyle, J. D., Hill R. R., and Roberts, D. R. (eds) *Light, chemical change and life*, Open University Press, 1982.)

new cases of skin cancer a year in the US. This number seems intolerably large. However, as Hugh Ellsaesser has pointed out (see Bibliography), there is another way of interpreting the data. The statistics themselves come from cancer incidence as a function of latitude. A 1 per cent decrease in ozone corresponds to about a 12-mile (~19 km) displacement equatorward, or a 300-ft (100-m) increase in elevation. That is, from a purely statistical point of view, the chances of contracting skin cancer are increased identically by a perturbation leading to 7 per cent ozone reduction or by moving 84 miles south! Of course, one does not move populations around to compensate for controllable pollution, but Ellsaesser's ideas should put the cancer scare in perspective. In any case, human skin cancer would be fairly controllable if some natural disaster should deplete the ozone layer by several per cent. Only for a few hours during the middle of the day need one avoid exposure. Window glass gives sufficient protection even at midday. Actually, acute sunburn may be more of a hazard than skin cancer, and it is responsible for the loss of many man-hours of work.

A potentially hazardous response to increased ultraviolet exposure concerns the immune system of both animals and humans. Even a mild sunburn has been found to decrease the lymphocyte viability and function in humans for as long as 24 hours. Exposure of animals to otherwise tolerable doses of ultraviolet radiation can produce changes in allergic reaction, skin graft rejection rates, and so on; uninhibited growth also occurs of normally-rejected skin tumours. The notion that 'sunlight is good for you' and that a well-tanned skin is a sign of good health has certainly suffered a set-back!

Possible effects of increased UV-B levels on speceis other than humans are rather difficult to evaluate. Certainly, many plants and animals are sensitive to the ultraviolet radiation dose that they receive, but few experiments have been performed to examine response as a function of dosage. Higher-order species also depend on the results of competition among the lower orders, thus increasing the complexity of analysis. In addition, small climatic temperature changes could also well affect the competition. Ultraviolet irradiation of key cytoplasmic and nuclear constituents is definitely damaging, and animals and plants have developed both avoidance and repair mechanisms. Even the humble plankton may migrate to great depths during the day and rise towards the surface at night. Since plants do not enjoy mobility, ultraviolet radiation can be avoided only through protective shielding (e.g. by waxes and flavenoid pigments) or by changes in the orientation of the leaves. Photosynthesis may ultimately prove to be the most critical element if the ozone shield is depleted, because of its general dependence on cellular integrity and specific dependence on UV-vulnerable enzymes and other proteins.

Models for the prediction of the present-day climate are only approximate, and they are not sufficiently sensitive to estimate perturbations resulting from changes in stratospheric ozone concentrations. Atmospheric mass and energy are mostly contained in the troposphere, so that small changes in stratospheric temperature structure seem unlikely to force direct tropospheric response. However, some feedback mechanisms result from the intricate coupling of energy between the troposphere and the stratosphere. For example, since H_2O content in the lower stratosphere is at least partly controlled by the temperature of the tropical tropopause (p. 65), small temperature changes at this point could lead to large changes in water vapour mixing ratio. Quite apart from the coupling with biospheric responses, the possibility thus exists for stratospheric changes to either trigger or inhibit complex tropospheric, and hence climatic, interactions.

So far in this section, we have examined the deleterious consequences of perturbing stratospheric ozone levels. Instinct tells us that it would be foolish not to avoid any activity that is likely to alter our ozone shield. Curiosity, on the other hand, forces us to ask whether the present-day concentrations of ozone are the 'best' than can be envisaged. Ozone levels have certainly been much smaller than they are now, and probably they have been larger, during the time that life has existed on Earth. Even today, the average intensity of

UV-B reaching the surface varies by a factor of 50 in going from poles to the equator. Any effects of reduction in ozone are uncovered by presuming a relationship between response and exposure to ultraviolet radiation. Some 'bad' effects do show a correlation: could there be 'good' effects? One widely discussed 'good' effect of sunlight is the production of vitamin D in the skin. There is no doubt that vitamin D deficiency (rickets) is frequently seen in town-dwellers at relatively high latitudes, but the disease is also associated with an inadequate diet. More convenient sources exist for the treatment and prevention of rickets than exposure to sunlight. The real argument for avoiding what amounts to deliberate change in ozone levels seems to be concerned with the time-scale. Anthropogenic perturbations, such as those to be discussed in the next sections, can lead to changes over a few years comparable with those occurring naturally over evolutionary periods. Avoidance responses in living organisms have been learnt over the course of evolution of the species. Most important, those responses are generally triggered by particular combinations of *visible* light intensity and temperature. So long as UV-B and visible light bear a constant intensity relationship, the longer wavelength light is a measure of the ultraviolet. Depletion of the ozone layer changes the balance. Of all species, only humans could perhaps directly measure the UV-B intensities and adjust their life-style rapidly. Without making the measurement, enough humans at present get sunburnt by misjudging their ultraviolet exposure from the visible intensity (or temperature) on misty beaches and snowy mountains! Most flora and fauna have adapted to existence at particular latitudes, so that the 50-fold greater UV dose at the equator is of little interest to polar species: a factor of two increase might be intolerably large.

The conclusion that has to be drawn from this section is that rather little is yet known about the physical effects or the ecological consequences of stratospheric ozone changes in either downward or upward direction. It *may* well be that negative feedback mechanisms make the atmosphere 'self-healing' and it *may* be that life can cope with changed ultraviolet intensities. However, the identification of areas where damage might be done, coupled with our lack of knowledge about them, demand that we do not make unwarranted perturbations of the ozone layer. In the sections that follow we look at some of the ways in which man is now presumed capable of modifying his ultraviolet screen. We start by considering a series of individual perturbations. These perturbations are hypothetical in the sense that they do not represent any change to the real atmosphere, in which several parameters vary simultaneously. They do, however, provide a basis for realistic 'scenarios' of simultaneous multiple perturbations.

4.5.2 Stratospheric aircraft (SST)

High-temperature combustion in air leads to formation of NO by 'fixing' of atmospheric N_2 and O_2, as well as the ordinary combustion products CO_2,

Table 4.4   Emission indices for some products of combustion in subsonic and supersonic aircraft engines.

Species	Altitude (km)	Emission index (g/kg fuel)	
		Subsonic	Supersonic
NO_x	9–12	10	10
(as NO_2)	12–15	7	18
	15–21	–	8
CO		3	3
Hydrocarbons		0.5	0.5
H_2O		1250	1250
CO_2		3220	3220
SO_2		1	1
Soot, oil		0.12	0.12

Data from *Aircraft emissions: Potential effects on ozone and climate, a review and progress report*. Report FAA-EQ-77-3, US Dept. of Transportation, Washington, DC, 1977.

CO, and H_2O. Injection of these species, especially the catalytically active NO, into the stratosphere could destroy ozone. Because of the vertical stability of the stratosphere, physical lifetimes against removal of injected species are considerable, ranging from 1–2 years at 17 km to 2–4 years at 25 km altitude. Stratospheric aircraft envisaged in 1970 included the Concorde type flying at 17 km, the US SST flying at ~20 km, and a conceptual hypersonic transport with a flight level of about 25 km.

Emission indices are often used in discussions of fuel utilization. They give the number of grams of a particular combustion product for each kilogram of fuel consumed. Table 4.4 lists some emission indices for typical current subsonic and supersonic (Concorde) jet aircraft engines. While NO_x is a minor combustion product in comparison with H_2O and CO_2, the emission index must be placed in the context of fuel burned. For the 100-passenger Concorde, about 19 100 kg h^{-1} of fuel is consumed during cruise, corresponding to 344 kg h^{-1} of NO_x emitted, or about 8.7×10^4 molecule cm^{-2} s^{-1} averaged over the globe. Early estimates of growth of aircraft anticipated several hundred Concorde and US SSTs in use by 1990. Such a fleet was perhaps an unreasonable projection, since it would have been capable of transporting the entire population of London to New York in a week or so! The oil and economic crises led to revised estimates of growth rates, but even so, 1976 forecasts suggested up to 144 Concorde-like supersonic aircraft in service by 1990. Table 4.5 shows the contributions of subsonic and supersonic aircraft to annual NO_x production at various altitudes calculated in the 1976 studies. The aircraft estimates themselves were certainly not reached by 1990. As an upper limit, they suggest that maximum NO_x emission will correspond to

Table 4.5   Emission of NO_x at various altitudes from the subsonic and supersonic aircraft fleet forecast for 1990

Altitude	Subsonic	Supersonic	Total
	$10^{-8}[NO_x]/$kg NO_2 yr^{-1}		
6–8	2.27	0.07	2.34
8–9	2.63	0.06	2.69
9–10	5.97	0.06	6.03
10–11	9.21	0.05	9.26
11–12	5.95	0.04	5.99
12–13	0.77	0.05	0.82
13–14	0.07	0.08	0.15
14–15	–	0.08	0.08
15–16	–	0.16	0.16
16–17	–	0.26	0.26
17–18	–	0.21	0.21
18–19	–	0.09	0.09
		Grand Total	2.81×10^9 kg yr^{-1}

Data from *Aircraft emissions: Potential effects on ozone and climate, a review and progress report*. Report FAA-EQ-77-3, US Dept. of Transportation, Washington, DC, 1977, for high end of range of predictions. Corrected for underestimated subsonic emissions.

2.6×10^7 kg yr^{-1} at 16–17 km altitude. That release rate corresponds to about 2×10^6 molecule cm^{-2} s^{-1} distributed globally, or a volume addition rate of 20 molecule cm^{-3} s^{-1} if the injection is confined to a 1-km-thick layer of the atmosphere. By way of contrast, most of the model calculations have been based on release rates of 1 to 2×10^8 molecule cm^{-2} s^{-1} at altitudes of 17 and 20 km (corresponding to Concorde and defunct US SST flight altitudes). In Concorde terms, the release rate is equivalent to 2245 aircraft flying 4.4 h day^{-1} every day of the year. Since less than twenty production aircraft have been built, it seems unlikely that the modelled release rates will be reached for some time to come.

Regardless of the clear overestimate of NO_x production rates, the SST has shown in a most interesting way the development of stratospheric models. The ozone controversy was triggered by the fear that SSTs would destroy ozone, and the earliest models indeed showed large ozone depletions for NO_x injection. Laboratory measurements of kinetic parameters, and changes to the models, subsequently led to a range of results. To show what happened, we examine the predictions of a single model (the Lawrence Livermore National Laboratory 1-D model) for a fixed NO_x injection (2×10^8 molecule cm^{-2} s^{-1}). Figure 4.17 is a rather unusual diagram that shows predicted depletion of ozone as a function of year of research: it is sometimes called a *Wuebbles*

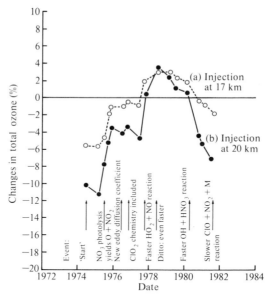

Fig. 4.17. Time history of calculations of the expected change in total ozone concentration from NO$_x$ injection at (a) 17km; (b) 20km. All calculations are from the Lawrence Livermore National Laboratory one-dimensional model. Major changes in model parameters are labelled as 'events'. Unlabelled alterations generally result from minor changes in the chemistry or kinetic parameters. (From Hudson, R., ed.-in-chief, *The stratosphere 1981*, World Meteorological Organization, Geneva, 1981.)

diagram, after its originator. Two curves are shown for NO$_x$ injection, at the 'standard' altitudes of 17 km and 20 km. As will be seen, by mid-1978, NO$_x$ had the effect not of *destroying* ozone but of *making* it!

The reason, as indicated on the diagram, was mainly the discovery of a high rate coefficient for reaction (4.23), between HO$_2$ and NO, a process discussed many times before in this chapter. At the time of writing, the results continue to show ozone column increases for NO$_x$ injections below 17 km, but now predict decreases again for injections at altitudes above 17 km. The most recent alteration to the reaction kinetic data base has resulted in reduced HO$_x$ levels, and hence a reduced [NO$_2$]/[NO] ratio. Understanding of the processes that determine the ozone distribution in the stratosphere has greatly improved over the last decade, but it is by no means obvious that further changes in model sensitivity may not occur in the future. In any case, our ultimate interest in the model predictions must be for the combined perturbations to be discussed later.

An interesting feature of the model calculations for SSTs has been the way in which they have shown the importance of *subsonic* aircraft operating in the

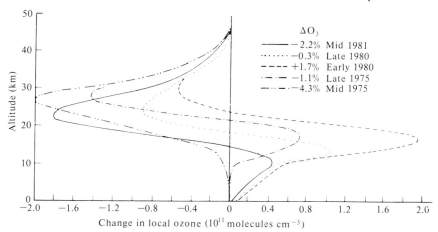

Fig. 4.18.   Changes in local ozone for several chemical schemes from early 1975 to mid-1981, with 2×10^8 cm^{-2} s^{-1} injections of NO$_x$ at an altitude of 17 km. The changes in model parameters are identified in Fig. 4.17. (From Hudson, R., ed.-in-chief, *The stratosphere 1981*, World Meteorological Organization, Geneva, 1981.)

upper troposphere and lower stratosphere. Local ozone between 8 and 10 km is *increased* by up to 20 per cent by the fleet of subsonic aircraft originally predicted for 1990 (Table 4.5). The total column ozone is then increased by about one per cent, even if the depletions due to SSTs predicted by the 'latest' chemistry in Fig. 4.17 are allowed for. Altitude variability of ozone depletion is, in fact, the key to the fickle nature of the model predictions. Figure 4.18 shows predicted altitude variations of ozone change for five of the critical changes in chemistry explained in Fig. 4.17. In every case there is a sizeable depletion of ozone between 20 and 30 km, near the peak of the ozone layer. The difference between the five curves lies in how much ozone is generated at altitudes below 20 km.

Two-dimensional modelling confirms the essential features established by the one-dimensional models, although both the atmospheric lifetimes of injected odd nitrogen and the predicted ozone depletions are very roughly two-thirds of those obtained with the one-dimensional model for the same chemistry and background gas levels. Maximum ozone depletions are produced by NO$_x$ injections at 28 km. Tropical injection causes the greatest depletion, but the largest ozone column reductions occur in polar regions. On the basis of these calculations, a fleet of 500 aircraft operating at an altitude of 22 km might be expected to give global ozone reductions of about 20 per cent if the engines had the properties of those used in the subsonic aircraft of 1988. However, it must be recognized that engine developments might cut nitric oxide emissions by as much as a factor of ten (although those developments might themselves depend on sufficiently good prospects for the building of several hundred SSTs).

4.5.3 Rockets and the space shuttle

Space shuttle launch and re-entry operations generate NO_x, although the amounts are too small to produce a noticeable change in stratospheric ozone. Ammonium perchlorate solid-fuel boosters on the shuttle launch vehicles release HCl vapour to the stratosphere as an exhaust constituent, and could therefore influence the ClO_x catalytic cycle for destruction of ozone. Original estimates suggested a one to two per cent global ozone depletion with 60 launches a year each releasing 50–100 tons (5–10×10^4 kg) of HCl, with up to five per cent depletion in the launch corridor. Later estimates have been much lower, with a maximum depletion of 0.3 per cent (and no corridor effect) now seeming reasonable. About 8×10^7 kg yr^{-1} of water are injected into the stratosphere by the space shuttle, while the proposed Heavy Lift Launch vehicle could inject 2×10^9 kg yr^{-1} for 300 launches a year. Natural H_2O injection rates at the tropics are probably 10^{12} kg yr^{-1}, so that the rocket contribution is very small.

4.5.4 Halocarbons: basic chemistry

Fluorinated chlorocarbons were developed in 1930 by the General Motors Research Laboratories in a search for a non-toxic non-flammable refrigerant to replace sulphur dioxide and ammonia then in use. Dichlorodifluoromethane, CF_2Cl_2, is a typical member of the class of compounds, and is identified[a] as CFC-12. 'Freon' (Du Pont, US) and 'Arcton' (ICI, UK) are trade names for the CFCs. Chemical inertness has made the CFCs valuable as aerosol propellants, as blowing agents for plastic foam production, and as solvents, in addition to their use as refrigerants. CFC production is a world-wide industry with an estimated 2.8×10^8 kg of CFC-11 and 3.7×10^8 kg of CFC-12 being manufactured in 1985. The uses of the CFCs all lead ultimately to atmospheric release, since even 'hermetically sealed' refrigerators and closed-cell foams finally leak to the air. About 90 per cent of all the CFC-11 and CFC-12 produced is believed to have been released.

In 1973, Jim Lovelock[b] and his collaborators reported the presence of halogenated hydrocarbons in the troposphere. It soon became apparent that

---

[a] CFC (chlorofluorocarbon) is followed by a coded two or three digit number. The hundreds digit is the number of carbon atoms in the molecules *less* one, the tens is the number of hydrogen atoms *plus* one, the units is the number of fluorine atoms, and the residue is assumed to be chlorine. If the first digit is zero it is dropped. Thus CFC-11 is $CFCl_3$, CFC-12 is CF_2Cl_2 and CFC-115 is CF_3CF_2Cl.

[b] Lovelock, J. E., Maggs, R. J., and Wade, R. J., Halogenated hydrocarbons in and over the Atlantic. *Nature, Lond.* **241**, 194 (1973).

the quantities of the CFCs were equal, within experimental error, to the total amount ever manufactured. Tropospheric inertness of the CFCs was thus confirmed, and lifetimes of up to hundreds of years were indicated. Only one escape route is possible for the compounds: transport to the stratosphere followed by ultraviolet photolysis:

$$CF_2Cl_2 + h\nu \rightarrow CF_2Cl + Cl \qquad (4.44)$$

is the photolysis process for CFC-12. Space Shuttle-generated chlorine was in the minds of atmospheric scientists (see Section 4.5.3). Mario Molina and Sherry Rowland were quick to see[a] that the chlorine atoms from reaction (4.44) were a much more serious threat to the ozone layer, since the CFC source emissions were known to exist, and their magnitudes were easily calculated. Tropospheric measurements have since confirmed the presence of all the man-made CFCs, as well as chlorinated compounds such as carbon tetrachloride (CCl_4) and 1,1,1-trichloroethane (methyl chloroform, CH_3CCl_3) which are almost certainly solely of anthropogenic origin. Figure 4.19 shows how tropospheric concentrations of some halocarbons have been increasing recently. More tellingly, these compounds have subsequently been detected in the stratosphere, and the concentration–altitude profiles there are consistent with

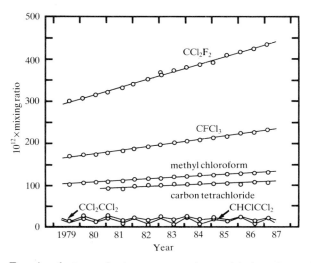

Fig. 4.19. Trends of atmospheric concentrations of halocarbons in the mid-latitude Northern Hemisphere observed in Hokkaido, Japan (42–45°N) in the years 1979–1987. Source: *Stratospheric Ozone 1988*. U.K. Stratospheric Ozone Review Group (Her Majesty's Stationery Office, London, 1988.)

[a] Molina, M. J. and Rowland, F. S., Stratospheric sink for chlorofluoromethanes: chlorine atom-catalysed destruction of ozone. *Nature, Lond.* **249**, 810 (1974).

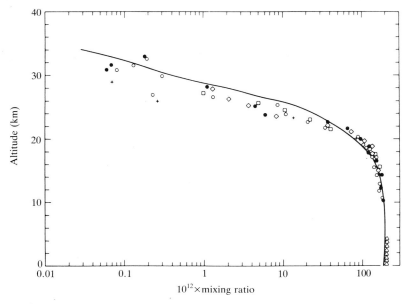

Fig. 4.20. Vertical distribution of CFCl$_3$ in the Northern Hemisphere. The different symbols are used to identify particular experiments performed over the period 1980–83, but have no significance here. Source: as for Fig. 4.6.

the photochemical loss mechanism (some additional loss, especially for the hydrogenated species, occurs *via* reaction with OH radicals). Figure 4.20 illustrates for a typical CFC how the concentration remains virtually constant throughout the troposphere, where the compound is inert and well mixed, but drops suddenly beyond the tropopause as photolysis destroys it and releases active chlorine. Measurements of the vertical distribution of CH$_3$Cl (the most important *natural* chlorine-bearing species) show that chlorine of anthropogenic origin now predominates in the stratosphere. The threat to the ozone layer is thus real and present, and some more detailed examination both of the catalytic chemistry and of the release of chlorinated hydrocarbons is in order.

Cycle 4:

$$Cl + O_3 \rightarrow ClO + O_2 \qquad (4.19)$$

$$ClO + O \rightarrow Cl + O_2 \qquad (4.20)$$

Net $$O + O_3 \rightarrow O_2 + O_2$$

is the major odd-oxygen loss cycle if ClO$_x$ alone is present as catalyst. At least three other cycles appear to be significant if HO$_x$ and NO$_x$ are present as well:

Cycle 12:

$$Cl + O_3 \rightarrow ClO + O_2 \qquad (4.19)$$

$$OH + O_3 \rightarrow HO_2 + O_2 \qquad (4.15)$$

$$ClO + HO_2 \rightarrow HOCl + O_2 \qquad (4.31)$$

$$HOCl + h\nu \rightarrow OH + Cl \qquad (4.45)$$

Net $\qquad O_3 + O_3 \rightarrow O_2 + O_2 + O_2$

Cycle 13:

$$Cl + O_3 \rightarrow ClO + O_2 \qquad (4.19)$$

$$ClO + NO \rightarrow Cl + NO_2 \qquad (4.24)$$

$$NO_2 + O \rightarrow NO + O_2 \qquad (4.18)$$

Net $\qquad O + O_3 \rightarrow O_2 + O_2$

Cycle 14:

$$Cl + O_3 \rightarrow ClO + O_2 \qquad (4.19)$$

$$NO + O_3 \rightarrow NO_2 + O_2 \qquad (4.17)$$

$$ClO + NO_2 + M \rightarrow ClONO_2 + M \qquad (4.33)$$

$$ClONO_2 + h\nu \rightarrow Cl + NO_3 \qquad (4.46)$$

$$NO_3 + h\nu \rightarrow NO + O_2 \qquad (4.47)$$

Net $\qquad O_3 + O_3 \rightarrow O_2 + O_2 + O_2.$

Each of these cycles involves species from the different HO_x, NO_x, and ClO_x families, so that it is difficult to attribute the odd-oxygen loss to a particular family. Reaction (4.18), for example, is the rate-limiting step in the simple NO_x catalytic process, cycle 3, as well as in cycle 13. In the contemporary atmosphere, with a Cl mixing ratio of 2×10^{-9}, reaction (4.24) is responsible for 20 per cent of NO oxidation at most; but in a perturbed atmosphere containing four times as much total chlorine, the reaction would become as important as the $NO + O_3$ path, reaction (4.17). Odd-oxygen loss rates due to the four cycles depend on altitude. Above 20 km, cycle 4 is always the major contributor, although the other cycles are operative. At lower altitudes, cycles 12–14 become dominant.

Bromine-containing compounds can also interfere with stratospheric ozone. Source gases are mainly CH_3Br and brominated CFCs (the 'halons'). Methyl bromide is largely of natural origin, although it is also used as a soil fumigant.

Tropospheric mixing ratios are typically about 10^{-11}, but may be higher in the marine environment. Concentrations fall rapidly above the tropopause, demonstrating that the compound can release its bromine. The halons are used as fire retardants and as fuel additives. At present CH_3Br dominates, but recent measurements suggest a 20 per cent annual growth rate for compounds such as CF_2BrCl (halon 1211).

The chemistry of bromine in the stratosphere is very interesting because of a chlorine–bromine synergism with respect to ozone depletion. A coupling reaction between BrO and ClO produces Br and Cl atoms that react with ozone, and complete a catalytic cycle

Cycle 15:

$$BrO + ClO \rightarrow Br + Cl + O_2 \qquad (4.48a)$$

$$Br + O_3 \rightarrow BrO + O_2 \qquad (4.49)$$

$$Cl + O_3 \rightarrow ClO + O_2 \qquad (4.19)$$

Net $\qquad O_3 + O_3 \rightarrow O_2 + O_2 + O_2.$

With a mixing ratio of 2×10^{-11} of bromine in the stratosphere, the depletion of ozone by the CFCs is enhanced by 5 to 20 per cent. There are no known efficient reservoirs for Br or BrO, in contrast to the case for Cl and ClO, because HBr and $BrONO_2$ are very rapidly photolysed. On a molecule for molecule basis, therefore, bromine is a more efficient catalyst for O_3 destruction than is chlorine. At increased bromine levels (mixing ratios $> 10^{-10}$), odd-oxygen loss increases rapidly because the fast reaction

$$BrO + BrO \rightarrow Br + Br + O_2 \qquad (4.50)$$

is second order in BrO.

There are two further channels for the BrO + ClO interaction that compete with reaction (4.48a)

$$BrO + ClO \rightarrow Br + OClO \qquad (4.48b)$$

$$BrO + ClO \rightarrow BrCl + O_2 \qquad (4.48c)$$

The branching ratios into the three channels are 0.45, 0.43, and 0.12 for reactions (4.48a), (4.48b), and (4.48c), respectively, at room temperature, although the OClO channel is favoured at low temperatures. Photolysis of OClO regenerates O, so that reaction (4.48b) does not lead to loss of odd oxygen. The BrCl produced in reaction (4.48c) acts as a temporary reservoir for Br but like the other Br reservoirs it is rapidly photolysed during the day. Reaction (4.48b) is the only known stratospheric source of OClO, so that observations of the molecule provide useful diagnostic information about stratospheric bromine chemistry.

4.5.5 Halocarbons: ozone depletion potentials

Several factors can influence the impact that releases of particular halocarbons might have on stratospheric ozone. They include the detailed chemistry of the halocarbon once in the stratosphere, and the fraction of the emission that reaches the stratosphere, as well as the absolute amount of the material released.

Although there are some remaining uncertainties in stratospheric models, they do indicate an important feature of perturbation of the stratosphere by CFCs. Since the relative importance of different chlorine-related cycles depends on altitude, the vertical distribution of chlorine sources from different halocarbons must be considered. Destruction of relatively small *fractions* of ozone at around 15–20 km are more important on an absolute scale than larger fractional destructions at higher altitudes. Some halocarbons are photolysed at shorter wavelengths, and thus at higher altitudes, than others. Other things being equal, substitution of chlorine by fluorine shifts absorptions to shorter wavelengths. For equal concentrations, $CFCl_3$ has a maximum photolysis rate at ~ 25 km, CF_2Cl_2 at 32 km, while $CClF_2CF_3$ does not produce its maximum contribution until 40 km. It follows that the more heavily chlorinated halocarbons are more active in destroying ozone for two reasons. First, they are photolysed at lower altitudes where their absolute impact is greater. Secondly, they can release more chlorine atoms per molecule to the catalytic cycle.

Atom for atom, bromine is potentially far more destructive towards stratospheric ozone than is chlorine. Not only does photolysis of the source gases occur lower in the stratosphere, but the synergistic effects described at the end of the last section lead to increased rates of ozone destruction. Estimates of the catalytic activity of Br relative to that of Cl range from 10 to 50.

The total atmospheric burden of a halocarbon is determined both by the rate of its release and by the atmospheric lifetime (itself directly related to the rates of loss processes). Photolysis in the stratosphere (and possibly the upper troposphere) constitutes one important sink for the halocarbons. The atmospheric lifetime against photolysis is shorter for molecules that contain relatively more chlorine than fluorine, and is even shorter if bromine is substituted for chlorine. Molecules that contain hydrogen in place of halogens are susceptible to attack by OH in the troposphere, as are molecules possessing double bonds. If a compound is removed rapidly in the troposphere, then it will have little impact on stratospheric ozone (although the degradation products might adversely affect the troposphere).

Table 4.6 puts these concepts on a quantitative basis. The atmospheric lifetimes given in the second column reflect the ideas about reactivity just presented. The hydrogenated species CHF_2Cl and CH_3CCl_3 are tropospherically much less inert than the other CFCs, and their lifetimes correspondingly small. Figure 4.19 shows that measured concentrations of CH_3CCl_3 were about three times lower than those of CF_2Cl_2 throughout the measurement period, even though the release rates were higher. The figure also shows the effect of unsaturated bonds on tropospheric reactivity. Both CCl_2CCl_2 and $CHClCCl_2$ are emitted in quantities larger than those of $CFCl_3$, yet the concentrations are much lower, and substantially constant: the atmospheric lifetimes for these compounds are only 5 months and 8 days, respectively.

One way of expressing quantitatively the efficiency of a source gas in depleting stratospheric ozone is in terms of an *Ozone Depletion Potential* (ODP), which is specified relative to the depletion by $CFCl_3$ on a kilogram-for-kilogram basis. The ODP thus incorporates the atmospheric lifetime, the stratospheric catalytic activity, and the relative molecular mass of each species. The third column of Table 4.6 lists some ODPs. It must be emphasized that exact values of these ODPs (as well as the atmospheric lifetimes) depend on the particular model used for their assessment. In addition, there is an increasing awareness that the ODP is based on steady-state concepts, comparison being made with the long-lived $CFCl_3$, while short-term releases of highly reactive compounds puts a transient burden on the atmosphere. The peak ozone depletions in such cases may exceed those predicted on the basis of ODPs. Nevertheless, the ODPs do appear to give a reasonable pointer to which compounds are likely to be least damaging.

We see from the table that, for every kilogram of halocarbon released, the ozone destroyed is greater the more chlorine there is in the molecule: $CFCl_3$ is more efficient at destroying ozone than CF_2Cl_2, and CCl_4 is the most efficient of all. Hydrogenated compounds have much lower ODPs, but brominated halocarbons possess much greater potential for damage. Compare, for example, the ODPs of CHF_2Cl and CF_2BrCl. Although the lifetimes of the two compounds are comparable, the ODP of the brominated species is 75 times more.

The last two columns of Table 4.6 show some historical release rates for the halocarbons and the resulting relative contribution to ozone depletion. As anticipated by the original theory, $CFCl_3$ and CF_2Cl_2 make the major contribution to ozone loss. The limiting magnitude of loss is appreciable, and should be measurable, although it will not be reached for several hundred years. Surprisingly, however, the *non-fluorinated* halocarbons CCl_4 and CH_3CCl_3 contribute about 9 and 7 per cent to the overall loss. Although the CFCs were first seen as the potential culprits in anthropogenic ozone destruction, it is now obvious that the enormous release of less inert species such as carbon tetrachloride and methyl chloroform can also have its effect.

4.5.6 Halocarbons: control, legislation, and alternatives

So long as the release rates are unaltered, a steady-state concentration of halocarbon must ultimately be reached. Both the time taken to reach the steady state and the value of the limiting concentration depend on the lifetime.

Table 4.6   Effect of different halocarbons on stratospheric ozone

Halocarbon	Atmospheric Lifetime (yr)[a]	Ozone depletion potential[a]	Release rate $(10^6 \text{ kg yr}^{-1})$[b]	Percentage contribution to ozone loss[c]
$CFCl_3$	60	1.0	281	30.6
CF_2Cl_2	105	0.9	370	36.3
$CF_2ClCFCl_2$	101	0.9	138	13.5
CF_2ClCF_2Cl	236	0.6	–	–
CF_2ClCF_3	522	0.4	–	–
CHF_2Cl	17.2	0.04	72	0.3
CF_3Br	72	7.8	3	2.5
CF_2BrCl	18	3.0	3	1.0
CH_3CCl_3	6.3	0.14	474	7.2
CCl_4	52.2	1.2	66	8.6

[a] Calculated using University of Oslo 2-dimensional model. Quoted in *Scientific assessment of stratospheric ozone: 1989*. Volume 1. World Meteorological Organization, Geneva, 1990.
[b] Source as for Fig. 4.19.
[c] From halogen compounds, based on those listed in the table.

The figures given in Table 4.6 suggest steady-state mixing ratios for CF_2Cl_2 of $\cong 1.8 \times 10^{-9}$ and for $CFCl_3$ of $\cong 7.2 \times 10^{-10}$; these values are very approximately four times the current atmospheric burden shown in Fig. 4.19.

Constant release, leading to steady state atmospheric concentrations, is only one of many scenarios that can be imagined for the CFCs. For a start, historical release rates have been anything but constant. From a few hundred thousand kilograms of CF_2Cl_2 produced in 1931, an all-time high of 4.74×10^8 kg were produced in 1974. (The cumulative production up to 1990 was about 10^{10} kg). Similar figures apply to $CFCl_3$. The Western nations produce 5–10 times as much of the CFCs as the rest of the world, but since concern about the ozone problem was expressed in 1974, production has been steady and then falling. Some legislation was passed in the 1970s forbidding certain uses of the CFCs, especially in the US, and production in 1981 was more than 20 per cent less than it had been in 1974. A much more wide-ranging control is embodied in the 'Montreal Protocol on Substances that Deplete the Ozone Layer' that was agreed in September 1987 and entered into force in January 1989. Each party to the Protocol will freeze, and then reduce according to an agreed timetable, its production and consumption of the first five CFCs listed in Table 4.6; it will also freeze consumption and production of the halons. Consumption in bulk of the five CFCs will be frozen at 1986 levels from 1989, and is to be reduced to 80 per cent (1994) and then 50 per cent (1999) of those levels. Production is less severely restricted in order to meet the domestic needs of developing countries, but production must not exceed 110 per cent of 1986

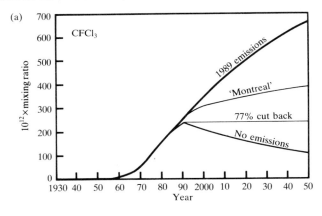

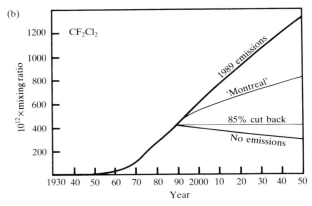

Fig. 4.21. Model predictions of the average concentrations of CFC11 and CFC12 in the atmosphere resulting from four different emission scenarios. Source: as for Fig. 4.19.

levels initially, and must drop back to 90 per cent (1994) and then 65 per cent (1999).

It is possible to calculate, using the atmospheric lifetimes, the future atmospheric concentrations of different halocarbons to be expected with different emission scenarios. Figure 4.21 shows the results of one such calculation for $CFCl_3$ and CF_2Cl_2 for four different situations. In the first, emissions are continued at assumed 1989 levels; in the second, CFC emissions are reduced in line with the demands of the Montreal protocol. The third and fourth scenarios envisage a cut-back of 85 per cent and a complete cessation of emission in 1989. The most striking result of these calculations is immediately evident. Although adherence to the Montreal protocol will alleviate the problem, it would require a much larger reduction (by 85 per cent) to stabilize atmospheric concentrations at their 1989 level. Even were releases of the CFCs to stop altogether now, it appears that it would be well into the 22nd century before atmospheric concentrations would be restored to the values they had before 1960.

The Montreal protocol allows for consumption or production of any CFC to be switched with that of another in such a way that the total potential for depletion of ozone remains unchanged (exchange between brominated compounds and non-brominated ones is *not* permitted). According to the ODPs listed in Table 4.6, that means, for example, that every tonne of $CFCl_3$ could be substituted by 2.5 tonnes of CF_2ClCF_3. Some flexibility in the use of CFCs is thus gained without increasing the potential damage to stratospheric ozone. Reductions in the use of CFCs can be achieved by avoiding their use altogether where possible, or by improving recovery and recycling of them. However, there are instances where halocarbons offer outstanding advantages in particular applications, such as use as refrigerants. A search is on, therefore, for halocarbons that are 'alternatives' to the conventional ones. The requirements are that the alternative compound should have a low ODP (which means, in essence, that it should have a short tropospheric lifetime) while retaining the desired physical and thermodynamic properties (such as boiling point or heat of vaporization). Examples are CF_3CHCl_2 (ODP $\cong 0.013$) for $CFCl_3$ (ODP $= 1$) as a foam-blowing agent; CF_3CH_2F (ODP $\cong 0$) for CF_2Cl_2 (ODP $= 0.9$) as a refrigerant, and CH_2ClCF_2Cl (ODP < 0.05) for $CF_2ClCFCl_2$ (ODP $= 0.9$) as a solvent in the electronics industry. All these compounds are non-flammable, but in some cases the toxicity has not yet been established.

4.5.7 Halocarbons: future ozone depletions

Projected changes in atmospheric loadings of various halocarbons, such as those described in the last section, can be used in conjunction with numerical models of the atmosphere to predict how atmospheric ozone concentrations might alter in the future. Different models, possessing different physical formulations and making different assumptions, produce different predictions. A fine balance must exist between the various influences on stratospheric ozone, thus making it difficult to make precise forecasts of the impact of the halocarbons on atmospheric ozone.

The exact extent predicted for ozone depletion by CFCs has, of course, changed as the chemical input data to the models have been refined, and it will doubtless continue to change in the future. Figure 4.22 shows how the predicted effects of CFC perturbation have evolved with time in some of the earlier calculations; the diagram corresponds to Fig. 4.17 for NO_x catalytic loss. Changes in chemistry at each date-step are the same as those marked on Fig. 4.17. Although the predicted magnitude of the effect has altered, the sense has always been the same: ozone is *lost*. In view of the strong likelihood of an

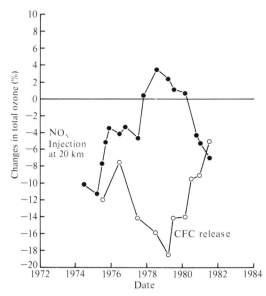

Fig. 4.22. Time history of calculations of the expected change in total ozone from chlorofluorocarbon release as predicted by the one-dimensional Lawrence Livermore National Laboratory model. Changes in model parameters are identified in Fig. 4.17. (From Hudson, R., ed.-in-chief, *The stratosphere 1981*, World Meteorological Organization, Geneva 1981.)

atmospheric response, and of the time lag for that response, it is evident that CFCs should be regarded as a potential danger.

As a representative example of the results of the more recent numerical calculations, Fig. 4.23 shows the changes of ozone column concentrations (global means) predicted by a 2-D model (University of Oslo) in 1988. The 'no control' scenario allows CFC and halon release rates to grow (at roughly 3–4 per cent annually for the CFCs, but reducing to 2.5 per cent by the year 2000), while the 'Montreal' line uses the reduced emissions described in the last section.

The predicted depletion of ozone depends on altitude, season, and latitude. Integrated column ozone concentrations are generally depleted more at high latitudes than at the equator, and the high-latitude depletion is greatest during winter. However, the changes in column abundances are all generally much smaller than the local changes at high altitudes. Some models show depletions for high latitudes of more than 50 per cent at altitudes of 45 km, where the effects of CFC release are a maximum. Depletion of ozone in the upper stratosphere appears to be partially compensated by increased penetration of solar ultraviolet radiation, which leads to an increased rate of generation of ozone in the lower stratosphere. This compensation is sometimes referred to

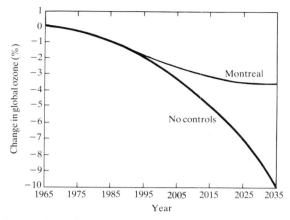

Fig. 4.23. Change in global mean column ozone predicted by the Oslo model. Allowance is made for growth of CO_2, N_2O and CH_4 (at annual rates of ca. 0.6%, 0.2%, and 1.0%, respectively). Source as for Fig. 4.19.

as 'self-healing'. A further complication arises because changes in concentrations of other atmospheric gases, such as N_2O, CH_4, and CO_2, also affect both the distribution and the total column density of ozone. We shall see in the next three sections how these gases acting alone or in combination might influence stratospheric ozone. Perhaps more significant than the uncertainties about the future evolution of the trace gases is the absence from the models of any heterogeneous chemistry. Depletions of ozone far greater than those predicted by the models for the middle of the next century have been seen already in the Antarctic ozone 'holes' (Section 4.7); the models could not have predicted those depletions, because the phenomenon involves polar stratospheric cloud chemistry that was not incorporated in the models.

4.5.8 Nitrous oxide (N_2O): agriculture

Nitrous oxide from the troposphere is the principal source of stratospheric NO_x. Possible perturbations to its concentration because of human activities could therefore affect stratospheric ozone concentrations. Intensive use of fertilizers could lead to increased N_2O production in the biosphere, the nitrogen 'fixed' artificially in the manufacture of the fertilizer being returned to the atmosphere through nitrification or denitrification processes (Section 1.5.3). While perturbations due to the use of SSTs or halocarbons may be regarded as discretionary, agricultural use of fertilizers may well be essential if populations continue to grow. The response of stratospheric ozone to increased emission of N_2O to the atmosphere is evidently of considerable interest. Nitrous oxide concentrations in the atmosphere are known to be increasing slowly (about 0.2 per cent annually between 1976 and 1980), and a simple projection places the concentration by the year 2000 at 5 to 7 per cent above the 1980 value, even if fertilizer use does not increase. With a world population of 6.5×10^9 and a grain consumption of 400 kg per person per year (about $1\frac{1}{2}$ times the present value), N_2O production in the biosphere could double by the turn of the century. The atmospheric lifetime of N_2O seems to be as long as a hundred years, so that any deleterious effects due to increased N_2O release will manifest themselves decades after the perturbation is first applied.

Until recently, it was thought that anaerobic bacterial denitrification (reduction of NO_3^-) was the microbial source of N_2O, a few per cent of the nitrogen converted being released as the oxide. Lowered pH increases the fraction released as N_2O, but decreases the overall rate of denitrification. Denitrification is completely inhibited below a pH of 5.5. Natural N_2O production was assumed to be augmented in proportion to the amount of fixed nitrogen added to the soil as fertilizer. However, normal agricultural practice is to drain and aerate the soil, particularly that soil to which fertilizer is applied. Intensification of agriculture might, according to these ideas, decrease, rather than increase, N_2O output. Incidentally, acid rain (a common form of pollution: see Chapter 5) could inhibit N_2O formation *via* denitrification since the pH of raindrops saturated by CO_2 is already near the limiting value of 5.5.

Observations in oceans, rivers, ponds, and soils have provided direct evidence for a quantitative relation between oxidation of organic material and production of N_2O. That is, *nitrification* (oxidation of NH_4^+) rather than denitrification is implicated as the major source of natural N_2O. The fractional yield of N_2O from nitrification of fertilizer nitrogen appears to be linear at low rates of application (10^{-3} moles N_2O per mole of NH_4^+) but increases rapidly at high rates. Where they are heavily loaded with reduced nitrogen, aquatic and soil systems both appear to exhibit non-linear amplification of the N_2O flux in response to increased loading. For both systems also, enhanced N_2O production is triggered by oxygen depletion. It seems clear, then, that fertilization of soils with ammonium or organic nitrogen, and disposal of human and animal wastes, strongly stimulate nitrification, and are likely to lead to globally significant efflux of N_2O to the atmosphere.

According to 2-D model calculations, response of ozone to a uniform increase in N_2O varies considerably with latitude, and is most important in polar regions in winter or spring. Altitude distributions show the maximum depletion at 20–30 km near the poles. The depletions are only partially balanced by tropospheric increases, which are in the equatorial zone. Although such large increases in N_2O are improbable in the near future, a doubling in N_2O concentration is predicted to give a global ozone depletion of between 9 and 16 per cent.

4.5.9 Direct and indirect action: CH_4, CO, CO_2, and 'greenhouse gases'

This section serves as an introduction to our discussion of 'combined influences' (Section 4.5.10) by looking at how trace gases can affect stratospheric ozone concentrations both by direct involvement in catalytic cycles and by indirect alteration of stratospheric chemistry. Methane affords an excellent first example of a compound exerting several types of influence, since it can be a source of catalytically active hydroxyl radicals, it can lead to *production* of ozone, it can play a part in the formation of reservoir compounds, and, in turn, its concentration in the troposphere can influence the amounts of other trace gases that reach the stratosphere.

We have already seen (Section 4.4.3) that reaction of excited atomic oxygen, $O(^1D)$, with both H_2O and CH_4 is a source of stratospheric OH, and, further, that oxidation of CH_4 accounts for at least some of the stratospheric H_2O. Hydroxyl radicals enter into several stratospheric cycles, both as participants in catalytic ozone destruction (Section 4.4.1), and in determining the partitioning of other catalysts between active and reservoir forms (Section 4.4.2). Methane also acts in another significant way by converting Cl atoms to the reservoir HCl.

Not only may methane influence the detailed chemistry of the catalytic cycles for ozone destruction, but it may also act as a source of ozone in the lower stratosphere. The chemistry concerned is described in connection with tropospheric ozone production (Section 5.3.3). The essential idea is that peroxy radicals derived from the CH_4 can convert NO to NO_2. Nitrogen dioxide can be photolysed at relatively long wavelengths to yield atomic oxygen, and is thus a source of odd oxygen in altitude regions where there is only feeble penetration of radiation able to dissociate O_2. It is evident, then, that increased $[CH_4]$ could increase $[O_3]$ in the lower stratosphere both by sequestering active chlorine and by converting NO to NO_2.

The concentrations of many species in the stratosphere are in large part determined by processes occurring in the troposphere. As we shall see in Chapter 5, reaction with OH is the major tropospheric sink for important trace gases such as CO, CH_4, H_2, CH_3Cl, NO_2, and many others. Any decrease in tropospheric [OH] would lead to increased global concentrations of naturally-occurring species such as CH_4 and CH_3Cl, as well as of anthropogenic substances such as CH_3CCl_3, known to be released in large quantities (cf. p. 161 and Table 4.6). Lowered tropospheric [OH] could thus have an indirect impact on stratospheric ozone through enhanced catalytic destruction. Reactions with carbon monoxide and methane

$$OH + CO \rightarrow H + CO_2 \qquad (4.34)$$

$$OH + CH_4 \rightarrow CH_3 + H_2O \qquad (4.51)$$

are the major sinks for OH in the troposphere (Section 5.3.2), so that increases of either [CO] or $[CH_4]$ can act to reduce [OH]. The detail is complicated, because increased concentration of either carbon species will decrease [OH], and thus lead to increased concentrations of the other species; furthermore, CO is itself one end product of the oxidation of CH_4 (see Sections 1.5.1 and 5.3.2).

There is, in fact, already evidence that $[CH_4]$ and [CO] are increasing in the troposphere, by very roughly one per cent per year (for CO the increase is higher in the Northern Hemisphere and negligible in the Southern). Isotopic data suggest that over half the 'new' methane is of fossil origin (e.g. from combustion, natural gas leakage, oil extraction, and possibly from frozen tundra bogs) and less than half from biogenic sources (e.g. rice agriculture and animal husbandry).

Carbon dioxide concentrations are increasing in the atmosphere largely as a result of the burning of fossil fuels. Aspects of this increase are discussed in more detail in Chapter 9. The potential effect on ozone comes about because CO_2 is the principal atmospheric constituent that contributes to stratospheric cooling. Infra-red radiation from CO_2 escapes from the stratosphere (where the CO_2 is optically thin) rather than being trapped as in the lower atmosphere (Section 2.2.4). Increased CO_2 thus leads to lower stratospheric temperatures, which can alter chemical reaction rates and atmospheric dynamics. Similar remarks apply to changes in concentration of all other 'greenhouse' gases such as CO, N_2O, CH_4, and halocarbons that may be present.

A temperature decrease in the stratosphere is converted to an *increase* in ozone concentration. As we pointed out in Sections 4.4.1 and 4.4.5, the simple Chapman $O + O_3$ process for the loss of odd oxygen, reaction (4.4), has a relatively large activation energy. In the absence of catalytic chains, therefore, loss rates decrease—and O_3 concentrations increase—with lowered temperature. The indirect catalytic routes for odd-oxygen loss are often effective precisely because they have small activation energies. To that extent, catalytic destruction of ozone is less affected by temperature changes than is the direct, oxygen only, route. Nevertheless, cooling the lower stratosphere increases the ozone there by reducing ozone loss resulting from reactions with OH and HO_2. Doubling the present-day CO_2 content would lead to a maximum temperature decrease of 7 to 10 K near 40 km, and a corresponding local ozone increase in the unpolluted stratosphere of 20 to 30 per cent. Column abundances would be increased by ~ 6 per cent. However, if, for example, the CFCs already make a substantial contribution to ozone loss, then the increase of ozone due to increased CO_2 is much smaller. The NO_x catalytic chain is affected in a more subtle way by temperature changes. Loss processes for NO in the upper stratosphere include reaction with atomic nitrogen, and the atom concentration increases with decreasing temperature. Lowered temperatures therefore lead to lower NO_x concentrations, less catalytic loss of odd oxygen, and higher ozone concentrations.

Two-dimensional models can include the alterations to dynamical processes caused by changes in atmospheric temperature structure. Global ozone increases of about 9 per cent are predicted for an approximate doubling of atmospheric CO_2 content, in rough agreement with the 1-D results quoted earlier. Greatest column changes occur at high latitudes, as seems to be indicated for all perturbations of the ozone layer.

Bringing together the indirect influences we have suggested in this section, we can see that the greenhouse gases can influence tropospheric chemistry *via* temperature changes, for example by increasing evaporation of surface water. Increased infra-red opacity due to CO_2 or CFC increases can alter the average H_2O content of the troposphere, and hence the OH concentration. Trace constituent concentrations will then be affected as we described earlier in this section.

4.5.10 Combined influences

A study of the perturbations produced by a single source allows examination of the details of the perturbing influence and of the corresponding feedback processes. The single-source scenario does not, however, reflect reality, and we have already seen that the various perturbations are most surely not additive. Even over the past few decades, CO_2, N_2O, NO_x from subsonic aircraft, and the halocarbons have all increased in concentration. To interpret long-term ozone trends over the same period, and to predict the effects of future perturbations, we must clearly evaluate how the perturbations operate when they are present simultaneously. Comparison of Figs. 4.17 and 4.22 provides a nice demonstration of the interaction between NO_x and ClO_x perturbations. Since 1976, each change of chemistry incorporated in the model has had an opposite effect on the magnitude of ozone depletion due to the two catalytic chains. One might begin to suspect that, with a well-judged balance of NO_x pollution from SSTs and CFC pollution from the ground, the model could be insensitive to the values of rate parameters employed!

Two kinds of model calculation have been employed to look at combined perturbations. In the first, the historical release rates, or measured atmospheric concentrations, are used for the source species (N_2O, NO_x, CFCs, as well as CO_2), and the calculations are run over (say) a decade up to a date in the immediate past. The second kind of calculation looks into the future by continuing the model with a combined scenario of trace gases based on reasonable expectations of growth or decline of the individual sources.

Carbon dioxide cooling (Section 4.5.9) has a major effect in offsetting ozone destruction due to release of catalytically active pollutants. Combined scenario models have made it very apparent that carbon dioxide growth rates must be included in any realistic attempt at predicting ozone changes. The CO_2-related increase in O_3 is not algebraically additive with CFC-related decreases, because of the reduced temperature sensitivity of catalytic processes (cf. p. 174). Nevertheless, inclusion of CO_2 effects in models greatly reduces

Table 4.7 Effect of coupled anthropogenic perturbations on stratospheric ozone.[a]

Species present	Percentage ozone change 1970–80	
	Total column	Local at 40 km
$CFC + NO_x + N_2O + CO_2$	$+0.13$	-3.8
$CFC + NO_x + N_2O$	-0.26	-4.5
$CFC + CO_2$	-0.11	-3.3
CFC only	-0.48	-3.9

[a] LLNL model. Data from Wuebbles, D. J., Luther, F. M., and Penner, J. C. *J. geophys. Res.* **88,** 1444 (1983).
Temperature feedback effects are incorporated.

the impact of a given CFC release. When two catalytic species are considered simultaneously, the dominant cycle alone seems to affect the total ozone, although the altitude distribution of depletion may be shifted by the less important cycle. For example, release of CFC until a steady-state concentration is reached depletes ozone by about 5 per cent in one model (May 1981 point on Fig. 4.22). Doubling N_2O alone leads to 12.5 per cent depletion of O_3, but doubling N_2O *and* releasing CFCs is predicted to destroy 12.9 per cent of the total ozone. On the other hand, most of the change is at ~ 22 km for N_2O alone, while with $N_2O + CFC$ the peak additional ozone destruction rate is at ~ 35 km, almost the same altitude as for CFC alone. With $CO_2 + CFC + NO_x$ from aircraft the model predicts almost no change in total ozone between 1970 and the turn of the century, whereas with CFC alone increasing, there would be a 2 per cent depletion by the year 2000.

Some more detailed combined scenario results for changes over the decade January 1970 to January 1980 are given in Table 4.7. The calculations include temperature feedback effects from ozone heating and CO_2 cooling. Note particularly that with all anthropogenic sources present, total column ozone concentrations are now expected to have increased very slightly over the decade. That the feedback effects are important is shown by calculations without them: for all sources present, the total ozone column is then predicted to have *decreased* by 0.37 per cent. Figure 4.24 shows how the ozone changes are distributed locally with altitude. The difference between having CFCs alone present, and having NO_x, N_2O and CO_2 present in addition, is virtually all a result of ozone *production* below 25 km in the latter case. Projections for dates in the future appear in Fig. 4.25. Local ozone increases in the upper troposphere (~ 10 km) are almost constant after 1990. Quite considerable *percentage* decreases in local ozone occur at around 40 km, but they are sufficiently small in absolute terms to be partially offset by the increases in the lower stratosphere (15–25 km).

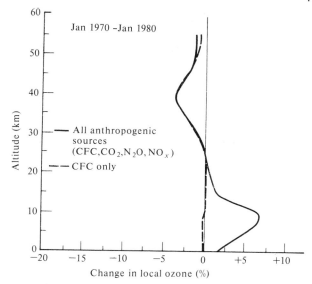

Fig. 4.24. Calculated change in Northern Hemisphere local ozone from 1971 to 1980 for release of chlorofluorocarbons only, and for release of chlorofluorocarbons together with other anthropogenic species. The model stratosphere contains a mixing ratio of 6×10^{-10} of ClO_x in 1970. [From Wuebbles, D. J., Luther, F. M. and Penner, J. E. *J. geophys. Res.* **88**, 1444 (1983).]

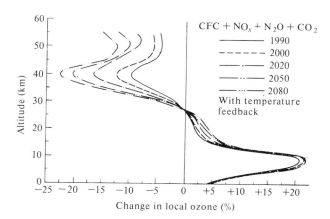

Fig. 4.25. Changes in Northern Hemisphere local ozone relative to background projected into the future for perturbation containing all anthropogenic sources. Background atmosphere contains a mixing ratio 6×10^{-10} of ClO_x. [From Wuebbles, D. J., Luther, F. M., and Penner, J. E. *J. geophys. Res.* **88**, 1444 (1983).]

Similar conclusions are reached about the need to consider future changes in CH_4 concentrations in predictions of ozone depletions. For example, in one model the depletions by the year 2040 with the releases controlled by the Montreal protocol are four times larger if probable increases in $[CH_4]$ are neglected than if they are included in the calculation. These interactions between different influences make quantitative prediction of ozone depletions more difficult than ever, particularly since the atmospheric trends of species such as CH_4 and N_2O have been recognized only relatively recently, and the reasons for the increases are not yet fully understood. In particular, if the trends in $[CH_4]$ are less than anticipated, or those in $[N_2O]$ are greater, then greater ozone destruction will occur than models currently predict.

4.6 Ozone variations and trends

Model predictions of present-day ozone concentrations are mainly of interest in so far as they adequately explain experimental observations and thus give credence to the theory on which they are based. Predicted ozone concentrations are not more 'reliable' than measured ones, although there has been some tendency to interpret observations in the light of expectations from some model or other. Increases in atmospheric CO_2, N_2O, and CFCs over the last 10–20 years were sufficient to have had quite marked effects on ozone according to early models. It so happens that the most recent model, at the time of writing, predicts rather small changes in average column ozone abundances. But that is immaterial. The question we really have to ask is: have ozone concentrations actually changed to a detectable and measurable degree?

Quite severe problems interfere with providing a direct answer to our question. Each measurement of ozone concentration is subject to random errors as well as calibration errors that may be regarded as noise. On top of the measurement noise are real random fluctuations of ozone concentration and longer term variations such as those due to seasonal variations. Two natural phenomena are known to affect ozone concentrations. Variations in solar ultraviolet intensities over the eleven-year solar cycle produce changes of up to two per cent. Tropical winds in the lower stratosphere switch from being easterly to westerly in a cycle that has a period of 26 months (the *quasi-biennial oscillation* or QBO). Column ozone concentrations alter in sympathy with the QBO, typical changes being about 2.7 per cent for mid-latitudes. The observed data series must exist over a long enough period that all the noise and variability factors can be filtered out to reveal possible long-term trends.

Figure 4.26 illustrates some of the techniques used for measuring ozone concentrations in the stratosphere. Ground-based and satellite-platform techniques have different advantages and drawbacks. Ground-based instruments can be calibrated periodically, and the series of measurements dates back to

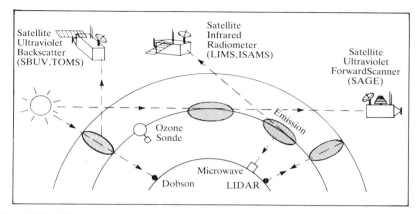

Fig. 4.26. Some of the techniques that can be used for measuring stratospheric ozone concentrations from the ground, from airborne platforms, and from satellites.

1957 in some cases. However, the distribution of stations is not adequate to give a truly global average. Satellite observations can overcome the averaging problem but so far have been available only for relatively short periods. The backscatter UV (BUV) instrument on the Nimbus 4 satellite was in operation for seven years from April 1970, but suffered from an apparent drift in sensitivity. A solar backscatter UV (SBUV) and total ozone mapping spectrometer (TOMS) on the Nimbus 7 satellite have been returning ozone measurements since 1978, and UV and IR photometers on the Solar Mesosphere Explorer have been measuring ozone concentrations for the altitude region above ca. 48 km since late 1981 (cf. p. 141). Both the SBUV and TOMS instruments are calibrated by reflecting sunlight into the instruments with a diffuser plate which is likely to degrade with time, thus altering the calibration.

An elaborate analysis of the available data was undertaken by the NASA/WMO Ozone Trends Panel, and reported in 1988. In this work, the satellite data and ground-based measurements were used to provide cross checks. For example, the TOMS column ozone values as the satellite passed over each ground station were used for intercomparison and quality control of the ground data. The acceptable ground-based measurements (about 50 to 65 stations in any one month) were then used to back-calibrate the satellite instruments. Allowance was made for increases in *tropospheric* ozone (which affect the ground instruments more than the satellite ones) and for the effects of pollutants. In this way, the drifts in the SBUV and TOMS instruments can then be assessed. Finally, the changes in ozone ascribed to natural phenomena (solar cycle and QBO: see above) can be removed and any long-term trends identified.

The analysis (restricted to the non-tropical region of the Northern Hemisphere) did reveal some trends over the period 1969 to 1986. Changes were greatest in the latitude band 53–64°N during winter where decreases reach as much as 8.3 per cent. During summer at these latitudes, however, the change over the seventeen years was not significant. Further south, the changes are smaller, but more evenly distributed over the seasons. Measured stratospheric temperature changes provide additional support for such ozone depletions. Qualitatively, predictions of the University of Oslo model (cf. Section 4.5.7) incorporating increased trace gas concentrations and allowing for the effects of the solar cycle are consistent with these results. On the other hand, it is evident that the full 'combined scenario' of the model for which results are reported in Table 4.7 does not correctly reflect the real changes if the period 1970–1980 is representative of the longer period of the experimental observations.

Data for comparisons of ozone profiles, or local ozone concentrations, are even more limited than those for column abundances. The Ozone Trends Panel inferred long-term changes in the profiles from both ground-based and satellite measurements. For example, the satellite data suggest a decline of about 2.5 ± 1 per cent at about 40 km in the latitude band 20–50° in both hemispheres over the period 1979/1981 to 1984/1987, with the depletion dropping to near-zero at 30 km. These results are consistent, within the limits of uncertainty, with the changes in vertical profiles calculated by recent models and, with certain assumptions, with the predictions for higher altitudes given in Table 4.7 and Fig. 4.24.

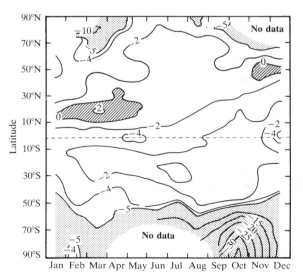

Fig. 4.27. The percentage difference in two-year averages of column ozone (1986/87 minus 1979/80), derived from corrected TOMS data. The dotted areas represent regions suffering losses of more than 5 per cent, while the shaded areas show slight gains in ozone over the period. Source as for Fig. 4.19.

Seasonal trends for column ozone concentrations can also be extracted from the satellite data. For example, a comparison of the averages for 1986/87 with 1979/80 (years chosen to reduce the effect of the QBO) is shown in Figure 4.27. Depletions polewards of 60°S are in excess of five per cent throughout the year; depletions exceeding six per cent are found at northern high latitudes during the winter months. But the outstanding feature is a change of more than 30 per cent over the Antarctic during spring. The decrease in average springtime ozone over the seven years and the even more precipitous fall in ozone between August and October mark changes so unequivocal that there are few concerns about the statistical significance of the observations. It is these enormous and deepening seasonal depletions of ozone in polar regions that are the subject of the last section of this chapter.

4.7 The Antarctic ozone hole

4.7.1 Discovery of abnormal depletion

The chemistry presented in earlier sections of this chapter does not suggest that man's activites can *yet* have had a large effect on stratospheric ozone concentrations, although it indicates that the build-up of CFCs in the future might lead to substantial ozone depletions (Section 4.5.7). In general terms, models based on known chemistry and the dynamics of the atmosphere seemed to explain stratospheric chemical behaviour well. Indeed, the 'Montreal protocol' (Section 4.5.6) that seeks to control CFC emissions was based on just such models. It is thus ironic that the protocol was being established in the autumn of 1987, just as experimental evidence was emerging for the largest ever depletions of Antarctic ozone, and for the connection between these depletions and man's release of the CFCs.

The use of the word 'hole' to describe what happens to ozone in the Antarctic is perhaps an exaggeration, because the total column is depleted rather than removed completely. Nevertheless, column depletions of up to 50 per cent have been observed. At some altitudes the ozone can be almost completely removed. For example, Fig. 4.28 shows that in 1989 about 95% of the ozone that was present between 14 and 18 km in mid-August had gone by mid-October. The first line shows the altitude distribution of ozone on August 23, 1989, late winter in the South, and the second the distribution on October 20, in the spring. It is now apparent that this thinning of the ozone layer in the Antarctic spring has been going on for more than a decade. Scientists of the British Antarctic Survey (BAS) have been measuring ozone concentrations regularly from their base at Halley Bay at 76°S for many years now. The BAS team believed in 1982 that they had detected a decline in springtime ozone concentrations since 1977, and by October 1984 they were sure, with something like 30% total ozone loss, and the depletion apparently increasing over the years 1982, 1983, and 1984. They published their conclusions in *Nature* in

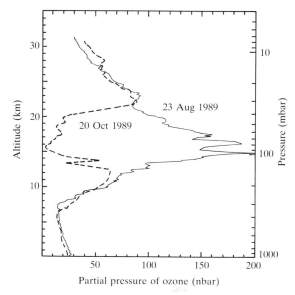

Fig. 4.28. Ozone concentrations over McMurdo Station in August and in October, 1989. These altitude profiles show that the concentrations dropped drastically in the layer between 14 and 22 km, the reduction in ozone being 98% at 15 km. Source: Deshler, T., Hoffmann, D. J., Hereford, J. V., and Sutter, C. B., *Geophys. Res. Letts.* **17**, 151 (1990).

1985. One worrying feature was that rather more sophisticated ozone instruments (see Section 4.6 and Fig. 4.26) such as the Total Ozone Mapping Spectrometer (TOMS) and the Solar Backscattered UltraViolet (SBUV) instrument on the Nimbus 7 satellite had apparently not detected the ozone depletion. However, it subsequently emerged that ozone concentrations as low as those observed by BAS were being rejected from the satellite data on the grounds that they lay outside what was thought to be a 'reasonable' range. When the error was recognized the satellite results were reprocessed, and they confirm the BAS findings.

The satellite TOMS results give a better overall view of ozone over the region, but it is instructive that the depletion of ozone in the ozone hole was recognized by the British scientists working with relatively simple equipment from the ground. The TOMS data show that the depleted region has grown much deeper since 1979, and it was larger than the Antarctic continent in 1987 and 1989. In some years (e.g. 1988) the depletions are less severe than others, but the overall trend of decreasing ozone concentrations is clear. The behaviour of ozone in the atmosphere can be seen even more dramatically in moving picture images that NASA has produced, in which the time-dependent evolution and fluctuations of the hole can be followed from the period when the

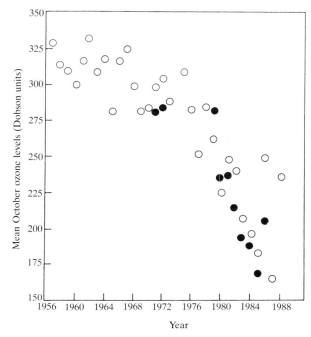

Fig. 4.29.   Decline in October ozone concentrations over Halley Bay during the period 1956–1988. The open circles are results from ground-based measurements using Dobson spectrophotometers, while the closed circles show TOMS measurements where they are available.

hole begins to form until it finally dissipates in late November or early December.

A further illustration that the springtime Antarctic ozone levels have been declining since the late 1970s is given in Fig. 4.29. The open circles show the results from Halley Bay of October-average ozone columns from 1956–1988. The large trend superposed on fluctuations from year to year is obvious. The closed circles show the measurements obtained from the satellite experiments where they are available, and they make it clear that there is now no real conflict between the measurement techniques.

As Susan Solomon, one of the scientists closely connected with interpretation of the new phenomenon, has said, a period of complacency about the ozone layer thus came to an end with the discovery of unprecedented and completely unanticipated depletions of atmospheric ozone over the Antarctic. The magnitudes of the changes were not predicted by any model of future stratospheric composition, even for 50 or 100 years ahead. The depletions were the more startling because they occurred in the present-day atmosphere. The seasonal nature of the depletions heightened the mystery, but demonstrated that the effect was real, rather than being a result of slow degradation of the instruments used to measure ozone in the atmosphere.

4.7.2 Special features of polar meteorology

There are two features of polar stratospheric meteorology and dynamics that appear to have a close bearing on the interpretation. The very low temperatures (below $-80°C$ in the Antarctic) lead to the formation of high-altitude clouds: the so-called polar stratospheric clouds. They are much commoner in the colder Antarctic than in the Arctic regions. There is much suspicion that the polar stratospheric clouds are involved in polar ozone destruction.

The second, and related, feature of polar meteorology is that a vortex forms as air cools and descends during the winter. A westerly circulation is set up as illustrated in Fig. 4.30, with very high wind speeds of perhaps 100 metres per second or more by spring. The vortex develops a core of very cold air,

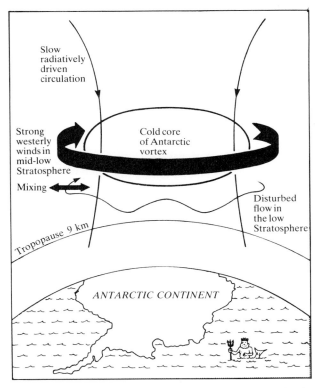

Fig. 4.30. The winter vortex over Antarctica. The cold core is almost isolated from the rest of the atmosphere, and acts as a reaction vessel in which the constituents may become chemically 'preconditioned' during the long polar night.

and it is these low temperatures that allow the polar stratospheric clouds to form in the lower stratosphere. When the Sun returns in September, temperatures rise, the winds weaken, and the vortex breaks down in November. But in the winter and early spring, the stability of the vortex is so great that air at polar latitudes is almost sealed off from that at lower latitudes. That is, the air over the pole is more or less confined to what is effectively a giant reaction vessel. There is a slow downward circulation which drives the polar air through the cold core of the vortex that contains the polar stratospheric clouds. As we shall see shortly, these clouds seem to be involved in unusual chemistry, so that the downward circulation could allow the core region to act as a 'chemical processor'.

4.7.3 Anomalous chemical composition

The confirmation of the British Antarctic Survey observations of ozone depletion naturally generated a great deal of scientific activity. An ozone expedition was mounted in 1986, and it was repeated, in a greatly expanded form, in 1987. The 1987 campaign involved not only ground- and satellite-based measurements and balloons, but two airborne laboratories which flew into the depleted region several times to obtain detailed information about the extent of the perturbed region, and about the chemistry in it. The instrumented aircraft used in the 1987 campaign were a DC8 and what used to be the spy plane, the U2, but now called the ER2, that can fly at such great heights that it is within the stratosphere. The airborne laboratories were able to measure concentrations of ozone; particulate matter, aerosols, and condensation nuclei; the oxides of nitrogen and of the halogens; water vapour concentrations; and temperatures, pressures, and other meteorological parameters. Many of the measurements were obtained by two or more methods. An enormous collaborative effort was involved. Some 150 scientists and support staff from 19 organizations and four nations co-ordinated their efforts in a most effective manner. The wealth of data obtained has permitted much interpretation of the ozone hole phenomenon.

The stratospheric aircraft measurements of 1987 have confirmed, beyond all doubt, that there is anomalous chemistry going on within the vortex region. First consider the concentrations of ozone and of catalytically active ClO radicals. Figure 4.31 shows measurements obtained at altitudes of around 18 km in late August. Concentrations of ozone are normal. But the chlorine oxide concentrations show a sharp rise at a latitude of about 65°S. The concentration goes up by a factor of 10 over a few hundred kilometres. In the perturbed region, the concentrations are, in fact, more than 100 times greater than they are at lower latitudes. It is worth remembering that almost all the chlorine in the Earth's atmosphere is a consequence of man's release of compounds containing it! Only a few weeks later, in mid-September, there is a dramatic change in the behaviour of ozone (Fig. 4.32). This is a time at which

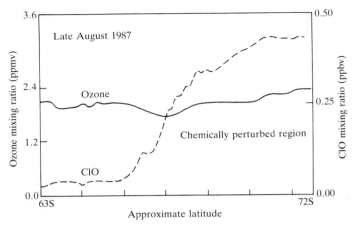

Fig. 4.31. Latitude dependence of ozone and chlorine monoxide (ClO) on entering the chemically perturbed region: late August 1987.

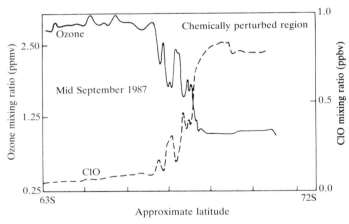

Fig. 4.32. Latitude dependences as for Figure 4.31, but now in mid-September.

the hole has developed fully. Ozone concentrations decrease over just that latitude range where the chlorine monoxide concentration increases on entering the chemically perturbed region within the polar vortex. The strong anti-correlation between the ozone and chlorine oxide concentrations is a strong indication that chlorine chemistry is somehow responsible for the ozone depletion. This anti-correlation is exhibited not only by the large-scale changes but also by the smaller fluctuations in concentration seen on traversing the critical latitude region.

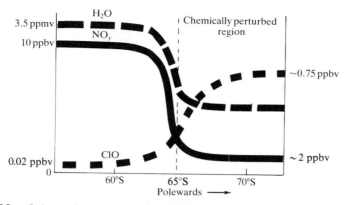

Fig. 4.33. Schematic representation of the changes in concentration of some of the species discussed in the text on entering the chemically perturbed region.

Other measurements, summarized in Fig. 4.33, show that the stratosphere within the disturbed region is abnormally dry, and that it is highly deficient in the oxides of nitrogen, the drop from normal stratospheric concentrations occurring at just those latitudes where the chlorine monoxide concentration increases and the ozone concentrations decrease. The dehydration and denitrification is explained by the condensation of water and the conversion of the oxides of nitrogen to nitric acid in the polar stratospheric clouds. The clouds are made up of particles of two sizes, the smaller of which (about 1 μm diameter) condense at a temperature of about $-80°$C, and the larger (about 10 μm diameter) condense at about $-85°$C. The smaller sized cloud particles are, in fact, composed of nitric acid trihydrate, but the larger ones are made up of water-ice, with substantial amounts of nitric acid dissolved in them, and are large enough to sink slowly through the atmosphere, and thus remove both water and active nitrogen compounds. It is this removal of the oxides of nitrogen which leads to the anomalous chlorine chemistry. Only the core of the vortex is cold enough for the larger particles to form, so we see again that it is the combination of low temperature and special dynamics in the atmosphere that set up the conditions needed for perturbed chemistry.

4.7.4 Perturbed chemistry

The central feature of the perturbed chemistry of the polar stratosphere is the conversion of reservoir compounds (Section 4.4.2) to catalytically active species (or their precursors) on the surface of the polar stratospheric clouds (PSCs). Most of the chlorine in the stratosphere is usually bound up in the reservoir molecules hydrogen chloride and chlorine nitrate, as a result of the reactions

$$Cl + CH_4 \longrightarrow HCl + CH_3 \tag{4.29}$$

$$ClO + NO_2 \xrightarrow{\ M\ } ClONO_2. \tag{4.33}$$

Liberation of the active chlorine from the reservoirs is normally rather slow. But it turns out that the two reservoir molecules can react together in the presence of ice particles, such as those that make up the polar stratospheric clouds

$$ClONO_2 + HCl \rightarrow Cl_2 + HNO_3. \tag{4.52}$$

The outcome is that molecular chlorine is released as a gas, and the nitric acid remains in the ice particles, which can ultimately transport water and nitric acid out of the vortex. The molecular chlorine is photodissociated to atoms

$$Cl_2 + h\nu \rightarrow Cl + Cl \tag{4.53}$$

if sunlight is present. That the surface reaction (4.52) occurs has been demonstrated recently in several laboratory studies. The PSCs thus disturb the balance between active and reservoir chlorine in two related ways. They provide surfaces on which unusual chemical change can occur, and they also transport active nitrogen out of the stratosphere, in the form of HNO_3, thus reducing the amount of $ClONO_2$ reservoir that can be formed in the first place.

Surface reactions such as reaction (4.52) can obviously play an important role in atmospheric chemistry when particles are present, but their involvement in atmospheric chemical transformations has frequently been neglected. Ignorance of the reaction is a major reason why the ozone hole was not predicted by any atmospheric model. Other surface reactions that have subsequently been demonstrated to occur in the laboratory, and that may be involved in polar chemistry are

$$ClONO_2 + H_2O \rightarrow HOCl + HNO_3 \tag{4.54}$$

$$N_2O_5 + HCl \rightarrow ClNO_2 + HNO_3. \tag{4.55}$$

Reaction (4.55) converts another important reservoir molecule, N_2O_5. The reactions occur on ice particles (with dissolved HCl in the second case), and HNO_3 remains in the ice particle after reaction. The gas-phase products HOCl and $ClNO_2$ (nitryl chloride) are both easily photolysed by near ultraviolet and visible light to yield ClO_x radicals.

Because the conversion can occur on the surface of the polar stratospheric clouds, release of molecular chlorine from the major reservoir molecules can continue in the chemically perturbed region throughout the polar winter and early spring. The vortex largely isolates the air within it, so that this part of the stratosphere can become chemically altered, or 'pre-conditioned', over the long polar night.

Consider, as an example, the molecular chlorine released by reaction (4.52). As we have noted already, when the Sun finally returns again in the Spring, the Cl_2 generated from the reservoir gases as a result of the pre-conditioning is rapidly split into chlorine atoms which can destroy ozone, and at the same time liberate chlorine monoxide.

$$Cl_2 + h\nu \rightarrow Cl + Cl \qquad (4.53)$$

$$Cl + O_3 \rightarrow ClO + O_2. \qquad (4.19)$$

The beginnings of an explanation for elevated concentrations of chlorine monoxide accompanying ozone depletions are already apparent. However, on their own, reactions (4.53) followed by (4.19) cannot lead to much ozone loss. It will be recalled from Section 4.4.1 that, in the ordinary atmosphere, a *chain* process

$$Cl + O_3 \rightarrow ClO + O_2 \qquad (4.19)$$

$$ClO + O \rightarrow Cl + O_2 \qquad (4.20)$$

destroys hundreds of thousands of ozone molecules for each chlorine atom made available. Something of the sort must be going on in the perturbed Antarctic stratosphere, but it cannot be this chain, because the concentration of oxygen atoms in the lower stratosphere is far too low when the Sun first appears.

Alternative catalytic cycles are required to explain substantial ozone depletion in the polar stratosphere in early spring. The most important of these cycles seem to involve the ClO dimer, $(ClO)_2$, formed in the self-reaction of ClO. These dimers have been shown experimentally to be readily photolysed to yield, by an indirect route, two free chlorine atoms. The cycle is thus

Cycle 16:

$$ClO + ClO + M \rightarrow (ClO)_2 + M \qquad (4.56)$$

$$(ClO)_2 + h\nu \rightarrow Cl + ClOO \qquad (4.57)$$

$$ClOO + M \rightarrow Cl + O_2 + M \qquad (4.58)$$

$$2(Cl + O_3 \rightarrow ClO + O_2) \qquad (4.19)$$

$$\overline{}$$

Net $2O_3 + h\nu \rightarrow 3O_2$

Dimers such as $(ClO)_2$ are only formed at low temperatures, so that, once again, the low Antarctic polar temperatures are an essential component of another part of the perturbed chemistry. High concentrations of chlorine monoxide radicals also favour the formation of the dimer, and such high concentrations are a feature of the chemically perturbed region of the Antarctic vortex. Recent experiments have shown that for conditions appropriate to the polar stratosphere, dimer formation in reaction (4.56) is the only channel of importance in the self-reaction of ClO. The photolysis of $(ClO)_2$ in reaction

(4.57) was another process completely unknown when the ozone hole was first found. Experimental proof that the reaction liberates atomic chlorine gives some confidence in the validity of the explanation for the Antarctic ozone hole.

Other cycles may supplement cycle 16 in the catalytic destruction of ozone in the polar spring. It has been argued that if bromine were present with mixing ratios in the range $15-30 \times 10^{-12}$, the coupled reactions of BrO with ClO

$$BrO + ClO \rightarrow Br + Cl + O_2 \tag{4.48a}$$

could make a significant contribution to depletion of ozone in the Antarctic stratosphere *via* cycle 15 (shown on p. 164). Measurements by the ER-2 aircraft of BrO (which is probably the major bromine-containing species within the perturbed region) showed mixing ratios of about 5×10^{-12} at an altitude of 18 km, and ground-based measurements suggest a maximum mixing ratio of 10×10^{-12} in the lower stratosphere, so that the BrO–ClO cycle probably only contributes 5–10 per cent to the depletion of ozone. Because of the denitrification of the Antarctic stratosphere, concentrations of HO_x species are elevated. Yet another cycle (cycle 12, p. 163) then becomes possible based on the reactions

$$ClO + HO_2 \rightarrow HOCl + O_2 \tag{4.31}$$

$$HOCl + h\nu \rightarrow OH + Cl. \tag{4.45}$$

Recent calculations suggest, however, that this cycle contributes much less than the others to the net ozone loss.

Efforts are now being made to include the effects of PSCs and heterogeneous chemistry into atmospheric models, but this area of stratospheric modelling is in its infancy and the premises on which the models are based are not well quantified. Models of the gas-phase chemistry that follows after the heterogeneous preconditioning can explain the atmospheric observations reasonably well, although the detailed results suggest that new reservoirs, such as the higher oxides of chlorine (Cl_2O_4, Cl_2O_6, and Cl_2O_7) may have to be invoked to reproduce the behaviour during early spring.

4.7.5 Origin of chlorine compounds; dynamics

All the unusual chemistry that leads to the Antarctic ozone hole depends on the presence of chlorine compounds, and most of those chlorine compounds have been released by man in the form of the chlorofluorocarbons. The concentration of the CFCs in the *lower* atmosphere near the South Pole is almost exactly the same as it is in rural areas of Britain or the United States. In the Antarctic, the compounds have been transported in from populated regions of the Earth. However, at higher altitudes over Antarctica, as the stratosphere is reached, the concentrations of CFCs drop very abruptly, as shown already in Fig. 4.20 for one of the compounds. The absence of CFCs in the stratosphere means, of course, that the chlorine atoms have already almost all been released by photochemical decomposition, and are available to destroy ozone. This release must occur during the period that the polar stratosphere is illuminated, and the species formed then become trapped as old air in the winter vortex, initially mostly in the form of the reservoir gases. Simultaneous analyses of CO_2 and CFCs suggest that the 'age of air' in the vortex may be as much as five years. Details of dynamics and transport will thus determine how deep the hole is each year, because these factors determine the supply of the reservoir gases as well as the efficiency of conversion of the reservoir gases to active chlorine species. The rather shallow hole found in 1988 is a result of just such meteorological factors. It most certainly does not mean that the problem has gone away, and, indeed, the depletion in September 1989 was as great as that in 1987. The variability of the depth of the hole from year to year illustrates just how sensitive the phenomenon is to changes in dynamical activity, which are superposed on a general deepening trend resulting from the increased atmospheric burden of the chlorofluorocarbons. Incidentally, the concentration of the chlorine monoxide dimer depends on the *square* of the active chlorine concentration, thus partially explaining both the sensitivity to dynamics and the way in which the hole has shown such a rapid deepening over the last decade while the CFC concentrations have built up more slowly.

4.7.6 The Arctic stratosphere

It is of obvious interest to discover if the anomalous chemistry seen in the Antarctic atmosphere could also be of significance in the Arctic. Because the winter stratosphere is generally warmer over the Arctic than over Antarctic regions, polar stratospheric clouds are far less abundant and persistent in the North than in the South polar regions. No drastic depletions of ozone have been detected yet near the North Pole. However, there *is* now much evidence that the chemistry in the lower stratosphere is perturbed. After preliminary campaigns in the winter of 1987–88, a major expedition was mounted during 1988–89 in the Arctic, which started at the end of December 1988, and reached a climax in February 1989. It mirrored the Antarctic expedition of 1987, and used the airborne instrumentation on the ER2 and DC8, which were based at Stavanger, in Norway. The objectives were once again to study the mechanisms of production and loss of ozone in the North Polar stratosphere, and to study the effect of the Arctic polar vortex and stratospheric clouds on ozone distribution. One particularly interesting aspect of this campaign was that it was carried out earlier in the season than the Antarctic experiments, just as widespread PSCs were beginning to form. It was thus possible to study the conversion of 'unprocessed' to processed air and its subsequent photochemistry.

The evidence for perturbed chemistry in the Arctic stratosphere is very similar to that obtained in the Antarctic. The abundance of ClO was enhanced by a factor of about 50 over predictions based on gas-phase chemistry, and enhancements in [OClO] provide corroboration. Abrupt increases in [ClO] were associated in a number of cases with the edges of PSCs, thus once again implicating cloud processing of precursors. Abundances of $ClONO_2$ were elevated near the edge of the vortex but decreased towards the interior. The ratio of [HCl]/[HF] was reduced by a factor of two or three inside the vortex compared with outside it, indicating chemical conversion of HCl to other chlorine-containing species (the reduction in the perturbed Antarctic atmosphere is even higher, approaching a factor of six).

Extremely low column abundances of NO_2 were found within the Arctic vortex, once again suggesting that cloud processing had brought about denitrification. In retrospect, it is now evident that earlier observations, dating back to 1979, had revealed similarly low winter column NO_2 abundances that possessed very steep latitudinal gradients associated with stratospheric flow from polar regions (a phenomenon known as the Noxon 'cliff' after its discoverer). Column concentrations of HNO_3 in the Arctic, on the other hand, were much greater than in the Antarctic, and these and other observations of nitrogen-containing compounds suggest that, although denitrification occurs at altitudes around 20 km, the PSCs may have evaporated in the lower stratosphere, thus limiting the reduction in integrated column $[HNO_3]$. Another contrast with the Antarctic was the small extent of dehydration accompanying the denitrification. The observations are not yet understood, but may indicate that denitrification can occur at higher temperatures than was supposed on the basis of evidence from the Antarctic, so that the associated effects on chlorine chemistry might possibly also take place in warmer situations than previously expected.

Whatever the details of the denitrification mechanism, the measurements of chlorine and nitrogen compounds in the Arctic stratosphere provide evidence of perturbed chemistry with many similarities to that seen in the Antarctic. However, in the Arctic campaign of 1988–89, the total ozone losses did not exceed a few per cent, and maximum depletions, at altitudes of around 20 km, were not more than 15–20 per cent. Detection of these losses was complicated by an apparent increase in ozone concentration resulting from ozone-rich air from higher altitudes sinking into the vortex. Comparison of O_3 concentrations with those of N_2O (treated as a relatively stable tracer) was necessary to establish the apparent real photochemical losses of ozone. Some further evidence links patches of ozone depletion with altitude regions that carried PSCs. Nevertheless, it is evident that the Arctic ozone losses were much less than those observed in the Antarctic. One possible explanation is that the winter was unusual in that temperatures remained very low until mid-February, when there was a sudden warming and the PSCs disappeared. Thus air that was sufficiently cold for rapid loss of O_3 through the ClO-dimer mechanism did not, in the event, receive much photolytic solar radiation. In addition, the warming may have brought about mixing of air containing active nitrogen species and thus have restored much of the chlorine to its normal reservoirs. Antarctic ozone is lost mainly in September, when polar temperatures are still very low, but illumination has returned. The corresponding month in the Arctic would be March, so that significant ozone depletion might be confined to winters where the stratosphere remains cold until March. Interannual differences are thus likely to have a large effect on ozone depletion in the northern vortex, but the anomalous chemistry seems primed to take effect whenever the meteorological conditions are appropriate.

4.7.7 Implications of the polar phenomena

A major reason for concern about stratospheric ozone depletion is the biological danger posed by increased solar ultraviolet radiation reaching the Earth's surface (Section 4.5.1). In that context, the creation of Antarctic ozone holes must increase the amount of UV that can penetrate to the ground, especially in the significant UV-B region ($\lambda = 280$–320 nm). Springtime biologically-active irradiances can exceed those experienced in the (unperturbed) Antarctic midsummer. For example, on 5 November 1987 the calculated UV-B irradiance averaged over the day exceeded the solstice value by over 50 per cent, and the irradiance at noon was relatively even more enhanced. Nevertheless, the irradiances should be placed in context: for all realistic ozone depletions, the UV exposure in the Antarctic spring is still less than typical values at a mid- to low-latitude location such as Miami. The real threat is posed to indigenous organisms rather than to human visitors to the Antarctic. One area of particular concern is the possible sensitivity of the spring bloom of phytoplankton living in the surface waters around Antarctica. The UV dose is already greatest in spring, when the ice is much more transparent than in the summer.

Quite apart from the local effects of Antarctic ozone depletion, it is important to know if the processes occurring within the chemically perturbed region can influence ozone concentrations at lower latitudes. Although the vortex is conveniently pictured as a 'containment vessel' that isolates the chemically perturbed region, there is no seal around the polar air and the containment provided by the vortex is somewhat leaky. Ozone-deficient or chemically processed air can thus be transported to lower latitudes. Later in spring, when the vortex breaks up, more extensive lateral mixing takes place. Export of ozone-poor air from polar regions to middle latitudes might lead to a dilution of ozone that could persist for up to a year because of the relatively slow photochemical replacement of ozone, and transport of processed air rich in active chlorine could enhance the effect. If the deficit lasts from one spring to another, there could be a cumulative, permanent depletion of ozone. The TOMS data displayed in Fig. 4.27 are not inconsistent with a dilution effect following seasonal Antarctic ozone depletion, although it must be noted that volcanic aerosols and other factors might be responsible for the changes. Reductions of four per cent or more between 1979/80 and 1987/88 are apparent throughout the year at 50–60°S, and the detailed seasonal changes at these latitudes also support the dilution concept. There is some direct evidence that ozone-poor air even reached 32°S in the Australia–New Zealand area in December 1987 following the enormous depletions over Antarctica of two to three months earlier.

The understanding that heterogeneous chemistry is central to Antarctic ozone depletion has brought a heightened awareness of the importance of aerosol processes in stratospheric chemistry as well as physics. Recent laboratory experiments have shown that reactions can proceed on sulphuric acid particles that are similar to those taking place on the surfaces of PSCs. Sulphuric acid aerosol is widely distributed in the stratosphere, and might thus be responsible for ozone depletion in non-polar regions. Such activity would obviously be enhanced in periods following volcanic eruptions that deposit sulphate aerosol in the stratosphere. Stratospheric ozone trends in middle-to-high latitudes in winter (Section 4.6) are hard to explain by gas-phase mechanisms, but could be accounted for if the sulphate aerosol injected into the stratosphere by the El Chichón explosion (see Section 1.3) was responsible for heterogeneous loss. Ozone measurements show record lows in late 1982 and most of 1983 following the eruption. Total column NO_2 abundances also show a marked decrease, in the same way as in the Antarctic perturbed chemistry, coincident with the poleward movement of the El Chichón cloud in late 1982. This circumstantial evidence is at least suggestive of chemical processing on aerosol surfaces.

One very important lesson learned from the surprise discovery of the Antarctic ozone hole was that the gas-phase stratospheric chemistry thought to be essentially 'complete' in the mid-1980s was, in reality, missing a vital part. The possibility that heterogeneous chemistry might significantly affect ozone throughout the stratosphere emphasizes the need to characterize the reactions through reliable laboratory experiments and to develop atmospheric models that incorporate surface processes with the requisite degree of sophistication.

Bibliography

Books containing introductions to the study of atmospheric chemistry in general and to the investigation of stratospheric ozone chemistry in particular.

Aeronomy of the middle atmosphere: chemistry and physics in the stratosphere and mesosphere. Brasseur, G. and Solomon, S. 2nd edn. (D. Reidel, Dordrecht, 1986).

Dynamics of the middle atmosphere. Holton, J. R. and Matsuno, T. (eds.). (Terrapub, Tokyo, 1984).

Light, chemical change and life. Coyle, J. D., Hill, R. R., and Roberts, D. R. (eds.). (Open University Press, Milton Keynes, 1982.)

> *These three documents are major sources of information concerning the chemistry of the stratosphere, models of it, the sources of trace gases, and their concentrations. Much of the quantitative information in Chapter 4 of the present book is gleaned from these publications.*

Scientific assessment of stratospheric ozone: 1989. World Meteorological Organization, Global ozone research and monitoring project: report no. 20. (WMO, Geneva, 1990).

Atmospheric ozone 1985. World Meteorological Organization, Global ozone research and monitoring project: report no. 16. (WMO, Geneva, 1986).

The stratosphere 1981. Theory and measurement. World Meteorological Organization ozone research and monitoring project: report no. 11 (WMO, Geneva, 1981).

> *The UK Stratospheric Ozone Research Group reports are another important source of information. The report for 1988 is particularly useful in its coverage of models, source gases, trends in ozone concentrations, and its overview of the Antarctic 'ozone hole' phenomenon.*

Stratospheric Ozone 1988. Stratospheric Ozone Research Group (SORG) (Her Majesty's Stationery Office, London, 1988).

Section 4.2

> *Historical introductions to the study of atmospheric ozone*

Forty years' research on atmospheric ozone at Oxford: a history. Dobson, G.M.B. *Applied Optics,* **7**, 401 (1968).

Pioneers of ozone research. A historical survey. Schmidt, M. (Max Planck Institute, Katlenburg-Lindau, 1988).

Sections 4.3 and 4.4

> *Stratospheric ozone and the influence of catalytic destruction cycles*

Chemistry of the stratosphere. Thrush, B. A. *Rep. Prog. Phys.* **51**, 1341 (1988).

The photochemistry of the stratosphere. Turco, R. P. in *The photochemistry of atmospheres.* Levine, J. S. (ed.). (Academic Press, Orlando, 1985).

Photochemistry and dynamics of the ozone layer. Prinn, R. G., Alyea, F. M., and Cunnold, D. M. *Annu. Rev. Earth & planet. Sci.* **6**, 43 (1978).

Section 4.4.3

> *Budgets of catalytically active species.*

Chemical budgets of the stratosphere. Crutzen, P. J. and Schmailzl, U. *Planet. Space Sci.* **31**, 1009 (1983).

The photochemical time constants of minor constituents and their families in the middle atmosphere. Shimazaki, T. *J. atmos. terr. Phys.* **46**, 173 (1984).

Production of odd nitrogen in the stratosphere and mesosphere. An intercomparison of source strengths. Jackman, C. H., Frederick, J. E., and Stolarski, R. S. *J. geophys. Res.* **85**, 7495 (1980).

The role of NO and NO$_2$ in the chemistry of the troposphere and stratosphere. Crutzen, P. J. *Annu. Rev. Earth & planet. Sci.* **7**, 443 (1979).

Release of active chlorine from NaCl particles.

Formation of chemically active chlorine compounds by reactions of atmospheric NaCl particles with gaseous N$_2$O$_5$ and ClONO$_2$. Finlayson-Pitts, B. J., Ezell, M. J., and Pitts, J. N., Jr. *Nature* **337**, 241 (1989).

Section 4.4.5

Observations of ozone and of minor constituents. The papers are representative of a large number describing in situ, satellite, and ground-based investigations. See also the papers reporting work from the AAOE described under Section 4.7.2. In addition to providing the primary source of information about the anomalous composition in the Antarctic springtime stratosphere, the papers listed in that section incidentally provide a cross-section of contemporary methods for the remote and in situ investigation of trace constituents of the atmosphere.

Ozone measuring instruments for the stratosphere. Grant, W. B. (ed.). (Optical Society of America, Washington, DC, 1989.)

Remote sounding of atmospheres. Houghton, J. T., Taylor, F. W., and Rodgers, C. D. (Cambridge University Press, Cambridge, 1984).

Minor constituents in the stratosphere and mesophere. Hudson, R. D. *Rev. Geophys. & Space Phys.* **17**, 467 (1979).

The middle atmosphere as observed from balloons, rockets and satellites. *Phil. Trans. R. Soc.* **A296**, 1, (1980).

In situ measurements of stratospheric trace constituents. Ehhalt, D. H. *Rev. Geophys. & Space Phys.* **16**, 217 (1978).

Infrared absorption spectroscopy applied to stratospheric profiles of minor constituents. Louisnard, N., Fergant, G., Girard, A., Gramont, L., Lado-Bordowsky, O., Laurent, J., Le Boiteux, S., and Le Maitre, M. P. *J. geophys. Res.* **88**, 5365 (1983).

Remote sensing by IR heterodyne spectroscopy. Kostiuk, T. and Mumma, M. J. *Applied Optics* **22**, 2644 (1983).

Solar mesosphere explorer [summary of results]. *Geophys. Res. Lett.* **10**, 237–68 (1983).

Heterodyne spectrophotometry of ozone in the 9.6 μm band using a tunable diode laser. McElroy, C. T., Goldman, A., Fogal, P. F., and Murcray, D. G. *J. geophys. Res.* **95**, 5567 (1990).

UV rocket spectroscopy measurement of the night time ozone distribution. Lean, J. L. *J. geophys. Res.* **88**, 1468 (1983).

A rocket-borne photometric measurement of O$_3$ and NO$_2$ in the equatorial stratosphere. Schlyter, P. *J. geophys. Res.* **87**, 7228 (1982).

The mixing ratio of stratospheric hydroxyl radical from far infrared emission measurements. Carli, B., Carlotti, M., Dinelli, B. M., Mencaraglia, F., and Park, J. H. *J. geophys. Res.* **94**, 11049 (1989).

Progress in stratospheric hydroxyl measurements by balloon-borne lidar. Heaps, W. S., and McGee, T. J. *J. geophys. Res.* **90**, 7913 (1985).

Stratospheric hydroperoxyl measurements. Traub, W. A., Johnson, D. G., and Chance, K. V. *Science* **247**, 446 (1990).

A measurement of stratospheric HO_2 by ground-based millimeter-wave spectroscopy. de Zafra, R. L., Parrish, A., Solomon, P. W., and Barrett, J. W. *J. geophys. Res.* **89**, 1321 (1984).

Evidence for stratospheric hydrogen peroxide. Chance, K. V., and Traub, W. A. *J. geophys. Res.* **92**, 3061 (1987).

Measurements of odd nitrogen compounds in the stratosphere by the ATMOS experiment on Spacelab 3. Russell, J. M., III, Farmer, C. B., Rinsland, C. P., Zander, R., Froidevaux, L., Toon, G. C., Gao, B., Shaw, J., and Gunson, M. *J. geophys. Res.* **93**, 1718 (1988).

Measurements of stratospheric NO_2 from the solar mesosphere explorer. 1. An overview of the results. Mount, G. H., Rusch, D. W., Noxon, J. F., Zawodny, J. M., and Barth, C. A. *J. geophys. Res.* **89**, 1327 (1984).

On the distribution of nitrogen dioxide in the high altitude stratosphere. Solomon, S., and Garcia, R. R. *J. geophys. Res.* **88**, 5229 (1983).

Stratospheric NO_2 at night from balloons. Naudet, J. P., Rigaud, P., and Huguenin, D. *J. geophys. Res.* **89**, 2583 (1984).

In situ stratospheric measurements of HNO_3 and HCl near 30 km using the balloon-borne laser in situ sensor tunable diode laser spectrometer. May, R. D. and Webster, C. R. *J. geophys. Res.* **94**, 16343 (1989).

Simultaneous measurements of HNO_3, NO_2, HCl, O_3, N_2O, CH_4, H_2O and CO, and their latitudinal variations as deduced from airborne infrared spectrometry. Karcher, F., Amodei, M., Armand, G., Besson, C., DuFour, B., Froment, G., and Meyer, G. P. *Ann. Geophys.* **6**, 425 (1988).

Atmospheric NO_3. 1. Measurement technique and the annual cycle at 40 N. Solomon, S., Miller, H. L., Smith, J. P., Sanders, R. W., Mount, G. H., Schmeltekopf, A. L., and Noxon, J. F. *J. geophys. Res.* **94**, 11041 (1989).

Atmospheric NO_3. Observations in polar regions. Solomon, S., Sanders, R. W., Mount, G. H., Carroll, M. A., Jakoubek, R. O., and Schmeltekopf, A. L. *J. geophys. Res.* **94**, 16423 (1989).

Simultaneous measurements of vertical distributions of NO_3 and O_3 at different periods of the night. Rigaud, P., Naudet, J. P., and Huguenin, D. *J. geophys. Res.* **88**, 1463 (1983).

Infrared emission measurements of morning stratospheric N_2O_5. Blatherwick, R. D., Murcray, D. G., Murcray, F. H., Murcray, F. J., Goldman, A., Vanasse, G. A., Massie, S. T., and Cicerone, R. J. *J. geophys. Res.* **94**, 18337 (1989).

Stratospheric N_2O_5 profiles at sunrise and sunset from further analysis of ATMOS/Spacelab 3 solar spectra. Rinsland, C. P., Toon, G. C., Farmer, C. B., Norton, R. H., and Namkung, J. S. *J. geophys. Res.* **94**, 18341 (1989).

Latitudinal and temporal changes of stratospheric HCl and HF. Mankin, W. G. and Coffey, M. T. *J. geophys. Res.* **88**, 10776 (1983).

In-situ northern mid-latitude observations of ClO, O_3 and BrO in the wintertime lower stratosphere. Brune, W. H., Toohey, E. H., Anderson, J. G., Starr, W. L., Vedder, J. F., and Danielsen, E. F. *Science,* **242**, 588 (1988).

ClO and O_3 stratospheric profiles: balloon microwave measurements. Waters, J. W.,

Stachnik, R. A., Hardy, J. C., and Jarnot, R. F. *Geophys. Res. Letts.* **15**, 780 (1988).

Stratospheric ClO: Insitu detection with a new approach. Brune, W. H., Weinstock, E. M., Schwab, J. J., Stimpfle, R. M., and Anderson, J. G. *Geophys. Res. Letts.* **12**, 441 (1985).

A critical analysis of ClO and O_3 in the mid-latitude stratosphere. Froidevaux, L., Allen, M., and Yung, Y. L. *J. geophys. Res.* **90**, 12999 (1985).

Diurnal variation of ClO. Implications for the stratospheric chemistries of $ClONO_2$, HOCl, and HCl. Ko, M. K. W., and Sze, N. D. *J. geophys. Res.* **89**, 11619 (1984).

Measurement of stratospheric HOCl: concentration profiles, including diurnal variation. Chance, K. V., Johnson, D. G., and Traub, W. A. *J. geophys. Res.* **94**, 11059 (1989).

Infrared spectroscopic measurements of halogenated sink and reservoir gases in the stratosphere with the ATMOS instrument. Raper, O. F., Farmer, C. B., Zander, R., and Park, J. H. *J. geophys. Res.* **92**, 9851 (1987).

Particles and aerosols are an important component of the stratosphere, and may influence its chemistry. These papers offer introductory reviews of the subject. Additional references can be found in connection with Aerosols and Volcanic activity at the end of Chapter 1. See also the last entries for Section 4.7.7 for some specific references to the influence of volcanic aerosol on stratospheric ozone.

Stratospheric aerosols: observation and theory. Turco, R. P., Whitten, R. C., and Toon, O. B. *Rev. Geophys. & Space Phys.* **20**, 233 (1982).

Particles above the tropopause. Toon, O. B., and Farlow, N. H. *Annu. Rev. Earth & planet. Sci.* **9**, 19 (1981).

Possible effects of volcanic eruption on stratospheric minor constituent chemistry. Stolarski, R. S. and Butler, D. M. *Pure & appl. Geophys.* **117**, 486 (1978).

Increases in the stratospheric background sulfuric acid aerosol mass in the past ten years. Hofmann, D.J. *Science* **248**, 996 (1990).

Representative photochemical models of the stratosphere are presented in the references that follow. Some further models are described in references given in connection with Section 4.5.

Theoretical predictions. Chapter 3, Volume 1 of *Scientific assessment of stratospheric ozone: 1989*. World Meteorological Organization, Global ozone research and monitoring project: report no. 20. (WMO, Geneva, 1990.)

Upper atmosphere models and research. Ryecroft, M. J., Kasting, G. M., and Rees, D., eds. *Adv. Space Res.* **10**, No. 6 (1990).

The ozone budget in the stratosphere: results of a one-dimensional photochemical model. Cariolle, D. *Planet. Space Sci.* **31**, 1033 (1983).

A numerical model of the zonally averaged dynamical and chemical structure of the middle atmosphere. Garcia, R. R. and Solomon S. *J. geophys. Res.* **88**, 1379 (1983).

A two-dimensional model of stratospheric chemistry and transport. Miller, C., Filkin, D. L., Owens, A. J., Steed, J. M., and Jesson, J. P. *J. geophys. Res.* **86**, 12039 (1981).

A numerical response of the middle atmosphere to the 11-year solar cycle. Garcia, R. R., Solomon, S., Roble, R. G., and Rusch, D. W. *Planet. Space Sci.* **32**, 411 (1984).

Effect of recent rate data revisions on stratospheric modeling. Ko, M. K. W. and Sze, N. D. *Geophys. Res. Lett.* **10**, 341 (1983).

The seasonal and latitudinal behavior of trace gases and O_3 as simulated by a two-dimensional model of the atmosphere. Ko, M. K. W., Sze, N. D., Livshits,

M., McElroy, M. B., and Pyle, J. A. *J. atmos. Sci.* **41**, 2381 (1984).

Chemistry and transport in a three-dimensional stratospheric model: chlorine species during a simulated stratospheric warming. Kaye, J. A., and Rood, R. B. *J. geophys. Res.* **94**, 1057 (1989).

A three-dimensional model of chemically active trace species in the middle atmosphere during disturbed winter conditions. Rose, K., and Brasseur, G. *J. geophys. Res.* **94**, 16387 (1989).

The roles of dynamical and chemical processes in determining the stratospheric concentration of ozone in one-dimensional and two-dimensional models. Ko, M. K. W., Sze, N. D., and Weisenstein, D. *J. geophys. Res.* **94**, 9889 (1989).

A plea that the predictions of models should be allowed neither to replace observations nor to influence their interpretation.

Should we trust models or observations? Ellsaesser, H. W. *Atmos. Environ.* **16**, 197 (1982).

Ozone concentrations may have been affected by nuclear tests in the past.

Effects of atmospheric nuclear explosions on total ozone. Bauer, E. and Gilmore, F. R. *Rev. Geophys. & Space Phys.* **13**, 451 (1975).

The atmospheric nuclear tests of the 50s and 60s: a possible test of ozone depletion theories. Chang, J. S., Duewer, W. H., and Wuebbles, D. J. *J. geophys. Res.* **84**, 1755 (1979).

Nuclear war might be expected to severely damage the Earth's ozone shield.

Photochemical war on the stratosphere. Hampson, J. *Nature, Lond.* **250**, 189 (1974).

Possible ozone depletions following nuclear explosions. Whitten, R. C. Borucki, W. J., and Turco, R. P. *Nature, Lond.* **257**, 38 (1975).

An additional atmospheric effect of nuclear war that has recently been identified concerns the smoke emitted by wide-ranging fires. Indirect effects on ozone concentrations might follow.

The atmosphere after a nuclear war: twilight at noon. Crutzen, P. J. and Birks, J. W. *Ambio* **11**, 114 (1982).

Atmospheric effects of a nuclear war. Birks, J. B. and Crutzen, P. J. *Chem. Br.* **19**, 927 (1983).

Beyond Armageddon. Raloff, J. *Science News* **124**, 314 (1983).

Nuclear winter: global consequences of multiple nuclear explosions. Turco, R. P., Toon, O. B., Ackerman, T. P., Pollock, J. B., and Sagan, C. *Science* **222**, 1293 (1983).

Global atmospheric effects of massive smoke injections from a nuclear war: results from general circulation model simulations. Covey, C., Schneider, S. H., and Thompson, S. L. *Nature, Lond.* **308**, 21 (1984).

Solar protons, arriving at Earth, enhanced during a 'solar proton event', may produce oxides of nitrogen in the atmosphere. The papers listed describe measurements of the associated depletion of ozone.

Possible composition and climate changes due to past intense energetic particle precipitation. Hauglustaine, D., and Gérard, J.-C. *Ann. Geophys.* **8**, 87 (1990).

Effect of solar proton events in 1978 and 1979 on the odd nitrogen abundance in the middle atmosphere. Jackman, C. H. and Meade, P. E. *J. geophys. Res.* **93**, 7084 (1988).

Solar proton events: stratospheric sources of nitric oxide. Crutzen, P. J., Isaksen, I. S. A., and Reid, G. C. *Science* **189**, 457 (1975).

Solar proton event: influence on stratospheric ozone. Heath, D. F., Krueger, A. J., and Crutzen, P. J. *Science* **197**, 886 (1977).

Analysis of the August 1972 solar proton event including chlorine chemistry. Solomon, S. and Crutzen, P. J. *J. geophys. Res.* **86**, 1140 (1981).

Observations of ozone depletion associated with solar proton events. McPeters, R. D., Jackman, C. H., and Stassinopoulos, E. G. *J. geophys. Res.* **86**, 12071 (1981).

Effect of solar proton events in the middle atmosphere during the past two solar cycles as computed using a two-dimensional model. Jackman, C.H., Douglass, A.R., Rood, R.B., and McPeters, R.D. *J. geophys. Res.* **95**, 7417 (1990).

Section 4.5

Ozone in the stratosphere is an essential filter of ultraviolet radiation that protects life on the Earth's surface. One effect of reduced ozone concentrations could therefore be an increase in biologically damaging radiation, a subject explored in the references that follow.

Living with our sun's ultraviolet rays. Giese, A. C. (Plenum Press, New York, 1976).

Stratospheric ozone reduction, solar ultraviolet radiation, and plant life. Worrest, W. C., and Calwell, M. M. (Springer-Verlag, Berlin, 1986).

Indication of increasing solar ultraviolet-B radiation flux in alpine regions. Blumthaler, M. and Ambach, W. *Science* **248**, 206 (1990).

The budget of biologically active radiation in the earth–atmosphere system. Frederick, J. E., and Lubin, D. *J. geophys. Res.* **93**, 3825 (1988).

Possible long-term changes in the biologically active ultraviolet radiation reaching the ground. Frederick, J. E., and Lubin, D. *Photochem. Photobiol.* **47**, 571 (1988).

Biological effects of ultraviolet radiation. Beddard, G. S., in *Light, chemical change and life* (eds. Coyle, J. D., Hill, R. R., and Roberts, D. R.). (Open University Press, Milton Keynes, 1982.)

Possible ozone reductions and UV changes at the Earth's surface. Pyle, J. A. and Derwent, R. G. *Nature, Lond.* **286**, 373 (1980).

Biologically damaging radiations amplified by ozone depletions. Gerstl, S. A. W., Zardecki, A., and Wiser, H. L. *Nature, Lond.* **294**, 352 (1981).

Many activities of man can potentially affect stratospheric ozone concentrations. The first two papers in this set consider the influence that man may have on atmospheric chemistry in general, while the others are devoted more specifically to the problem of stratospheric ozone. The book by Dotto and Schiff describes the political and scientific controversies surrounding the question of ozone depletion

Changing composition of the global stratosphere. McElroy, M. B., and Salawitch, R. J. *Science* **243**, 763 (1989).

Atmospheric chemistry: response to human influence. Logan, J. A., Prather, M. J., Wofsy, S. C., and McElroy, M. B. *Phil. Trans. R. Soc.* **A290**, 187 (1978).

Pollution of the stratosphere. Johnston, H. S. *Annu. Rev. phys. Chem.* **26**, 315 (1975).

Man and stratospheric ozone. Bower, F. A. and Ward, R. B. (eds). (CRC, Cleveland, 1980.)

Causes and effects of stratospheric ozone reduction: an update. (National Academy Press, Washington, DC, 1982).

Causes and effects of changes in stratospheric ozone: update 1983. (National Academy Press, Washington, DC, 1984.)

A catalog of perturbing influences on stratospheric ozone, 1955–1975. Bauer, E. *J. geophys. Res.* **84**, 6929 (1979).

Transport processes and ozone perturbations. Solomon, S., Garcia, R. R., and Stordal, F. *J. geophys. Res.* **90**, 12981 (1985).

The ozone war. Dotto, L. and Schiff, H. I. (Doubleday, Garden City, New York, 1978.)

Ozone crisis: the 15-year evolution of a sudden global emergency. Roan, S. L. (John Wiley, Chichester, 1989).

Some specific perturbing influences are now considered, following the order of subdivision of Section 4.5.

Section 4.5.2

Environmental impact of stratospheric flight. (US Nat. Acad. Sci., Washington, DC, 1975.)

Aircraft emissions: potential effects on ozone and climate. Oliver, R. C., Bauer, E., Hidalgo, H., Gardner, K. A. and Wasylkiwskyj, W. Report FAA-EQ-77-3, US DoT, Washington, DC (1977).

Nitrogen oxides from high-altitude aircraft: an update on potential effects on ozone. Johnston, H. S., Kinnison, D. E., and Wuebbles, D. J. *J. geophys. Res.* **94**, 16351 (1989).

Section 4.5.3

An assessment of the impact on stratospheric chemistry and ozone caused by the launch of the space shuttle and Titan IV. Prather, M. J., Garcia, M. M., Douglass, A. R., Jackman, C. H., Ko, M. K. W., and Sze, N. D. *Report to US Congress from the upper atmosphere program*, 1990.

Environmental effects of the space shuttle. Potter, A. E. *J. Env. Sci.* **21**, 15 (1978).

Section 4.5.4

Some references concerned with the nature of the problem and the sources of the chloro (fluoro) carbons.

Chlorofluoromethanes in the environment. Rowland, F. S. and Molina, M. J. *Rev. Geophys. & Space Phys.* **13**, 1 (1975).

Halogens in the atmosphere, Cicerone, R. J. *Rev. Geophys. & Space Phys.* **19**, 123 (1981).

Chlorofluoromethanes in the environment: the aerosol controversy. Sugden, T. M. and West, T. F. (eds.). (John Wiley, New Jersey, 1980.)

Chlorocarbon emission scenarios: potential impact on stratospheric ozone. Wuebbles, D. J. *J. geophys. Res.* **88**, 1433 (1983).

How have the atmospheric concentrations of halocarbons changed? Prinn, R. G. in *The changing atmosphere*, Rowland, F. S. and Isaksen, I. S. A., eds. (John Wiley, Chichester, 1988).

Atmospheric concentration of halocarbons in Japan 1979–1986. Makide, Y., Yokohara, A., Kubo, Y., and Tominaga, T. *Bull. Chem. Soc. Japan* **60**, 571 (1987).

Sources of stratospheric gaseous chlorine. Ryan, J. A. and Mukherjee, N. R. *Rev. Geophys. & Space Phys.* **13**, 650 (1975).

Selected man-made halogenated chemicals in the air and oceanic environments. Singh, H. B., Salas, L. J., and Stiles, R. E. *J. geophys. Res.* **88**, 3675 (1983).

Measurements of atmospheric methyl bromide and bromoform. Cicerone, R. J., Heidt, L. E., and Pollack, W. H. *J. geophys. Res.* **93**, 3745 (1988).

Chemistry of organic traces in air. VIII. Sources and distribution of bromo- and bromochloro-methanes in marine air and surface water of the Atlantic ocean. Class, T., and Ballschmiter, K. *J. atmos. Chem.* **6**, 35 (1988).

Man's impact on stratospheric ozone as a result of release of halogen-containing species will depend on the emission rates, the concentrations, and the atmospheric lifetimes of the pollutants. Several experiments and models address these questions. The 'Atmospheric Lifetime Experiment', in particular, is designed to determine accurately the atmospheric concentrations of the four halocarbons $CFCl_3$, CF_2Cl_2, CCl_4, and CH_3CCl_3, and also of N_2O (see Section 4.5.8), with emphasis on measurement of their long term trends in the atmosphere. The paper to which the reference is given is one of six consecutive papers describing the techniques, methodology, and results.

The atmospheric lifetime experiment. Prinn, R. G., Simmonds, P. G., Rasmussen, R. A., Rosen, R. D., Alyea, F. N., Cardelino, C. A., Crawford, A. J., Cunnold, D. M., Fraser, P. J., and Lovelock, J. E. *J. geophys. Res.* **88**, 8353 (1983).

Global distribution and southern hemispheric trends of atmospheric CCl_3F. Fraser, P. J., Hyson, P., Enting, I. G., and Pearman, G. I. *Nature, Lond.* **302**, 696 (1983).

A 2-D model calculation of atmospheric lifetimes for N_2O, CFC-11 and CFC-12. Ko, M. K. W. and Sze, N. D. *Nature, Lond.* **297**, 317 (1982).

The transient response of atmospheric ozone to injections of catalytically active species may be the most valid way of assessing the impact of releases of halogen-containing species.

Stratospheric ozone depletion and future levels of atmospheric chlorine and bromine. Prather, M. J. and Watson, R. T. *Nature* **344**, 729 (1990).

Some models used to investigate the effect of chlorofluorocarbons on stratospheric ozone concentrations.

Two-dimensional modelling of potential ozone perturbation by chlorofluorocarbons. Steed, J. M., Owens, A. J., Miller, C., Filkin, D. L., and Jesson, J. P. *Nature, Lond.* **295**, 308 (1982).

A two-dimensional photochemical model of the atmosphere. 1: Chlorocarbon emissions and their effect on stratospheric ozone. Gidel, L. T., Crutzen, P. J., and Fishman, J. *J. geophys. Res.* **88**, 6622 (1983).

Non-linear response of stratospheric ozone column to chlorine injections. Cicerone, R. J., Walter, S., and Lin, S. C. *J. geophys. Res.* **88**, 3647 (1983).

Revised predictions of the effect on stratospheric ozone of increasing atmospheric N_2O and chlorofluoromethanes: a two-dimensional model study. Whitten, R. C., Borucki, W. J., Woodward, H. T., Capone, L. A., and Reigel, C. A. *Atmos. Environ.* **17**, 1995 (1983).

Section 4.5.5

Ozone depletion potentials. Chapter 4, Volume I and Chapter VIII, Volume II of *Scientific assessment of stratospheric ozone: 1989*. World Meteorological Organization, Global ozone research and monitoring project: report no. 20. (WMO, Geneva, 1990.)

Section 4.5.6

Montreal Protocol on substances that may deplete the ozone layer–final act. United Nations Environment Programme (UNEP, Nairobi, Kenya, 1987).
Future emission scenarios for chemicals that may deplete stratospheric ozone. Hammitt, J. K., Camm, F., Connell, P. S., Mooz, W. E., Wolf, K. A., Wuebbles, D. J., and Bemazai, A. *Nature* **330**, 711 (1987).
Alternative fluorocarbon environmental acceptability study (AFEAS) Volume II of *Scientific assessment of stratospheric ozone: 1989*. World Meteorological Organization, Global ozone research and monitoring project: report no. 20. (WMO, Geneva, 1990).
Model calculations of the relative effects of CFCs and their replacements on stratospheric ozone. Fisher, D. A., Hales, C. H., Filkin, D. L., Ko, M. K. W., Sze, N. D., Connell, P. S., Wuebbles, D. J., Isaksen, I. S. A., and Stordal, F. *Nature* **344**, 508 (1990).
The CFC-ozone issue: progress on the development of alternatives to CFCs. Manzer, L.E. *Science* **249**, 31 (1990).

Section 4.5.8

Nitrous oxide: trends and global mass balance over the last 3000 years. Khalil, M. A. K., and Rasmussen, R. A. *Ann. Glac.* **10**, 73 (1988).
Ozone calculations with large nitrous oxide and chlorine changes. Kinnison, D. E., Johnston, H. S., and Wuebbles, D. J. *J. geophys. Res.* **93**, 14165 (1988).
Effects of nitrogen fertilizers and combustion on the stratospheric ozone layer. Crutzen, P. J. and Ehhalt, D. H. *Ambio* **6**, 112 (1977).
Factors influencing the loss of fertilizer nitrogen into the atmosphere as N_2O. Conrad, R., Seiler, W., and Burse, G. *J. geophys. Res.* **88**, 6709 (1983).
The nitrogen cycle: perturbations due to man and their impact on atmospheric N_2O and O_3. McElroy, M. B., Wofsy, S. C., and Yung, Y. L. *Phil. Trans. R. Soc.* **B277**, 159 (1977).
Nitrous oxide emissions from fossil fuel combustion. Linak, W.P., McSorley, J.A., Hall R.E., Ryan, J.V., Srivastava, R.K., Wendt, J.O.L., and Mereb, J.B. *J. geophys. Res.* **95**, 7533 (1990).

Section 4.5.9

Carbon monoxide in the Earth's atmosphere: indications of a global increase. Khalil, M. A. K. and Rasmussen, R. A. *Nature* **332**, 242 (1988).

Continuing worldwide increases in tropospheric methane, 1978–1987. Blake, D. R. and Rowland, F. S. *Science* **239**, 1129 (1988).

Increased carbon dioxide and stratospheric ozone. Groves, K. S., Mattingley, S. R., and Tuck, A. F. *Nature, Lond.* **273**, 711 (1978).

On the relationship between the greenhouse effect, atmospheric photochemistry and species distribution. Callis, L. B., Natarajan, M., and Boughner, R. E. *J. geophys. Res.* **88**, 1401 (1983).

Influence of stratospheric cooling from CO_2 on the ozone layer. Isaksen, I. S. A., Hesstvedt, E., and Stordal, F. *Nature, Lond.* **283**, 189 (1980).

Effect of water vapour on the destruction of ozone in the stratosphere perturbed by Cl_x or NO_x pollutants. Lin, S. C., Donahue, T. M., Cicerone, R. J., and Chameides, W. L. *J. geophys. Res.* **81**, 3111 (1976).

Section 4.5.10

Impact of coupled perturbations of atmospheric trace gases on Earth's climate and ozone. Nicoli, M. P. and Visconti, G. *Pure & appl. Geophys.* **120**, 626 (1982).

Effect of coupled anthropogenic perturbations on stratospheric ozone. Wuebbles, D. J., Luther, F. M., and Penner, J. E. *J. geophys. Res.* **88**, 1444 (1983).

Section 4.6

Global trends. Chapter 2, Volume 1 of *Scientific assessment of stratospheric ozone: 1989*. World Meteorological Organization, Global ozone research and monitoring project: report no. 20. (WMO, Geneva, 1990.)

Several papers presented at the Quadrennial Ozone Symposium, Göttingen, FRG, 1988 deal with long-term trends in stratospheric ozone levels, and address the problems of drift and calibrations in satellite instruments. The papers are published in

Proceedings of the Quadrennial Ozone Symposium: Göttingen, 1988. Bojkov, R., and Fabian, P., (eds.). (Deepak Publishing Co., Hampton, Virginia, 1989.)

The Ozone Trends Panel Report listed next describes the detailed findings of the Panel.

Ozone Trends Panel Report, 1988. World Meteorological Organization, Global ozone research and monitoring project: report no. 18. Watson, R. T. (ed.). (WMO, Geneva, 1989.)

Other papers deal with some of the factors leading to ozone variability and to the methods for extracting trends from the data.

A statistical trend analysis of revised Dobson data over the Northern Hemisphere. Bojkov, R., Bishop, L., Hill, W.J., Reinsel, G.C., and Tino, J.C. *J. geophys. Res.* **95**, 9785 (1990).

An analysis of the 7-year record of SBUV satellite ozone data: global profile features and trends in total ozone. Reinsel, G. C., Tiao, G. C., Ahn, S. K., Pugh, M., Basu, S., DeLuisi, J. L., Mateer, C. L., Miller, A. J., Connell, P. S., and Wuebbles, D. J. *J. geophys. Res.* **93**, 1689 (1988).

Interannual variability in Antarctic ozone and the quasi-biennial oscillation. Garcia, R. R., and Solomon, S. *Geophys. Res. Letts.* **14**, 848 (1987).

A two-dimensional model of the quasi-biennial oscillation of ozone. Gray, L. J., and Pyle, J. A. *J. Atmos. Sci.* **46**, 203 (1989).

Associations between the 11-year solar cycle, the QBO and the atmosphere. Part I: the troposphere and stratosphere in the northern hemisphere in winter. Labitzke, K., and van Loon, H. *J. Atmos. Terr. Phys.* **50**, 197 (1988).

Ozone variability. Dütsch, H. U. *Planet. Space Sci.* **31**, 1053 (1983).

The distribution and annual cycle of ozone in the upper stratosphere. Frederick, J. E., Huang, F. T., Douglass, A. R., and Raber, C. A. *J. geophys. Res.* **88**, 3819 (1983).

The quasi-biennial oscillation in atmospheric ozone. Oltmans, S. J. and London, J. *J. geophys. Res.*, **87**, 8981 (1982).

Semiannual oscillation of stratospheric ozone. Maeda, K. *Geophys. Res. Lett.* **11**, 583 (1984).

Global ozone long-term trends from satellite measurements and the response to solar activity variations. Keating, G. M., Lake, L. R., Nicholson III, J. Y., and Natarajan, M. *J. geophys. Res.* **86**, 9873 (1981).

Section 4.7

The hole in the sky. Gribbin, J. (Corgi Books, London, 1988)

Punching a hole in the stratosphere. Wayne, R. P. *Proc. Royal Inst.* **61**, 13 (1990).

Polar ozone. Chapter 1, Volume 1 of *Scientific assessment of stratospheric ozone: 1989.* World Meteorological Organization, Global ozone research and monitoring project: report no. 20. (WMO, Geneva, 1990.)

The Antarctic ozone hole, Stolarski, R. S. *Sci. Amer.* **258**, 20 (Jan 1988).

The Antarctic ozone hole. Gardiner, B. G. *Weather*, **44**, 291 (1989).

Ozone depletion at the poles: the hole story emerges. *Physics Today*, p17 (July 1988).

The mystery of the Antarctic ozone 'hole'. Solomon, S. *Rev. Geophys.* **26**, 131 (1988).

Section 4.7.1

The 1987 ozone hole: a new record low. Krueger, A. J., Schoeberl, M. R., Stolarski, R. S., and Sechrist, F. S. *Geophys. Res. Letts.* **15**, 1365 (1988).

The 1988 Antarctic ozone depletion. Comparison with previous year depletions. Schoeberl, M. R., Stolarski, R. S., and Krueger, A. J. *Geophys. Res. Letts.* **16**, 377 (1989).

Ozone and temperature profiles over McMurdo station Antarctica in the spring of 1989. Deshler, T., Hofmann, D. J., Hereford, D. J., and Sutter, C. B. *Geophys. Res. Lett.* **17**, 151 (1990).

Section 4.7.2

The most extensive data on chemical composition of the Antarctic atmosphere came from the campaign of the Airborne Antarctic Ozone Experiment (AAOE) in August–September 1987. Some 52 papers reporting the results are published in two special issues of

Journal of geophysical Research, Volume 94 (Number D9, 30 Aug 1989; Number D14, 30 Nov 1989).

Polar stratospheric clouds inferred from satellite data, Austin, J., Remsberg, E. E., Jones, R. L., and Tuck, A. F. *Geophys. Res. Letts.* **13**, 1256 (1986).

Characteristics of polar stratospheric clouds during the formation of the Antarctic ozone hole. Hamill, P., Toon, O. B., and Turco, R. P. *Geophys. Res. Letts.* **13**, 1288 (1986).

On the growth of nitric and sulfuric acid aerosol particles under stratospheric conditions. Hamill, P., Toon, O. B., and Turco, R. P. *J. Atmos. Chem.* **7**, 287 (1988).

Section 4.7.4

Progress towards a quantitative understanding of Antarctic ozone depletion. Solomon, S. *Nature* **347**, 347 (1990).

Chemistry of the Antarctic stratosphere. McElroy, M. B., Salawitch, R. J., and Wofsy, S. C. *Planet. Space Sci.* **36**, 73 (1988).

Heterogeneous physicochemistry of the polar ozone hole. Turco, R. P., Toon, O. B., and Hamill, P. *J. geophys. Res.* **94**, 16493 (1989).

Antarctic stratospheric chemistry of chlorine nitrate, hydrogen chloride, and ice: release of active chlorine. Molina, M. J., Tso, T. L., Molina, L. T., and Wang, F. C. Y. *Science* **238**, 1253 (1987).

The role of chlorine chemistry in Antarctic ozone loss: implications of new kinetic data. Rodriguez, J. M., Ko, M. K. W., and Sze, N. D. *Geophys. Res. Lett.* **17**, 255 (1990).

Stratospheric nitric acid vapour measurements in the cold arctic vortex—implications for nitric acid condensation, Arnold, F. and Knop, G. *Nature* **338**, 746 (1989).

Nitric acid cloud formation in the cold Antarctic stratosphere: a major cause for the springtime 'ozone hole'. Crutzen, P. J. and Arnold, F. *Nature* **324**, 651 (1986).

Heterogeneous reactions on nitric acid trihydrate. Moore, S. B., Keyser, L. F., Leu, M.-T., Turco, R. P., and Smith R. H. *Nature* **345**, 333 (1990).

Heterogeneous reactions of N_2O_5 with H_2O and HCl on ice surfaces: implications for Antarctic ozone depletion Hanson, D. and Mauersberger, K. *Geophys. Res. Letts.* **15**, 855 (1988).

Heterogeneous chemical reaction of chlorine nitrate and water on sulfuric acid surfaces at room temperature. Rossi, M. R., Malhotra, R., and Golden, D. M. *Geophys. Res. Letts.* **14**, 127 (1987).

Heterogeneous interactions of $ClONO_2$, HCl, and HNO_3 with sulfuric acid surface at stratospheric temperatures. Tolbert, M. A., Rossi, M. J., and Golden, D. M. *Geophys. Res. Letts.* **15**, 851 (1988).

Antarctic ozone depletion chemistry: reactions of N_2O_5 with H_2O and HCl on ice surfaces. Tolbert, M. A., Rossi, M. J., and Golden, D. M. *Science* **240**, 1018 (1988).

The stability and photochemistry of dimers of the ClO radical and implications for Antarctic ozone depletion. Cox, R. A. and Hayman, G. D. *Nature* **322**, 796 (1988).

Role of the ClO dimer in polar stratospheric chemistry: rate of formation and implications for ozone loss. Sander, S. P., Friedl, R. J., and Yung, Y. K. *Science* **245**, 1095 (1989).

Direct ozone depletion in springtime Antarctic lower stratospheric clouds. Hofmann, D. J. *Nature* **337**, 447 (1989).

Section 4.7.5

Effects of initial active chlorine concentration on the Antarctic ozone spring depletion. Henderson, G. S., Evans, W. F. J., and McConnell, J. C. *J. geophys. Res.* **95**, 1899 (1990).

Section 4.7.6

Results from the 1989 Airborne Arctic Stratospheric Expedition are summarized in a special issue (Supplement to issue of March 1990)

Geophys. Res. Lett. **17**, 313–564 (April 1990).

Observations of denitrification and dehydration in the winter polar stratospheres. Fahey, D. W., Kelley, K. K., Kawa, S. R., Tuck, A. F., Loewenstein, M., Chan, K. R., and Heidt, L. E. *Nature* **344**, 321 (1990).

Stratospheric clouds and ozone depletion in the Arctic during January 1989. Hofmann, D. J., Aimedieu, P., Matthews, W. A., Johnston, P. V., Kondo, Y., Sheldon, W. R., and Byrne, G. J. *Nature* **340**, 117 (1989).

Observations of stratospheric NO_2 and O_3 at Thule, Greenland. Mount, G. H., Solomon, S., Sanders, R. W., Jakoubek, R. O., and Schmeltekopf, A. L. *Science* **242**, 555 (1988).

Section 4.7.7

Ultraviolet levels under sea ice during the Antarctic spring. Trodahl, H. J. and Buckley, R. G. *Science* **245**, 194 (1989).

A general circulation model simulation of the springtime Antarctic ozone decrease and its impact on mid latitudes. Cariolle, D., Laserre-Bigorry, A., Royes, J. -F., and Geleyn, J. -F. *J. geophys. Res.* **95**, 1883 (1990).

Global impact of the Antarctic ozone hole: dynamical dilution with a three-dimensional chemical tansport model. Prather, M., Garcia, M. M., Suozzo, R., and Rind, D. *J. geophys. Res.* **95**, 3449 (1990).

Global impact of the Antarctic ozone hole: chemical propagation. Prather, M. and Jaffe, A. H. *J. geophys. Res.* **95**, 3473 (1990)

Antarctic ozone hole: possible implications for ozone trends in the southern hemisphere. Sze, N. D., Ko, M. K. W., Weisenstein, D. K., Rodriguez, J. M., Stolarski, R. S., and Schoeberl, M. R. *J. geophys. Res.* **94**, 11521 (1989).

Observations of the decay of the El Chichón stratospheric aerosol cloud in Antarctica. Hofmann, D. J., Rosen, J. M., Harder, J. W., and Rolf, S. R. *Geophys. Res. Letts.* **14**, 614 (1987).

Ozone destruction through heterogeneous chemistry following the eruption of El Chichón. Hofmann, D. J., and Solomon, S. *J. geophys. Res.* **94**, 5029 (1989).

El Chichón volcanic aerosols: impact of radiative, thermal, and chemical perturbations. Michelangeli, D. V., Allen, M., and Yung, Y. L. *J. geophys. Res.* **94**, 18429 (1989).

5 The Earth's troposphere

5.1 Introduction

Hydroxyl radicals dominate the daytime chemistry of the troposphere in the same way that oxygen atoms and ozone dominate the chemistry of the stratosphere. High reactivity of the OH radical with respect to a wide range of species leads to oxidation and chemical conversion of most trace constituents that have an appreciable physical lifetime in the troposphere. Free radical chain reactions oxidize hydrogen, methane and other hydrocarbons, and carbon monoxide to CO_2 and H_2O. The reactions thus constitute a low-temperature combustion system. Other species, notably NO_x and sulphur compounds, participate in the reactions to modify the course of the combustion processes[a]. Species which survive both physical loss and chemical conversion in the troposphere (e.g. N_2O, some CH_4 and CH_3Cl) are transported to the stratosphere where they yield the NO_x, HO_x, and ClO_x radicals that destroy ozone in the catalytic cycles discussed in Chapter 4. Hydroxyl radical chemistry in the troposphere provides an efficient chemical scavenging mechanism for both natural and man-made trace constituents, and has a major influence not only on tropospheric composition, but also on stratospheric behaviour.

At night, the nitrate radical, NO_3, takes over from OH as the dominant oxidant in the troposphere. Although NO_3 is generally much less reactive than OH, its peak tropospheric concentration is higher, so that it plays an important role in atmospheric chemical transformations. The diurnal impact of OH and NO_3 is complementary, because OH is generated photochemically only during the day, while NO_3 is readily photolysed, and so can survive only at night.

About 90 per cent of the total atmospheric mass resides in the troposphere, and the bulk of the minor trace gas burden is also found there. The Earth's surface acts as the main source of the trace gases, although some NO_x and CO may be produced in thunderstorms. Surface emissions include natural and 'pollution' sources, with the latter being concentrated in urban and industrial areas of the Northern Hemisphere. Natural sources are more evenly distributed, but since the land area is twice as great in the Northern as in the Southern Hemisphere, some asymmetry remains.

Radical-chain processes in the troposphere are photochemically driven,

[a] See footnote [a] on page 126 for a definition of the representations NO_x and NO_y.

although stratospheric ozone limits the solar radiation at the Earth's surface to wavelengths longer than 280 nm (Chapter 4). In 1961, P.A. Leighton wrote his classic work on atmospheric chemistry. Its title, *The Photochemistry of Air Pollution* (see Bibliography) hardly does justice to its coverage, which goes far beyond pollution matters. Leighton noted the need to characterize carefully the flux of solar photons as a function of wavelength as these photons pass through the atmosphere and are absorbed and scattered.

The most important species that are photochemically labile at $\lambda > 280$ nm are O_3, NO_2, and HCHO (formaldehyde). As we shall see in Section 5.3, all three of these species can indirectly lead to OH (or HO_2) formation, and thus initiate the oxidation chains. Ozone photolysis is a critical step, since the other photolytic processes owe either their origin or their importance to it. Although only 10 per cent of the total atmospheric ozone is found in the troposphere, all *primary* initiation of oxidation chains in the natural atmosphere depends on that ozone. The origin of tropospheric ozone is therefore of great interest. Mechanisms exist in tropospheric chemistry that potentially can generate O_3 at low altitudes, but their efficiencies are related to absolute and relative concentrations of hydrocarbons and, especially, of NO_x. Ozone is also transported, *via* stratospheric–tropospheric exchange, from the stratospheric ozone layer. The relative importance of stratospheric and locally-produced ozone continues to be a controversial subject.

5.2 Sources, sinks, and transport

Some of the natural and artificial sources of various trace species have already been identified in Chapters 1 and 4, and we shall consider further certain specific pollutants in Section 5.5. Our interest here is to follow the general path of trace components from their sources to their sinks. Physical removal processes are divided into *dry deposition* in which the species are absorbed irreversibly on soil, water, or plant surfaces, and *wet deposition* in which the constituents are incorporated into precipitation elements (clouds, rain droplets, and aerosols). In between release to and removal from the atmosphere, the species may be transported horizontally and vertically. The source–transport–sink sequence defines a physical lifetime for the constituent. However, chemical changes may occur on a time-scale comparable with, or smaller than, this lifetime. Degradation of one trace gas frequently generates another which survives long enough to have a separate existence. For example, released methane is not rapidly removed by physical processes, and is oxidized slowly (with a lifetime of several years). Formaldehyde is an intermediate in the oxidation, but is rapidly photolysed (lifetime of a few hours). The products are hydrogen and carbon monoxide, which have lifetimes of several years and several months respectively. Dry deposition acts efficiently only where a specific chemical or biological interaction is available, and even then only when the trace gases are close to the surface. Wet deposition by incorporation into falling precipitation ('washout') or cloud droplets ('rainout') is only significant for those species that are water soluble. Chemical conversions may play an essential part in the overall scavenging process by generating some new compound that *can* be removed physically. A tendency exists for natural trace gases of biogenic origin to be reduced (CH_4, terpenes, H_2S, etc.) or only partially oxidized (CO, N_2O). These species are also of only modest solubility in water. By way of contrast, man-made species, which frequently involve combustion sources, are often more highly oxidized (CO_2, NO_2, SO_2) and somewhat more soluble. However, chemical transformations may be precursors of deposition even for the latter species. For example, washout removal times for sulphur dioxide in moderate rainfall are estimated as several hours. Most of the sulphate found in precipitation, however, is a result of rainout of hygroscopic cloud condensation nuclei (Sections 1.3 and 2.5) involving SO_3 or H_2SO_4: that is, the SO_2 has already been oxidized before nucleation occurs. Table 5.1 is a summary[a] of the most important natural and man-made sources of tropospheric trace gases. Of the species listed, only a handful undergo direct deposition. Dry deposition mechanisms exist for SO_2, O_3, CO_2, and SO_3, while microbiological sinks are known for soil removal of CO and H_2. Acids such as HCl, HF, and HNO_3 are readily soluble, as are NH_3, SO_2, and NO_x after conversion to aerosol species: wet deposition is possible. Nitrous oxide, carbonyl sulphide (COS), and the chlorofluorocarbons are sufficiently inert, physically and chemically, to survive exchange with the stratosphere. All other species undergo tropospheric photo-oxidation with OH, or direct photolysis (e.g. O_3, NO_2, HCHO).

The lower limit of the troposphere is often envisaged not as the Earth's surface, but rather as a boundary layer of the atmosphere. Some hint of this idea has appeared in earlier chapters, but for simplicity was left unexplained. The boundary layer is the region of highly turbulent mixing next to the ground; it is generally confined to the first 0.5–2 km by day, and less at night. By repeatedly bringing air parcels in contact with the surface, deposition of any local source gas, or its oxidation products, is greatly enhanced. Over a time-scale of days, any remaining constituents are incorporated into the free troposphere, and transported by the atmospheric circulation. Once in the free troposphere, the trace gases move over much larger distances than in the boundary layer. One consequence of the existence of a boundary layer is that local surface releases (natural or artificial) of highly reactive species tend to be geographically confined to the source region.

In the free troposphere, and at mid-latitudes, wind velocities are typically 10–30 m s^{-1} in the E–W direction, but much less in both N–S and vertical directions. An air parcel is therefore moved around a circle of latitude in a

---

[a] Cox, R. A. and Derwent, R. G. *Gas-phase chemistry of the minor constituents of the troposphere. Gas kinetics and energy transfer*, Specialist Periodical Reports Chem. Soc. **4**, 189 (1981).

Table 5.1 Natural and man-made sources of the minor trace gases of the troposphere

Compound	Natural sources	Man-made sources
Carbon-containing trace gases		
Carbon monoxide (CO)	Oxidation of natural methane, natural C_5, C_{10} hydrocarbons; oceans, forest fires	Oxidation of man-made hydrocarbons; incomplete combustion of wood, oil, gas, and coal, in particular motor vehicles, industrial processes; blast furnaces
Carbon dioxide (CO_2)	Oxidation of natural CO; destruction of forests; respiration by plants	Combustion of oil, gas, coal, and wood; limestone burning
Methane (CH_4)	Enteric fermentation in wild animals; emissions from swamps, bogs, *etc.*, natural wet land areas; oceans	Enteric fermentation in domesticated ruminants; emissions from paddies; natural gas leakage; sewerage gas; colliery gas; combustion sources
Light paraffins, C_2—C_6	Aerobic biological source	Natural gas leakage; motor vehicle evaporative emissions; refinery emissions
Olefins, C_2—C_6		Motor vehicle exhaust; diesel engine exhaust
Aromatic hydrocarbons		Motor vehicle exhaust; evaporative emissions; paints, petrol, solvents
Hemiterpenes, C_5H_8 Terpenes, $C_{10}H_{16}$ Diterpenes, $C_{20}H_{32}$	Trees, broadleaves, and conifers; plants	
Nitrogen-containing trace gases		
Nitric oxide (NO)	Forest fires; anaerobic processes in soil; electric storms	Combustion of oil, gas, and coal
Nitrogen dioxide (NO_2)	Forest fires; electric storms	Combustion of oil, gas and coal; atmospheric transformation of NO
Nitrous oxide (N_2O)	Emissions from denitrifying bacteria in soil; oceans	Combustion of oil and coal

Table 5.1 (*continued*)

Compound	Natural sources	Man-made sources
Peroxyacetyl nitrate (PAN)	Degradation of isoprene	Degradation of hydrocarbons
Ammonia (NH_3)	Aerobic biological source in soil	Coal and fuel oil combustion; waste treatment
	Breakdown of amino acids in organic waste material	
Sulphur-containing trace gases		
Sulphur dioxide (SO_2)	Oxidation of H_2S; volcanic activity	Combustion of oil and coal; roasting sulphide ores
Hydrogen sulphide (H_2S)	Anaerobic fermentation; volcanoes and fumaroles	Oil refining; animal manure; Kraft paper mills; rayon production; coke oven gas
Carbon disulphide (CS_2)	Anaerobic fermentation	Viscose rayon plants; brick making; fish meal processing
Carbonyl sulphide (COS)	Oxidation of CS_2; slash and burn agriculture; volcanoes and fumaroles	Oxidation of CS_2; brick making; effluent from Kraft mills; blast furnace gas; coke oven gas; shale and natural gas
Sulphur trioxide (SO_3)		Combustion of S-fuel
Methyl mercaptan (CH_3SH)	Anaerobic biological sources	Animal rendering; animal manure; pulp and paper mills; brick manufacture; oil refining
Dimethyl sulphide (CH_3SCH_3)	Aerobic biological sources	Animal rendering; animal manure; pulp and paper mills
Dimethyl disulphide (CH_3SSCH_3)	Anaerobic biological sources	Animal rendering; fishmeal processing
Other organic sulphur compounds: C_2—C_4 mercaptans, dialkyl disulphides, dimethyl trisulphide, alkyl thiophenes, benzothiophenes		Animal rendering; fishmeal processing; brick making

Chlorine-containing trace gases

Hydrogen chloride (HCl)	Volcanoes and fumaroles; degradation of CH_3Cl	Coal combustion; degradation of chlorocarbons
Methyl chloride (CH_3Cl)	Slow combustion of organic matter; marine environment; algae	PVC and tobacco combustion
Methylene dichloride (CH_2Cl_2)		Solvent
Chloroform ($CHCl_3$)		Pharmaceuticals; solvent; combustion of petrol; bleaching of wood pulp; degradation of C_2HCl_3
Carbon tetrachloride (CCl_4)		Solvent; fire extinguishers; degradation of C_2Cl_4
Methyl chloroform (CH_3CCl_3)		Solvent; degreasing agent
Trichloroethylene (C_2HCl_3)		Solvent; dry cleaning agent; degreasing agent
Tetrachloroethylene (C_2Cl_4)		Solvent; dry cleaning agent; degreasing agent
Other chlorofluorocarbons: CCl_3F, CCl_2F_2, $C_2Cl_3F_3$, $C_2Cl_2F_4$, C_2ClF_5		Aerosol propellants; refrigerants; foam blowing agents; solvents

Other minor trace gases

Hydrogen	Oceans: soils; oxidation of methane, isoprene and terpenes via formaldehyde	Motor vehicle exhaust; oxidation of methane via formaldehyde
HF	Volcanoes and fumaroles	
Ozone	Stratosphere: natural $NO-NO_2$ conversion	Man-made $NO-NO_2$ conversion
H_2O	Evaporation from oceans	Insignificant
SF_6		Electrical insulator
CF_4		Aluminium industry
CH_3Br	Aerobic biological source	Fumigation of soil and grain
CH_3I	Aerobic biological source	Insignificant

Reproduced from Cox, R. A. and Derwent, R. G. *Gas kinetics and energy transfer, Specialist Periodical Reports Chem. Soc.* **4**, 189 (1981).

few tens of days, while motions in the N–S direction are much slower. Transfer between hemispheres across the equator is inhibited (cf. p. 66) and occurs only at certain seasons of the year, N → S in the upper troposphere and S → N in the lower troposphere; the time-scale is about one year. Because of the E–W averaging, two-dimensional models (altitude, latitude) are generally adequate for describing the tropospheric distribution of trace gases with lifetimes of more than about one month.

5.3 Oxidation and transformation

5.3.1 Photochemical chain initiation

In the real troposphere, several species are present that are capable of absorbing solar radiation and hence initiating radical-chain oxidation. More insight may be gained if we take our starting point as the artificial situation where no CH_4 has yet been oxidized. Ozone is then the photochemical precursor of hydroxyl radicals. As we discussed at some length in Sections 3.2.1 and 3.2.2, ozone is photolysed at wavelengths less than ~ 310 nm to yield an excited, ^1D, oxygen atom that is energetically capable of reacting with water vapour to yield OH:

$$O_3 + h\nu \rightarrow O^*(^1D) + O_2^*(^1\Delta_g) \tag{5.1}$$

$$O^*(^1D) + H_2O \rightarrow OH + OH. \tag{5.2}$$

Since H_2O is itself a minor component of the atmosphere, reaction (5.2) is a minor fate of $O^*(^1D)$ atoms compared with quenching

$$O^*(^1D) + M \rightarrow O + M \tag{5.3}$$

$$M = N_2, O_2.$$

However, virtually all ground state O atoms will regenerate ozone

$$O + O_2 + M \rightarrow O_3 + M, \tag{5.4}$$

so that quenching does not constitute a loss of odd oxygen. Although the recycling of atomic oxygen maintains the mass balance of odd oxygen, the rate of OH production and of chain initiation *does* depend on the relative rates of reactions (5.2) and (5.3), and, in a similar way, on the quantum yield for $O(^1D)$ production in reaction (5.1) as a function of wavelength.

Ozone of stratospheric origin (see Section 5.1) can well be that needed for $O(^1D)$ production, but if NO_2 is also present in the troposphere, then NO_2 photolysis (at $\lambda \leqslant 400$ nm),

$$NO_2 + h\nu \rightarrow O + NO, \tag{5.5}$$

followed by the combination reaction (5.4) can provide a tropospheric source. As we shall see shortly, NO itself can be oxidized back to NO_2, so that the formation of O_3 is not stoicheiometrically limited by the number of NO_2 molecules initially present.

5.3.2 Oxidation steps

Hydroxyl radicals formed in reaction (5.2) react mainly with CO and with CH_4:

$$OH + CO \rightarrow H + CO_2, \tag{5.6}$$

$$OH + CH_4 \rightarrow CH_3 + H_2O. \tag{5.7}$$

Roughly 70 per cent of the OH reacts with CO, and 30 per cent with CH_4, in the unpolluted atmosphere. The reaction of OH with CO is kinetically and mechanistically unusual, and we shall return to it in Section 5.3.5. For the time being, we note that in both processes an active species is formed that is capable of adding molecular oxygen to produce a peroxy radical

$$H + O_2 + M \rightarrow HO_2 + M, \tag{5.8}$$

$$CH_3 + O_2 + M \rightarrow CH_3O_2 + M. \tag{5.9}$$

In tropospheric regions where NO concentrations are very low, the peroxy radicals HO_2 and CH_3O_2 are consumed mainly in the reactions

$$HO_2 + HO_2 \rightarrow H_2O_2 + O_2 \tag{5.10}$$

$$CH_3O_2 + HO_2 \rightarrow CH_3OOH + O_2. \tag{5.11}$$

(Self-reaction of two CH_3O_2 radicals is rather slow at ambient temperatures). One fate of the hydrogen peroxide (H_2O_2) and methyl hydroperoxide (CH_3OOH) is that they can dissolve in cloud droplets and be removed from the troposphere in the form of rain. In this respect, therefore, reactions (5.10) and (5.11) are loss or terminating steps, although alternative fates of the peroxides, such as photolysis or reaction with OH, regenerate radicals.

If oxides of nitrogen are present in the atmosphere, then a quite different course of events can follow the formation of the peroxy radicals in reactions (5.8) and (5.9). We know already from our discussion of stratospheric chemistry (Section 4.4) that HO_2 reacts rapidly with NO, and the analogous reaction occurs with methylperoxy radicals,

$$HO_2 + NO \rightarrow OH + NO_2 \tag{5.12}$$

$$CH_3O_2 + NO \rightarrow CH_3O + NO_2. \tag{5.13}$$

Reaction (5.10) regenerates OH, while (5.11) produces a methoxy radical that can in turn react with O_2,

$$CH_3O + O_2 \rightarrow HCHO + HO_2, \tag{5.14}$$

to yield formaldehyde. Formaldehyde itself is photochemically labile; the

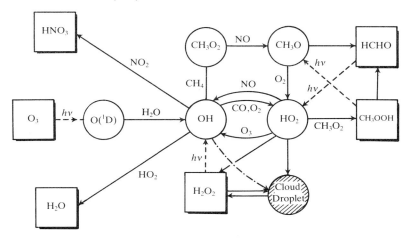

Fig. 5.1. Chemistry of the troposphere. This schematic diagram emphasizes the processes that create and destroy the OH and HO_2 radicals that are key intermediates in the oxidation steps.

major photolytic pathway at $\lambda < 338$ nm produces two radical fragments

$$HCHO + h\nu \rightarrow H + HCO, \tag{5.15}$$

and both radicals re-enter the HO_x chain, *via* reaction (5.8) for H, or reaction (5.16)

$$HCO + O_2 \rightarrow CO + HO_2, \tag{5.16}$$

for HCO.

The essential feature is the conversion of NO to NO_2 while preserving active radicals capable of oxidizing further molecules of CO and CH_4 or converting more NO to NO_2. This formation of NO_2 is essential to the formation of ozone in the troposphere, as we shall discuss in Section 5.3.3.

Figure 5.1 presents in schematic form the chemistry discussed so far, with the emphasis laid on processes creating and destroying the radical intermediates OH and HO_2.

The oxidation steps that we have written for methane obviously have their analogues for higher hydrocarbons, but in all cases depend on the switch between RO_2 (peroxy-) radicals and RO (oxy-) radicals in the interaction with NO, represented for the general case as

$$RO_2 + NO \rightarrow RO + NO_2. \tag{5.17}$$

We turn now to a rather more detailed consideration of the oxidation of simple alkanes and alkenes. To keep the discussion reasonably compact, we must assume that the initial attack is always by hydroxyl radicals, although it must be understood that attack by NO_3 is important at night (Section 5.3.6), and that HO_2, O, and O_3 can play a minor role.

Reaction with OH is, in fact, almost exclusively the loss process for alkanes in the troposphere with H-atom abstraction from the C–H bond yielding an alkyl, R, radical. The radicals react initially with O_2,

$$R + O_2 \rightarrow RO_2, \tag{5.18}$$

a third body probably being needed in the case of $R = CH_3$ [cf. reaction (5.9)]. For ethane and propane, the subsequent steps are probably

$$RO_2 + NO \rightarrow RO + NO_2 \tag{5.17}$$

$$RO \rightarrow R' + R''CHO \tag{5.19}$$

$$RO + O_2 \rightarrow R'R''CO + HO_2, \tag{5.20}$$

where R' and R'' are daughter alkyl radicals or groups. The aldehydes and ketones (R''CHO, R'R''CO) can be photolysed, as we have seen above for formaldehyde in reaction (5.15), or become further oxidized in thermal reactions. For example, OH abstracts H from acetaldehyde to form CH_3CO, which itself adds oxygen to yield the acetylperoxy radical, $CH_3CO.O_2$:

$$CH_3CHO + OH \rightarrow CH_3CO + H_2O, \tag{5.21}$$

$$CH_3CO + O_2 \rightarrow CH_3CO.O_2. \tag{5.22}$$

As we shall see in Section 5.3.4, one important reaction of $CH_3CO.O_2$ is addition of NO_2, but the usual [cf. reactions (5.12) and (5.17)] oxidation of NO is possible, and in that case the resulting $CH_3CO.O$ radical can fragment:

$$CH_3CO.O_2 + NO \rightarrow CH_3CO.O + NO_2, \tag{5.23}$$

$$CH_3CO.O \rightarrow CH_3 + CO_2. \tag{5.24}$$

In the reactions described, then higher alkyl radicals are degraded to lower ones, and ultimately, *via* the reactions of CH_3, to CO and CO_2.

Ethene and propene, C_2H_4 and C_3H_6, appear to react with OH predominantly by an addition mechanism:

$$OH + C_2H_4 \rightarrow HOC_2H_4, \tag{5.25}$$

$$OH + CH_3CHCH_2 \rightarrow CH_3CHCH_2OH. \tag{5.26}$$

For ethene, at least, the reaction appears to require stabilizing collisions with a third body, and shows intermediate-order kinetics (see Section 3.4.2). Both reactions show a negative temperature coefficient of rate, characteristic of an addition process, and a collisionally stabilized adduct has been explicitly observed in the case of reaction (5.26). The formation of the OH-ethene adduct is ~ 134 kJ mol^{-1} exothermic, and elimination of an H atom is endothermic by ~ 30 kJ mol^{-1}, so that the only favourable decomposition path for the adduct is back to the reactants. For propene, the excess energy in the newly-formed adduct is similar to that in the ethene case; here elimination of CH_3

is thermochemically possible, although it seems not to be important for the CH_3CHCH_2OH isomer.

The alkene-OH adducts appear to undergo reactions analogous to those of alkyl radicals. Thus, for terminal addition of OH to propene, the major atmospheric oxidation steps are

$$CH_3CHCH_2OH + O_2 \rightarrow CH_3CH(O_2)CH_2OH, \tag{5.27}$$

$$CH_3CH(O_2)CH_2OH + NO \rightarrow CH_3CH(O)CH_2OH + NO_2, \tag{5.28}$$

$$CH_3CH(O)CH_2OH \rightarrow CH_3CHO + CH_2OH, \tag{5.29}$$

which should be compared with reactions (5.18), (5.17) and (5.19) for the R, RO_2, and RO radicals. In the atmosphere, the CH_2OH reacts exclusively with O_2 to yield $HCHO + HO_2$:

$$CH_2OH + O_2 \rightarrow HCHO + HO_2, \tag{5.30}$$

so that the products from attack of OH on propene are the aldehydes HCHO and CH_3CHO, which are ultimately oxidized or photolysed as described earlier. The HO_2 radical is also a product, and it is interesting to add reaction (5.12)

$$HO_2 + NO \rightarrow OH + NO_2 \tag{5.12}$$

to the sequence of reactions (5.26) to (5.30) to emphasize that the oxidation scheme is a chain process. Chain carriers are not consumed overall, even though C_3H_6 has been oxidized to CH_3CHO and HCHO. Of course, two molecules of NO are converted to NO_2 for each OH radical cycle, and that is a matter of importance for ozone production as we shall now see in the next section.

5.3.3 Tropospheric ozone production

Photolysis of NO_2 is the only known way of producing ozone in the troposphere. Reactions (5.12), (5.13) or (5.17) achieve not only the RO_2 to RO radical conversions, but also the oxidation of NO to NO_2. They thus provide the link required for catalytic generation of O_3 *via* reactions (5.5) and (5.4). One cyclic process for tropospheric ozone production can now be written:

$$NO_2 + hv \rightarrow O + NO \tag{5.5}$$

$$O + O_2 + M \rightarrow O_3 + M \tag{5.4}$$

$$OH + CO \rightarrow H + CO_2 \tag{5.6}$$

$$H + O_2 + M \rightarrow HO_2 + M \tag{5.8}$$

$$HO_2 + NO \rightarrow OH + NO_2 \tag{5.12}$$

Net $\qquad CO + 2O_2 + hv \rightarrow CO_2 + O_3.$

Very similar chain reactions can be written that involve the RO_2 (e.g. CH_3O_2) species. In all cases, NO_x must be present.

Below a certain critical value of the ratio $[NO]/[O_3]$, ozone loss through the sequence

$$HO_2 + O_3 \rightarrow OH + 2O_2 \tag{5.31}$$

$$CO + OH \rightarrow H + CO_2 \tag{5.6}$$

$$H + O_2 + M \rightarrow HO_2 + M \tag{5.8}$$

Net $\qquad CO + O_3 \rightarrow CO_2 + O_2$

dominates over the generation sequence. Current models suggest that production (via HO_2 and RO_2 oxidation of NO) balances or exceeds loss of O_3 for atmospheric mixing ratios of $NO \geqslant 3 \times 10^{-11}$. Regions of the Earth characterized by extremely low concentrations of NO, such as the remote Pacific, are thus likely to provide a net photochemical sink for odd oxygen, while the continental boundary layer at mid-latitudes, where concentrations of NO are relatively high, is likely to provide a net source. So long as NO_x is available, production of ozone in the troposphere is ultimately limited by the supply of CO, CH_4, and other hydrocarbons. Each molecule of CO can generate one molecule of O_3, while it is estimated that as many as 3.5 O_3 molecules could be formed from the oxidation of each CH_4.

Tropospheric loss processes for O_3 include photolysis in the visible and ultraviolet regions, and reaction with species such as NO_2, HO_2, and unsaturated hydrocarbons. Model calculations suggest that the globally averaged tropospheric sources and sinks for ozone are roughly in balance. This view of the *in situ* balance is consistent with estimates that the surface sink is equal to the stratospheric injection rate.

The tropospheric abundance of ozone is obviously critically important in determining the oxidizing capacity of the troposphere, because O_3 is the primary source of OH radicals (as well as being an oxidizing species in its own right). Background concentrations of NO_x and of the peroxy radicals derived from CO, CH_4, and other hydrocarbons determine the rate of formation of ozone. Future trends of tropospheric oxidizing capacity will thus be tied closely to future atmospheric burdens of all these species.

5.3.4 The importance of NO_x

The discussion of the previous two sections has shown how important NO_x, and especially NO, is to the overall oxidation rates and to the ozone distribution in the troposphere. The influence of NO_x is emphasized in Fig. 5.2, which is a restatement of the essential steps of Fig. 5.1. in another form. The left-hand semicircle shows the oxidation steps that do not require the presence of NO. In this case, the sequence of transformations following the heavy arrows leads

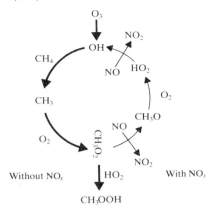

Fig. 5.2. Essential steps in tropospheric methane oxidation. The heavier arrows on the left-hand side of the diagram indicate the steps that can occur in the absence of NO_x. With NO_x present, the processes on the right-hand side can close a loop, with regeneration of OH and oxidation of NO to NO_2 (from a diagram devised by M. E. Jenkin.)

from O_3 to CH_3OOH. If, however, NO is present, then the transformations of the right-hand semicircle close the loop, with regeneration of OH and the concomitant production of two molecules of NO_2 (and thus, potentially, of O_3).

Model estimates of tropospheric ozone production obviously depend critically on the altitude–concentration profiles adopted for NO_x, so that we must consider the atmospheric budgets for these species. Natural sources of NO_x include microbial actions in the soil, which produce NO as well as N_2O (cf. Sections 1.5.3 and 4.5.8). Controlled laboratory experiments suggest that NO might even be a slightly more abundant initial product than N_2O, but NO has a far shorter chemical lifetime. Oxidation of biogenic NH_3, initiated by OH radicals, would be another significant natural source of NO_x. Lightning appears to be responsible for less than 10 per cent of the total NO_x budget, while biomass burning and high-temperature combustion (i.e. 'anthropogenic' sources) are perhaps responsible for 50 per cent globally. The ratio of $[NO]/[NO_2]$ depends on the rate of photolysis (reaction 5.5) and the rates of NO oxidation in processes (5.12), (5.13) and (5.17), together with the reaction

$$NO + O_3 \rightarrow NO_2 + O_2. \tag{5.32}$$

This latter reaction has a significant activation energy (~ 13 kJ mol^{-1}) so that $[NO]/[NO_2]$ is larger at higher tropospheric altitudes where the temperature is lower.

It is important to note that all the sources mentioned, except lightning, release NO_x to the boundary layer. Since the major loss for NO_x involves nitric acid formation,

$$OH + NO_2 + M \rightarrow HNO_3 + M, \tag{5.33}$$

followed by wet deposition, the NO_x release to the free troposphere may be appreciably smaller than the surface source strengths suggest. Some workers have argued that injection of NO_x from the stratosphere could provide a more important source for the upper troposphere than the surface emissions. Mixing ratios of NO, in the *lower* troposphere, over the Pacific Ocean, where there is no surface source, are as low as 4×10^{-12}, which would suggest that the stratospherically-derived NO_x is alone important and that the upper troposphere is the dominant region for photochemical production of ozone. However, several other observations in the remote troposphere indicate mixing ratios in the range 10^{-11} to 2×10^{-10}, and ozone production in the lower (<5 km) troposphere would then be important. The question remains in dispute at the time of writing.

Measurements in regions remote from man-made sources have generally been thought to be representative of NO_x in the 'natural' atmosphere because the chemical lifetime of NO_x is small compared with the times taken for geographical redistribution. The situation is rather more involved than appears at first sight, because a reservoir species for NO_x is now known that can extend the lifetime of NO_x and provide a source when others are absent. This reservoir species is an adduct of the peroxyacetyl radical, $CH_3CO.O_2$, with nitrogen dioxide, of formula $CH_3CO.O_2NO_2$. It is known universally as peroxyacetyl nitrate, or PAN[a]. PAN has been recognized for many years as a component of photochemical air pollution (Section 5.5.7), but is now known to be present also in the clean air of remote oceanic regions. Concentrations of 10 to 400 parts in 10^{12} of PAN were observed over the Pacific Ocean where NO_2 itself constitutes less than 30 parts in 10^{12} of the air. The importance of PAN is that it is in thermal equilibrium with its precursors,

$$CH_3CO.O_2 + NO_2 \rightleftharpoons CH_3CO.O_2NO_2, \tag{5.34}$$

and that the equilibrium is shifted to the right-hand side at lower temperatures. Above the boundary layer, temperatures are sufficiently low for PAN to be relatively stable, but the molecule is unstable close to the surface. PAN will release NO_2 as it is transferred from cooler to warmer regions.

Formation of PAN, and of other peroxyacyl nitrates, requires the initial generation of the peroxyacyl radicals. As we saw in Section 5.3.2, many hydrocarbons, both saturated and unsaturated, produce aldehydes as oxidation intermediates, and it is attack of OH on these carbonyl compounds that

---

[a] A proper systematic name for PAN might be ethane peroxoic nitric anhydride as the compound is a mixed anhydride of two acids; the compound is not an ester of HNO_3 so the 'nitrate' in the name is not really appropriate.

yields the acyl, and ultimately the peroxyacyl, radicals. Several other oxidation mechanisms can lead to alkoxy radicals, RO. So long as R contains more than two carbon atoms, oxidation can yield a carbonyl fragment $R'CO$ and thence, *via* reactions analogous to (5.21) and (5.22), a peroxyacyl species. PAN is therefore a compound expected when hydrocarbons are oxidized in the presence of NO_2. It is now apparent that the species must be included with NO, NO_2, and HNO_3 when considering NO_x or NO_y budgets and transport in the troposphere.

5.3.5 The reaction OH + CO

Oxidation of CO by OH,

$$OH + CO \rightarrow H + CO_2, \tag{5.6}$$

is one of the most important reactions in tropospheric chemistry. The process exhibits some interesting kinetic features that must be taken into account in constructing tropospheric models. First, the reaction shows a temperature dependence that deviates markedly from the Arrhenius form (p. 98). At temperatures below ~ 300 K, the activation energy is near zero. Secondly, the overall rate of reaction is pressure dependent, the apparent rate coefficient increasing with pressure.

A possible explanation of the kinetic behaviour is that the reaction involves discrete steps with an HOCO intermediate:

$$OH + CO \rightleftharpoons HOCO^*, \tag{5.35}$$

$$HOCO^* \rightarrow H + CO_2, \tag{5.36}$$

$$HOCO^* + M \rightleftharpoons HOCO + M, \tag{5.37}$$

$$HOCO + O_2 \rightarrow HO_2 + CO_2. \tag{5.38}$$

Reaction (5.37) followed by (5.38) offers a pressure-dependent pathway that does not lead to the $H + CO_2$ products of reaction (5.6), but that particular path can operate only when O_2 is present, which it certainly is in the Earth's atmosphere! The reversible association reaction (5.35) will be favoured at low temperatures, and thus account for the slight negative temperature dependence observed for the overall rate.

5.3.6 The nitrate radical

Radical intermediates in chemical reactions are most positively identified through their optical absorption spectra. The nitrate radical, NO_3, was observed in this way over 100 years ago, in 1881, and may thus have been the first transient species to be detected directly. Over the past two decades, it has become evident that the NO_3 radical plays a significant part in chemical transformations in both the stratosphere (see Sections 4.4.2 and 4.4.5) and the troposphere of the Earth.

In the Earth's atmosphere, NO_3 is formed by the reaction

$$NO_2 + O_3 \rightarrow NO_3 + O_2 \tag{5.39}$$

in both stratosphere and troposphere. Dissociation of N_2O_5

$$N_2O_5 + M \rightarrow NO_3 + NO_2 + M \tag{5.40}$$

is apparently an additional source, but since N_2O_5 is formed by the reaction

$$NO_3 + NO_2 + M \rightarrow N_2O_5 + M \tag{5.41}$$

it is ultimately dependent on the occurrence of reaction (5.39). Dinitrogen pentoxide, N_2O_5, is itself an important product, since it can react heterogeneously with H_2O (Section 5.3.9) to yield HNO_3 and thus contribute to atmospheric acidification (cf. Section 5.5.6). Incidentally, NO_3 does not itself appear to react with H_2O, although it does react with aqueous negative ions to yield nitrate ions and other products.

During the day, the NO_3 radical is rapidly photolysed: the product channels may be $NO + O_2$ or $NO_2 + O$, the relative and absolute yields being dependent on wavelength

$$NO_3 + h\nu \rightarrow NO + O_2 \tag{5.42a}$$

$$NO_3 + h\nu \rightarrow NO_2 + O. \tag{5.42b}$$

In the stratosphere, therefore, reaction (5.39) followed by photolysis contributes to the detailed balance of the chemistry of odd nitrogen compounds. Reactions of NO_3 in the troposphere present a different picture because of the multitude of organic compounds available. Although the hydroxyl radical is usually the main agent of attack on organic species during the day, the nitrate radical may be the most important oxidizing species in the troposphere at night. For some species, such as CH_3SCH_3, the reaction at night with NO_3 may even dominate over the daytime reaction with OH. Two main kinds of initial step can be envisaged: hydrogen abstraction and addition to unsaturated bonds, as typified by the reactions

$$NO_3 + RH \rightarrow HNO_3 + R \tag{5.43}$$

$$NO_3 + {>}C{=}C{<} \rightarrow {>}C(ONO_2)C(.){<} \tag{5.44}$$

Nitric acid is thus a direct product of hydrogen abstraction by the radical. Furthermore, the radical produced in reaction (5.43) is likely to add O_2 in air to form a peroxy radical, RO_2. In the special case that RH is formaldehyde, HCHO, the HCO radical will ultimately yield the HO_2 radical; for other aldehyde precursors, the acyl radical products of reaction (5.43) yield acylperoxy, $R.CO.O_2$, and are thus potential sources of peroxyacylnitrates.

The initial adduct formed in reaction (5.44) can eliminate NO_2 to yield an epoxide. However, the species is a radical, and in the presence of air it is therefore expected to add oxygen. For example, for the radical derived from propene, the reaction is

$$CH_3\dot{C}HCH_2 + O_2 \rightarrow CH_3CHCH_2 \qquad (5.45)$$
$$\quad\quad |\qquad\qquad\qquad\quad |\ \ |$$
$$\quad ONO_2 \qquad\qquad\quad O\ ONO_2$$
$$\qquad\qquad\qquad\qquad\qquad\ \ |$$
$$\qquad\qquad\qquad\qquad\qquad\ \ O^\bullet$$

Several products are observed in the system when O_2 and NO_x are present, including CH_3CHO, $HCHO$, 1,2-propanediol dinitrate (PDDN), nitroxyperoxypropyl nitrate (NPPN), and α-(nitrooxy)acetone

$$CH_3-CH-CH_2 \qquad CH_3-CH-CH_2 \qquad CH_3CCH_2ONO_2$$
$$\quad\ \ |\qquad\quad |\qquad\qquad\quad |\qquad\quad |\qquad\qquad\qquad \|$$
$$\ \ O_2NO\quad\ ONO_2 \qquad\ O_2NO\quad\ OONO_2 \qquad\qquad O$$

$$\qquad\ \text{PDDN} \qquad\qquad\qquad \text{NPPN} \qquad\qquad \alpha\text{-(nitrooxy)acetone}$$

The formation of the nitrated acetone from the peroxy radical product of reaction (5.45) is illustrated by the sequence of processes

$$CH_3-CH-CH_2 \xrightarrow[-NO_2]{+NO} CH_3CH-CH_2 \xrightarrow[-HO_2]{+O_2} CH_3C-CH_2$$
$$\quad\ \ |\qquad\quad |\qquad\qquad\qquad |\qquad\quad |\qquad\qquad\qquad \|\quad\ \ |$$
$$\quad\ \ O\qquad\ ONO_2 \qquad\qquad O\bullet\quad\ ONO_2 \qquad\qquad O\ \ ONO_2$$
$$\quad\ \ |$$
$$\quad\ \ O\bullet \qquad\qquad\qquad\qquad\qquad\qquad\qquad\qquad\qquad\qquad (5.46)$$

which is clearly the analogue of (5.17) and (5.20). Conversion of NO to NO_2 and the generation of HO_2 are both effected, but the possible formation of α-(nitrooxy)acetone in ambient air is of interest, since the compound has been reported to be a mutagen. The other products are also of concern because dinitrates have been shown to produce several adverse health reactions.

The preceding discussion suggests that involvement of NO_3 in tropospheric chemistry has four significant consequences:

1. Primary organic pollutants can be oxidized and removed during the night.
2. Nitric acid can be formed, either by hydrolysis of N_2O_5 from reaction (5.41) or as a product of the hydrogen-abstraction process (5.43).
3. Free radicals, probably undergoing rapid conversion to HO_2 and RO_2, can participate in further steps, so that reaction of organic compounds with NO_3 may initiate oxidation chain reactions (and provide a source of hydroxyl radicals at night).
4. Toxic or otherwise noxious compounds, including peroxyacyl nitrates and other nitrates and oxidized compounds, may be formed.

Rate constants for the reactions of NO_3 with hydrocarbons are generally much lower than for the reactions of OH. Typical relative reactivities of OH and NO_3 towards hydrocarbons such as butane and 1-butene are roughly 10^5

and 3000 at ambient temperature. Against these intrinsic reactivities must be set the greater concentration of NO_3 during the night, which may be about 10^9 molecule cm^{-3}, compared with that of OH during the day, for which 10^6 molecule cm^{-3} is of the right order of magnitude (cf. Section 5.4). Reaction of NO_3 with the alkenes can thus be comparable in extent with the attack by OH during the day. Indeed, for some naturally occurring terpenes, the reactions with NO_3 are particularly rapid (rate constants several hundredths of those for the corresponding reactions with OH), so that the night-time NO_3 processes dominate. We shall see in the next section that NO_3 may play a similarly important role in the oxidation of some sulphur compounds. Reaction of NO_3 with alkanes is probably never more than a minor contributor to the total diurnally-averaged oxidation; from the point of view of HNO_3 formation, however, reaction (5.43) may add substantially to other sources, such as hydrolysis of N_2O_5.

Although the reactions of NO_3 with alkanes and simple alkenes are relatively slow, the same is not true of the reactions with radical species. Interactions of NO_3 with radicals such as HO_2 and RO_2 that persist at night may be of atmospheric significance, particularly since NO, so important in the reactions of the peroxy radicals in reactions (5.12) and (5.17), falls to low concentrations during the night. More research into the rates, products, and mechanisms of these reactions is needed to elucidate the participation of the radical–radical reactions in tropospheric chemistry.

5.3.7 Compounds of sulphur

In Section 1.5.4, we presented the main features of the natural biogeochemical cycle of sulphur in the atmosphere. Biogenic processes generally emit sulphur species in the reduced forms of H_2S, CS_2, COS, and the organic compounds CH_3SH (methyl mercaptan), CH_3SCH_3 (dimethyl sulphide, DMS) and CH_3SSCH_3 (dimethyl disulphide, DMDS). The importance of these compounds in producing aerosols in the troposphere (Section 1.6) and stratosphere (Sections 1.5.4 and 4.7.7) has been mentioned already; oxidation products of these compounds act either as the main components or as condensation nuclei in the aerosols. Carbonyl sulphide, COS, is rather uniformly distributed throughout the troposphere, at high concentrations (mixing ratios = 5–6 × 10^{-10}). These observations are both attributable to the long atmospheric lifetime, which is probably in excess of 50 years. Carbonyl sulphide, from natural or anthropogenic sources, can therefore reach the stratosphere and there be oxidized to the stratospheric sulphuric acid or sulphate aerosol (see Section 4.7.7 and the Bibliography to Chapter 4). Oceanic emissions of biogenic sulphur gases provide an important flux to the atmosphere, and constitute a source that is believed to be responsible for the background levels of SO_2, methanesulphonic acid, and non-sea-salt sulphate. The dominant sulphur compound released from the oceans is DMS, which is produced by metabolic processes in certain algae. Oceanic surface waters are super-saturated with DMS with respect to atmospheric concentrations and thus supply a net flux to the atmosphere. In a recent cruise from Germany to Brazil of the research vessel 'Polarstern', both DMS and CS_2 in the marine boundary layer were found to exhibit a marked spatial variability. Concentrations of DMS decreased closer to land, a result that might reflect removal of DMS in processes involving oxidants present in continental rather than marine air. Sulphur dioxide, SO_2, may be a product (perhaps minor) of the oxidation of the biogenic sulphur compounds, but man makes an especially large contribution to the atmospheric sulphur burden in urban and industrial regions almost entirely in the form of SO_2. For this reason, it is convenient to divide sulphur chemistry into two parts. First, we treat in this section the oxidation of the reduced compounds, released naturally, perhaps through six oxidation states to SO_2. Then, in Section 5.5.4, we consider the oxidation of anthropogenic and natural SO_2 to H_2SO_4.

For the inorganic reduced sulphur species, oxidation is driven by the hydroxyl radical, as with so many tropospheric processes. In each case SH radicals are formed

$$OH + CS_2 \rightarrow COS + SH \tag{5.47}$$

$$OH + COS \rightarrow CO_2 + SH \tag{5.48}$$

$$OH + H_2S \rightarrow H_2O + SH. \tag{5.49}$$

Reaction (5.47) with CS_2 probably proceeds via an intermediate complex $HOCS_2$ that can react with O_2 to yield $COS + SH$. This process is analogous to the reaction of OH with CO (Section 5.3.5), and it possesses a negative temperature coefficient of rate as well as a dependence on total pressure. In the presence of O_2, O_3, or NO_2, the SH radicals are then further oxidized to SO_2, *via* the intermediate radicals SO and possibly HSO and S atoms. Details of the mechanism are not known, but the oxidation steps are probably of the type

$$SH + O_2 \rightarrow OH + SO, \tag{5.50}$$

$$SH + OH \rightarrow S + H_2O, \tag{5.51}$$

$$S + O_2 \rightarrow SO + O, \tag{5.52}$$

$$SO + O_2 \rightarrow SO_2 + O. \tag{5.53}$$

The disproportionation reaction

$$SH + SH \rightarrow H_2S + S, \tag{5.54}$$

creates one S atom that can be oxidized in reaction (5.52) followed by (5.53), but also provides a pathway for the formation of H_2S from the more oxidized SH radicals. Reaction of SH with HO_2, HCHO, H_2O_2, and CH_3OOH

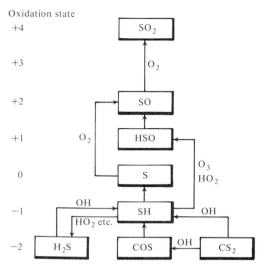

Fig. 5.3. Conversions of sulphur containing species in the troposphere, showing the progression towards more oxidized compounds.

likewise generates H_2S, so there exist photochemical pathways that link CS_2 and COS to H_2S. Figure 5.3 illustrates the main transformations that lead to SO_2 from reduced inorganic sulphur compounds.

Oxidation of the organic sulphur compounds can be initiated by attack either of OH or of NO_3. For CH_3SCH_3 (DMS) and CH_3SH, the probable atmospheric lifetime against attack by NO_3 at night is less than 1.1 hr, as against 4 days and 17 hr for daytime attack by OH radicals. CH_3SSCH_3 seems to react predominantly with OH, although NO_3 still contributes about 10 per cent to the loss of the compound. In marine environments, where the large organic sulphur source is almost entirely DMS, an additional possible oxidation reaction involves IO radicals generated photochemically from CH_3I (see Section 5.3.8).

The types of oxidation process that the organic sulphides undergo can be illustrated with reference to DMS. Initial attack of OH on DMS proceeds via two parallel steps

$$OH + CH_3SCH_3 \rightarrow H_2O + CH_2SCH_3 \qquad (5.55a)$$

$$OH + CH_3SCH_3 \rightarrow CH_3S(OH)CH_3. \qquad (5.55b)$$

In the presence of air, the adduct reacts further with O_2, perhaps to yield methane sulphonic acid, CH_3SO_3H, one of the observed products. Secondary reactions of either the CH_2SCH_3 radical product of reaction (5.55a) or of the adduct of reaction (5.55b) can lead to the methyl thiyl radical, CH_3S.

The reaction of NO_3 with DMS is more rapid than many interactions of this radical with organic compounds, and it possesses a small negative temperature coefficient, which is indicative of an addition mechanism. Although the products have not been identified, it seems possible that they lead to CH_3S as in the oxidation of DMS by OH.

Methyl thiyl radicals may be able to react with O_2 to yield SO_2

$$CH_3S + O_2 \rightarrow CH_3 + SO_2, \qquad (5.56)$$

although they may react preferentially in steps that lead via the CH_3SO radical to other products

$$CH_3S + NO_2 \rightarrow CH_3SO + NO \qquad (5.57)$$

$$CH_3S + O_3 \rightarrow CH_3SO + O_2 \qquad (5.58)$$

$$CH_3SO + O_2 \rightarrow CH_3SO(O_2) \rightarrow \text{products} \qquad (5.59)$$

$$CH_3SO + NO_2 \rightarrow CH_3SO_2 + NO. \qquad (5.60)$$

The end-product of reaction (5.59) is predominantly SO_2, but at relatively high $[NO_x]$ reaction (5.60) becomes the preferred pathway, and methane sulphonic acid is the major product. A quite different product for the reaction between IO and DMS is proposed: dimethyl sulphoxide is apparently formed in a rapid process

$$IO + CH_3SCH_3 \rightarrow I + CH_3SOCH_3. \qquad (5.61)$$

The species that result from the atmospheric oxidation of DMS are of considerable importance. It has been proposed (see Sections 1.5.4 and 1.6) that DMS may be a link in an ecosystem regulation of cloudiness over Earth. Sulphate aerosols are very efficient cloud condensation nuclei. If oxidation of DMS leads to SO_2, then SO_4^{2-} can be formed in the aqueous phase (see Section 5.5.4), and the activity of the marine biota could influence cloud cover. In turn, backscattering by clouds influences both illumination and temperature at the surface, and thus affects the rate at which DMS is generated. The elements of a 'planetary thermostat' are thus in place (cf. Section 1.6). Quite apart from this regulatory mechanism, oxidation of DMS could contribute as much as 25 per cent to total acid deposition (Section 5.5.6), at least over Europe.

5.3.8 Natural halogen-containing species

Some natural halogenated compounds are found in the troposphere. The methyl halides, CH_3Cl, CH_3Br, and CH_3I, are produced predominantly in the oceans. Methyl chloride and methyl bromide both have tropospheric lifetimes towards attack by OH in excess of one year, so that they are able to reach the stratosphere and play a role there as catalytic species in ozone chemistry (Section 4.4). The behaviour of methyl iodide is, in some ways, more interesting. Unlike the chlorine and bromine compounds, CH_3I can be photolysed by the radiation that penetrates to the troposphere, and it generates iodine atoms

$$CH_3I + h\nu \rightarrow CH_3 + I \tag{5.62}$$

which could participate in tropospheric chemistry. For example, the reactions

$$I + O_3 \rightarrow IO + O_2 \tag{5.63}$$

$$IO + IO \rightarrow 2I + O_2 \tag{5.64}$$

provide a catalytic route to ozone destruction that might play a minor part in ozone loss in the troposphere. However, the rather low tropospheric abundances of CH_3I and the existence of alternative channels for the interaction of two IO radicals (to yield $I_2 + O_2$ or an IO dimer) limit the importance of the chain. As discussed in Section 5.3.7, IO is proposed as an agent of initial attack on dimethyl sulphide in marine environments. It has also been suggested that IO radicals might influence NO_x and HO_x chemistry in the troposphere by forming $IONO_2$ and HOI (species to be compared with the analogous 'reservoir' compounds of the stratosphere, $ClONO_2$ and HOCl, Section 4.4.2). Although the participation of IO in the 'mainstream' chemistry of the troposphere is uncertain at present, the oxidizing potentiality of the radical should clearly be kept under review.

5.3.9 Heterogeneous processes and cloud chemistry

The discussion of Sections 1.3 and 4.7.4 has already highlighted the possibility that solid particles and liquid droplets in the atmosphere may participate in chemical processes as well as modifying the radiation balance and temperature. Reactions involving condensed phase species are of great importance within the troposphere. Heterogeneous chemical reactions are usually thought of as those occurring at the interface between two phases. Some important atmospheric processes, however, involve transfer of reactants from the gaseous to a condensed phase followed by homogeneous chemical change within the condensed phase. For our purposes, therefore, it is convenient to include in the category of heterogeneous reactions all processes that involve more than one phase.

The heterogeneous processes of interest in tropospheric chemistry include the formation of secondary particulate matter (as aerosols of solids or liquids) as well as reactions within aerosols. The condensation of a single gaseous component to form a new suspended particle is *homogeneous, homomolecular* nucleation, and it is obviously central to cloud formation. The thermodynamics of this process were discussed in Section 2.5. The analogous process for reacting species involves two or more gases that form a condensable product species in a *homogeneous, heteromolecular* process. For example, NH_3 and HNO_3 in the gas phase can form NH_4NO_3 that ultimately produces particles. A more important route to aerosol NH_4NO_3 may, however, be *heterogeneous, heteromolecular* reaction on the surface of pre-existing particles. Another example of the heterogeneous, heteromolecular process is afforded by the reaction of NO_2 or HNO_3 with sea-salt particles (NaCl) to form $NaNO_3$. Finally, chemical reactions within the aerosol itself can form particles of changed composition, as in the oxidation of SO_2 to sulphate ions in clouds (Section 5.5.4).

The heteromolecular reactive condensation of gas-phase molecules on pre-existing particles is sometimes called *aerosol scavenging*. It can have an impact on bulk tropospheric chemistry by providing a sink for nitrogen and hydrogen species such as HNO_3, NO_3, N_2O_5, H_2O_2, and HO_2, as well as organic nitrates and peroxides. Clouds and raindrops have a major effect on gas-phase species through the scavenging mechanism. Rainout (removal of gases by cloud droplets) is believed to be more important than washout (removal of gases by raindrops) because of the longer lifetime and greater surface area of cloud droplets compared with raindrops. Water-soluble species, such as the acids, acid anhydrides, and peroxides are obviously particularly susceptible to removal by these mechanisms.

As well as providing a liquid phase in which soluble gases can be dissolved, clouds also offer an active chemical medium for aqueous-phase reactions that can affect the distribution of active species in the troposphere. The particular case of SO_2 oxidation is deferred until Section 5.5.4, but similar considerations apply to the photochemistry of background air. Clouds occupy only about 15 per cent of the volume of the lower troposphere, and the volume fraction of liquid water in the clouds is less than one-millionth. Nevertheless, the direct chemical influence of clouds on the tropospheric reactants may be more important in atmospheric photochemistry than the scattering and reflection of actinic ultraviolet radiation.

One of the most important conversions in water droplets is likely to be the formation of HNO_3 from N_2O_5

$$[N_2O_5 + H_2O \rightarrow HNO_3 + HNO_3]_{aq}. \tag{5.65}$$

The gas-to-aqueous phase transfer of N_2O_5 is limited by gas-phase diffusion and transfer through the interface, while reaction (5.65) within the aqueous phase is so fast as to be essentially 'instantaneous', so that the dissolution of N_2O_5 is irreversible. Cloud reactions further influence the oxides of nitrogen by removal of NO_3 radicals (Section 5.3.6). Because the radicals have a long lifetime against gas-phase removal processes, they can be incorporated readily into cloud water as $(NO_3)_{aq}$. The dissolved radical reacts rapidly with chloride ion

$$[NO_3 + Cl^- \rightarrow NO_3^- + Cl]_{aq}. \tag{5.66}$$

Conversion times are a few tens of seconds, and the effect is not only to remove oxides of nitrogen from the gas phase, but also to enhance the concentration of NO_3^- dissolved in the cloud water.

Another heterogeneous reaction can convert HO_2 to H_2O_2 by producing first the superoxide ion which then reacts with more HO_2

$$[HO_2 \rightarrow H^+ + O_2^-]_{aq} \qquad (5.67)$$

$$[HO_2 + O_2^- \xrightarrow{\text{H}^+} H_2O_2]_{aq}. \qquad (5.68)$$

Evaporation of clouds containing high concentrations of dissolved H_2O_2 could represent a net source of H_2O_2 to the atmosphere. However, the H_2O_2 can also react with OH in the aqueous phase

$$[OH + H_2O_2 \rightarrow H_2O + HO_2]_{aq} \qquad (5.69)$$

so that the sequence of reactions (5.67) + (5.68) + (5.69) constitutes a free-radical sink

$$[OH + HO_2 \rightarrow H_2O + O_2]_{aq}. \qquad (5.70)$$

Thus the overall oxidizing capacity of the troposphere can be strongly influenced by cloud chemistry. The reduction in HO_x radical concentrations is compounded by both the destruction of hydrated HCHO and inhibition of its photolysis to give HO_2 in reactions (5.15) and (5.16). Indeed, there are even aqueous-phase reactions that destroy ozone itself, such as

$$[O_2^- + O_3 + M \rightarrow 2O_2 + OH + OH^-]_{aq}. \qquad (5.71)$$

Various reaction sequences can be written, of which the most obvious is reactions (5.67) + (5.71) that then acts as the aqueous-phase analogue of the gas-phase sink reaction (5.31) between HO_2 and O_3. Despite the relatively poor solubility of O_3 in water, reaction (5.71) also turns out to be the dominant source of dissolved hydroxyl radicals.

The disproportionate influence of the aqueous-phase reactions derives in part from the concentration of the reagents in solution, which especially enhances the rate of second-order reactions. Added to the concentration effect, the solution reactions often have lower activation energies than the corresponding gas-phase processes, thus further increasing the rates. But a special feature in the tropospheric reactions is high solubility of certain key compounds such as HO_2 or N_2O_5; the reactions of these species within the droplets ensure that dissolution is irreversible. The partitioning into the aqueous phase is thus strongly favoured, and partially offsets the small relative volume of the water droplets. Because NO is relatively *insoluble*, HO_2 and NO are separated and prevented from participating in the reaction

$$HO_2 + NO \rightarrow OH + NO_2 \qquad (5.12)$$

that converts NO to NO_2 (and thus promotes ozone formation) in the gas phase.

Two-dimensional cloud–climate models have been used to assess the influence that cloud chemistry might have on the tropospheric ozone budget. Initial calculations suggest that ozone production rates might be decreased by up to 40 per cent, and destruction rates increased by 30 to 270 per cent compared with the cloud-free estimates. Substantial calculated decreases in the concentrations of OH, HO_2, HCHO, and oxides of nitrogen are also brought about by the cloud chemistry.

A quantitative interpretation of heterogeneous chemistry requires a method for introducing the coupling of the chemistry that occurs in the two phases, as well as a knowledge of the solubilities, interfacial mass-transport kinetics, and the reaction kinetics of the species in the condensed phase. The theoretical background to the coupling needs further development, and the whole study of heterogeneous reactions in particles and droplets is in its infancy. However, in view of the likely involvement of reactions in and on surfaces in both the troposphere and the stratosphere, this field is one that merits the most detailed attention.

5.4 Observations and models

Although the broad outline of tropospheric behaviour now seems clear enough, there remain many problems of detail. The variability in concentration of several species is one cause of uncertainty. Relatively short atmospheric lifetimes mean that geographical distributions are frequently determined by local release patterns and meteorological conditions. Furthermore, tropospheric measurements of concentrations often prove to be much more difficult and imprecise than those made in the stratosphere. Chemical reactions are themselves more complex, involving a wide variety of relatively large molecules, with the additional possibility of heterogeneous processes on surfaces on the planet or in the atmosphere.

Is it possible, with the uncertainties and complexities, to compare the predictions of models with field observations of concentrations in the troposphere? Such comparisons, it will be remembered, provide one basis for the validation of stratospheric models (Section 4.4.5). Two obvious candidate species for study are the radical OH, which is central to tropospheric oxidation chemistry, and CO, which is both formed and destroyed in OH-mediated reactions (Section 5.3.2). Reliable values for OH concentrations are needed to estimate tropospheric lifetimes for many trace gases, including CO and CH_4 whose sources are extremely difficult to quantify. The analysis is complicated since OH itself is lost mainly in reaction with these two gases (see Section 5.3.2), both of whose tropospheric concentrations are slowly increasing. It is evident that verification of the theories of tropospheric OH would benefit from good measurements of the spatial and temporal distribution of concentrations of the radical. Validation of a time-dependent model requires simultaneous determination not only of [OH] itself, but also of all the parameters that control the local OH chemistry. The challenge has not yet been met entirely satisfactorily.

Three methods currently exist for the time-resolved measurement of local hydroxyl radical concentrations. Laser induced fluorescence (LIF) is a modification of the resonance fluorescence method described in Section 4.4.5, in which lasers are used as the source of radiation resonant in frequency with the $OH(A^2\Sigma^+ \leftarrow X^2\Pi)$ transition. Although the weight of evidence is against it, there exists the possibility of interference with the measurements with high intensity laser sources. Artificial OH can be produced from atmospheric H_2O and O_3 by the laser pulse itself. Long-path absorption spectroscopy appears to give more reliable results. In a recent high-resolution instrument, several absorption features of OH in the ultraviolet region are monitored simultaneously in order to overcome interference problems. The method gives absolute [OH] without the need for calibration. A further technique for determining [OH] relies on ^{14}C isotopic measurements: the principle of the method is outlined later in this section.

Daytime [OH] ranges from 0.4 to 9×10^6 molecule cm^{-3}, in line with the general predictions of realistic models of the troposphere. Where simultaneous supporting data are available that allow full model calculations to be performed for the expected local [OH], the predictions and measurements agree reasonably well when NO_2 is high (mixing ratio $> 2 \times 10^{-9}$). However, for mixing ratios of NO_2 below about 10^{-9}, corresponding to relatively clean air, the calculated concentrations exceed those measured by a factor of two. This result suggests that there is some deficiency in the way that the model deals with losses of HO_x species, and emphasizes the need for further measurement of [OH], especially in remote regions, as well as the need for refinement of the models.

Although direct measurements of [OH] at particular times in specific locations are essential for the characterization of models, another body of data refers to OH abundances obtained as a global average. One technique for obtaining such averages uses measured atmospheric concentrations of some trace gas to infer [OH] indirectly. The idea is that if the gas has a known source strength, and is removed from the atmosphere by reaction with OH alone, then the value of [OH] can be derived so long as the rate coefficient for the reaction is available for appropriate temperatures and pressures. Certain of the man-made halocarbons (Sections 4.5.4 and 4.5.5) meet some of the requirements for a test compound. For example, CH_3CCl_3 (methyl chloroform) seems to fulfil the criteria quite well, the global emission rates probably being known to within 5 per cent. Uncertainties in the absolute atmospheric concentrations of CH_3CCl_3 pose one of the greatest obstacles to accurate derivation of [OH]. According to values accepted at the time of writing, the atmospheric lifetime of CH_3CCl_3 is 6.3 years (cf. Table 4.6), and the corresponding globally averaged [OH] is about 7×10^5 molecule cm^{-3}. The data imply that [OH] is somewhat higher in the Northern Hemisphere than in the Southern. Halocarbons with a shorter lifetime than CH_3CCl_3 potentially provide a more sensitive indicator of latitudinal and seasonal gradients of OH concentration. Chemically suitable halocarbons include 1,2-dichloroethane (CH_3ClCH_3Cl) or tetrachloroethene (CCl_2CCl_2), but unfortunately the absolute and seasonally varying source strengths are too poorly known to allow reliable conclusions about the OH field to be drawn.

One major objective of obtaining globally averaged values of [OH] has been in assessing the rate of tropospheric scavenging of man-made pollutants. For example, the determination of the impact of halocarbons on stratospheric ozone (Section 4.5.5) first requires calculation of the atmospheric lifetime, a quantity strongly dependent on OH concentrations. However, the use of a global average of [OH] in the calculations is of dubious validity first because of the lack of uniformity of the abundances both of the OH itself and of the molecules with which it is to react and secondly because of differences in temperature dependences of the rates of the reactions being compared. One procedure that has been used to avoid some of these difficulties is first to calculate a three-dimensional tropospheric OH field that allows for observed trace-gas mixing ratios, and climatological factors such as temperatures and sunlight intensities. Relatively simple chemistry is used but the predicted and observed mixing ratios for CH_3CCl_3 are subsequently matched by scaling the absolute OH concentrations.

The preceding two paragraphs have described the measurement of mixing ratios of a trace gas of known source strength to determine [OH] in the troposphere. The same concept lies behind the use of measurements of ^{14}CO in obtaining [OH] as a global average or to calculate its distribution in a more detailed way. Cosmic-ray bombardment of atmospheric nitrogen is the source of new ^{14}C. Natural biospheric processes liberate and consume a small but significant amount of ^{14}CO, but human combustion processes do not, since they burn ancient fossil carbon in which the radioactive ^{14}C component has decayed. Figure 5.4 shows annual variations of ^{14}CO measured at a fixed location. The concentrations are extremely small, always being less than about 30 molecule cm^{-3}. The solid lines on the figure represent ^{14}CO calculated from a two-dimensional model. Hydroxyl radical concentrations for the lower curve were double those used in calculating the upper one. Observed seasonal variations are clearly reproduced correctly, and the OH distribution appears to lie between the two limits modelled. An optimization procedure was used to refine the fit, and thus provide the absolute concentration field for OH radicals. Figure 5.5 presents the results of the 2-D model for OH concentrations in July. The global mean $[OH] = 6.5 \pm 2.5 \times 10^5$ molecule cm^{-3}, which corresponds well with the values obtained by other techniques. Although the calculations were first conducted in the early 1980s, a re-evaluation nearly a decade later suggested that the average [OH] derived from the ^{14}CO measurements was unaffected by the improved data for kinetic and other factors, probably as a result of fortuitous cancellation of effects!

Determinations of [^{14}CO] provide information not only about tropospheric OH, but also about CO budgets. Recent studies of CO indicate that there are

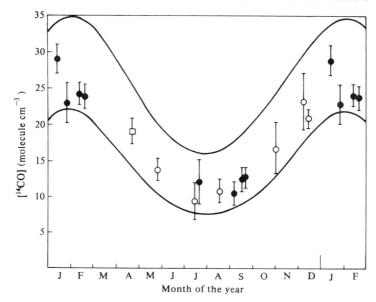

Fig. 5.4. Annual variations in the observed and calculated concentrations of ¹⁴CO at 51°N. The upper calculated line was obtained using the standard OH distribution calculated by the AERE Harwell two-dimensional model; the lower line was obtained by doubling those OH concentrations. [From Cox, R. A. and Derwent, R. G. *Gas kinetics and energy transfer*, Specialist Periodical Reports Chem. Soc. **4**, 189 (1981).]

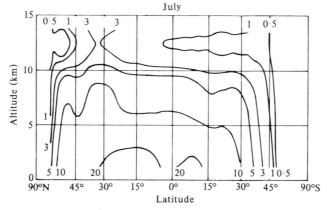

Fig. 5.5. Contours of the OH distribution used in Fig. 5.1. Units are 10^5 molecule cm^{-3}. (Source as Fig. 5.1).

large sources for the gas other than the oxidation of CH_4 and the combustion of fossil fuels. Possible additional sources include oxidation of non-methane hydrocarbons (NMHC), especially natural isoprene and terpenes, together with biomass burning. Part of the problem in quantitative calculation arises because of large localized production of CO from man's activities. However, the carbon monoxide releases required to balance the observed latitudinal CO distribution and the calculated OH distribution are considerably greater than the injection rate from all known man-made sources. The ^{14}CO source term used in the calculations of [OH] described in the last paragraph is the sum of the (known) cosmic-ray production rate and the biospheric injection. This latter term can be chosen at will in the calculations. However, the ^{14}C to ^{12}C isotopic ratio is known for the biospheric source, so that there is an additional constraint if ^{14}CO and ^{12}CO cycles are to be balanced simultaneously. In principle, at least, both OH distributions and biospheric CO sources can be balanced simultaneously. More direct measurements of the CO field have been made with satellite instruments, for example in the 'measurement of air pollution from satellite' (MAPS) experiment. These data show clearly the expected enhancements over industrial source regions, but there were also regions of elevated [CO] in the tropics that are possibly associated with agricultural burning. Seasonal variations are marked, with a typical modulation of ± 25 per cent in Europe between minimum [CO] in late summer and maximum values in late winter. The secular increases in [CO] (perhaps as much as one per cent per year) reveal an interesting aspect of the coupling between OH, CO, and CH_4 in the troposphere. The present-day concentrations of CO (and of CH_4, whose concentration is also increasing) are not in equilibrium with the sources because of the coupling: there can be a positive feedback since oxidation of CH_4 is an important source of CO, and CO and CH_4 are the dominant sinks for the OH radical.

Ozone is one of the most important trace gases in the troposphere since it is the ultimate initiator of photochemical chain oxidation (Section 5.3.1). Tropospheric ozone is produced both by the *in situ* chemistry described in Section 5.3.3 and by transfer from the stratosphere. Model calculations suggest that sources and sinks for ozone within the troposphere are nearly balanced on a global scale, and other estimates indicate that the stratospheric injection is roughly equal to the surface sinks. However, both observations and analysis of regional budgets show that $[O_3]$ is increasing in the Northern Hemisphere, probably because of a shift in the balance of *in situ* production and loss. The balance depends on the abundance of NO_x in the background atmosphere (see Section 5.3.4). The rate of conversion of NO to NO_2 is also a key factor, and it is directly influenced (Section 5.3.2) by the concentrations of peroxy radicals that are a consequence of oxidation of CO, CH_4, and the non-methane hydrocarbons.

We have emphasized, in this section, determinations of [OH], [CO], and $[O_3]$ to illustrate the possibilities and problems of tropospheric measurements.

Much effort is being put into the investigation of other trace gases, and the interpretation of secular, seasonal, and geographical trends of concentrations will further our understanding of tropospheric chemistry. Man's agricultural and industrial activities may be altering tropospheric chemistry on a global scale, as concentrations of important trace species, and the tropospheric oxidizing capacity itself, undergo change. It is not yet clear what effects these global changes might have. On a more local scale, the consequences of pollution of the atmosphere by man is all too evident, and it is with such pollution that the rest of this chapter is concerned.

5.5 Air pollution

5.5.1 Clean and polluted air

Air is never perfectly 'clean' in the sense of containing only N_2, O_2, CO_2, H_2O, and the inert gases. Even before man existed, there were many sources of the trace constituent species, some of which would now be classified as pollutants. 'Natural air pollution' is caused by volcanic eruptions, breaking waves, pollens, dust and terpenes from plants, windblown dust, and forest fires. Man has accentuated the frequency and intensity of release of some of these pollutants. Agricultural practices, in particular, have been a source of additional pollutant species ever since man has been on Earth. 'Slash-and-burn' clearing of land adds greatly to the fire source; the removal of natural vegetation cover itself increases the rate of dust erosion.

Nitrogen fixed as fertilizer is in part returned to the atmosphere as N_2O, and alterations in the types and utilization of crops also affect the N_2O budget. Increased population of livestock and of rice paddy cultivation imply an increase in CH_4 production. Destruction of forests, and their replacement by grasslands, may provide suitable habitats for termites that are a potentially significant source of atmospheric methane. Human-accentuated forms of pollution, although significant, may seem minor compared with the releases of man-*made* pollutants in industrial societies. Man seems capable of releasing— by accident or design—almost all known gaseous species to the atmosphere. Even substances, such as lead, not commonly regarded as gaseous, can be produced as aerosols by man's activities. It is not our purpose to catalogue here all the pollutants that have been, or might be, released. Rather, we wish to concentrate on pollution processes which are of rather widespread occurrence, with respect both to geographical location and to frequency. We have already seen, in Chapter 4, that some trace gases whose release is due to man (e.g. CF_2Cl_2), or accentuated by him (e.g. N_2O, CH_4), can have global consequences because of stratospheric interactions. In the troposphere, the most important forms of pollution seem to be related to the combustion of fossil fuels. It is these types of pollution that will be used for illustration in this chapter.

Combustion of fossil fuels affects the atmosphere in three main ways: by the formation of carbon dioxide, by the release of substances such as sulphur dioxide and partially oxidized or unburnt fuel, and by the high-temperature fixation of atmospheric N_2 and O_2 to yield oxides of nitrogen. The carbon dioxide problem arises because, in the course of a few hundred years, appreciable fractions of the Earth's fuel deposits have been burnt to release carbon dioxide, thus reversing the chemistry that photosynthesis took millions of times as long to achieve. At present, atmospheric CO_2 concentrations are increasing by about one part in 300 a year, corresponding to about one-fifth of the annual release of carbon dioxide from combustion of fossil fuels (see Fig. 1.5). The potential climatic effects of increases in a 'greenhouse' gas are considered in relation to the evolving atmosphere (Chapter 9); we have already seen that the changes in stratospheric temperature consequential on CO_2 increase can affect ozone chemistry (Section 4.5.9). Coal and petroleum products contain appreciable quantities of sulphur compounds (up to 2.5 per cent of coal by weight, and as much as 1.8 per cent in residual fuel oil). Combustion then leads to sulphur dioxide release. Particulate matter (e.g. soot), carbon monoxide, and unburnt hydrocarbons are also important minor by-products of the combustion process. We shall discuss sulphur dioxide chemistry in Section 5.5.4, and examine some consequences of SO_2 release in the following two sections. Sulphur dioxide and the oxides of nitrogen, formed when combustion is supported by air, contribute to the problems of acid rain (Section 5.5.6), and the oxides of nitrogen also play a special part in 'photochemical air pollution' (Section 5.5.7).

Pollutants released to the boundary layer may be quite rapidly removed by wet or dry deposition, and so degrade air quality only near the source. Soot, SO_2, and malodorous compounds emitted from factories often fall into this category, and contribute to *local air pollution*. More widespread dispersal within the troposphere can lead to *regional* pollution of quite large geographical areas. If the entire atmosphere is affected, as is the case for trends discussed at the end of Section 5.4 or for the long-term CO_2 increases, then the pollution is *global* in extent. Attempts to reduce the local impact of released pollutants can sometimes backfire by increasing the impact on the larger scale. For example, by building very tall chimney stacks for factories, local pollution can be much reduced. However, the removal of pollutants at higher altitudes may be a far slower process than near the surface, and on the regional scale the pollution problem may be exacerbated.

5.5.2 Effects of pollution

Air pollution is seen as a growing threat to our welfare, and especially to human health, as ever-increasing emission of contaminants is made into the constant-sized atmosphere. An average adult male gets through about 13.5 kg of air a day, compared with about 1.2 kg of food and 2 kg of water. The quality of the air breathed is therefore at least as important as the cleanliness of our food and water. Air pollutants can exert an influence on health in two ways. First, they may be physiologically toxic, and secondly they may possess a nuisance value. Thus pollution can, by being smelly or dirty, have a health effect far beyond that of simple poisoning.

One of the major difficulties we face in assessing the impact of a pollutant on *human* health concerns the experiments that are possible. It is obviously out of the question to perform clinical tests of human response to any substance at concentrations that could conceivably be toxic. We have, therefore, to fall back either on the effects of accidental exposure of small groups to high doses, or to statistical correlations between mortality or morbidity and (chronic) exposure to low doses: such studies form the basis of *epidemiology*. By far the most hazardous pollution is that inflicted on self (and neighbours) by tobacco smokers, and the difficulties of air pollution epidemiology are compounded because of this factor. Indeed, the differences in virtually all health indicators are so large as between smokers and non-smokers that it is often impossible to detect whether other forms of air pollution are presenting a health hazard. Cigarette smoke contains, in addition to carcinogens and nicotine, sufficient carbon monoxide to clearly worsen coronary heart disease and respiratory problems. Toxicological trials on animals are of only limited relevance, because we cannot be sure that human responses will match those of the species tested, even after due allowance has been made for the size of the species.

We shall consider the health effects of individual pollutants in subsequent sections, but there are some comments of general applicability which can be made here. The main route of entry into the body of air pollutants is *via* the respiratory system, although some solid material deposited there may subsequently be swept to the gastrointestinal system. Because of the mode of access, respiratory symptoms are the most common response produced by pollutants. Low concentrations of oxidants such as O_3 or NO_2 produce nasal and throat irritation, while slightly higher concentrations impair mechanical lung function. In healthy subjects, the change is reversible within a few hours. So long as the change in function is unaccompanied by symptoms or by decreased work capacity, the exposure probably does not correspond to a real impairment of health. Of course, in persons with underlying respiratory illness, the situation is altered, since the *normal* function may already limit activity. Further increases in irritant concentration (say >0.3 p.p.m. of ozone) lead to sufficient discomfort to restrict normal activity. Even in the healthy, the effects experienced depend on whether the subjects are exercising or at rest. Presumably, when exercising, the expiratory flow rates are increased, and breathing is through the mouth, so that the pollutant is delivered deeper within the respiratory tree.

Living species other than man may be at risk from man-made pollutants. As we shall see in Section 5.5.6, fish seem to be highly sensitive to the pH of their environment, and release of pollutants tending to increase acidity can have serious consequences for freshwater fish populations. Acute and chronic injury to the leaves of trees and plants can be caused by a number of pollutant gases. Sulphur dioxide, and the oxidants O_3 and NO_2, all cause damage: they are said to be *phytotoxic*. Combinations of the gases seem particularly harmful. Economic consequences in terms of lost foodstuffs or forests can be very great indeed. The indirect effects of reduced nitrogen fixation and CO_2 turnover, and of interference with entire ecosystems, could be of even greater long-term importance.

Economic loss can also be sustained by damage to inanimate materials. Many organic substances are susceptible to attack by oxidants such as ozone, especially if they contain double bonds. Elastomers such as natural and synthetic rubber, paints, and dyes are attacked. Chemical measures introduced to prevent damage may themselves constitute a major cost in production. Structural damage can be caused to buildings by species such as sulphur dioxide or sulphuric acid, which attack carbonates present in the stone. Compounds of larger volume are formed, which lead to flaking of the stonework. Disfigurement by blackening of stone surfaces may be a result of simple deposition of particles, but it may also involve a more complex organic process which first requires the absorption of SO_2.

Aerosols reduce visibility in their own right; in addition, particles may act as condensation nuclei to cause or aggravate water fogs. Apart from the 'nuisance' aspect of reduced visibility, it is also evident that limited visual range is a contributing cause of automobile and aircraft accidents.

Release of species such as CO or CH_4, or alteration by man of their natural production rates, seems to exert its main effect in an indirect way. Of course, carbon monoxide is toxic at high concentrations, and may be implicated in traffic accidents at rush-hour periods when urban atmospheric concentrations (up to 85 p.p.m.) exceed the amounts known to degrade behavioural performance. However, average tropospheric concentrations (e.g. ~ 0.2 p.p.m. even in the industrialized Northern Hemisphere) are too low to elicit a direct response. Much more important is the reduction in tropospheric [OH] that is likely to follow increased CO emission rates. Reaction (5.6) between CO and OH (cf. Section 5.3.5) is a major loss process for OH (~ 70 per cent) so that [OH] is very sensitive to [CO]. Since the remaining 30 per cent of OH reacts with CH_4 in the 'clean' atmosphere, and CO is the product of oxidation (Section 5.3.2), increases in atmospheric CH_4 also reduce hydroxyl radical levels in the troposphere. Smaller tropospheric [OH] in turn has consequences for the concentrations of a wide variety of natural and man-made trace species, since reaction with OH is often the principal scavenging mechanism. As we have pointed out frequently, the possible responses then include not only changes in the troposphere but also modifications to stratospheric chemistry and global climate.

5.5.3 Primary and secondary pollutants

Pollutants may be grouped into two categories: primary and secondary. *Primary pollutants* are the chemical species emitted directly from identifiable sources. *Secondary pollutants*, on the other hand, are species formed from the primary pollutants by chemical transformation. Adverse effects of pollution are often associated more with the secondary than with the primary pollutants. For example, although atmospheric sulphur dioxide has itself many harmful effects, the sulphuric acid formed as a secondary pollutant by oxidation of SO_2 is even more damaging to the environment.

Table 5.2 shows estimates of some man-made emissions for the year 1976. For comparison, a rough idea is given of the natural source strengths for the chemical species. (We note here that different assumptions about natural sources lead to very different values: the figures quoted in Table 5.2 are intended to be self-consistent with those for the man-made emissions. Man-accentuated processes are not included, so that the total contribution of man's activities to species such as CH_4, CO, or NO_x may be much higher.)

Examination of Table 5.2 shows immediately that man releases very large quantities of material to the atmosphere. The units are 10^9 kg yr^{-1}—that is, millions of tonnes annually. For most species, the 'anthropogenic' emissions are a minor, but significant, contribution to the global total budget. In the case of sulphur dioxide, man produces more than the natural sources. Overall sulphur emissions include H_2S, COS, CS_2 and the organic sulphides, as well as SO_2. As a world average, man contributes about 70 per cent compared with biogenic sources, sea-spray, and volcanoes, but in the industrial Northern Hemisphere he dominates over the sum of natural sulphur emissions.

Table 5.2   Global man-made and natural emissions of various species (estimates for 1976)[a]

Species	Emission estimate (10^9 kg yr^{-1})	
	Man-made	Natural
CO_2	2×10^4	10^6
CH_4, hydrocarbons	188	1800
CO	600	2500
SO_2	207	~ 10
H_2S	2	50
NO_x (as NO)	90	1200
NH_3	7	1200

[a] Cullis, C. F. and Hirschler, M. M. *Global cycles—2. Man's activities*, Education in Chemistry **17**, 40 (1980).

Table 5.3   Emissions in the United States (estimates for 1977)[a]

Source	Particulate matter	Emission estimate (10^9 kg yr^{-1})			
		Sulphur oxides	Nitrogen oxides	Carbon monoxide	Hydrocarbons
Fuel combustion					
Static (heating, power generation)	5.3	24.3	14.3	1.3	0.3
Transport	1.2	0.9	9.7	91.7	11.8
Industrial and chemical processing	7.1	4.7	0.8	8.2	12.8
Fires (forest and agricultural)	0.6	~ 0	0.1	5.3	0.8
Waste disposal	0.5	0.05	0.1	3.3	0.9
Solvent evaporation	0	0	0	0	3.0
Total, US	14.7	29.9	25.0	109.8	29.6

[a] Source: 1977 National Emissions Report. US Environmental Protection Agency Publication No. EPA-450/4-80-005 (1980).

Table 5.3 gives a more detailed breakdown of the emission sources (for the United States) in order to show the importance of combustion-related processes to atmospheric pollution. Industrial operations include chemical manufacturing, metal smelting and refining, and mineral extraction, and so contain a combustion element. Even without that contribution, fuel combustion is by far the largest source of the oxidized species (SO_x, NO_x, CO). Transport (especially private cars and light duty vehicles using petroleum fuel) is clearly the single most polluting activity in the USA; the quantity of CO and unburnt hydrocarbons emitted reveals something about the efficiency of the internal combustion engine! That the sulphur emission is not higher is a result of desulphurization of the fuel. Motor spirit ('gasoline') contains between 0.026 per cent (USA Premium grade) and 0.040 per cent (UK) of sulphur, compared with several per cent in the crude oil.

5.5.4 Sulphur dioxide chemistry

Sulphur dioxide is present at mixing ratios of less than 10^{-9} in the free troposphere of remote areas (eg the South Pacific), up to 30×10^{-9} in rural areas of the Northern Hemisphere, and as high as 2×10^{-6} in heavily polluted environments. The results suggest an atmospheric lifetime measured in weeks, and a largely anthropogenic source. Natural contributions to sulphur dioxide are generally a result of oxidation of reduced sulphur compounds as discussed in Section 5.3.7.

In the troposphere, SO_2 is almost all oxidized to H_2SO_4, in the form of aerosol, and the atmospheric sulphur cycle is closed by wet precipitation of the sulphuric acid. The aerosol particles, once formed, are rapidly incorporated into water droplets, and the particles may, indeed, act as condensation nuclei. Gas-phase, aqueous-phase, and surface oxidation steps all seem to contribute to the overall conversion of SO_2 to H_2SO_4. Meteorological conditions such as the relative humidity and the presence or absence of clouds or fogs are likely to control the relative importance of the homogeneous and heterogeneous (Section 5.3.9) processes. Enhanced oxidation rates at high relative humidities or when condensed water is present indicate the participation of liquid-phase reactions. Rates of wet and dry deposition of SO_2 itself are five to ten times faster than the rate of homogeneous oxidation in the boundary layer, thus suggesting that heterogeneous conversion of SO_2 to H_2SO_4 must normally constitute the major oxidation path. Some reaction may occur on solid aerosol surfaces, but within a cloud most of the oxidation takes place in the liquid phase. However, there are uncertainties about the rates for many of the reactions possible in the gas and liquid phases. In highly polluted environments, species present other than sulphur compounds may act as oxidants, and various metals can act as catalysts. Quantitative estimates of the contributions of the different routes to SO_2 oxidation are not, therefore, available. Instead, we must examine the processes most likely to effect homogeneous and heterogeneous oxidation.

Photodissociation of SO_2 has a threshold wavelength of about 210 nm, corresponding to the O–SO bond energy of ca. 548 kJ mol^{-1}. Radiation of this wavelength does not penetrate to the troposphere, and the process cannot, therefore, play a part in the tropospheric oxidation of SO_2. Significant quantities of sunlight *are* absorbed to generate excited singlet and triplet states, but physical deactivation is the major fate, and direct photo-oxidation of gaseous SO_2 to H_2SO_4 in air is negligible. The ubiquitous hydroxyl radical seems once again to initiate the dominant oxidation route by addition to SO_2

$$OH + SO_2 + M \rightarrow HOSO_2 + M. \qquad (5.72)$$

The $HOSO_2$ radical is further oxidized to H_2SO_4, although the mechanism is probably more complex than just the addition of a further OH radical. Kinetic studies of the OH–SO_2 reaction, using long-path Fourier transform infra-red spectroscopy to follow chemical changes, have begun to help in further elucidation of the oxidation mechanism. In irradiated mixtures of HONO (source of OH), CO, SO_2, NO_x and O_2/N_2, the [OH] is insensitive to the concentration of SO_2 in the mixture, suggesting that SO_2 termination of HO–HO_2-chain reactions is unimportant. The overall reaction seems most nearly represented by the equation

$$OH + SO_2(+O_2, H_2O) \rightarrow H_2SO_4 + HO_2. \tag{5.73}$$

Other experiments have shown that aerosol can be formed when OH reacts with SO_2 in the presence of O_2 and H_2O. The direct reaction of O_2 with $HOSO_2$

$$O_2 + HOSO_2 \rightarrow HO_2 + SO_3 \tag{5.74}$$

would be followed by rapid hydrolysis of SO_3 to yield H_2SO_4. Reaction (5.74) is probably slightly endothermic, so it may be that O_2 adds to hydrated $HOSO_2$ radicals, $HOSO_2 \cdot H_2O$, the enthalpy of hydration then being available to promote reaction to the final products, $H_2SO_4 + HO_2$. Experimental determinations of the rate constant for reaction (5.74) suggest that the lifetime of the $HOSO_2$ adduct would be only 0.5 μs at the Earth's surface, so that regardless of whether hydration of the adduct occurs before reaction, H_2SO_4 formation seems assured following addition of OH to SO_2. An interesting consequence of HO_2 formation is that the oxidation is catalytic in the presence of NO, since OH is regenerated in the process

$$HO_2 + NO \rightarrow OH + NO_2. \tag{5.12}$$

As an alternative to the exchange reaction with O_2, it has been suggested that the oxidation of $HOSO_2$ may first involve addition of O_2 to form HSO_5 ($HOSO_2O_2$) that has some of the properties of other peroxy radicals, but which may also become hydrated:

$$HOSO_2 + O_2 \overset{M}{\rightleftharpoons} HSO_5 \tag{5.75}$$

$$HSO_5 + nH_2O \rightarrow HSO_5(H_2O)_n. \tag{5.76}$$

Hydrated HSO_5 is seen as a strong oxidizing agent that can convert SO_2 to SO_3 in a reaction that is described as 'quasi-heterogeneous' because, with large n, the hydrated radicals are virtually aerosol particles. Sulphuric acid formation is possible, as indicated by the equation

$$HSO_5(H_2O)_n + SO_2 \rightarrow HSO_4(H_2O)_nSO_3$$
$$\rightarrow HSO_4(H_2O)_{n-m} + H_2SO_4(H_2O)_{m-1}. \tag{5.77}$$

Reactions analogous to (5.77) can also oxidize NO and HO_2. Nitric acid is a potential product of the reaction with NO:

$$HSO_5(H_2O)_n + NO \rightarrow HSO_4(H_2O)_nNO_2$$
$$\rightarrow H_2SO_4(H_2O)_{n-m} + HNO_3(H_2O)_{m-2}, \tag{5.78}$$

a matter of considerable interest since nitric and sulphuric acids are frequently found together in polluted environments (see Section 5.5.6). The hydrated HSO_4 radical which appears as a product in reaction (5.77) is likely to be a highly reactive species that could react with itself, or with HO_2 or NO_2.

Peroxy radicals may play a minor role in the atmospheric oxidation of SO_2. HO_2 itself reacts rather slowly with SO_2, but the reaction of methylperoxy (CH_3O_2) is relatively rapid, and in the boundary layer, especially of polluted atmospheres, oxidation *via* this route may be significant. Ozone cannot react significantly with SO_2 *in the gas phase*. However, ozone reacts with olefins to produce 'Criegee biradicals' such as CH_3CHOO (from propene) that are thought to oxidize SO_2

$$CH_3CHOO + SO_2 \rightarrow CH_3CHO + SO_3. \qquad (5.79)$$

These reactions cannot be a major route for SO_2 oxidation under most atmospheric conditions, but they are distinguished from many other SO_2 oxidation mechanisms in that they can occur at night.

We consider now the oxidation from the S(IV) oxidation state of SO_2 to S(VI) of SO_4^{2-} in the aqueous droplets that constitute aerosols, clouds, fogs, and rain. Solution of SO_2 in water proceeds first through hydration of the neutral SO_2 and then by ionization to yield HSO_3^- and SO_3^{2-} ions

$$SO_2 + H_2O \rightleftharpoons H^+ + HSO_3^-, \qquad (5.80)$$

$$HSO_3^- \rightleftharpoons H^+ + SO_3^{2-}. \qquad (5.81)$$

These ions are finally oxidized to SO_4^{2-}. Hydrogen peroxide and ozone dissolved from the gas phase both oxidize HSO_3^-. The reaction with ozone probably proceeds *via* an ionic mechanism that can be expressed as

$$HSO_3^- + OH^- + O_3 \rightarrow SO_4^{2-} + H_2O + O_2. \qquad (5.82)$$

The rate constant for this reaction increases with increasing pH, as does the solubility of the S(IV) reactants, so that oxidation by ozone is probably important only for pH > 4.5. Oxidation of S(IV) by hydrogen peroxide, on the other hand, has a rate coefficient that decreases with increasing pH, which offsets the increasing solubility of S(IV), and makes the production of S(VI) only slightly dependent on pH. The great solubility of H_2O_2 in water droplets means that equilibrium concentrations of H_2O_2 in solution would be about six orders of magnitude greater than those of O_3 expected for atmospheric conditions. It thus seems likely that H_2O_2 is a major oxidant for S(IV) in solution. Molecular oxygen plays a more equivocal role, since it is not clear if any oxidation occurs at all in the absence of catalysts; several metals have, however, been shown to promote the reaction.

Free radicals in the liquid phase, such as OH and HO_2, may play a role in the aqueous oxidation of S(IV). The radicals can be scavenged from the gas phase, or formed by reactions in the droplets, as described in Section 5.3.9. Aqueous OH reacts with HSO_3^- and SO_3^{2-} to yield SO_4^{2-}, possibly in processes involving SO_3^- and SO_5^- as intermediates. Hydroperoxyl radicals act as a source of H_2O_2: the process may be represented by the equation

$$HO_2 + HO_2 \rightarrow H_2O_2 + O_2 \qquad (5.83)$$

although its mechanism may consist of the two steps of reactions (5.67) and (5.68) presented in Section 5.3.9.

Whatever the oxidation mechanism, H^+ ions become dissolved in cloud-water and lower its pH. Precipitation thus brings with it acidity as well as dissolved sulphate. Any SO_2 that is not oxidized remains to exert its own irritant and phytotoxic effects. It is clear that SO_2 is a particularly noxious pollutant, and effects associated with it are discussed in the next two sections.

5.5.5 Smoke and sulphur pollution

Air pollution from the burning of coal has been a problem for centuries. At the beginning of the fourteenth century, Edward I forbade the use of coal because of the smell and smoke it produced: and a violator of that law was executed! London, although not unique in suffering from the combination of smoke and SO_2 pollution produced by coal combustion, has had severe problems well into the twentieth century. London 'pea-souper' fogs have certainly become part of the folk-myth about Victorian London, although the Hollywood version includes Sherlock Holmes and Jack the Ripper prowling through the nearly impenetrable gloom. The word *smog* was coined in 1905 to describe the combination of smoke and fog that was so disastrous.[a] Almost all heavily industrialized cities suffered to some extent. Probably the special place of London in smog history arose because of the size of the city and the scale of industrialization at a time when possible control measures were unknown. British bituminous coal is high in sulphur content, and the tars and hydrocarbons make for a high smoke yield. Such a combination is dramatically effective in fog nucleation in a climate already humid and possibly supersaturated. In 1952, a tragic air pollution episode occurred in London, as a result of which more than 4000 people died.

The deaths that followed acute smog episodes usually involved those with pre-existing heart and respiratory problems, primarily the elderly. It is recognized, however, that human health can suffer as a result of slower subtle effects, especially on the respiratory system, produced at lower air pollution levels. Sulphur dioxide is itself a respiratory irritant, the effects appearing at concentrations above about 1 p.p.m. (SO_2 levels can still reach values as high as 1.5 p.p.m. in some cities, such as New York). Below about 25 p.p.m., irritation is confined to the upper respiratory tract. The situation is greatly altered if particles (e.g. soot) are also present in the pollution. The lower part of the respiratory tract may then be involved at the lower concentration levels. A three- or four-fold potentiation of the irritant response to SO_2 results from the presence of particulate matter. Chronic and acute bronchitis, pleurisy, and

---

[a] *Smog* is now used to describe any smoky or hazy pollution of the atmosphere, and includes the conditions encountered in Los Angeles (Section 5.5.7). A suitable qualifier, such as 'London', 'classical', or 'Los Angeles', 'photochemical', is often used to provide greater clarity of expression.

emphysema are all produced by SO_2-containing smoke such as that generated in the combustion of bituminous coal. Part of the potentiation may be caused by the delivery and retention in the respiratory system of substances, including SO_2, absorbed on soot particles. In addition, as we saw in Section 5.5.4, SO_2 oxidation to H_2SO_4 (or SO_3) may occur efficiently on aerosol surfaces; in heavily polluted atmospheres, the presence of metals may enhance catalytic effects.

Shortly after the 1952 London smog disaster, decisive action was taken in Britain to alleviate pollution. Controls were placed on the type of fuel burnt and the kinds of smoke that might be emitted. A ban on all but 'smokeless' fuels in urban areas has been particularly effective in reducing the emissions of particulate matter if not of sulphur dioxide. Although there were a few further serious episodes in the period immediately following the legislation, pollution disasters related to smog now appear to have become a thing of the past in London. Most other cities report a similar success with smoke control. Sulphur dioxide, however, remains a problem, as the emission inventories of Tables 5.2 and 5.3 show. The most serious consequences may be due in part to the acidity of precipitation, and we turn now to this question.

5.5.6 Acid rain

Natural precipitation—rain and snow—is slightly acid because carbon dioxide is dissolved in the falling droplets. Indeed, the acidity forms part of the geochemical weathering cycle, as discussed in Chapter 1. However, because 'carbonic acid' is a weak acid, the pH in natural rain-water is limited to minimum values of ~ 5.6. Over the last few decades, rain-water of much greater acidity (lower pH) has been of widespread occurrence, and the acids involved have mostly been 'strong' ones such as sulphuric, nitric, and hydrochloric. Acidic rain is potentially damaging to the environment, the two most serious influences appearing to be on freshwater fish and on forest ecology. Strong acids such as H_2SO_4 and HNO_3 have their origin in gaseous SO_2 and NO_2, while HCl may be produced by the reaction of H_2SO_4 with atmospheric NaCl of marine origin. We shall show shortly that most of the strong acid load is a consequence of fossil fuel combustion.

Two regions have been particularly badly affected by the acid rain problem. They are the north-eastern United States and neighbouring parts of Canada, and the Scandinavian countries, particularly Sweden and southern Norway. 'Fossil' precipitation is sometimes preserved as glacial ice, and it indicates that the pH was generally greater than 5 outside urban areas prior to about 1930. Direct records in north-west Europe have been kept for three decades now, and show that precipitation has become increasingly acid and that this acidity is more widespread geographically (Fig. 5.6). In some parts of Scandinavia, the H^+ concentration in precipitation has increased by a factor of more than 200 over the last two decades. Average pH in the north-eastern United States

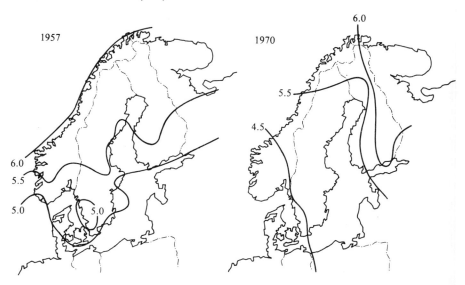

Fig. 5.6. Increase in acidity of precipitation in Scandinavia between 1957 and 1970. The solid lines mark contours of equal pH. [Source: Likens, G. E., *Chem. Engng. News* **54**, No. 48, 29 (Nov. 1976)].

is now between 4.0 and 4.2, but values as low as 2.1 have been recorded for individual storms. The greatest increase in acidity of precipitation in the US appears to have taken place some time between 1930 and 1950. On the other hand, acid rain is no new phenomenon. In the early part of this century, acidity was recognized in the rain-water of the industrial northern cities of England. What is new is the commonplace nature of acid precipitation in regionally widespread and remote areas.

One important question to be answered is whether or not the increased acidity in precipitation is causally connected with combustion-related releases of SO_2 and NO_x. An alternative possibility, which would require quite different control strategies, would be that changes in biogenic sources (perhaps related to fertilizer use) had arisen over the last 20 to 30 years. On the *global* scale, man contributes so little to NO_x (Table 5.2) that only small effects might be anticipated. For sulphur, man already contributes more *dioxide* than nature, and—depending on the estimate adopted for biogenic sources—one-quarter to one-half of the total sulphur. Several pieces of evidence plainly implicate combustion of fossil fuels. Potential sources of biogenic sulphur emission, such as the salt marshes along the eastern seaboard of the US, or in the Baltic sea, generally lie *downwind* of the areas experiencing highest acidity. Major industrial areas lie upwind, and there is a high correlation between low pH in precipitation and storm tracks that have passed over large emission sources. Most (about two-thirds) of the acidity in the affected rain-water is due to H_2SO_4, but the concentrations of NO_3^- have increased dramatically over the last ten years. A fourfold increase, for example, is reported for New York State. Over this same period, both SO_2 and NO_x emissions from fuel combustion have increased sharply, but the fractional contribution of NO_x has risen even more (probably because improved de-sulphurization techniques have prevented the full load of SO_2 from reaching the atmosphere). The increase and geographical extension of acid precipita-tion came during a period when local industrial pollution generally declined. As we pointed out at the end of Section 5.5.1, smoke-stacks have been heightened to disperse pollutants and thereby reduce concentrations locally at ground level. Dispersal means, however, that the pollutants are delivered to the free troposphere, in which they can be transported over long distances. One infamous stack in Ontario, Canada, is over 400 m tall: its *visible* plume can be detected for distances of up to 200 km.

Various types of 'tracer' experiment bear out the idea that the acids, or their oxidic precursors, are transported over long distances. Sulphur hexafluoride (SF_6) has been injected into the waste gases leaving power station chimneys, and its dispersion followed by an instrumented aircraft. Release from the east coast of England can be followed well on the way to Scandinavia. Detection of the element vanadium in Arctic air suggests as a source the burning of fuel oil, which is rich in vanadium, in middle latitudes. Manganese is also found, and this element is released by metal processing and coal burning. Measure-ments of the two elements together pinpoint the source of the metals in the Arctic as the central Soviet Union and western Europe. Similar analysis of the [Mn]/[V] 'signature' for the north-eastern US indicates sources in the Midwest, where the ratio is high (1–10), rather than the east, where less coal and more oil is burned (ratio 0.1–0.2). Caution must be exercised in extending these data to acid precipitation, since the metals leave the air as deposited aerosols while the acidic components remain as gases. None the less, trace element analysis, perhaps utilizing isotopic abundances of arsenic and selenium, may prove to be a valuable method for identifying distant pollution sources.

The data presented do suggest anthropogenic sources of pollutants, coupled with quite long-range transport, as the cause of acid precipitation. Meteorological factors may lead to deposition, in a small area, of pollutants picked up over widespread densely urbanized and industrialized areas. Such a situation is thought to exist in Scandinavia, where the surface features favour precipitation of moisture from the air. More than 70 per cent of the sulphur in the atmosphere over Sweden is thought to be anthropogenic, and, of this load, 77 per cent may originate outside Sweden. Britain and the Ruhr Valley, together with some Eastern Bloc countries, are cited as the sources of the foreign sulphur and the associated increase in acid precipitation.

Let us now consider the damage done by acidity. Large areas of southern

Scandinavia and the north-eastern US—those very regions most affected by acid rain—are underlain by granite-type rocks. Surface waters in such areas contain little dissolved matter, and are poorly buffered (in distinction to those waters underlain by chalk or limestone). The lakes and rivers are thus particularly sensitive to the prevalent acid precipitation, and have undergone extensive ecological damage. Fish are sensitive to acidity, both as a direct toxic response, and because a variety of other aquatic organisms (e.g. algae) in the food web are adversely altered. Freshwater fish have become extinct, or have declined in number, in Sweden, Norway, Canada, and the US. About 10 000 Swedish lakes have been acidified to a pH below 6.0, and 5000 to below 5.0. Fish populations have been seriously affected, with losses of trout and salmon being particularly heavy. Those two types of fish spawn in rivers and streams where pulses of acidity may occur just at the same period as the vulnerable stage of egg hatching. Ion separation in the freezing and thawing process can lead to the acidity being concentrated in the first portion of melt-water liberated in spring. Figure 5.7 indicates at least a statistical

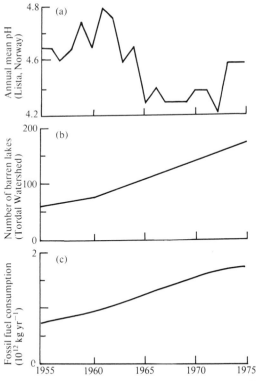

Fig. 5.7. Correlations between increases in (a) rainwater acidity, and (b) barren lakes in Norway, with (c) increases in European fossil fuel consumption. (Source as for Fig. 5.6).

relationship between the barrenness of lakes (curve (b)), rain-water pH (curve (a)) and European fossil fuel consumption (curve (c)). Similar effects have been observed in the Adirondack mountains of New York State. Between 1929 and 1937 only 4 per cent of 217 mountain lakes were devoid of fish, and only 4 per cent had a pH under 5.0. By 1976, half the lakes had a pH below 5.0, and, of these, 90 per cent contained no fish.

Additional information about the effects of acidity on aquatic ecosystems has been gathered from an area of natural airborne acidification. Spontaneous burning of bituminous shales at the Smoking Hills in the Canadian Arctic has been in progress for at least several hundreds, and perhaps thousands, of years. Sulphur dioxide, and H_2SO_4 aerosol, are released, the pH of summer rain-storms being as low as 2.0. Thirty per cent of the ponds studied have a pH in the range 2.5 to 3.5, so that the region provides a valuable opportunity of assessing biological and chemical responses to long-term acid inputs. Other ponds in the region, buffered by the HCO_3^-/CO_2 system, have water of much higher pH (6.5 to 10.5), and represent the normal Arctic environment. One major conclusion of the studies is that the acid ponds have become barren, and have then been repopulated. The biota of the alkaline ponds are typically Arctic in character, while those of the acidic ponds resemble the biota found world-wide in highly acidic waters. The acid-dwelling biota are not a subset of the much more diverse biota of alkaline ponds, but, rather, seem to be pre-adapted invaders.

The other effect of acid rain that we shall consider is that on land vegetation. Interpretation of response is difficult because the gaseous pollutants are themselves phytotoxic. Soil may also be affected, *via* leaching of inorganic ions, reduced nitrogen availability, and decreased soil respiration. Conversely, acid rain may actually supply needed sulphur or nitrogen to soils deficient in these species. Adverse effects can therefore be masked by nutritional benefits. Beyond doubt, forest growth, in particular, has been reduced in southern Scandinavia and the north-eastern US over the last two or three decades. Whether this decline is caused by acid precipitation, or the gaseous pollutants, acting alone or in combination, is not yet known.

Reduced growth of grasses and trees is produced by exposure to SO_2 at low doses, and larger doses lead to cellular leaf injury. Comparison of growth rates of, for example, grasses in purified or in ambient air has shown that the air of British cities is sufficiently polluted to inhibit growth. Such inhibition of growth is being produced by lower concentrations of SO_2 (0.011–0.026 p.p.m.) than those needed to reduce growth ($\gtrsim 0.038$ p.p.m.) if SO_2 is added to pure air. A mixture of SO_2 and NO_2 seems to elicit additive, or more-than-additive, responses, and the effects are markedly affected by season. Grass plants exposed over winter behave in different ways to different exposure regimes. Sulphur dioxide has little effect on its own at concentrations of 0.062 p.p.m., while NO_2 alone *stimulates* growth at the same concentration. A combination of NO_2 and SO_2, however, produces a substantial fall in yield.

Acids themselves can leach nutrients from plant foliage, accelerate cuticular erosion of the leaves, and directly damage the leaves at pH values below 3.5. Germination of seeds is reduced, and the establishment of seedlings made less secure under acid conditions. Finally, other organisms associated with the plant systems (e.g. pathogens) may change in quantity and type in response to increased acidity. The evidence that acid rain has an effect on plants, and especially on forests, is thus not unambiguous, but it does remain probable that a link exists between reduced growth and increased acidity.

The economic, as well as the ecological, costs of acidic precipitation are potentially enormous. Control of SO_2 and NO_x emission is unfortunately also costly. Energy conservation, resulting in reduced fuel consumption, is an attractive strategy, especially in view of the shortage of fossil fuel resources. On the other hand, smaller energy requirements mean a reduction in the supply of material goods, which conflicts with industry's view of the needs of the community. Fuels can be desulphurized, or low-sulphur fuels used. Effluent stack gases can have their SO_2 and NO_x content reduced before release. But these processes are expensive: an estimate has been suggested of \$250 per ton of SO_2 removed. Reduction by one-half of 207×10^9 kg yr^{-1} (Table 5.2) would cost the staggering sum of about $\$2.5 \times 10^{10}$ annually. Put another way, the cost of electric power would increase by at least 10 per cent, which might prove crippling to the already-burdened consumer. Alternative energy sources, such as nuclear fuels, are currently unpopular with the general public precisely because of the pollution dangers. Until fusion power becomes commercially available, some conflict seems bound to arise between economic and environmental pressures. Of course, combustion of fossil fuels will last for only a brief period in man's history. No energy conservation plan could reasonably envisage large fractions of our energy needs being met by fossil fuels alone for more than a few hundred years. By that time, an alternative energy source or a completely new life-style must have been found. Until then, acid rain—often produced by one nation to the detriment of another—seems to be one of the most serious problems in atmospheric pollution.

5.5.7 Photochemical smog

The London type of air pollution, characterized by particulate matter and SO_x (Section 5.5.5) has been recognized for centuries. In the mid-1940s, the effects of a new kind of oxidizing air pollution, which caused eye irritation, plant damage, and visibility degradation, became evident in Los Angeles. The oxidants include ozone, nitrogen dioxide, and peroxyacetyl nitrate (PAN), and they are formed photochemically by the action of solar radiation on mixtures of NO_x and hydrocarbons (HC) in air. Because of the region where it was first observed, and because of its photochemical origin, oxidizing pollution of the kind described is called *Los Angeles Smog* or *Photochemical Smog*. Very many other cities, especially in the south-west USA, but also in Israel, Japan, Australia, and even in the UK, have subsequently been found to suffer from photochemical pollution. Photochemical smog has, however, been a particularly serious problem in Los Angeles and Southern California, and there are several contributing factors. The Los Angeles basin faces the Pacific Ocean to the south-west, and is otherwise almost enclosed by mountain ranges. This topography results in frequent temperature inversions in the boundary layer and lower troposphere, and pollutants are trapped within the basin. Intensely sunny days are frequent, thus promoting photochemical processes. Last, but not least, there is a very high density of automobiles that are thought to be the most important source of primary pollutants. Curiously, it turns out that photochemical 'pollution' is not really a new phenomenon at all! San Pedro Bay was named the 'Bay of Smokes' in 1542, and eye irritation was first recorded in Los Angeles by 1868. The blue haze of the Smoky Mountains—a tourist attraction—probably owes its existence to a biogenic variant of the automobile problem, as we shall mention later. The severity and incidence of pollution have, however, grown out of all recognition from the early precursors of photochemical smog, and the growth has paralleled the vast expansion of use of the internal combustion engine (especially for light motor vehicles) over the last fifty or sixty years.

What is observed on a smoggy day? Nitric oxide (NO) concentrations build up during the night and during the early-morning period of heavy commuter traffic. After dawn, NO becomes replaced by NO_2, and ozone is generated. By noon there are high concentrations of ozone and nitrogen dioxide in the atmosphere, there is a brown haze because particles are present, and the eyes run because PAN, a powerful lachrymator, is formed. Figure 5.8 shows the

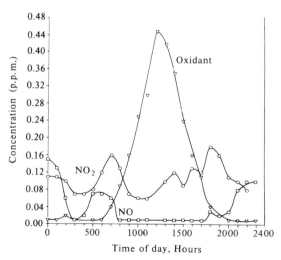

Fig. 5.8. Variations in concentration of oxidant (mainly ozone) and oxides of nitrogen during the course of a smoggy day in Southern California. [Source: Finlayson-Pitts, B. J. and Pitts. J. N., Jr. *Adv. Environ. Sci. Technol.* **7**, 75 (1977).]

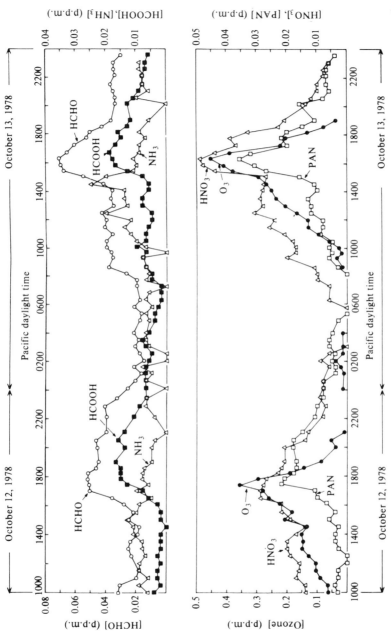

Fig. 5.9. Long-path Fourier Transform Infrared study of the build-up and decay of ozone, peroxyacetyl nitrate (PAN) and other species during two successive polluted days in Southern California. [From Tuazon, E. C., Winer, A. C., and Pitts, J. N., Jr., *Environ. Sci. Technol.* **15**, 1232 (1981).]

inter-relations between $NO-NO_2-O_3$ concentrations during an air pollution episode, while Fig. 5.9 indicates clearly the diurnal variations of several key species during two days of severe pollution in Southern California. The measurements on which the second figure is based were obtained by infra-red spectroscopy, using a multipass cell of total optical path of up to 1 km, and an interferometric spectrometer ('Fourier Transform Infra-red Spectroscopy', FTIR) to provide high speed and high resolution. FTIR determinations have placed *in situ* measurement of pollutants on a secure footing in recent years.

Because of the complexity of the chemical reactions involved in photo-chemical smog formation, one experimental strategy for studying smog processes has been to use large test chambers ('*smog chambers*') in which is contained a 'surrogate' gas mixture intended to simulate the composition of polluted air. Sunlight or artificial light is then used to irradiate the mixture, and photochemical change is followed by a variety of suitable analytical techniques, including long-path FTIR spectroscopy. The temporal behaviour of pollutant species in the atmosphere is well-mimicked by the smog chamber experiments. Figure 5.10 demonstrates the effects of irradiating an olefin–NO_x–air mixture with a 25 kW xenon lamp as source. Trends observed in the

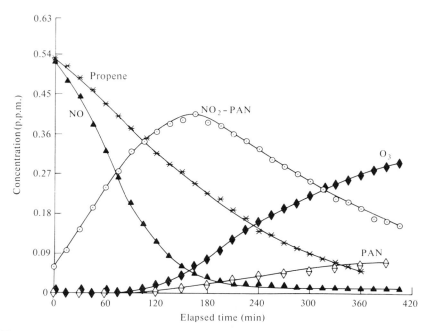

Fig. 5.10. Concentration–time profiles of the major primary and secondary pollutants during irradiation of 0.53 p.p.m. propene and 0.59 p.p.m. NO_x in 1 atm of purified air in an evacuable smog chamber (Source as for Fig. 5.8).

pollution episodes (Figs. 5.8 and 5.9) show up clearly: NO_2 builds up at the expense of NO, and O_3 at the expense of NO_2, and propene is consumed.

We must now consider the chemistry that gives rise to the observed behaviour. The primary pollutants in automobile exhaust are NO_x (mainly NO) from the high-temperature combustion, carbon monoxide, partially oxidized and unburnt hydrocarbons, and sulphur dioxide from sulphur-containing fuels. It is these species that then undergo photochemical transformation to ozone, nitrogen dioxide, aldehydes, ketones and acids, PAN, and inorganic and organic aerosols. As we have already emphasized (Section 5.3.3), NO_2 photolysis followed by $O + O_2$ combination is the only known tropospheric source of O_3:

$$NO_2 + hv \rightarrow O + NO. \tag{5.5}$$

$$O + O_2 + M \rightarrow O_3 + M. \tag{5.4}$$

A route must therefore be found for conversion of the primary NO to NO_2, before ozone can be formed. Inorganic chemistry on its own seems unable to bring about the oxidation. Three-body reaction

$$NO + NO + O_2 \rightarrow 2NO_2, \tag{5.84}$$

is far too slow at the concentrations of NO present (its rate is proportional to $[NO]^2$). Reaction of NO with O_3,

$$NO + O_3 \rightarrow NO_2 + O_2, \tag{5.32}$$

would be fast enough to convert NO to NO_2, but, paradoxically, requires ozone to be available already (and the sequence (5.32), (5.5), (5.4) neither creates nor destroys ozone). Obviously, the organic species released in the exhaust gases must play a part in the oxidation process and in ozone formation. This result is confirmed in test chamber experiments where the organic components are omitted from the gas mixture: no ozone is formed. The part played by reaction (5.32) seems to be to prevent O_3 build-up until almost all free NO has been consumed (and converted to NO_2) as suggested by the curves of Fig. 5.10.

We do, of course, already know processes that convert NO to NO_2 in the unpolluted troposphere. They include the reactions with peroxy radicals

$$HO_2 + NO \rightarrow OH + NO_2, \tag{5.12}$$

$$CH_3O_2 + NO \rightarrow CH_3O + NO_2, \tag{5.13}$$

$$RO_2 + NO \rightarrow RO + NO_2. \tag{5.17}$$

These reactions do, indeed, seem to be the critical ones for oxidizing NO in photochemical smog. Smog chemistry is, then, a grotesquely exaggerated form of the oxidation and transformation chemistry (Section 5.3) of the unperturbed troposphere. Higher concentrations of primary species (e.g. NO_x, HCs) are present in the polluted atmosphere, and perhaps a wider variety of saturated, unsaturated, and aromatic HCs is liberated. But the oxidation chain is still carried by OH, HO_2, and organic oxy- and peroxy-radicals, as in the natural troposphere. We may emphasize the conversion of NO to NO_2 in the chain process by writing the reaction sequence following attack of OH on an alkane, RCH_3 [reactions (5.86) to (5.88) are more detailed forms of eqns (5.18), (5.17), and (5.20)]:

$$OH + RCH_3 \rightarrow H_2O + RCH_2 \qquad (5.85)$$

$$RCH_2 + O_2 \rightarrow RCH_2O_2 \qquad (5.86)$$

$$RCH_2O_2 + NO \rightarrow RCH_2O + NO_2 \qquad (5.87)$$

$$RCH_2O + O_2 \rightarrow RCHO + HO_2 \qquad (5.88)$$

$$HO_2 + NO \rightarrow OH + NO_2 \qquad (5.12)$$

Net $\qquad RCH_3 + 2NO + 2O_2 \rightarrow RCHO + 2NO_2 + H_2O.$

Attack of OH on RCHO continues the hydrocarbon oxidation by yielding carbonyl radicals

$$OH + RCHO \rightarrow RCO + H_2O, \qquad (5.89)$$

that can lead, directly or indirectly, to carbon monoxide and the radical R possessing one less carbon atom than the starting hydrocarbon. Acids, RCOOH, are a minor product from RCO radicals, but they, together with the aldehydes, are found in photochemical smog (Fig. 5.9). The degradation from RCH_2 to R radicals is accompanied by aldehydes and acids down to HCHO and HCOOH.

Carbon monoxide is the final oxidation product of the organic chain, but is itself oxidized by OH (Section 5.3.5). The sequence

$$OH + CO \rightarrow H + CO_2 \qquad (5.6)$$

$$H + O_2 + M \rightarrow HO_2 + M \qquad (5.8)$$

$$HO_2 + NO \rightarrow OH + NO_2 \qquad (5.12)$$

Net $\qquad NO + O_2 + CO \rightarrow NO_2 + CO_2$

can be written to show again how this oxidation of CO can be accompanied by the oxidation of NO to NO_2.

Although we have illustrated the oxidation chain with an alkane as fuel, olefins, in fact, react with OH radicals even faster, at rates approaching the collision- or diffusion-controlled limit. The initial attack appears to be the addition of OH to the double bond, with the major products being the appropriate aldehydes and ketones (e.g. CH_3CHO from C_2H_4, and $C_2H_5CHO + CH_3COCH_3$ from C_3H_6). Aromatic compounds constitute a significant fraction of the reactive hydrocarbons in automobile exhaust gases, but little is known of their atmospheric chemistry. Hydroxyl radical attack is rapid, and the products are of particular interest because they may include long-chain oxygenated compounds that can be involved in aerosol formation (see below).

The general mechanism of hydrocarbon oxidation, and of NO to NO_2 conversion, in a free radical chain reaction seems well established. Hydroxyl radicals (and probably HO_2) attack the organic 'fuel' to propagate the chain. We must now consider the origin of radicals in polluted atmospheres. The ozone photochemical source (Section 5.3.1) that is important in the natural troposphere may be supplemented by several other processes. Of these, two are of particular interest, since they involve species detected in photochemical smog. Nitrous acid (HONO) can be formed in the process

$$NO + NO_2 + H_2O \rightarrow 2HONO, \tag{5.90}$$

by either a homogeneous or a heterogeneous route. The molecule is photolysed at relatively long wavelengths ($\lambda < 400$ nm) that reach ground level

$$HONO + hv \rightarrow OH + NO. \tag{5.91}$$

One oxidized molecule, NO_2, is lost in (5.90), but two HONO molecules, and thus potentially two chain-initiating OH radicals, are created. Aldehydes may provide an important entry into the radical chain. One channel for the photodissociation of formaldehyde (at $\lambda < 340$ nm) yields H and HCO radicals. Both these radicals are converted to HO_2 in the presence of O_2, so that the photochemical initiation steps can be represented by the sequence

$$HCHO + hv \rightarrow H + HCO \tag{5.15}$$

$$HCO + O_2 \rightarrow CO + HO_2 \tag{5.16}$$

$$H + O_2 + M \rightarrow HO_2 + M \tag{5.8}$$

Net $\qquad HCHO + 2O_2 + hv \rightarrow 2HO_2 + CO.$

Higher aldehydes are probably less significant sources of chain carriers, since they require shorter wavelength ultraviolet light for photolysis. Formaldehyde itself is emitted directly into the air from automobile exhausts.

Smog chamber experiments support the idea that aldehydes are important. Addition of formaldehyde to a hydrocarbon–NO_x–air mixture causes NO to be converted much more rapidly into NO_2, and ozone appears earlier and at higher concentrations. Concentration–time profiles are shown in Fig. 5.11 for irradiation of butane–NO_x mixtures with and without added aldehyde. Even the relatively small amount of HCHO used has a dramatic effect on the conversion rates.

Aldehydes are also implicated in the formation of peroxyacetyl nitrate, an important component of photochemical smog. Carbonyl compounds, either emitted as primary pollutants, or produced *via* processes such as (5.88) as

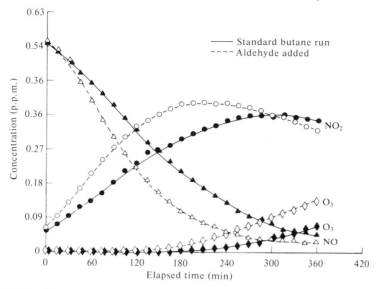

Fig. 5.11. Concentration–time profiles for some primary and secondary pollutants during irradiation of 2.2 p.p.m. *n*-butane and 0.61 p.p.m. NO$_x$ in an evacuable
smog chamber without added aldehyde (——) and with 0.13 p.p.m. HCHO added
(- - -). (Source as for Fig. 5.8).

oxidation intermediates, can be converted to acyl radicals in reaction (5.89)
and thence to peroxyacyl radicals

$$OH + RCHO \rightarrow RCO + H_2O, \tag{5.89}$$

$$RCO + O_2 \rightarrow RCO.O_2. \tag{5.92}$$

Addition of NO$_2$ to RCO.O$_2$ then yields a peroxyacyl nitrate:

$$R.CO.O_2 + NO_2 \rightarrow RCO.O_2.NO_2. \tag{5.93}$$

Formation of PAN itself from acetaldehyde is shown in reactions (5.21), (5.22)
followed by (5.34) (p. 217 and p. 221). Peroxyacyl radicals behave in a similar
way to peroxyalkyl radicals in oxidizing NO to NO$_2$:

$$R.CO.O_2 + NO \rightarrow RCO.O + NO_2, \tag{5.94}$$

[cf. reactions (5.17), (5.23) and (5.87), p. 217 and p. 257], and so play an
additional part in smog chemistry.

Many of the undesirable effects of photochemical smog arise from the
presence of suspended particulate matter. Nearly half the aerosol mass can
be organic in severe photochemical smog, and of this organic fraction, 95
per cent is secondary in origin. A variety of long-chain aliphatic and aromatic

compounds (including the carcinogen benzpyrene) is found, together with oxygenated species such as acids, esters, aldehydes, ketones, and peroxides. The mechanisms leading to the formation and growth of particles are not yet clear. It has, however, long been known that oxidation by ozone of hydrocarbons such as terpenes leads to polymerization and aerosol formation. Indeed, the 'natural' photochemical smog of California or the Smoky Mountains, alluded to at the beginning of this section, has been ascribed to the reaction between oils from pine forests or citrus groves with ozone naturally present in the troposphere (Section 5.3.3).

Inorganic aerosol in photochemical smog includes sulphate, nitrate, and ammonium ions, as well as a variety of trace metals. Sulphuric acid is formed (Section 5.5.4) from SO_2 released by combustion of sulphur-containing fuels, while nitric acid involves the usual reaction with OH [process (5.33) p. 221]. Ammonia is assigned a prominent role in the neutralization of the acids, especially when there are high local concentrations produced by primary sources (e.g. cattle stations) as in California. The time dependence of NH_3 and HNO_3 concentrations shown in Fig. 5.9 is suggestive of reaction between the two species, with $[NH_3]$ decreasing as $[HNO_3]$ builds up. Sulphur dioxide to H_2SO_4 conversion seems to be accelerated by photochemical smog. Observation shows that, in the Los Angeles area, oxidation of SO_2 may proceed up to 100 times faster than the simple photo-oxidation rate. Recent laboratory investigations have also shown that an intermediate produced in ozone–olefin reactions in air can rapidly oxidize gaseous SO_2 to sulphate (cf. p. 245).

Photochemical air pollution degrades the 'quality' of the environment in all the ways outlined in Section 5.5.2. Human health is affected primarily by the oxidant species such as ozone, but PAN, NO_2, and the aerosols are also harmful. Impairment in physical performance has been demonstrated at oxidant levels (ozone + PAN) above 0.15 p.p.m. Attacks in asthmatics are exacerbated at 0.25 p.p.m. of oxidant (but the same level may have no effect on healthy persons). In general, it seems that discomfort can be perceived—as chest pains, cough, and headache—for concentrations of oxidant beyond 0.25 to 0.3 p.p.m. The US 'air quality standard' quotes a limiting ozone concentration of 0.08 p.p.m. over one hour. Reference to Figs 5.8 or 5.9 shows that the ozone concentrations in the episodes represented exceeded the standard throughout the period 10 a.m. to 6 p.m., and that the levels around midday were sufficient to have noticeable effects on health. In Great Britain, meteorological conditions rarely favour serious smog formation, although atmospheric ozone has exceeded the US air quality standard on several occasions since regular monitoring began in 1970. During the exceptionally hot summer of 1976, photochemical pollution was enhanced, and between 22 June and 17 July rural hourly-mean ozone levels exceeded 0.25 p.p.m. At one one rural site, ozone concentrations were in excess of 0.1 p.p.m. over at least eight hours for 18 consecutive days of the 21 days of the episode.

Peroxyacetylnitrate, PAN, is a powerful *lachrymator*, as well as having the effects on the respiratory system of the oxidants. That is, it causes intense irritation of the eye, with consequent tear formation. Irritation increases steadily for oxidant concentrations between 0.1 and 0.45 p.p.m., although ozone on its own is *not* an eye irritant. We cannot be clear whether eye irritation constitutes a real impairment of public health since it is reversible, and there is no proven association between pollution-induced irritation and chronic eye damage. Nevertheless, the effect is undoubtedly unpleasant, and is perhaps the most obviously perceivable nuisance aspect of being exposed to photochemical air pollution.

Vegetation is easily harmed by photochemical air pollution. Once again, the main agents of damage are PAN, which is one of the most phytotoxic substances known, and ozone. Plants respond to the oxidants by first increasing their cell membrane permeability. Higher doses lead to cellular and biochemical changes with visible leaf injury, leaf drop and reduced vigour and growth, and finally death. The classic injury caused by PAN is a glaze, followed by bronzing of the lower leaf surface, although the upper leaf surface is affected in plants such as tobacco. Serious leaf damage to experimental tobacco crops in England was reported in early July 1976, following the June–July pollution episode. Growth experiments in field chambers have shown that there can be up to 50 per cent loss in yield in citrus fruits, grapes, potatoes, and tobacco, and up to 30 per cent loss in cotton. As little as 0.05 p.p.m. of oxidant can cause significant effects, so that the implications for agricultural economy are very serious.

Can anything be done to reduce and control the formation of photochemical smog? The State of California, one of the worst afflicted areas, has taken the lead in recommending and enforcing legislation. One of the first moves, in 1961, was to require the installation of positive crankcase ventilation on new and used cars, followed by the approval of catalytic converters in 1964–6 to reduce hydrocarbon and carbon monoxide emissions. In 1966, an alternate, 'lean-burn' method of reducing HC and CO emissions was implemented, that did not require catalytic devices or afterburners. Unfortunately, this approach led simultaneously to large increases in NO_x emissions. By 1975, catalytic converters were virtually universal, and stringent standards for HC and CO emissions could be met without greatly impairing engine performance or economy. An essential requirement of catalytic afterburners is that they have a long enough lifetime, and that they should be effective against partially oxidized species. Indeed, incomplete oxidation of hydrocarbons over an inefficient catalyst could aggravate the pollution problem by producing aldehydes, which are more reactive in initiating and promoting smog than the parent hydrocarbons.

Control of NO_x emissions is no less essential than that of HCs, but there is considerable debate as to the degree of control that is appropriate for any particular area. The nature of the problem can be seen from the consequences of the 1966 California legislation, which led to increased NO_x levels. The legislation did have the required effect of reducing average ozone levels in downtown Los Angeles, but, unfortunately, oxidant levels *downwind* of the central area actually increased. Ozone concentrations near the release area were decreased because NO—the main component of the increased NO_x emission—rapidly destroys O_3 [reaction (5.32)]. While the air is being transported downwind, the NO_2 is photolysed, and ultimately produces ozone. Similar pollution patterns to those of the Los Angeles basin were seen in the British smog episodes of 1976. Rural sites downwind of the primary pollutant source (in this case, presumably, Greater London) experienced the highest oxidant concentrations. Ozone measurements made at five different sites during the episode suggest that the Greater London source was augmented by secondary pollutant formation over north-west Europe as a whole, and brought to the UK by the prevailing winds.

The complexities of pollution response to NO_x control are well illustrated by the smog chamber results set out in Fig. 5.12. Ozone concentrations were measured for fixed irradiation periods of different mixtures of HC and NO_x with air. Reduced reactive [HC], at constant $[NO_x]$ leads to decreased $[O_3]$ in all cases. However, if one is on the right-hand side of the maximum in the curves (a region typical of urban ambient air), reducing $[NO_x]$ at constant [HC] will *raise* the maximum ozone level. These results, and the atmospheric observations described in the preceding paragraph, emphasize an important aspect of pollution reduction. That is, the most effective strategy may differ according to whether it is areas upwind or downwind of the primary source that are to be protected; what time of day release occurs; and what the relative and absolute HC and NO_x concentrations are.

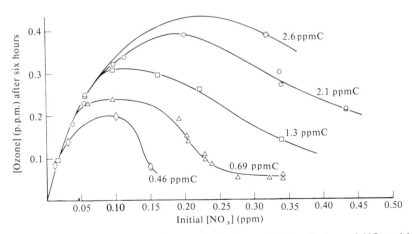

Fig. 5.12. Ozone concentration reached after a 6 h irradiation of NO_x with a surrogate mixture of hydrocarbons simulating ambient air, with varying initial concentrations and ratios of HC/NO_x. (Source as for Fig. 5.8).

Primary emission reduction is the obvious approach to pollution control, but an alternative is to interfere chemically with secondary oxidant formation *after* the primary pollutants have been released. Inhibition of chain reactions by radical scavengers is a well-established phenomenon in the laboratory. The proposal is that some suitable inhibitor be dispersed in polluted atmospheres to interrupt the radical oxidation steps. Much attention has been given to diethylhydroxylamine (DEHA, $(C_2H_5)_2NOH$) as a possible chain inhibitor. Several objections have to be answered before such a control strategy could be adopted. There is the straightforward question of cost and practicability, for the amounts of material required are enormous, and dispersion would have to be fairly even over wide areas. Perhaps more important, though, is the question of toxicity of the inhibitor and its major and minor products: one possible side product from DEHA is diethylnitrosamine, a potent carcinogen. Delayed conversion of NO to NO_2 achieved by an inhibitor may not, in fact, be particularly useful, since it might serve only to increase the pollution downwind of the source later in the day.

Rather than deliberately introduce yet more chemicals into the air we breathe, it seems that the reduction of primary emissions is a better approach. The economic costs of emission control can be offset by changes in life-style. Small decreases in legal speed limits not only conserve fuel, but significantly reduce NO_x emission with negligible increase in hydrocarbon. Increased use of public transport rather than the private car further reduces total emissions. Use of fuels such as hydrogen or methanol eliminates at once the hydrocarbon problem, but it is not yet commercially feasible. The parts of this chapter concerned with pollution will have shown that, in the long term, the cleanliness of the air we breathe, as well as the climate of our planet, will depend on the development of non-combustion energy sources.

Bibliography

As explained in the reference list for Chapter 3, the book by Leighton is not concerned solely with pollution topics. Although some details of the chemistry have been revised since the book was written, it still provides a satisfactory overview of tropospheric chemical processes in unpolluted as well as polluted atmospheres.

Atmospheric chemistry. Finlayson-Pitts, B. J. and Pitts, J. N., Jr. (John Wiley, Chichester, 1986.)

Atmospheric chemistry and physics of air pollution. Seinfeld, J. L. (John Wiley, Chichester, 1986.)

Chemistry of the natural atmosphere. Warneck, P. (Academic Press, London, 1988.)

Gas-phase tropospheric chemistry of organic compounds: a review. Atkinson, R. *Atmos. Env.* **24A**, 1 (1990).

Global tropospheric chemistry: a plan for action. National Research Council. (National Academy Press, Washington, D. C., 1984.)

Photochemistry of air pollution. Leighton, P. A. (Academic Press, New York, 1961.)

The references that follow are to some general accounts of tropospheric chemistry. The first two are concerned primarily with sources, budgets, and cycles of minor species, while the other references consider also details of the chemical transformation processes. The SCOPE reports discussed in connection with the bibliography for Chapter 1 provide much important additional information.

Tropospheric chemistry: a global perspective. Logan, J. A., Prather, M. J., Wofsy, S. C., and McElroy, M. B. *J. geophys. Res.* **86**, 7210 (1981).

The global troposphere: biogeochemical cycles, chemistry and remote sensing. Levine, J. S. and Allario, F. *Environ. Monitg & Assessm.* **1**, 263 (1982).

Chemistry in the troposphere. Chameides, W. L. and Davis, D. D. *Chem. Engng News* **60**, No. 40, 39 (1982).

Gas-phase chemistry of the minor constituents of the troposphere. Cox, R. A. and Derwent, R. G. *Gas kinetics and energy transfer*. Specialist Periodical Reports of the Chemical Society **4**, 189 (1981).

Chemistry of the unpolluted and polluted troposphere. Georgii, H. W. and Jaeschke, W. (eds). (D. Reidel, Dordrecht, 1982.)

Photochemistry of the troposphere. Levy, H., II, *Adv. Photochem.* **9**, 364 (1974).

Organic chemistry of the Earth's atmosphere. Isidorov, V.A., trans. Kovoleva, E.A. (Springer Verlag, Berlin, 1990).

Much tropospheric chemistry involves processes occurring at surfaces or within aerosols and raindrops. Some further general references to this very important, but rather poorly understood, topic can be found in the list for Chapter 1.

Particulate matter in the atmosphere: primary and secondary particles. Chapter 12 in *Atmospheric chemistry*. Finlayson-Pitts, B. J. and Pitts, J. N., Jr. (John Wiley, Chichester, 1986.)

Importance of heterogeneous processes to tropospheric chemistry. Turco, R. P., Toon, O. B., Whitten, R. C., Keesee, R. G., and Hamill. P. *Geophys. Monograph Ser.* **26**, 231 (1982).

Section 5.2

These references are concerned mainly with 'natural' sources of trace atmospheric constituents, but include also some discussion of man's influence (cf. Section 5.5).

Sources, atmospheric lifetimes, and chemical fates of species in the natural troposphere. Chapter 14 in *Atmospheric chemistry*. Finlayson-Pitts, B. J. and Pitts, J. N., Jr. (John Wiley, Chichester, 1986.)

Tropospheric ozone: regional and global scale interactions. Isaksen, I.S.A. (ed.) NATO ASI Series C, (D. Reidel, Dordrecht, 1988.)

Ozone in the troposphere. Fishman, J., in *Ozone in the free atmosphere*, Whitten, R. C. and Prasad, S. S. (eds.). (Van Nostrand Reinhold, New York, 1985.)

Tropospheric ozone: the role of transport. Levy, H., II, Mahlman, J. D., Moxim, W. J., and Liu, S. C. *J. geophys. Res.* **90**, 3753 (1985).

Tropospheric ozone: seasonal behavior, trends and anthropogenic influence. Logan, J. A. *J. geophys. Res.* **90**, 10463 (1985).

Nitrogen oxides in the troposphere: global and regional budgets. Logan, J. A. *J. geophys. Res.* **88**, 10785 (1983).

Nitrogen oxides produced by lightning. Franzblau, E., and Popp, C. J. *J. geophys. Res.* **94**, 11089 (1989).

Production of nitrogen oxides by lightning discharges. Tuck, A. F. *Q. J. R. Meteorol. Soc.* **102**, 749 (1976).

Production of NO and N_2O by soil nitrifying bacteria. Lipschulz, F., Zafiriou, O. C., Wofsy, S. C., McElroy, M. B., Valois, F. W., and Watson, S. W. *Nature, Lond.* **294**, 641 (1981).

Production of nitrous oxide and consumption of methane by forest soils. Keller, M., Goreau, T., Wofsy, S. C., Kaplan, W. A., and McElroy, M. B. *Geophys. Res. Lett.* **10**, 1156 (1983).

Sources and seasonal cycles of atmospheric methane. Khalil, M. A. and Rasmussen, R. A. *J. geophys. Res.* **88**, 5131 (1983).

The influence of termites on atmospheric trace gases: CH_4, CO, $CHCl_3$, N_2O, CO, H_2, and light hydrocarbons. Khalil, M. A. J., Rasmussen, R. A., French, J. R. J., and Holt, J. A. *J. geophys. Res.* **95**, 3619 (1990).

Global production of methane by termites. Rasmussen. R. A. and Khalil, M. A. K. *Nature, Lond.* **301**, 700 (1983).

Biogenic hydrocarbon contribution to the ambient air of selected areas. Arnts, R. R. and Meeks, S. A. *Atmos. Environ.* **15**, 1643 (1981).

Organic material in the global troposphere. Duce, R. A., Mohnen, V. A., Zimmerman, P. R., Grosjean, D., Cautreels, W., Chatfield, R., Jaenicke, R., Ogren, J. A., Pellizzari, E. D., and Wallace, G. T. *Revs. Geophys. Space Phys.* **21**, 921 (1983).

Measurements of atmospheric hydrocarbons and biogenic emission fluxes in the Amazon boundary layer. Zimmerman, P. R., Greenberg, J. P., and Westberg, C. E. *J. geophys. Res.* **93**, 1407 (1988).

Biomass burning appears to be an important source of atmospheric trace gases

Emission of some trace gases from biomass fires. Hegg, D. A., Radke, L. F., Hobbs, P. V., Rasmussen, R. A., and Riggan, P. J. *J. geophys. Res.* **95**, 5669 (1990).

Importance of biomass burning in the atmospheric budgets of nitrogen-containing gases. Lobert, J.M., Scharffe, D.H., Hao, W.M., and Crutzen, P.J. *Nature* **346**, 552 (1990).

Enhanced biogenic emissions of nitric oxide and nitrous oxide following surface biomass burning. Anderson, I. C., Levine, J. S., Poth, M. A., and Riggan, P. J. *J. geophys. Res.* **93**, 3893 (1988).

Carbon monoxide and the burning earth. Newell, R. E., Reichle, H. G., Jr., and Seiler, W. *Sci. Amer.* **261** (4), 58 (Oct 1989).

Biomass burning as a source of atmospheric gases CO, H_2, N_2O, NO, CH_3Cl and COS. Crutzen, P. J., Heidt, L. E., Krasnec, J. P., Pollock, W. H., and Seiler, W. *Nature, Lond.* **282**, 253 (1979).

There is considerable evidence that the concentrations of many trace gases in the troposphere are increasing, probably as a result of man's activities. The oxidizing capacity of the troposphere, and hence the ability of the atmosphere to process added substances, may thus be reduced. See also the references for Sections 1.5 and 5.4.

The changing atmosphere. Rowland, F. S. and Isaksen, I.S.A. (eds.). (John Wiley, Chichester, 1988.)

The changing atmosphere. Graedel, T. E. and Crutzen, P. J. *Scient. Am.* **261**, 28 (Sept 1989).

Trends of atmospheric methane during the 1960s and 1970s. Khalil, M.A.K., Rasmussen, R. A., and Shearer, M. J. *J. geophys. Res.* **94**, 18279 (1989).

Causes of increasing atmospheric methane: depletions of hydroxyl radicals and the rise of emissions. Khalil, M. A. K., and Rasmussen, R. A. *Atmos. Environ.* **19**, 397 (1985).

Constraints on the global sources of methane and an analysis of recent budgets. Khalil M.A.K. and Rasmussen, R.A. *Tellus* **42B**, 229 (1990).

Increased tropospheric concentrations of trace gases can increase radiation trapping and thus influence climate. This topic is explored further in Section 9.6.

Trace gas trends and their possible role in climate change. Ramanathan, V., Cicerone, R. J., Singh, H. B., and Kiehl, J. T. *J. geophys. Res.* **90**, 5547 (1985).

Biogenic sulfur in the environment. Saltman, E. S., and Cooper, W. J. (eds). ACS Symposium Series No. 393, American Chemical Society, Washington, DC, 1989.

Photochemical production of carbonyl sulphide in marine surface waters. Ferek, R. J. and Andreae, M. O. *Nature, Lond.* **307**, 148 (1984).

Oxidation of CS_2 and COS: sources for atmospheric SO_2. Logan, J. A., McElroy, M. B., Wofsy, S. C., and Prather, M. J. *Nature, Lond.* **281**, 185 (1979).

The contribution of volcanoes to the global atmospheric sulfur budget. Berresheim, H. and Jaeschke, W. *J. geophys. Res.* **88**, 3732 (1983).

Section 5.3

The photochemistry of the troposphere. Graedel, T. E., in *The photochemistry of atmospheres*, Levine, J. S. (ed.). (Academic Press, Orlando, Fla, 1985.)

Photochemistry of biogenic gases. Levine J. S., in *Global ecology: towards a science of the biosphere*, Rambler, M. B., Margulis, L., and Fester, R. (Academic Press, Orlando, Fla, 1989.)

Sections 5.3.1 and 5.3.2

Oxidation and chemistry of non-methane hydrocarbons.

Atmospheric chemistry of ethane and ethylene. Aikin, A. C. *J. geophys. Res.* **87**, 3105 (1982).

Permutation reactions of organic peroxy radicals. Madronich, S. and Calvert, J. G. *J. geophys. Res.* **95**, 5697 (1990).

On the mechanism of the interaction between hydroxyl radicals and olefins, and the detection of a stabilized adduct.

Photoionization mass spectrometer studies of the collisionally stabilized product distribution in the reaction of OH radicals with selected alkenes at 298 K. Biermann, H. W., Harris, G. W., and Pitts, J. N., Jr. *J. phys. Chem.* **86**, 2958 (1982).

The direct reaction between ozone and olefins is probably of minor importance in the natural troposphere.

Atmospheric ozone-olefin reactions. Niki, H., Maker, P. D., Savage, C. M., and Breitenbach, L. P. *Environ. Sci. Technol.* **17**, 312A (1983).

Section 5.3.3

An important question, to which some uncertainty is still attached, is that of how much ozone in the troposphere is of tropospheric origin, and how much has been transported down from the stratosphere. The last papers of this group investigate whether tropospheric ozone can have a climatic influence.

Photochemistry of tropospheric ozone. Stewart, R. W., Hameed, S., and Pinto, J. P. *J. geophys. Res.* **82**, 3134 (1977).

Vertical profiles of tropospheric gases: chemical consequences of stratospheric intrusions. Bamber, D. J., Healey, P. G. W., Jones, B. M. R., Penkett, S. A., Tuck, A. F., and Vaughan, G. *Atmos. Environ.* **18**, 1759 (1984).

Stratospheric ozone in the lower troposphere. II. Assessment of downward flux and ground level impact. Viezee, W., Johnson, W. B., and Singh, H. B. *Atmos. Environ.* **17**, 1979 (1983).

On the origin of tropospheric ozone. Liu, S. C., Kley, D., McFarland, M., Mahlman, J. D., and Levy, H., II, *J. geophys. Res.* **85**, 7546 (1980).

Tropospheric ozone, its changes and possible radiative effect. Bojkov, R. D., World Meteorological Organization Special Environment Report No. 16 (1983).

Tropospheric ozone and climate. Fishman, J., Ramanathan, P. J., Crutzen, P. J., and Liu, S. C. *Nature, Lond.* **282**, 818 (1979).

Section 5.3.4

The oxides of nitrogen play a central role in tropospheric chemistry. Nitric oxide (NO) converts peroxy radicals to oxy radicals, and is itself oxidized to NO_2. Nitrogen dioxide (NO_2) is a photochemical source of O atoms. The nitrate radical (NO_3) provides a potential source of night-time oxidation (see Section 5.3.6). Peroxyacetyl nitrate has been identified as a species that could be involved in the spatial transport of NO_x.

The role of NO and NO_2 in the chemistry of the troposphere and stratosphere. Crutzen, P. J. *Annu. Rev. Earth & Planet. Sci.* **7**, 443 (1979).

The atmospheric chemistry of organic nitrates. Roberts, J. M. *Atmos. Environ.* **24A**, 243 (1990).

The global distribution of peroxyacetyl nitrate. Singh, H. B., Salas, L. J., and Viezee, W. *Nature*, **321**, 588 (1986).

Atmospheric measurements of peroxyacetyl nitrate (PAN) in rural southeast England: seasonal variations, winter photochemistry and long range transport. Brice, K. A., Penkett, S. A., Atkins, D. H. F., Sandalls, F. J., Bamber, D. J., Tuck, A. F., and Vaughan, G. *Atmos. Environ.* **18**, 2691 (1984).

Influence of peroxyacetyl nitrate (PAN) on odd nitrogen in the troposphere and lower stratosphere. Aikin, A. C., Herman, J. R., Maier, E. J. R., and McQuillan, C. J. *Planet. Space Sci.* **31**, 1075 (1983).

Peroxyacetyl nitrate in the free troposphere. Singh, H. B. and Salas, L. J. *Nature, Lond.* **302**, 326 (1983).

Section 5.3.5

The first paper is representative of current thinking about the OH + CO reaction, while the second one looks at the indirect atmospheric consequences of the reaction.

The pressure dependence of the rate constant of the reaction of OH radicals with CO. Paraskevopoulos, G. and Irwin, R. S. *J. chem. Phys.* **80**, 259 (1984).

Correlative nature of ozone and carbon monoxide in the troposphere: implications for the tropospheric ozone budget. Fishman, J. and Seiler, W. *J. geophys. Res.* **88**, 3662 (1983).

Section 5.3.6

The nitrate radical: physics, chemistry and the atmosphere. Wayne, R. P. (ed.). Atmos. Environ. **25A**, 1 (1991).

Gaseous nitrate radical: possible night-time sink for biogenic organic compounds. Winer, A. M., Atkinson, R., and Pitts, J. N., Jr. *Science* **224**, 156 (1984).

Section 5.3.7

Human influence on the sulphur cycle. Brimblecombe, P., Hammer, C., Rohde, H., Ryaboshapko, A., and Boutron, C. F. SCOPE **39**, 77 (1989).

Are global cloud albedo and climate controlled by marine phytoplanckton? Schwartz, S. E. *Nature* **336**, 441 (1988).

Atmospheric dimethyl sulphide and the natural sulphur cycle. Lovelock, J. E., Maggs, R. J., and Rasmussen, R. A. *Nature* **237**, 452 (1972).

Global ocean-to-atmosphere dimethyl sulfide flux. Erickson, D.J., III, Ghan, S.J., and Penner, J.E. *J. geophys. Res.* **95**, 7543 (1990).

Section 5.3.8

Impact of halogen oxides on dimethyl sulfide oxidation in the marine atmosphere. Barnes, I., Becker, K. H., Martin, D., Carlier, P., Mouvier, G., Jourdain, J. L., Laverdet, G., and Le Bras, G., in *Biogenic sulfur in the environment*. Saltman, E. S., and Cooper, W. J. (eds.). ACS Symposium Series No 393, American Chemisty Society, Washington, DC, 1989.

Chlorine-hydrocarbon photochemistry in the marine troposphere and lower stratosphere. Singh, H. B., and Kasting, J. F. *J. atmos. Chem.* **7**, 261 (1988).

Section 5.3.9

Influences of cloud photochemical processes on tropospheric ozone. Lelieveld, J. and Crutzen, P. J. *Nature* **343**, 227 (1990).

Section 5.4

Tropospheric measurements
Spectroscopic determinations are naturally part of the armoury of the atmospheric chemist measuring concentrations of trace gases. Fourier Transform Infrared (FTIR) methods have become increasingly popular (see Section 5.5.7). The first paper

referred to below describes an alternative new method using diode lasers. Subsequent references are to papers describing the results of measurement campaigns. First we consider O_3 and NO_x.

Tunable diode laser systems for measuring trace gases in tropospheric air. Hastie, D. R., Mackay, G. I., Iguchi, T., Ridley, B. A., and Schiff, H. I. *Environ. Sci. Technol.* **17**, 352A (1983).

Global tropospheric experiment (GTE). Special section of *J. geophys. Res.* **95**, 10047–10247 (1990).

Distribution of tropospheric ozone determined from satellite data. Fishman, J., Watson, C. E., Larsen, J. C., and Logan, J. A. *J. geophys. Res.* **95**, 3599 (1990).

Tropospheric NO_x and O_3 budgets in the equatorial Pacific, Liu, S. C., McFarland, M., Kley, D. Zafiriou, O., and Huebert, B. *J. geophys. Res.* **88**, 1360 (1983).

NO_3 and NO_2 in the mid-Pacific troposphere. Noxon. J. F. *J. geophys. Res.,* **88**, 11017 (1983).

Measurements of nitrogen oxides in the Arctic. Hanrath, R. E. and Jaffe, D. A. *Geophys. Res. Lett.* **17**, 611 (1990).

Atmospheric implications of simultaneous nighttime measurements of NO_2 radicals and HONO. Pitts, J. N., Jr., Biermann, H. W., Atkinson, R., and Winer, A. M. *Geophys. Res. Lett.* **11**, 557 (1984).

Direct measurements of ozone and nitrogen dioxide photolysis rates in the troposphere. Dickerson. R. R. *J. geophys. Res.* **87**, 4933 (1982).

Optical methods may also be used to measure concentrations of the OH radicals that dominate the chemistry of the troposphere.

Measurement of tropospheric OH concentrations by laser long-path absorption spectroscopy. Dorn, H.-P., Callies, J., Platt, U., and Ehhalt, D. H. *Tellus* **40B**, 437 (1988).

New tropospheric OH measurements. Platt, U., Rateike, M., Junkermann, W., Rudolph, J., and Ehhalt, D. H. *J. geophys. Res.* **93**, 5159 (1988).

Tropospheric OH concentrations: a comparison of field data with model predictions. Perner, D., Platt, U., Trainer, M., Huebler, G., Drummond, J. W., Junkermann, W., Rudolph, J., Schubert, B., Volz, A., Ehhalt, D. H., Rumpel, K. J., and Helas, G. *J. atmos. Chem.* **5**, 185 (1987).

The next papers present general assessments of OH concentrations, and the interrelation between OH, CH_4, and CO. Methyl chloroform concentrations provide another way of calculating [OH] in the troposphere (see text).

The tropospheric lifetime of halocarbons and their reactions with OH radicals: an assessment based on the concentrations of ^{14}CO. Derwent, R. and Volz-Thomas, A. *Alternative fluorocarbon environmental acceptability study (AFEAS)* Volume II of *Scientific assessment of stratospheric ozone: 1989.* World Meteorological Organization, Global ozone research and monitoring project: report no. 20. (WMO, Geneva, 1990.)

Tropospheric hydroxyl concentrations and the lifetimes of hydrochlorofluorocarbons (HCFCs). Prather, M. *Alternative fluorocarbon environmental acceptability study (AFEAS)* Volume II of *Scientific assessment of stratospheric ozone: 1989.* World Meteorological Organization, Global ozone research and monitoring project: report no. 20. (WMO, Geneva, 1990.)

Methane, carbon monoxide and methyl chloroform in the southern hemisphere. Fraser, P. J., Hyson, P., Rasmussen, R. A., Crawford, A. J., and Khalil, M. A. K. *J. atmos. Chem.* **4**, 3 (1986).

On the indirect determination of atmospheric OH radical concentrations from reactive hydrocarbon measurements. McKeen, S.A., Trainer, M., Hsie, E.Y., Tallamraju, R.K., and Liu, S.C. *J. geophys. Res.* **95**, 7493 (1990).

Ammonia is of interest as the only naturally occurring alkaline trace gas. It can be converted (via OH oxidation) to the oxides of nitrogen.

The concentration of ammonia in southern ocean air. Ayers, G. P. and Gras, J. L. *J. geophys. Res.* **88**, 10655 (1983).

Methane and the non-methane hydrocarbons are of particular significance as the 'fuels' that are oxidized in tropospheric chemical conversions.

Measurement of selected light hydrocarbons over the Pacific ocean: latitudinal and seasonal variations. Singh, H. B. and Salas, L. J. *Geophys. Res. Lett.* **9**, 842 (1982).

Non-methane hydrocarbons in remote tropical, continental, and marine atmospheres. Greenberg, J. P. and Zimmerman, P. R. *J. geophys. Res.* **89**, 4767 (1984).

Aldehydes, and especially formaldehyde, are capable of yielding hydroxyl radicals at the wavelengths of light penetrating to the troposphere. Formaldehyde is also an expected product of the oxidation of methane in the natural troposphere. The identification of formaldehyde in air far removed from human sources of pollutants in thus of considerable interest.

Formaldehyde (HCHO) measurements in the nonurban atmosphere. Lowe, D. C. and Schmidt, U. *J. geophys. Res.* **88**, 10844 (1983).

These discussions of trends in tropospheric trace-gas concentrations complement those listed in the references for Section 5.2; they are generally oriented to explanations of the changes.

Calculation of trends in the tropospheric concentrations of O_3, OH, CO, CH_4, and NO. Isaksen, I. S. A. and Hov, O. *Tellus* **39B**, 271 (1987).

Indications and causes of ozone increase in the troposphere. Penkett, S. A., in *The changing atmosphere*. Rowland, F. S. and Isaksen, I. S. A. (eds.). (John Wiley, Chichester, 1988.)

Correlative nature of ozone and carbon monoxide in the troposphere: implications for the tropospheric ozone budget. Fishman, J., and Seiler, W. *J. geophys. Res.* **88**, 3662 (1983).

Secular increase of the vertical column abundance of methane derived from IR solar spectra recorded at the Jungfraujoch station. Zander, R., Demoulin, P., Ehhalt, D. H., and Schmidt, U. *J. geophys. Res.* **94**, 11029 (1989).

Secular increase of the total vertical column abundance of carbon monoxide above central Europe since 1950. Zander, R., Demoulin, P., Ehhalt, D. H., Schmidt, U., and Rinsland, C. P. *J. geophys. Res.* **94**, 11021 (1989).

Is the oxidizing capacity of the atmosphere changing. Isaksen, I. S. A., in *The changing atmosphere*. Rowland, F. S. and Isaksen, I. S. A. (eds.). (John Wiley, Chichester, 1988.)

Sensitivity of tropospheric oxidants to global chemical and climate change. Thompson,

A. M., Stewart, R. W., Owens, M. A., and Heruche, J. A. *Atmos. Environ.* **23**, 519 (1989).

Tropospheric models: some representative recent descriptions.
The first paper describes the chemical kinetic and mechanistic basis for a sophisticated model of the troposphere. The oxidation of over 100 alkanes, alkenes, aromatic hydrocarbons, and other organic molecules is incorporated into the scheme.

A detailed mechanism for the gas-phase atmospheric reactions of organic compounds. Carter, W. P. L. *Atmos. Environ.* **24A**, 481 (1990).

A comparison of chemical mechanisms used in atmospheric models. Dunker, A. M., Kumar, S., and Berzins, P. H. *Atmos. Environ.* **18**, 311 (1984).

Modelling potential ozone impacts from natural hydrocarbons. I. Development and testing of a chemical mechanism for the NO_x-air photo-oxidations of isoprene and α-pinene under ambient conditions. Lloyd, A. C., Atkinson, R., Lurmann, F. W., and Nitta, B. *Atmos. Environ.* **17**, 1931 (1983).

Modelling potential ozone impacts from natural hydrocarbons. II. Hypothetical biogenic HC emission scenario modelling. Lurmann, F. W., Lloyd, A. C., and Nitta, B. *Atmos. Environ.* **17**, 1951 (1983).

Deviations from the O_3-NO-NO_2 photostationary state in tropospheric chemistry. Calvert, J. G. and Stockwell, W. R. *Can. J. Chem.* **61**, 983 (1983).

A two dimensional photochemical model of the atmosphere. 2. The tropospheric budgets of the anthropogenic chlorocarbons, CO, CH_4, CH_3Cl and the effect of various NO_x sources on tropospheric ozone. Crutzen, P. J. and Gidel, L. T. *J. geophys. Res.* **88**, 6641 (1983).

Does the photochemistry of the troposphere admit more than one steady state? White, W. H. and Dietz, D. *Nature, Lond.* **309**, 242 (1984).

The effects of clouds on photolysis rates and ozone formation in the unpolluted troposphere. Thompson, A. M. *J. geophys. Res.* **89**, 1341 (1984).

The first references are to general accounts of tropospheric air pollution and the physico-chemical behaviour of pollutants. National bodies, such as the National Academy of Sciences in the USA, issue reports from time to time. One such report from the UK is

Tackling pollution—experience and prospects. 10th Report of the Royal Commission on Environmental Pollution. Cmnd 9149, HMSO, London, 1984.

Social, economic, and political aspects are dealt with in addition to the chemical ones. Texts more specifically addressing the chemical problems include a series reporting conferences held under the auspices of the Commission of the European Communities, and entitled

Physico-chemical behaviour of atmospheric pollutants.

These first two books, listed also as general introductory texts, are particularly strong on the chemistry of air pollution.

Atmospheric chemistry. Finlayson-Pitts, B. J. and Pitts, J. N., Jr. (John Wiley, Chichester, 1986.)
Atmospheric chemistry and physics of air pollution. Seinfeld, J. H. (John Wiley, Chichester, 1986.)

Urban air pollution: state of the science. Seinfeld, J. H. *Science* **243**, 745 (1989).

Air pollution by particles. Shaw, R. W. *Sci. Amer.* **257** (2), 84 (Aug 1987).

Pollution: causes, effects and control. 2nd edn. Harrison, R. M. (ed.). (Royal Society of Chemistry, London, 1990.)

Atmospheric pollutants. Heicklen, J., Chapter IV in *Atmospheric chemistry.* (Academic Press, New York, 1976.)

Man's impact on the troposphere: Lectures in tropospheric chemistry, Levine, J. S. and Schryer, D. R. (eds). NASA Reference Publication 1022 (1978).

The following paper describes how the abundance ratios of elements provide a 'signature' that can help identify the source of pollution. Specific examples are discussed in Section 5.5.6.

Elemental tracers of distant regional pollution aerosols. Rahn, K. A. and Lowenthal, D. H. *Science* **223**, 132 (1984).

Section 5.5.2

Some effects of air pollution, excluding those on human health. A guide to the literature on health problems can be found in the textbooks on air pollution described earlier.

Non-health effects of airborne particulate matter. Lodge, J. P., Jr., Waggoner, A. P., Klodt, D. T., and Crain, C. T. *Atmos. Environ.* **15**, 431 (1981).

Effects of Gaseous Air Pollution in Agriculture and Horticulture. Unsworth, M. H. and Ormrod, D. P. (eds). (Butterworth, London, 1982).

Review: Atmospheric deposition and plant assimilation of gases and particles. Hosker, R. P., Jr. and Lindberg, S. E. *Atmos. Environ.* **16**, 889 (1982).

Section 5.5.4

Human influence on the sulphur cycle. Brimblecombe, P., Hammer, C., Rohde, H., Ryaboshapko, A., and Boutron, C. F., in *Evolution of the global biogeochemical sulphur cycle*, Brimblecombe, P. and Lein, A. Y. (eds) SCOPE 39, (John Wiley, Chichester, 1989.)

The homogeneous chemistry of atmospheric sulfur. Graedel, T. E. *Rev. Geophys. & Space Phys.* **15**, 421 (1977).

These references are concerned with mechanisms for the oxidation of SO_2 and with the formation of H_2SO_4. Conversion in solution and in cloud droplets is the subject of the later papers.

The mechanism of the $HO–SO_2$ reaction. Stockwell, W. R. and Calvert, J. G. *Atmos. Environ.* **17**, 2231 (1983).

Mechanism of the homogeneous oxidation of sulfur dioxide in the troposphere. Calvert, J. G., Su, F., Bottenheim, J. W., and Strausz, O. P. *Atmos. Environ.* **12**, 197 (1978).

SO_2 oxidation *via* the hydroxyl radical: atmospheric fate of the HSO_x radicals. Davis, D. D., Ravishankara, A. R., and Fischer, S. *Geophys. Res. Lett.* **6**, 113 (1979).

The Reaction of CH_3O_2 with SO_2. Sanhueza, E., Simonaitis, R., and Heicklen, J. *Int. J. chem. Kinet.* **11**, 907 (1979).

Kinetic flash spectroscopic study of the $CH_3O_2–CH_3O_2$ and $CH_3O_2–SO_2$ reactions.

Kan, C. S., McQuigg, R. D., Whitbeck, M. R., and Calvert, J. G. *Int. J. chem. Kinet.* **11**, 921 (1979).

Formation of sulfuric and nitric acids in acid rain and fogs. Chapter 11 in *Atmospheric chemistry*. Finlayson-Pitts, B. J. and Pitts, J. N., Jr. (John Wiley, Chichester, 1986.)

Aqueous oxidation of SO_2 by hydrogen peroxide. Kunen, S. M., Lazrus, A. L., Kok, G. L., and Heider, B. G. *J. geophys. Res.* **88**, 3671 (1983).

Measurements of the oxidation rate of sulfur (IV) by ozone in aqueous solution and their relevance to SO_2 conversion in non-urban tropospheric clouds. Maahs, H. G. *Atmos. Environ.* **17**, 341 (1983).

Kinetics and mechanism of the oxidation of S(IV) by ozone in aqueous solution with particular reference to SO_2 conversion in nonurban tropospheric clouds. Maahs, H. G. *J. geophys. Res.* **88**, 10721 (1983).

Section 5.5.6

Acid rain is regarded as one of the most serious problems of tropospheric pollution. The first papers deal with aspects of oxidative transformation and acid formation not specifically concerned with H_2SO_4 chemistry (Section 5.5.4).

Acid generation in the troposphere by gas-phase chemistry. Calvert, J. G. and Stockwell, W. R. *Environ. Sci. Technol.* **17**, 428A (1983).

A modelling study of SO_x–NO_x-hydrocarbon plumes and their transport to the background troposphere. Balko, J. A. and Peters, L. K. *Atmos. Environ.* **17**, 1965 (1983).

The free radical chemistry of cloud droplets and its impact upon the composition of rain. Chameides, W. L. and Davis, D. D. *J. geophys. Res.* **87**, 4863 (1982).

The mechanism of NO_3 and HONO formation in the nighttime chemistry of the urban atmosphere. Stockwell, W. R. and Calvert, J. G. *J. geophys. Res.* **88**, 6673 (1983).

Effects of heterogeneous processes on NO_3, HONO, and HNO_3 chemistry in the troposphere. Heikes, B. and Thomson, Anne M. *J. geophys. Res.* **88**, 10883 (1983).

Aqueous-phase source of formic acid in clouds. Chameides, W. L. and Davis, D. D. and Martin, L. R. *Atmos. Environ.* **17**, 2005 (1983).

Aqueous-phase source of formic acid in clouds. Chameides, W. L. and Davis, D. D. *Nature, Lond.* **304**, 427 (1983).

Naturally acidic water systems exist because of 'natural' pollution, and the ecology of these systems provides useful clues about the possible consequences of anthropogenic acidification.

The smoking hills: natural acidification of an aquatic ecosystem. Havas, M. and Hutchinson, T. C. *Nature, Lond.* **301**, 23 (1983).

The next references are to reports concerned directly with the causes and effects of acid rain.

Acid rain—cause and consequence. Mason, B. J. *Weather* **45**, 70 (1990).

Acid rain. Park, C. C. (Methuen, London, 1987.)

The acid rain controversy. Dudley, N., Barrett, M., and Baldock, D., Earth Resources Research, London, 1988.

Acid deposition: unravelling a regional phenomenon. Schwartz, S. E. *Science* **243**, 753 (1989).

Quantification of changes in lakewater chemistry in response to acidic deposition. Sullivan, T. J., Charles, D. F., Smol, J. P., Cumming, B. F., Selle, A. R., Thomas, D. R., Bernert, J. A., and Dixit, S. S. *Nature* **345**, 54 (1990).

Biological recovery of an acid lake after reductions in industrial emissions of sulphur. Gunn, J. M. and Keller, W. *Nature* **345**, 431 (1990).

Acid deposition. (National Academy of Sciences Press, Washington, DC, 1983.)

The menace of acid rain. Pearce, F. *New Scient.* **95**, 419 (1982).

Acid rain. Franks, J. *Chem. Br.* **19**, 504 (1983).

Acid precipitation. Likens, G. E. *Chem. Engng News* **54**, No. 48, 29 (1976).

Factors controlling the acidity of natural rainwater. Charlson, R. J. and Rodhe, H. *Nature, Lond.* **295**, 683 (1982).

Rainfall acidity in northern Britain. Fowler, D., Cape, J. N., Leith, I. D., Patterson, I. S., Kinnaird, J. W., and Nicholson, I. A. *Nature, Lond.* **297**, 383 (1982).

Acid rain on acid soil: a new perspective. Krug, E. C. and Frink, C. R. *Science* **221**, 520 (1983).

Air pollutants and forest decline. Tomlinson, G. H., II, *Environ. Sci. Technol.* **17**, 246A (1983).

Acid deposition and forest decline. Johnson, A. H. and Siccama, T. G. *Environ. Sci. Technol.* **17**, 294A (1983).

Certain of the ecological effects attributed to acid rain could have alternative causes. For example, forest damage can be a result of elevated ozone concentrations (see Section 5.5.7), and the recorded decline in fish populations may be a respsonse to changes more subtle than a simple increase in rainfall acidity. The following paper examines some of the alternative possibilities.

Acid Rain—the CEGB view. Howells, G. D. and Kallend, A. S. *Chem. Br.* **20**, 407 (1984).

Section 5.5.7

A better understanding of problems of composition of photochemical smog, and chemical transformations in it, has been provided by the use of tunable lasers and long-path Fourier Transform Infra-red Spectrometers (see also the first reference for Section 5.4).

Optical systems unravel smog chemistry. Pitts, J. N., Jr., Finlayson-Pitts, B. J., and Winer, A. M. *Environ. Sci. Technol.* **11**, 568 (1977).

The next references discuss the mechanisms of smog formation and possible control strategies in kinetic terms.

Photochemical air pollution: mechanisms of formation and chemical basis of control strategy options for oxidant and gaseous airborne toxic chemicals. Part 5 (Chapters 9 and 10) in *Atmospheric chemistry.* Finlayson-Pitts, B. J. and Pitts, J. N., Jr. (John Wiley, Chichester, 1986.)

Air pollution by photochemical oxidants: formation, transport, control, and effects on plants. Guderian, R. (ed.). (Springer-Verlag, Berlin, 1985.)

Mechanisms of photochemical air pollution. Wayne, R. P. in *Light, chemical change and life* (eds. Coyle, J. D., Hill, R. R., and Roberts, D. R.). (Open University Press, Milton Keynes, 1982).

Photochemistry of the polluted troposphere. Finlayson, B. J. and Pitts, J. N., Jr. *Science* **192**, 111 (1976).

The photochemistry of anthropogenic nonmethane hydrocarbons in the troposphere. Brewer, D. A., Augustsson, T. R., and Levine, J. S. *J. geophys. Res.* **88**, 6683 (1983).

Keys to photochemical smog control. Pitts, J. N., Jr. *Environ. Sci. Technol.* **11**, 456 (1977).

The chemical basis of air quality: kinetics and mechanisms of photochemical air pollution and application to control strategies. Finlayson-Pitts. B. J. and Pitts, J. N., Jr. *Adv. Environ. Sci. Technol.* **7**, 75 (1977).

The use of radical traps has been suggested as a pollution control technique, but many critics believe the method to be difficult in application, and potentially dangerous.

Control of photochemical smog by diethylhydroxylamine. Heicklen, J. *Atmos. Environ.* **15**, 229 (1981).

Models and computer simulations of smog formation have provided new insights into the chemical processes occurring.

Computer simulation of the rates and mechanisms of photochemical smog formation Calvert, J. G. and McQuigg, R. D. *Int. J. chem. Kin.* Symp. 1, 113 (1975).

Haziness and reduced visibility is an attribute of photochemical smog. Aerosols are formed in the oxidation of substances released by man's combustion of fossil fuels. Aerosols are, however, observed in many environments not polluted by man. They may be produced by processes analogous to those involved in photochemical smog, but using natural hydrocarbons such as terpenes as the precursor material.

Measurement and modeling of the concentrations of terpenes in coniferous forest air. Hov. Ø., Schjoldager, J., and Wathne, B. *J. geophys. Res.* **88**, 10679 (1983).

Natural organic atmospheric aerosols of terrestrial origin. Zenchelsky, S. and Youssefi, M. *Rev. Geophys. & Space Phys.* **17**, 459 (1979).

Blue hazes in the atmosphere. Went, F. W. *Nature, Lond.* **187**, 641 (1960).

Characterization of the aerosol in the Great Smoky Mountains. Stevens. R. K., Dzubay, T. G., Shaw, R. W., Jr., McClenny, W. A., Lewis, C. W., and Wilson W. E. *Environ. Sci. Technol.* **14**, 1491 (1980).

6 Ions in the atmosphere

6.1 Electrical charges in the atmosphere

Charged particles represent only a minute fraction of the total mass of the Earth's atmosphere, but they play a crucial part in many geophysical phenomena such as variations in the geomagnetic field, lightning, and auroras. Neutral species may be produced or destroyed in reactions involving ions or electrons, and ion chemistry would be important even in relation to neutral composition and behaviour alone. Communication by radio waves is now regarded as an essential feature of 'civilized' life, so that the reflection, absorption, and propagation through the charged atmosphere of radio-frequency electromagnetic waves is of vital importance. Much of the motivation for atmospheric research on charged species has derived from the need to understand how to provide secure and reliable communication: or even how to prevent it! The military implications of such communications have been regrettably clear, although civil interests are considerable. Electrons are now known to be the particles immediately involved in interactions with radio propagation. Atmospheric ions and their chemistry, which are properly the subjects of our study, play a central role in determining the concentration and distribution of electrons. Our survey of atmospheric chemistry would evidently be incomplete without some discussion of radio propagation, especially as radio experiments have given so much information about atmospheric ions. Section 6.1.3 is devoted to an explanation of the radio studies, and it follows a very brief introduction to two geophysical manifestations of charged species that are of particular historical interest.

Ions and electrons are most abundant at altitudes greater than ~ 60 km (i.e. within the mesosphere or above). Several factors contribute to this high altitude distribution. Most of the Sun's ionizing electromagnetic radiation, a major source of ions as discussed in Section 6.3, is absorbed at levels above about 60 km. Ions and electrons are highly reactive, because their charges can produce long-range attractive forces (Section 3.4.1). The rapid increase in mean free path with increasing altitude in itself favours longer lifetimes for the charged particles. But, in addition, loss processes are more efficient for complex molecular ions than for atomic ions (Section 3.4.2), and the atomic ions are found with greater relative abundance at higher altitudes (Section 6.2). Free electrons at concentrations sufficient to affect radio-wave propagation are certainly only located at levels above 60 km, and this region has conventionally been called the *ionosphere*. Peak instantaneous electron densities can exceed 10^6 electrons cm^{-3} at ~ 300 km, and drop to near-zero at 60 km. Total charged particle concentrations (positive ions + negative ions + electrons) also fall to a minimum of a few hundred per cm^3 at ~ 60 km, but increase a little in the stratosphere. Near ground level, the ion density is of the order of 10^3 cm^{-3}, but, of course, this value corresponds to a tiny mixing ratio (less than, say, 1 part in 10^{16}) while at 300 km several parts per thousand of the atoms and molecules are ionized. Stratospheric positive and negative ions are hard to study because of their small fractional abundance. Recent studies have, however, been successful, and some results are discussed in Section 6.5.

6.1.1 Aurora

Nature provides in the high atmosphere a phenomenon so obvious and striking that it has been the subject of scientific study for centuries. In regions not far from the geographic poles, a spectacular display of coloured lights can sometimes be seen at night. Complicated shapes are produced, which often move and change rapidly. The luminosity is given the name *aurora*, or 'dawn', with the qualifiers *borealis* ('northern') or *australis* ('southern') being added according to the polar region in which the display is seen.

Measurements of the angle of elevation of one auroral feature from two places simultaneously early showed that most of the light comes from altitudes of around 100 km. Such observations thus gave man some knowledge of the atmosphere at heights then inaccessible to *in situ* experiment. Spectroscopically, the auroral light is not dissimilar to that seen in the laboratory from electric discharges through low-pressure air. Atomic and molecular emission systems of oxygen and nitrogen predominate, the *auroral green line* of atomic oxygen, $O(^1S \rightarrow {}^1D)$, at $\lambda = 557.7$ nm being especially pronounced (see Section 7.4.1 for a further discussion of this line and of excitation mechanisms). The excitation seems to involve bombardment of atmospheric gases by particles from above and subsequent ionization. Neutralization by electron capture leads to excitation, and directly or indirectly to emission of visible radiation. If the bombarding particles were charged species (e.g. electrons or protons) projected from the Sun, then they would be deflected as they approached the Earth because of the latter's magnetic field. The charged beam would impinge equally on day and night sides, and would be concentrated in circular regions, one round each pole. Most auroral activity is, in fact, observed in an oval belt around the geomagnetic pole, between $15°$ and $30°$ from it. Violent changes in the observed intensity of auroral light, according to this picture, correspond to changes in the primary ionizing stream originating initially from the Sun. There is also a connection between *solar activity* and auroral phenomena. Solar 'activity' is often assessed in terms of the number of sudden 'storms' that disturb the Sun's surface. These storms reveal themselves as intense localized bursts of H-alpha light ($\lambda = 656.3$ nm; the $n = 3 \rightarrow n = 2$ transition). *Solar flares* of this kind are accompanied by the liberation of huge amounts of energy as electromagnetic radiation (especially in the X-ray region) and as kinetic energy of particles (protons mainly, but also α-particles and heavier nuclei). It is generally believed that the energy is produced by the annihilation of magnetic fields associated with sunspots. Sunspots are areas, dark only relative to the solar disc, that appear from time to time on the Sun's surface, and move across the disc in ~ 13.5 days (solar rotational period ~ 27 days). A definite *sunspot cycle* of the number of spots is known to repeat itself every eleven years, and is thought to represent some important oscillation in the structure of the Sun itself. One of several manifestations in our atmosphere of solar storms is the enhanced auroral activity mentioned earlier. A complication arises in that the ionizing particles are not simply the storm particles emitted by the Sun, partly because there is too great a time delay between the storm and the auroral response. One suggestion is that particles already trapped within the Earth's magnetic field are released and accelerated by the arrival of the storm particles. Whatever the detailed explanation, there is a statistical correlation between solar activity and frequency of auroras.

6.1.2 Geomagnetic fluctuations

Studies of the Earth's magnetism provided further early evidence that incident radiation might ionize the atmosphere at great altitudes. Detailed investigations had shown that the apparent direction of the Earth's magnetic field fluctuated regularly throughout a day, and that the fluctuations were smaller in winter than in summer. In 1882, Balfour Stewart suggested that electric currents circulating in the upper atmosphere might be responsible for diurnal changes in field. The atmospheric gases, *if conducting* and moved across the Earth's magnetic field, would generate and carry such currents, as in a giant dynamo. Currents of the order of tens of thousands of amperes are involved! If solar radiation were responsible for the ionization, then the day-to-night variations would be explicable, as would the seasonal dependence of the magnitude. Solar particles seem a less likely agent of ionization, since they would arrive at the atmosphere near the auroral zones, while the geomagnetic effect is spread over the entire earth. Nevertheless, the diurnal excursions of magnetic field do show a sympathy with the sunspot cycle, the swing being nearly twice as great at sunspot maximum as at minimum. This cyclic behaviour is now interpreted in terms of the increased ionization from solar X-rays at times of high sunspot activity.

As early as 1740, large, rapid, and irregular changes had been noticed as an occasional feature of the geomagnetic field. *Magnetic storms* of this kind are often seen when auroral displays are in progress, and they are most intense within the auroral zones. They are not, however, exclusively confined to the auroral regions, so that the excitation of auroras and the increase of atmospheric conductivity do not have the same *immediate* cause, even though both are somehow connected with the solar cycle.

6.1.3 Radio propagation

Marconi's success, in 1901, in sending signals from England to America is often regarded as heralding the birth of radio as a means of communication. Radio waves are electromagnetic radiation and travel in nearly straight lines as does light. Diffraction cannot possibly account for the travel of radio waves across the Atlantic, so that Marconi's feat demanded some explanation. An overhead reflector was suggested by Heaviside in England and by Kennelly in America: the hypothetical reflecting layer was called the *Heaviside Layer*. Commercial communication links were set up after Marconi's experiment, and it was soon noticed that signal strengths varied in a regular way throughout the day, the season, and the solar cycle. Magnetic storms were found to be associated with disturbances of the diurnal variability and sometimes with disruption of communication. When radio broadcasting started in the early 1920s, it was found that, *at night*, the signal strengths at distances of about 100 km varied widely over a few minutes, sometimes disappearing completely. This *fading* of the signals was ascribed to interference effects between a reflected *sky wave* and the direct *ground wave*. Reflections of the frequencies used for broadcasting thus seemed possible even for nearly vertical incidence, but only at night. The phenomenon is now very apparent with the large number of transmitting stations. At night, medium distance (up to a few thousand km) stations can be received only too readily, with resulting interference between stations. During the day, only relatively local signals can be received at the frequencies involved (~ 1 MHz). Such discoveries suggested that radio reflection is caused by ionized layers in the atmosphere and that the ionization is at least influenced, and probably caused, by solar radiation. As we shall see shortly, ionization at low altitudes leads also to absorption of broadcast frequencies, and prevents reflection from higher altitudes during the day. At night, the absorption is much diminished, and the various fading and interference effects become pronounced.

The early observations of the behaviour of radio waves were of great importance, because they indicated that radio waves could be used as a tool to explore the nature of the upper atmosphere and the ionization in it. Until rockets and satellites were able to enter the ionosphere or to examine it from above, radio studies provided much the most detailed information about our ionosphere. A crucial experiment concerned the polarization of reflected waves produced by the Earth's magnetic field. The results showed that the charged particles causing reflection were electrons and not ions. Sir Edward Appleton discovered that radio waves of frequency greater than 4–5 MHz penetrate the Heaviside layer, but, instead of escaping to space, they are reflected by a higher layer. This layer was named the *Appleton layer* after its discoverer. Appleton himself used the symbols E and F for the electric fields in the lower and upper layers, and suggested that the layers be called by the same letters in order to accommodate other possible layers by neighbouring

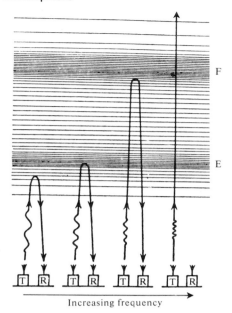

Increasing frequency

Fig. 6.1. Schematic representation of the operation of an ionosonde. T is the transmitter, and R the receiver. Horizontal shading indicates *rate of change* of ionization in the ionosphere. The four T-R pairs show, from left to right, what happens to radiowave pulses as the frequency increases. (Based on a diagram of Iribarne, J. V. and Cho, H.-R. *Atmospheric physics*, D. Reidel Co., Dordrecht, 1980).

letters of the alphabet. When discussing the ionosphere, it is universal practice to use this nomenclature, with D, E, and F layers or regions being particularly important. Electron densities in the ionosphere have been most commonly measured from the ground by a device called an *ionosonde*. A pulse of radiation of known frequency is directed vertically upwards, and the time delay of any reflection is recorded automatically. The frequency of the pulse train is increased to permit penetration to greater altitudes (see below). Figure 6.1 illustrates the principle of the ionosonde, and Fig. 6.2 is an *ionogram* obtained by day. At frequencies up to 4.4 MHz, reflection is by the E layer at an altitude of ~ 100 km. Wave penetration to the undisturbed F layer has a quite sharp onset at 4.4 MHz. Note that the F layer is partially split into two layers labelled F_1 and F_2: this is a day-time phenomenon, the F_1 'ledge' disappearing at night. The D region is more involved with absorption, rather than reflection, of radio waves, and so must be investigated by different techniques. *Topside sounding* by ionosondes in satellites now complements the data provided by ground-based ionosondes.

We have so far discussed the experimental results obtained on penetration, reflection, and absorption without explanation of why the phenomena occur, or why they are dependent on frequency, electron concentration, or altitude. Reflection itself is brought about by a process akin to the total internal reflection familiar in optics, although the detail is rather more subtle. Radio waves set atmospheric electrons into oscillation and cause them to radiate secondary wavelets. Since the electrons are randomly placed, the secondary wavelets coherently reinforce each other only in the original direction of propagation of the wave. (The *incoherent* or *Thomson* scatter in other directions produces a very weak wave that is also most valuable in ionospheric studies). Although the wavelets constituting the forward-scattered wave have traversed equal paths, the re-radiated wavelets are $\pi/2$ advanced in phase compared with the original driving wave: that is, they are one quarter of a cycle ahead. As a result, the composite combined wave also leads the driving wave by a phase angle dependent upon the relative contributions of scattered and unscattered radiation. Greater electron densities give rise to larger phase advances. In a very qualitative way, we can also say that higher frequencies are less able to force oscillation in the electrons, so that for the same electron density the phase advance is lower at high frequencies. The phase advance in the composite wave corresponds to an effective increase in the propagation velocity of the wave. An oblique wave is thus swung round and reflected as it enters regions of increasing electron density, since the higher parts of the wave move more rapidly than the lower. An extension of the argument then applies to vertically transmitted waves. The increase in velocity of propagation[a] corresponds to a decrease (below unity) of the refractive index, μ, of the medium. Total internal reflection becomes possible for angles of incidence, i, with the vertical such that $\mu = \sin i$. Reflection of vertically transmitted signals is possible if $\mu \to 0$. Theory shows that quantitatively

$$\mu^2 = 1 - (e^2/4\pi\varepsilon_0 mv^2)n, \qquad (6.1)$$

where n is the number density of electrons of charge e and mass m; ε_0 is the permittivity of free space, and v is the frequency of the radio wave. The reflection condition can thus be written in terms of the number density of electrons, n_{reflect}, needed to achieve zero refractive index

$$n_{\text{reflect}} = \text{constant} \times v^2 \qquad (6.2)$$

(constant $= 1.24 \times 10^{-8}$ s^2 cm^{-3} for n_{reflect} in electrons cm^{-3}).

The equation shows that, if electron densities increase with altitude, then at some critical altitude reflection may occur for a particular frequency. Higher

---

[a] To be more precise, the *phase velocity*; the group of waves in the transmitted pulse moves with the *group velocity*. Individual waves travel faster than the group in the ionosphere, so that when they reach the front of the group they die out and are replaced by new ones at the rear. At the turning point, the group velocity becomes zero for vertical reflection.

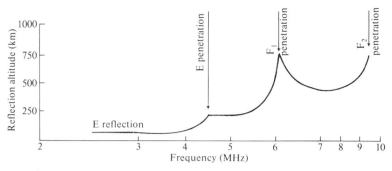

Fig. 6.2. Simplified and idealized ionogram obtained using the ionosonde technique of Fig. 6.1. Frequency of the emitted pulse of radiowaves is displayed on the horizontal scale, and the time delay for the return pulse is converted to effective altitude for the vertical scale. For frequencies less than about 4 MHz, reflection is from the E 'layer' at a height of about 120 km. At higher frequencies the E region is penetrated, and reflection is from the F_1 ledge, and (above $\sim$6–7 MHz) the F_2 peak. In reality, the traces tend to be doubled because the Earth's magnetic field splits the original wave into two 'characteristic' waves. (Derived from an actual ionogram trace reproduced in Ratcliffe, J. A., *Sun, earth and radio*, Weidenfeld & Nicolson, London, 1970.)

frequencies will, however, penetrate to higher altitudes before reflection occurs. A further important conclusion to be drawn from our discussion and the nature of eqn (6.2) is that *layers* of electrons are not necessary for reflection to occur, but rather only sufficient densities. Nevertheless, the rather sharp transitions between reflection altitudes at specific frequencies (Fig. 6.2) suggest that electron densities must themselves change rather sharply with altitude at those levels. Figure 6.3 gives some representative electron density profiles based on a variety of measurement techniques. The labels D, E, $F(F_1, F_2)$ are now indeed seen to be associated with regions rather than layers, the boundaries between regions being altitudes where electron densities increase particularly rapidly with height. Day–night variations are shown for the two extremes of the sunspot cycle. Night-time electron densities are orders of magnitude smaller than daytime ones, and the distinction between F_1 and F_2 regions is lost. Electron densities in the D region become near-zero at night (although negative ions continue to be present in this region).

The presence of electrons in the daytime D region is responsible for the attenuation of reflected waves of the medium frequencies ($\sim$1 MHz) used for broadcasting, as suggested earlier. For the frequencies concerned, eqn (6.2) predicts an electron density for reflection of $\sim 10^4$ electrons cm^{-3}, which is that found at altitudes of $\sim$80 km during the day. However, total gas densities—and hence collision rates—are relatively high in this low part of the ionosphere. Energy is removed from the oscillating electrons by collisions with

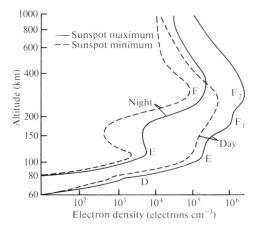

Fig. 6.3. Typical electron density–altitude profiles for the mid-latitude iono-sphere. (From a wallchart prepared by Swider, W., *Aerospace environment*, Air Force Geophysics Laboratory, Hanscom Air Force Base, Massachusetts.)

neutral molecules, so that the radio frequency signal is absorbed rather than reflected. Any process tending to increase ionization at low altitudes, where collision frequencies are high, tends also to bring with it enhanced absorption of radio waves. We have already seen how observations early established an association of solar activity and of magnetic storms with disturbed radio conditions, and we shall return, in passing, to this point several times again.

6.2 Ion chemistry in the atmosphere

The regions of the ionosphere have been introduced in this chapter in terms of the reflection of radio waves, and the electron densities that cause the reflections. An equally valid division of the ionosphere into its regions can be based on the chemical identity of the dominant ions and the chemical pro-cesses occurring at different altitudes.

Figure 6.4 shows ion compositions measured mass spectrometrically above about 100 km, together with total electron densities. The F region (say above 150 km) is characterized by atomic ions: O^+ and N^+ are dominant throughout most of the region, although H^+ and He^+ are more abundant at the very highest altitudes. Molecular ions (NO^+, O_2^+) are the most important ions in the E region (100–150 km), although total concentrations are lower than in the F region. At lower altitudes, in the D region, more complex molecular positive ions dominate, and negative ions are seen for the first time. We shall examine more carefully the composition and chemistry of each region in later

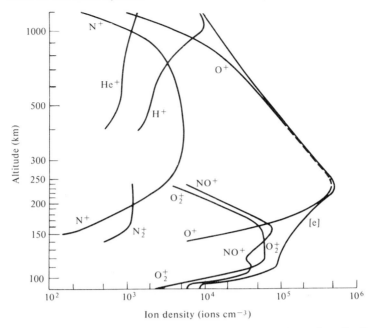

Fig. 6.4. Composition of positive ions in the E and F regions. Ion distributions were obtained by mass spectrometer experiments during daytime and at solar minimum; the data are normalized to the measured electron density distribution. [From Johnson, C. Y. *J. geophys. Res.* **71**, 330 (1966).]

sections. Here it is of interest to identify in outline some influences that lead to the compositions observed.

Gravitational separation (Section 2.1) of ionic species and their neutral precursors obviously favours atomic over molecular ions at high levels in the heterosphere. The abundances of the various atomic ions above the F peak of electron density at ~ 250 km fits this expectation well, with O^+ and N^+ being overtaken by the lightest ions H^+ and He^+ beyond about 1000 km. One point of interest concerns the scale heights (Section 2.1) for these ions. Electrons, being so light, could perhaps be expected to have near-infinite scale height, and to diffuse rapidly to the exosphere and so escape. Electrostatic attraction to the positive ions in reality prevents this escape, but the effect is to make the scale height for the ions *twice* what it would be for the corresponding neutral species. The scale height for the electrons themselves is determined by the concentration profiles of the most abundant ions: Fig. 6.4 shows very well that [e] follows $[O^+]$ closely where O^+ is dominant. Diffusion of the oppositely charged species is referred to as *ambipolar*, and it is of great importance in the F layer. The main distinction between the daytime F_1

and F_2 regions lies in the dominance of ambipolar diffusion over other loss processes in the F_2 region rather than in any fundamental chemical differences.

Ion production processes also favour atomic ion formation at high altitudes. Photoionization by extreme UV (EUV) is a major source of ions (Sections 3.2.1 and 6.3). Table 3.2 (p. 89) shows that the atoms N, O, and H have appreciably higher ionization potentials than the molecules NO or O_2. The shortest wavelength EUV is absorbed in the highest regions of the atmosphere, so that the remaining EUV can ionize only molecular species at lower altitudes (E, D regions). Energies of the photons available at high altitudes (F region) may be sufficient to dissociate a molecule and ionize one of the product atoms in a single step.

Altitude has a direct influence on chemical reactions through the pressure and hence the frequency of collisions. Ion-molecule reactions can convert the ions initially produced to different species that appear as *intermediate* and *terminal* ions. Although ion-molecule reactions are generally quite *efficient* because of long-range electrostatic attractions (Section 3.4.1), they are obviously *faster* for a given ion concentration at lower altitudes where the neutral reactant is more abundant. A number of ionic processes of atmospheric importance are brought together in Table 6.1. Some, such as the exchange processes (6.5) and (6.6), have clear counterparts in neutral chemistry, while others, such as charge transfer, (6.3), do not.

Conversion in the ionosphere of atomic ions to molecular ions is of critical importance in permitting charge neutralization, and thus in determining electron and ion concentrations throughout the atmosphere. Charge is lost, in the reactions shown, only in the recombination processes (6.7) to (6.9). As we discussed in Section 3.4.2, the ionization energy must be dissipated. Radiative recombination, process (6.7), is the only way in which atomic ions can be neutralized; it is extremely inefficient. Molecules can undergo dissociative recombination, as in reaction (6.8), where the fragments carry off the ionization energy as translational motion. Similarly, positive and negative molecular ions can interact [reaction (6.9)] to yield neutral particles that disperse the energy released as internal motions as well as kinetic energy of the products. A further mechanism for ion-ion recombination is recognized: a three-body association involving a neutral third-body, M, that can dissipate excess energy in the usual way. Such ion-ion charge annihilation processes are of particular importance at atmospheric altitudes less than about 60 km, since almost all the negative charge is carried by ions rather than electrons, although we shall see in Sections 6.4.3, 6.4.4 and 6.5 that the ions involved are considerably more complex than the NO^+ and NO_2^- represented in eqn (6.9).

Attachment of electrons to neutral molecules to form negative ions requires the removal of excess energy in the same way that neutralization does. Radiative attachment, reaction (6.10), is once again much less efficient than the three-body or reactive processes, (6.11) and (6.12), and is of little atmospheric

Table 6.1 Some types of ionic process of atmospheric importance

CHARGE TRANSFER		$N_2^+ + O_2 \rightarrow N_2 + O_2^+$	6.3
		$H^+ + O \rightarrow O^+ + H$	6.4
EXCHANGES		$N_2 + O^+ \rightarrow N^{(*)} + NO^+$	6.5
		$O_2^+ + N_2 \rightarrow NO + NO^+$	6.6
RECOMBINATIONS	Radiative	$O^+ + e \rightarrow O + h\nu$	6.7
	Dissociative	$NO^+ + e \rightarrow N^{(*)} + O$	6.8
	Ion–Ion	$NO^+ + NO_2^- \rightarrow NO + NO_2$	6.9
ATTACHMENTS	Radiative	$e + O \rightarrow O^- + h\nu$	6.10
	Three-body	$e + O_2 + M \rightarrow O_2^- + M$	6.11
	Dissociative	$e + O_3 \rightarrow O^- + O_2$	6.12
DETACHMENTS	Collisional	$O_2^- + M \rightarrow O_2 + e$	6.13
	Associative	$O_2^- + O \rightarrow O_3 + e$	6.14
	Radiative ("Photodetachment")	$O^- + h\nu \rightarrow O + e$	6.15
CLUSTERING		$O_2^+ + O_2 + M \rightarrow O_2^+.O_2 + M$	6.16
		$H^+(H_2O)_n + H_2O(+M) \rightarrow H^+(H_2O)_{n+1}(+M)$	6.17
		$NO_3^- + H_2O + M \rightarrow NO_3^-.H_2O + M$	6.18

Note—Products of exothermic processes may possess electronic as well as translational (and, where appropriate, vibrational and rotational) excitation. For example, the atomic nitrogen product of reactions (6.5) and (6.8) can be formed in the excited ^2D state.

significance. Much more important is the three-body reaction, but since its rate depends on the square of the pressure of neutral constituents, it is confined to the D region and below (as is the ozone reaction, (6.12)). Negative ions are consequently only found at low altitudes. Detachment processes, such as reactions (6.13) to (6.15), convert negative ions back to neutrals and electrons, so that the detailed balance between negative ion and electron densities depends on the competing rates of attachment, detachment and ion-ion recombination.

The attractive forces between neutral molecules are not usually strong enough to favour formation of long-lived 'Van der Waals' species except at very low temperatures. For ions, however, the interactions with polarizable, or especially dipolar, molecules are much stronger, and 'cluster ions' are a feature of ion chemistry at low altitudes, as we shall see later. Some typical clustering processes are shown in reactions (6.16) to (6.18).

6.3 Ionization mechanisms

Positive ions are formed in the atmosphere by three principal agencies: (i) solar radiation in the EUV and X-ray wavelength regions; (ii) galactic cosmic rays (GCR) and other galactic radiations; and (iii) precipitating energetic particles of solar origin and from the Earth's radiation belts. In the lower troposphere, radioactive emanation from rocks provides an ionization source at all times.

Photoionization, by EUV and X-rays, is the most important ion source above 60 km during the daytime with a 'quiet' Sun (i.e. in the absence of solar storms). Figure 6.5 shows the altitude by which the atmospheric components afford unity optical depth (p. 45) at different wavelengths. Expressed a different way, these altitudes correspond to the heights at which the incident solar

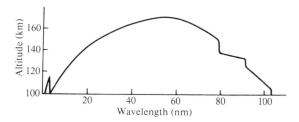

Fig. 6.5.   Height at which the remaining overhead atmospheric layer possesses unity optical depth, shown as a function of wavelength. This height corresponds to the altitude at which the solar intensity has been attenuated to a value $1/e$ times the intensity incident outside the atmosphere. The height also corresponds approximately to the altitude of maximum rate of absorption for any wavelength. (Adapted from Rishbeth, H. and Garriott, O. K., *Introduction to ionospheric physics*, Academic Press, New York, 1969.)

radiation has been attenuated by a factor e. Radiation of all wavelengths in the range reaches the F_2 region ($\gtrsim 200$ km), while the range 20–91 nm probably contributes to F_1 ionization (~ 140–200 km). E region (~ 90–140 km) ionization comes from the more deeply penetrating part of the spectrum, EUV radiation between roughly 80 and 103 nm and X-rays from 1 to 10 nm wavelength. Several strong atomic lines are superposed on the solar continuum in the EUV region: H Lyman-β (102.6 nm), C III (97.7 nm) and He I, II (58.4, 30.4 nm) are prominent. Lower down in the atmosphere, in the D region (~ 70–90 km), the only surviving ionizing radiations are EUV at $\lambda \gtrsim 103$ nm and 'hard' X-rays ($\lambda \sim 0.2$–0.8 nm). The only strong atomic feature is H Lyman-α at $\lambda = 121.6$ nm ($\equiv 10.19$ eV); rather curiously, O_2 has a gap in its spectrum, where absorption is relatively weak, at exactly this wavelength, providing a 'window' that allows the radiation to penetrate more deeply than it would otherwise (cf. p. 88). Primary ions formed depend on the available wavelengths and the neutral precursors at any altitude. In the F region, therefore, O^+, together with some N^+, are the dominant primary ions (see Table 3.2, p. 89, for photoionization limits). Molecular ions become the major primary products in the E region, O_2^+ being generated in particular by Ly-β and C III solar atomic lines. Soft X-rays ionize N_2 to make N_2^+ the second most important primary ion. Only minor constituents (cf. Table 3.2) such as NO or metal atoms can be ionized by the Ly-α that penetrates into the D region. In addition, metastable *excited* O_2, in the $^1\Delta_g$ state, can be ionized in the wavelength range 102.7–111.8 nm, as discussed in Section 3.2.1. Hard X-rays can ionize O_2 and N_2, while cosmic rays can ionize all constituents. Figure 6.6 shows the relative daytime rates of ionization in the D and lower E regions due to photons of different energies. In the 'quiet' D region, the primary ions are typically 80 per cent NO^+ and 20 per cent O_2^+. Solar intensities in the EUV, and especially X-ray, regions are markedly dependent on the 11-year

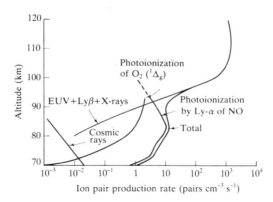

Fig. 6.6. Contribution of several ionization processes to ion pair formation in the 'quiet' daytime E and F regions. (From McEwan, M. J. and Phillips, L. F., *Chemistry of the atmosphere*, Edward Arnold, London, 1975.)

solar cycle (cf. Section 6.1.1 and Fig. 6.3), and they also show an oscillatory variation over the 27-day rotational period of the Sun.

Ionization does not cease entirely at night, even though removal of the main source, solar radiation, does drastically reduce the ionization rate. Radiation from the illuminated portion of the atmosphere into the night sector may be achieved by resonance scattering (resonance absorption followed by fluorescence). In the F region, $\lambda = 58.4$ and 30.4 nm lines scattered by He are a source of O^+; in the E region, $\lambda = 102.6$ nm Ly-β radiation, scattered by H, is an important source; and, in the D region, scattered Ly-α ($\lambda = 121.6$ nm) can ionize NO. Additional sources include starlight (stellar continuum radiation in the spectral interval $91.1-102.6$ nm) in the E region, and galactic cosmic rays (which continue to arrive at night, of course) in the D region and below.

Corpuscular ionization—that is, ionization by impact with energetic particles from, or released by, the Sun—is a potential source of night- or day-time ionization. During 'disturbed' solar conditions, especially just after the maximum of the sunspot cycle, solar storms and flares (Section 6.1.1) are fairly frequent. Increases in ionospheric electron density are associated with the storms, and lead to a variety of ionospheric disturbances. *Sudden ionospheric disturbances* (SID) lead to shortwave fadeout of radio signals due to absorption in the enhanced D region (Section 6.1.3). The SID is produced by an intensification of hard X-ray emissions by several orders of magnitude, and is thus a photon effect. Corpuscular effects are generally delayed by at least several hours after the solar flare, and can last for several days, in distinction to the flare that lasts for tens of minutes. One effect involves energetic electrons trapped in the Earth's radiation belts that are slowly precipitated in the days following the storm to produce increased ionization in latitudes from approximately $45°$ to $72°$. Solar particle events (SPE), on the other hand, lead to enhanced ionization at high magnetic latitudes ($>60°$), and are due to energetic particles, mainly protons, ejected from the Sun. The delay in arrival of the particles is a result of their having been guided into a long spiral path by the interplanetary magnetic field. Radio blackouts lasting several days can follow SPEs; because the D region enhancement covers the polar regions only, such radio disturbances are called polar cap absorption (PCA) events. In almost all disturbances (SID, electron, and SPE) of the D region, the primary ions are almost 100 per cent O_2^+, since a major constituent is capable of being ionized (NO^+, it will be remembered, is dominant in the quiet D region).

6.4 Chemistry of specific regions

6.4.1 F region processes

Chemistry of ions in the F region is relatively simple, and it centres on the conversion of O^+, formed as the primary ion (Section 6.3) to secondary molecular ions that can recombine with electrons (Section 6.2). Two pairs of processes illustrate the conversion and neutralization

$$O^+ + O_2 \rightarrow O + O_2^+, \qquad (6.19)$$

$$O_2^+ + e \rightarrow O^{(*)} + O, \qquad (6.20)$$

and

$$O^+ + N_2 \rightarrow N + NO^+, \qquad (6.21)$$

$$NO^+ + e \rightarrow N^{(*)} + O. \qquad (6.8)$$

Reaction (6.19) has the *effect* of charge transfer, although it may proceed partially through an exchange mechanism. Electronic excitation in one of the fragment atoms in reactions (6.20) or (6.8) can help to carry off excitation energy, and it is represented in the equations by the bracketed asterisk. Optical emission from the excited atoms is then sometimes observed as *airglow*, a subject we explore further in the next chapter. For example, $O^{(*)}$ from reaction (6.20) can be in the ^1D state, and the emission (cf. p. 317) is of the forbidden $^1D \rightarrow {}^3P$ atomic line in the red

$$O(^1D) \rightarrow O(^3P) + h\nu \ (\lambda = 630 \text{ nm}). \qquad (6.22)$$

At F region altitudes, $[N_2] \gg [O_2]$ because photodissociation of O_2 is almost complete. Thus, although the rate constant for reaction (6.21) is about an order of magnitude less than that for reaction (6.19), the NO^+ route to neutralization is the more important.

The significance of conversion from atomic to molecular ions is nicely illustrated by 'accidental' experiments in which rocket launches produce an artificial 'ionospheric hole'. For example, in the Atlas-F launch of a weather satellite in June 1982, the rocket burned to an altitude of 434 km, ejecting exhaust gases that included H_2, H_2O, and CO_2. As much as 10^{27} molecule s^{-1} of H_2O was released. Electron concentrations in the F region were approximately halved because the exhaust gases offer an efficient route from O^+ to molecular ions such as H_2O^+ and H_3O^+ which undergo particularly rapid recombination with electrons. The rate coefficient for reaction of O^+ with H_2 and H_2O is 100 times greater than that for reaction (6.19) and more than 1000 times greater than for reaction (6.21). An expanding shell of $O(^1D \rightarrow {}^3P)$ airglow at $\lambda = 630$ nm was seen, and appeared like a smoke ring. Studies of the emission can yield information about diffusion in the F region, as well as about the efficiency of production of $O(^1D)$ in plasma recombination.

The daytime peak in the F_2 region ion and electron concentrations might appear, at first sight, to be a result of Chapman layer formation (see Section 4.3.2). However, the peak concentrations occur at altitudes (~ 250–300 km) much greater than those (~ 100–150 km) where the rate of ionization is maximum. The peculiar behaviour is in part a result of the detailed charge neutralization (i.e. loss) mechanism. Since the rate-determining step in the loss processes is conversion of O^+ to molecular ions, mainly in reaction (6.21), the loss rate is proportional to $[N_2]$. Ionization rates are proportional to $[O]$, so that the steady state electron concentration is proportional to $[O]/[N_2]$. But the concentration of N_2 falls off more rapidly than $[O]$ with increasing altitude (gravitational separation, etc.), so that electron density in this steady state picture *increases* with height. In reality, the chemical destruction of electrons and ions from any altitude is supplemented by transport loss. Diffusion rates increase with decreasing pressure (increasing altitude) so that electron concentrations ultimately fall. The position of the F_2 peak will thus occur where the chemical and diffusive loss rates are identical.

Above the F_2 peak, the region dominated by O^+ gives way to the *protonosphere* dominated by H^+ (Fig. 6.4), the boundary being strongly influenced by the rapid and near-resonant charge-exchange reaction

$$O^+ + H \rightleftharpoons O + H^+, \tag{6.23}$$

of which we have represented the reverse step alone in eqn (6.4). Below the F_2 peak, concentrations of NO^+ and O_2^+ continue to increase, even though O^+ and electron concentrations decrease. In the lower F (or F_1) region, the molecular ion concentrations may even exceed $[O^+]$. The species are not, however, primary ions, but rather the products of reactions (6.21) and (6.19) favoured by the higher N_2 and O_2 concentrations and by the reduced electron densities.

6.4.2 E region processes

Molecular ions are the primary ionization products in the E region. As discussed in Section 6.3, both O_2^+ and N_2^+ are produced from the major neutral atmospheric constituents. *Observed* ion distributions (Fig. 6.4) show only a small contribution from N_2^+, so that it must be consumed rapidly in secondary reactions. In fact, these reactions that represent losses of N_2^+ are also sources of the terminal ions

$$N_2^+ + O \rightarrow N^{(*)} + NO^+, \tag{6.5}$$

and

$$N_2^+ + O_2 \rightarrow N_2 + O_2^+. \tag{6.3}$$

Exchange with atomic oxygen in the first reaction is favoured at high altitudes, and charge transfer with O_2 at lower levels, following the $[O]/[O_2]$ profile in the atmosphere.

Neutral nitric oxide begins to play an important part in atmospheric chemistry at E region altitudes. Ionic processes can themselves generate the molecule. For example, one source of excited atomic nitrogen, $N(^2D)$, is the dissociative recombination reaction (6.8), and the excited atom can react with O_2

$$N(^2D) + O_2 \rightarrow NO + O. \tag{6.24}$$

Reaction of O_2^+ with N_2,

$$O_2^+ + N_2 \rightarrow NO + NO^+, \tag{6.6}$$

has a rather small rate coefficient, but can be an important source of both NO and NO^+ because of the high values of $[N_2]$ relative to other species. Once present, NO tends to act as a charge transfer acceptor

$$O_2^+ + NO \rightarrow O_2 + NO^+, \tag{6.25}$$

because of its low ionization potential (Table 3.2). Reactions such as (6.6) and (6.25) thus tend to increase the NO^+ concentration at the expense of O_2^+. Dissociative recombination of NO^+ in reaction (6.8), rather than of O_2^+ in (6.20), is the most important charge neutralization step in the E region, since $[e] \ll [NO]$, and most O_2^+ ions will collide with NO to transfer charge before they meet an electron.

From time to time, unusual propagation of short-wave radio signals suggests the presence of local areas of increased ionization in the E region. These effects are known as *Sporadic E* phenomena. One manifestation of Sporadic E is long-distance reception of television pictures that have arrived by a reflected path that is usually absent. Narrow (1–3 km) localized layers of metal ions are often associated with Sporadic E. Peak metal ion concentrations can be two orders of magnitude higher than the molecular (NO^+, O_2^+) ion concentration in the vicinity of the layer. Increases in metal ion concentrations have been observed during meteor showers, and it seems likely that meteor ablation in the atmosphere is the source of the E region metals. Ionospheric chemistry is profoundly affected by the presence of metals, first because they have low ionization potentials ($\lesssim 7$ eV: cf. Table 3.2), and secondly because they are monatomic. The first factor means that they can become ionized by exothermic charge transfer from all molecular ions; e.g. with magnesium,

$$Mg + O_2^+(\text{or } NO^+) \rightarrow Mg^+ + O_2(\text{or NO}). \tag{6.26}$$

The second factor prevents the positive charge-carrying metal ions from being neutralized, since the only electron recombination route is the very inefficient radiative process (Section 6.2). Formation of (molecular) metal oxides does not assist the electron recombination rate much, since the regeneration of elemental ions in reactions such as

$$MgO^+ + O \rightarrow Mg^+ + O_2, \tag{6.27}$$

is very fast. Physical rather than chemical processes therefore seem to govern the distribution of the metal atoms and the formation of layers. Within the layers, electron (and ion) densities are abnormally high because there is no available loss pathway.

6.4.3 D region positive ion chemistry

Chemical complexity characterizes the D region! Low temperatures (the lowest in the entire atmosphere), relatively high pressures, and a wide range of minor trace reactants permit a multitude of reactions. At the same time, *in situ* experimental study of the D region is difficult because ion and electron concentrations are low and variable; high pressures hinder sampling into mass spectrometers; and the large ions encountered have a tendency to fragment while being sampled. Nevertheless, a quite considerable understanding of D region chemistry has emerged over the last 10–20 years, assisted by laboratory studies of the kinetics of potentially important ion-molecule reactions. Investigations of positive ion compositions (in the altitude range 64–112 km) date from 1965. These earliest mass spectrometric studies straight away showed that some unexpected ions were present at the lower altitudes, with mass numbers separated by 18 units: that is, the mass of H_2O. At first, there was much suspicion that contamination or some other artefact was responsible for the mass spectrometric peaks, but it is now certain that the ions are ionospheric hydrated protons, $H^+(H_2O)_n$, with n ranging from 2 to at least 8, and occasionally even 20: H_3O^+ itself seems at most to be of minor importance. Figure 6.7 shows some typical results.

A clear boundary, at ~ 82–85 km, is observed in most measured ion profiles, with the $H^+(H_2O)_n$ water cluster ions being dominant below the boundary, and NO^+ and O_2^+ above it. At the same altitudes, most daytime rocket flights find that the electron density decreases by almost an order of magnitude within a few kilometres height decrease. Reduced photoionization rates cannot account for the abrupt decrease in electron densities at this 'ledge', which means that the changes must have been brought about by greatly increased loss rates at lower altitudes. Negative-ion formation accounts for some of the electron loss,

$$e + O_2 + M \rightarrow O_2^- + M, \tag{6.11}$$

and, since it is a three-body process, its rate increases as the square of the pressure. Even more important, the large water cluster ions are exceptionally good at dissipating recombination energy, since several polyatomic fragments are formed

$$H^+(H_2O)_n + e \rightarrow H + nH_2O. \tag{6.28}$$

Dissociative recombination of $H^+(H_2O)_n$, with $n \sim 6$, is at least an order of magnitude faster than recombination with NO^+, reaction (6.8), at D region temperatures (~ 200 K). Reaction (6.28) is not only a loss process for electrons, but also for cluster ions. It is thus at least self-consistent that $H^+(H_2O)_n$ is absent when $[e]$ is relatively large ($\gtrsim 82$ km) and that $[e]$ is small when $[H^+(H_2O)_n]$ is relatively large ($\lesssim 82$ km). But we now have to ask what factors make the water ion clustering process so altitude dependent. The easiest way of answering this question is to turn to the results of laboratory experiments.

Many techniques exist for the study of ion-molecule reactions, including ion beam, ion cyclotron resonance, stationary discharge, high pressure mass spectrometer and drift tube methods. However, the largest contribution to the understanding of ionospheric chemistry has probably come from flow studies

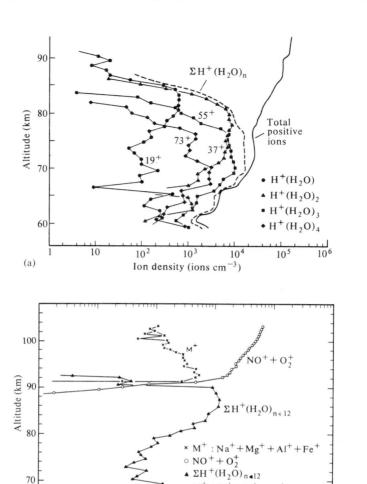

(a)

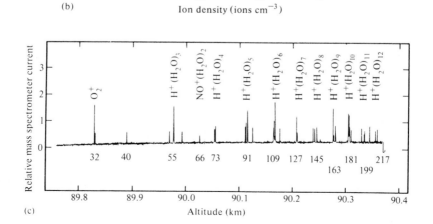

(b)

(c)

in which ions are generated upstream of the reaction region and sampled by a downstream mass spectrometer. The method as applied to studies of ion reactions is almost always referred to as the *flowing afterglow* technique. Its development by Eldon Ferguson and co-workers at Boulder, Colorado, has given us an understanding of D region chemistry formerly lacking, as well as providing detailed knowledge about ionic processes in the other regions. Variants of the flowing afterglow technique combine it with drift tubes or with upstream mass spectrometric selection (*selected ion flow tube*, 'SIFT') to sift the reactant ions and provide kinetic data of even greater sophistication. Studies using conventional cooling techniques to reach D region temperatures (150–200 K) are of limited applicability when water is a reactant because of the condensation problem. One solution is to rapidly expand gases containing the reactants, thus cooling them. By forcing the gases through an expansion nozzle, a flowing stream of cold gas is established which can be used for flow-kinetic studies. At supersonic flow velocities, temperatures less than 200 K can be attained from a room-temperature starting point: in essence, the three dimensional translational energy of the molecules has been rearranged to be one dimensional in the direction of flow. With such a system carrying ionized gases, and with a mass spectrometer as detector, D region ion reactions can be studied in the laboratory at the temperature appropriate to the atmosphere.

With O_2^+ ions present in ultra-dry O_2, the ions that are produced at low temperature are oxygen self-clusters: $O_2^+.O_2$ and even $O_2^+(O_2)_2$. Addition of a trace of water, however, prevents the formation of $O_2^+(O_2)_2$, and diminishes the yield of $O_2^+.O_2$. In their place, first the ion $O_2^+(H_2O)$ and then the ion $O_2^+(H_2O)_2$ appear. Further addition of water leads to the conversion of $O_2^+(H_2O)_2$ to $H^+(H_2O)_2$, and then to higher hydrates. Clusters with n up to 78 have been seen in the laboratory: they can almost be regarded as ice surrounding an H^+ ion. The reaction scheme envisaged to explain the observations is summarized below

$$O_2^+ + O_2 + M \longrightarrow O_2^+.O_2 + M \qquad \text{CLUSTER}, \qquad (6.16)$$

$$O_2^+.O_2 + H_2O \longrightarrow O_2^+(H_2O) + O_2 \qquad \text{SWITCH}, \qquad (6.29)$$

$$O_2^+(H_2O) + H_2O \xrightarrow{(M)} O_2^+(H_2O)_2 \qquad \text{CLUSTER}, \qquad (6.30)$$

◁ **Fig. 6.7.** (*opposite*). Results of some rocket-borne mass spectrometric determinations of ion composition in the D region: (a) above Red Lake, Ontario; the main proton hydrates are shown, and the decrease in concentration of these species above ~85 km is clearly marked; (b) above Kiruna, N. Sweden; the transition—here at ~90 km—between proton hydrates and NO^+, O_2^+ is even more evident; (c) a mass-spectral scan (from the Kiruna flight of (b)) just below the transition height, showing that $H^+(H_2O)_n$, $n = 3$ to 12, are the dominant ions. (From Kopp, E. and Herrman, U., *Ann. Geophysicae* **2**, 83, 1984.)

$$O_2^+(H_2O)_2 + H_2O \longrightarrow H^+(H_2O)_2 + O_2 + OH \qquad \text{EXCHANGE,} \quad (6.31)$$

$$H^+(H_2O)_2 + H_2O \xrightarrow{(M)} H^+(H_2O)_3 \qquad\qquad\qquad \text{CLUSTER,} \quad (6.32)$$

and so on. An essential feature of the scheme is that entry into the series of reactions leading to $H^+(H_2O)_n$ involves $O_2^+.O_2$, whose formation is a three-body process. That is, the rate of clustering is limited by a process whose rate is proportional to pressure squared, and is thus highly altitude dependent. Altitude dependence is further magnified by a reaction between atomic oxygen and $O_2^+.O_2$ that destroys the cluster

$$O + O_2^+.O_2 \rightarrow O_2^+ + O_3. \qquad (6.33)$$

Atomic oxygen concentrations increase sharply with altitude, so that as a consequence of reactions (6.33) and (6.16), $[O_2^+.O_2]$, and hence $[H^+(H_2O)_n]$, increases very rapidly with decreasing altitude.

Molecular oxygen ions, O_2^+, are dominant in the D region only in 'disturbed' conditions (cf. Section 6.3). However, reactions analogous to those just described can be written for the major (80 per cent) ion of 'quiet' conditions, NO^+. Direct addition of water to NO^+ is too slow to account for conversion to $H^+(H_2O)_n$. Rather, NO^+ must first cluster with one of the major neutral gases in a three-body process

$$NO^+ + N_2(\text{or } O_2, CO_2) + M \rightarrow NO^+.N_2(\text{or } O_2, CO_2) + M. \quad (6.34)$$

The resultant ion 'switches' in a *fast* reaction to form $NO^+(H_2O)$, which can then add a further water molecule: the dihydrate exchanges with H_2O to form the proton hydrates

$$NO^+(H_2O)_2 + H_2O \rightarrow H^+(H_2O)_2 + HNO_2. \qquad (6.35)$$

Once again, the laboratory experiments show no $H^+(H_2O)_n$ with $n < 2$, in substantial agreement with ionospheric observation. A summary of the reaction schemes is provided diagrammatically in Fig. 6.8.

An essential feature of the scheme presented is the formation of NO^+ cluster ions with N_2, O_2, and CO_2. Such ions are rather weakly bound, $NO^+.N_2$ having a bond dissociation energy of only 20 kJ mol^{-1}. Sampling into rocket-borne mass spectrometers is difficult, but some success has been achieved. Mass spectrometric peaks have been observed *in situ* that are assigned to the clusters $NO^+.N_2$, $NO^+.CO_2$ and $NO^+.H_2O.CO_2$. The rate of formation of the weakly bound cluster has a very strong *negative* temperature dependence. Accordingly, the conversion of NO^+ ions to water cluster ions is extremely sensitive to temperature changes, and the D region positive ion composition is expected to manifest strong seasonal, latitudinal, and even day-to-day variations as atmospheric temperatures fluctuate. Experimentally, such variability in the D region is well known, and the compositions and concentrations of Fig. 6.7 must be regarded only as representative.

Proton hydrates may be involved in the mesospheric phenomenon of

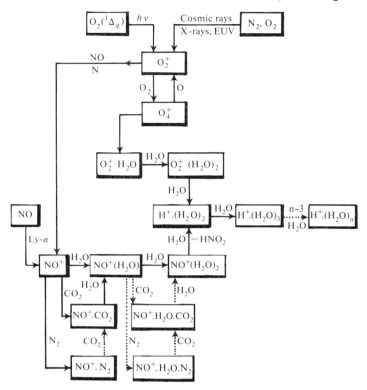

Fig. 6.8. Reaction scheme for positive ion chemistry in the D region showing the major steps in the conversion of the primary ions, O_2^+ and NO^+, to the terminal ions, $H^+ (H_2O)_n$. (Adapted from Ferguson, E. E., Fehsenfeld, F. C., and Albritton, D. L., in Bowers, M. T. (ed.), *Gas phase ion chemistry*, *Vol. 1*, pp. 45–82. Academic Press, New York, 1979).

noctilucent clouds. As the name implies, these clouds are only visible in the dark sky long after sunset: they are similar in appearance to thin cirrus clouds, but much higher and more tenuous. In order that they may still be lit up by the Sun, while the Earth's surface is in darkness, the noctilucent clouds must be situated at great height. The clouds lie at altitudes between 80 and 87 km, with the base most frequently near 82 km. Noctilucent clouds seem to be largely of summer-high latitude occurrence. That is, they are formed in the regions of the atmosphere and the seasons of the year where atmospheric temperatures are the lowest—perhaps even as low as 100–120 K. They are also, it will be noticed, formed at altitudes close to the D region transition in ionospheric composition and concentration. Various pieces of evidence, including direct rocket sampling, suggest that noctilucent clouds are composed of water ice. However, even at the low temperatures involved, it is hard to see how ice crystals could be formed by conventional nucleation mechanisms (Section 2.5). Dust from meteoric debris might conceivably be capable of acting as condensation nuclei. An alternative hypothesis is that the ionic water cluster species, known to be abundant at the cloud altitudes, are associated with the nucleation process. One possibility is that the dissociative recombination of $H^+(H_2O)_n$ in reaction (6.28) leaves a neutral cluster fragment $(H_2O)_m$ sufficiently large to start uninhibited growth (cf. Section 2.5).

$$H^+(H_2O)_n + e \rightarrow H + (H_2O)_m + (n - m)H_2O. \qquad (6.36)$$

Whatever the detailed formulation, the salient feature of the proton hydrate idea is that the energy barrier that exists to nucleation of neutral molecules can be overcome by the electrostatic ion-dipole attractive forces in the protonic species.

6.4.4 D region negative ion chemistry

Rather few detailed mass spectrometric measurements exist for D region negative ions. One reason is that negative ions exist as important species only where total (ion + electron) concentrations are low ($z < 78$ km). Further, the negative ion population consists of many different chemical species, so that any individual mass spectrometric peak is weaker than the less distributed positive ion peaks. Technical problems also interfere with the sampling and detection of negative ions. Notwithstanding the problems, an understanding, at least in outline, of D region negative ion chemistry has begun to emerge.

The terminal ions appear to be species such as CO_3^-, HCO_3^-, Cl^-, and the hydrates $NO_3^-(H_2O)_n$ and $CO_3^-(H_2O)_n$, while the primary negative ion is O_2^- produced by three-body electron attachment

$$e + O_2 + M \rightarrow O_2^- + M. \qquad (6.11)$$

The (pressure)^2 dependence of reaction rate thus confines negative ion production to altitudes below 70–80 km. Dissociative attachment of electrons to ozone,

$$e + O_3 \rightarrow O^- + O_2, \qquad (6.12)$$

is a minor source of negative ions; as with reaction (6.11), there is a strong altitude dependence because of the rapid increase in $[O_3]$ with decreasing altitude in this region. The transition from ions to electrons as dominant carriers of negative charge is sharpened by the steep increase in $[O]$ above 75–80 km. Atomic oxygen inhibits the growth of stable negative ions, an effect that can be expressed (with some simplification) as a result of the associative detachment reaction

$$O_2^- + O \rightarrow O_3 + e. \qquad (6.14)$$

The arguments for a sudden increase in negative ion concentrations below a boundary altitude are thus very similar to those used to explain $H^+(H_2O)_n$ emergence at the same altitudes, but with the feature of reduced dissociative recombination being absent.

Electron affinities of the neutral precursors of simple atmospheric negative ions are generally rather small, and much less than the ionization potentials of similar neutral species. For O_2 itself, the electron affinity is ~ 0.44 eV (42.5 kJ mol^{-1}). The negative ions are thus relatively unstable and have a tendency to participate in reactions, such as (6.14), that lead back to the release of electrons. In competition with the loss processes, ion-molecule reactions with minor molecular species can lead to the terminal ions that are stable enough to resist electron detachment. For example, the electron detachment energy of NO_3^- is ~ 3.9 eV ($\equiv 380$ kJ mol^{-1}), one of the largest known for a simple negative ion.

Photodetachment (the analogue of photoionization) is possible with visible and even infra-red radiation for ions with small electron affinity. With O_2^- itself, near-infra-red radiation can remove the electron

$$O_2^- + h\nu(\lambda \lesssim 2800 \text{ nm}) \rightarrow O_2 + e. \tag{6.37}$$

Daytime negative ion-to-electron concentration ratios are thus potentially affected by photodetachment processes, and there seems little doubt that the electron-ion transition boundary is shifted several kilometres to lower altitudes by day. A more detailed investigation of the rate at which the D region negative ions build up at twilight, and electrons build up at dawn, reveals further subtleties. Effects of illumination are particularly well revealed by absorption of radio-waves during PCA events (see Section 6.3). Absorption results from increased free electron concentrations in the D region. Negative ion formation therefore reduces the influence of the PCA. Very large day–night modulations of the absorption are observed, with the anomalous effects virtually disappearing at night. At dawn and twilight, however, the changes in electron density seem to require that the radiation causing photodetachment is screened by a layer for 30–40 km above the Earth's surface. Ozone is the obvious screening substance, but it filters out ultraviolet, and not visible radiation. Reaction (6.37) cannot, therefore, be the major detachment process. Instead, atomic oxygen from O_2 and O_3 photolysis (only ultraviolet is active) removes O_2^- by the associative detachment reaction

$$O_2^- + O \rightarrow O_3 + e. \tag{6.14}$$

Ozone photolysis

$$O_3 + h\nu \ (\lambda \lesssim 310 \text{ nm}) \rightarrow O(^1D) + O_2(^1\Delta_g), \tag{6.38}$$

provides not only atomic oxygen, but also an excited molecular fragment (Section 3.2.1) that is curiously efficient as an electron-detaching partner in the reaction

$$O_2^- + O_2(^1\Delta_g) \rightarrow 2O_2 + e. \tag{6.39}$$

The progression from O_2^- to the terminal ions (NO_3^- and its hydrates) proceeds *via* a series of increasingly complex ions. Two threads can be identified in the conversion, one of which involves the ions O_4^-, CO_4^-, NO_3^* ($O_2^-.NO$) and the other O_3^-, CO_3^-, NO_2^-. Clarification of the sequences can be achieved by writing the ions as though they were clusters, with a dot separating the initial ion and its cluster partner. Charge is, in reality, delocalized over the ion, but the cluster structure emphasizes the differences between, for example, the two nitrate ions

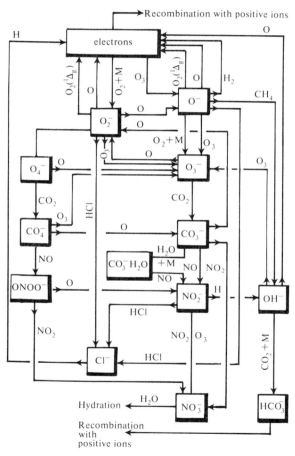

Fig. 6.9. Reaction scheme for negative ion chemistry in the D region. O_2^- and O^- are the primary ions, while the terminal ions are species such as NO_3^-, CO_3^-, and their hydrates. (Source as for Fig. 6.8).

$$NO_3^- \text{ or } \left(O-N\underset{O}{\overset{O}{<}}\right)^- \text{ and } NO_3^* \text{ or } (O-ONO)^-.$$

The initial step in each case is an addition, and each conversion represents a switching process with the appropriate neutral molecule:

$$O_2^- \rightarrow O_2^-.O_2; \quad O_2^-.CO_2; \quad O_2^-.NO \rightarrow O^-.NO_2, \tag{6.40}$$

or

$$O^- \rightarrow O^-.O_2; \quad O^-.CO_2; \quad O^-.NO \rightarrow O^-.NO_2. \tag{6.41}$$

The O_3^- ion ($O^-.O_2$) can be reached not only from the minor O^- primary species, but also from O_2^-

$$O_2^- + O_3 \rightarrow O_3^- + O_2. \tag{6.42}$$

At each stage of the conversion the intermediate negative ions can be removed by reaction with atomic species or by photodetachment. Figure 6.9 summarizes the steps leading to NO_3^- and HCO_3^-. Hydration appears to involve direct addition of H_2O to these ions. Conversion of CO_3^- to NO_2^- is the bottle-neck in the scheme, since the various routes from the primary ions to O_3^- are fast in comparison. Large concentrations of CO_3^- do not build up because of the rapidity of the 'back reactions', and the $O_2^- - CO_3^- - O_2^-$ loop occurs many times before an NO_2^- ion is produced. The slowness of the conversion step is a consequence of a small rate coefficient combined with a small concentration of neutral NO. A negative activation energy (cf. Section 3.4.1) characterizes the reaction, and atmospheric negative ion compositions may be very sensitive to temperature as well as variations in the minor species, O, O_3, NO, and H.

6.5 Ions in the stratosphere and troposphere

Although the D region is the bottom part of the 'conventional' ionosphere, speculation about underlying 'C', and lower regions has hardened into fact with the identification of ions in the stratosphere and troposphere. Rocket, balloon, and aircraft measurements have recently become feasible in these regions, and have revealed the presence of rather complex ions. Gaseous ions below the stratopause can play a role in trace gas processes and in aerosol formation. Long-range forces often make reaction cross-sections ('rate coefficients') large for ion-molecule processes. In addition, a newly-recognized class of process, *ion-catalysed reactions*, may occur in the stratosphere. Large cluster ions react with gas-phase species in a manner similar to surface-catalysed reactions: the reactant molecule can be regarded as being adsorbed on the 'surface' of the cluster ion. Ions promote nucleation *via* several processes, of which growth of large clusters (*ion nucleation*) and formation and growth of stable ion pairs by ion-ion recombination (*polyion nucleation*) are the most important. Gaseous ions are of additional interest in the stratosphere and troposphere since they can be used as powerful probes for neutral gas detection. Selective reactions of naturally occurring ions with neutral trace gases lead to characteristic ionic products. Mass spectrometric measurements of product and reactant ions allow concentrations of the trace species to be inferred with considerable accuracy. *Passive Chemical Ionization Mass Spectrometry* (PACIMS) of this kind has been supplemented by an *active* variant (ACIMS) in which ions are created in the stratospheric medium by an electron bombardment source.

Ionization in the stratosphere and troposphere is achieved largely by galactic cosmic rays, with some contribution from radioactive decay in the last kilometre above the Earth's surface. The primary processes give largely O_2^+ and electrons that attach to O_2 to yield O_2^-. Negative ions can also be generated in these regions by dissociative attachment reactions with species such as the chlorofluorocarbons that are released by man (cf. Section 4.5.4):

$$e + CF_2Cl_2 \rightarrow Cl^- + CF_2Cl \qquad (6.43a)$$

$$\rightarrow F^- + CFCl_2, \qquad (6.43b)$$

although three-body attachment reactions such as

$$e + O_2 + M \rightarrow O_2^- + M \qquad (6.11)$$

will always dominate primary negative ion production. Above about 35 km, the major terminal positive ions are water clusters, as in the lower D region. A fairly sharp transition is seen at 30–35 km to mixed and non-proton hydrate cluster ions, of which $H^+(CH_3CN)_l(H_2O)_{n-l}$ and $H^+(CH_3CN)_m$ are typical. Evolution of these ions can be envisaged as an extension of the D region chemistry reviewed in Fig. 6.8. Formation of $H^+(H_2O)_n$ ions is driven by the large proton affinity (~ 711 kJ mol^{-1}) of H_2O, and the strong bonding of H_2O molecules to the hydronium ion, H_3O^+. Displacement reactions

$$H^+(H_2O)_n + C \rightarrow H^+C(H_2O)_{n-1} + H_2O, \qquad (6.44)$$

are exothermic if the proton affinity of C is greater than that of water. Acetonitrile, CH_3CN, falls into this category, since its proton affinity is 777 kJ mol^{-1}. Tropospheric sources (perhaps biomass burning) seem most likely for CH_3CN, and the molecule survives transport to the stratosphere because it is neither attacked by OH nor photolysed efficiently (cf. Chapter 5). Other molecules, C, with high proton affinity include CH_3OH and HCHO from the methane oxidation chain (Section 5.3.2), and metal compounds such as NaOH or NaCl from meteor ablation and downward mixing from the mesosphere.

Stratospheric negative ions can be grouped into two families, NO_3^- $(HNO_3)_p$ ($p \sim 2$–3) and HSO_4^- $(H_2SO_4)_q$ $(HNO_3)_r$ ($q \lesssim 3$), with the former being dominant below about 25–30 km. The larger $HSO_4^-(H_2SO_4)_q$ ions are markedly hydrated, suggesting an increase in the H_2O bond energy for increasing q. Several other negative ions have been observed in smaller concentrations, most consisting of high electron affinity core molecules (e.g. CN, CO_3) and high gas phase acidity ligands (e.g. HCl, HNO_2, HOCl, HSO_3, H_2O). The major ions seem to be formed by a mechanism that starts from O_2^- as in the D region (Fig. 6.9). Nitric acid concentrations are sufficient in the stratosphere to displace water from $NO_3^-(H_2O)_n$ and establish a quasi-equilibrium size distribution of $NO_3^-(HNO_3)_p$. A second stage in the sequence involves reactions with S-containing gases (mostly H_2SO_4 and HSO_3), leading to HSO_4^- cores. Mixed cluster ions, $HSO_4^-(H_2SO_4)_q(HNO_3)_r$, are subsequently formed by the displacement of HNO_3 ligands by H_2SO_4. Strong co-operative bonding between H_2SO_4 and H_2O ligands leads to marked hydration of these mixed cluster ions, and the ions already resemble a small solution droplet composed of an H_2SO_4–H_2O mixture that has a large ΔH of mixing. Implications for stratospheric aerosol formation are apparent (cf. p. 149, and the Bibliography to Chapter 4).

The nature of tropospheric ions is at present just beginning to be unravelled, and the formidable problem of ion sampling from high pressures to be overcome. Aircraft flights at levels down to 3 km below the tropopause have returned mass spectrometric data on positive and negative ions. For example, the major negative ion is identified as $NO_3^-.(HNO_3)_2$ and there are sizeable contributions from $NO_3^-.HNO_3$ and NO_3^- itself. Negative ions up to mass 352 have been observed, with the highest mass ion being identified as $HSO_4^-(H_2SO_4)_2HNO_3$. A potential difference between tropospheric and stratospheric ions may arise from the small, and possibly strongly variable, abundances of trace gases that can be depleted by heterogeneous interaction with aerosols. Detailed *in situ* composition measurements of tropospheric ions will undoubtedly become refined in the near future, and investigation of the part played by ions in tropospheric chemistry will be a fruitful area of research.

Bibliography

Introductory texts dealing with the ionosphere and emphasizing the physical effects of charged particles in the atmosphere.

Sun, earth, and radio. Ratcliffe, J. A., Weidenfeld & Nicholson, London (1970).

An introduction to the ionosphere and magnetosphere. Ratcliffe, J. A. (Cambridge University Press, Cambridge, 1972.)

Physics and chemistry of the upper atmosphere. Rees, M. H. (Cambridge University Press, Cambridge, 1989.)

The Earth's ionosphere. Kelley, M. C. (Academic Press, San Diego, 1989.)

Introductions to the phenomenon of aurora.

The aurora. Akasofu, S-I., in *Atmospheric phenomena*, (ed. Lynch, D. K.) pp. 141–9. (Freeman, San Francisco, 1980.) (*Scient. Am.*, December 1965.)

The northern light. Brekke, A. and Engeland, A. (Springer, Berlin, 1983.)

Sections 6.2 and 6.3

> The nature and rates of reactions involving charged species are also discussed in Section 3.4.1 and the associated references. See Section 6.4.3 for experimental methods for the laboratory study of such reactions.
>
> The first two reviews are excellent general accounts of atmospheric ion chemistry, and that by Thomas also provides an introduction to the chemistry of neutral species in atmospheric regions above the stratosphere. The second two are concerned more particularly with the higher altitude regions (above about 100 km) and the last four with the lower altitudes.

Neutral and Ion Chemistry of the upper atmosphere. Thomas, L., in *Handbuch der Physik* (ed. Rawer, K.) XLIX/6, 7 Part VI. (Springer-Verlag, Berlin, 1982.)

Ion chemistry of the Earth's atmosphere. Ferguson, E. E., Fehsenfeld, F. C., and Albritton, D. L., in *Gas phase ion chemistry* (ed. Bowers, M. T.) Vol. 1, pp. 45–82. (Academic Press, New York, 1979.)

Upper atmosphere models and research. Ryecroft, M. J., Kasting, G. M., and Rees, D., (eds.). *Adv. Space Res.* **10**, No. 6 (1990).

Ionospheric Chemistry, Torr, D. G. *Rev. Geophys. & Space Phys.* **17**, 510 (1979).

Chemistry of the thermosphere and ionosphere. Torr, D. G. and Torr, M. R. *J. atmos. terr. Phys.* **41**, 797, (1979).

Modelling of the ion composition of the middle atmosphere. Thomas, L. *Ann. Geophysicae* **1**, 61 (1983).

Chemistry of middle atmospheric ionization—a review. Mitra, A. P. *J. atmos. terr. Phys.* **43**, 737 (1981).

On ionization of the lower mesosphere. Chakrabarty, D. K., Chakrabarty, P. and Witt, G. *J. atmos. terr. Phys.* **43**, 23 (1981).

Ion composition and electron and ion loss processes in the Earth's atmosphere. Arnold, F. and Krankowsky, D., in *Dynamical and chemical coupling* (ed. Grandal, B. and Holtet, J. A.) pp. 93–127. (D. Reidel, Dordrecht, 1977.)

> This paper examines quantitatively the effect of photoionization by starlight and scattered solar radiation at night.

Photoionization rates in the night-time E- and F-region ionosphere. Strobel, D. F., Opal, C. B., and Meier, R. R. *Planet. Space Sci.* **28**, 1027 (1980).

Sections 6.4.1 and 6.4.2

A study of the daytime E-F$_1$ region ionosphere at mid latitudes. Buonsanto, M.J. *J. geophys. Res.* **95**, 7735 (1990).

Relative abundances of the light ions in the winter topside ionosphere. Sanatini, S. and Breig, E. L. *J. geophys. Res.* **89**, 2918 (1984).

Modelling of the mid-latitude ionosphere. Taieb, C. and Poinsard, P. *Ann. Geophysicae* **2**, 197, 359 (1984).

> Dissociative recombination of ions such as O_2^+ or NO^+ can frequently lead to electronically excited fragment atoms. The efficiency of the process with O_2^+ is discussed in connection with atomic oxygen airglow in Section 7.4.1 and the associated references. Artificial injection of molecular species dramatically decreases electron densities by providing an efficient recombination route.

Ionospheric hole caused by rocket engine. Rycroft, M. J. *Nature, Lond.* **217**, 537 (1982).

Ablating meteors are usually thought to provide a source of metal atoms that inhibit recombination, so that meteor showers can enhance electron densities in the E region. The following paper suggests that sulphur may also be released on meteor ablation, and that SO and SO$^+$ ions become the principal carriers of sulphur following chemical conversion from S.

Sulfur chemistry in the E region. Swider, W., Murad, E., and Herrmann, U. *Geophys. Res. Lett.* **6**, 560 (1979).

Section 6.4.3

Chemistry and meteorology of the D region.

First ion composition measurements in the stratopause region using a rocket-borne parachute drop sonde. Pfeilsticker, K. and Arnold, F. *Planet. Space Sci.* **37**, 315 (1989).

Hydrogen constituents of the mesosphere inferred from positive ions: H_2O, CH_4, H_2CO, H_2O_2, and HCN. Kopp. E. *J. geophys. Res.* **95**, 5613 (1990).

Ion composition in the lower ionosphere. Kopp, E. and Herrman, U. *Ann. Geophysicae* **2**, 83 (1984).

Ion chemistry in the D region. Reid, G. C. *Adv. atomic molec. Phys.* **63**, 375 (1976).

Ion chemistry of the cold summer mesopause region. Reid, G. C. *J. geophys. Res.* **94**, 14653 (1989).

Meteorological control of the D region (Tutorial lecture). Taubenheim, J. *Space Sci. Rev.* **34**, 397 (1983).

NO and temperature control of the D region. Danilov, A. D. and Taubenheim, J. *Space Sci. Rev.* **34**, 413 (1983).

Identification of the mesospheric heavy ion ledge. Ganguly, S. *J. atmos. terr. Phys.* **46**, 99 (1984).

Cluster ions are of particular importance in the D region (and below). They may be implicated in the nucleation leading to formation of noctilucent clouds.

Nucleation and particle formation in the upper atmosphere. Keesee, R. G. *J. geophys. Res.* **94**, 14683 (1989).

Rapid growth of atmospheric cluster ions at the cold mesopause. Arnold, F. and Joos, W. *Geophys. Res. Lett.* **6**, 763 (1979).

Understanding the middle atmosphere via laboratory: ion cluster investigations. Keesee, R. G. and Castleman, A. W. *Ann. Geophysicae* **1**, 75 (1983).

Studies of ion clusters: relationship to understanding nucleation and solvation phenomena. Castleman, A. W., Jr., in *Kinetics of ion-molecule reactions* (ed. Ausloos, P.) pp. 295–321. (Plenum Publishing Corp., 1979.)

Nucleation and molecular clustering about ions. Castleman, A. W., Jr. *Adv. Colloid Interface Sci.* **10**, 73 (1979).

A reconsideration of nucleation phenomena in light of recent findings concerning the properties of small clusters, and a brief review of some other particle growth processes. Castleman, A. W., Jr. *Astrophys. Space Sci.* **65**, 337 (1979).

Noctilucent clouds. Gadsden, M. *Space Sci. Rev.* **33**, 279 (1982).

Dynamics, radiation, and photochemistry in the mesosphere: implications for the formation of noctilucent clouds. Garcia, R. R. *J. geophys. Res.* **94**, 14605 (1989).

Noctilucent clouds. Soberman, R. K., in *Atmospheric phenomena* (ed. Lynch, D. K.) pp. 131–40. (Freeman, San Francisco, 1980.) (*Scient. Am.*, June 1963.)

The effect of atmospheric screening on the visible border of noctilucent clouds. Taylor, M. J., Hapgood, M. A., and Simmonds, D. A. R. *J. atmos. terr. Phys.* **46**, 363 (1984).

On the diurnal variation of noctilucent clouds. Jensen, E., Thomas, G. E., and Toon, O. B. *J. geophys. Res.* **94**, 14693 (1989).

The references collected below discuss some of the laboratory techniques that have proved of particular value in the interpretation of ionospheric chemistry. The first three are reviews, while the remainder trace the development of flowing afterglow, drift, SIFT, and supersonic flow methods.

Flowing afterglow measurement of ion-neutral reactions. Ferguson, E. E., Fehsenfeld, F. C., and Schmeltekopf, A. L. *Adv. atomic. molec. Phys.* **5**, 1 (1969).

Recent advances in flow tubes: measurement of ion-molecule rate coefficients and product distributions. Smith, D. and Adams, N. G., in *Gas phase ion chemistry*, Vol. 1 (ed. Bowers, M. T.). (Academic Press, New York, 1979.)

Studies of ion-ion recombination using flowing afterglow plasmas. Smith, D. and Adams, N. G., in Brouillard, F. and McGowan, J. W. (eds.) *Physics of ion–ion and electron–ion collisions.* (Plenum Publishing Corp., New Jersey, 1983.)

Thermal energy ion-neutral reaction rates. I. Some reactions of helium ions. Fehsenfeld, F. C., Schmeltekopf, A. L., Goldan, P. D., Schiff, H. I., and Ferguson, E. E. *J. chem. Phys.* **44**, 4087 (1966).

Ion-molecule reaction studies from 300° to 600° in a temperature controlled flowing afterglow system. Dunkin, D. B., Fehsenfeld, F. C., Schmeltekopf, A. L., and Ferguson, E. E. *J. chem. Phys.* **49**, 1365 (1968).

Temperature dependence of some ionospheric ion-neutral reactions from 300° to 900°K. Lindinger, W., Fehsenfeld, F. C., Schmeltekopf, A. L., and Ferguson, E. E. *J. geophys. Res.* **79**, 4753 (1974).

Flow-drift technique for ion-mobility and ion-molecule reaction rate constant measurements. I. Apparatus and mobility measurements. McFarland, M., Albritton, D. L., Fehsenfeld, F. C., Ferguson, E. E., and Schmeltekopf, A. L. *J. chem. Phys.* **59**, 6610 (1973).

Effects of ion speed distributions in flow-drift studies of ion-neutral reactions. Albritton, D. L., Dotan, I., Lindinger, W., McFarland, M., Tellinghuisen, J., and Fehsenfeld, F. C. *J. chem. Phys.* **66**, 410 (1977).

Product-ion distributions for some ion-molecule reactions. Adams, N. G. and Smith, D. *J. Phys. B, Atom. molec. Phys.* **9**, 1439 (1976).

The selected ion flow tube (SIFT): a technique for studying ion-neutral reactions. Adams, N. G. and Smith, D. *Int. J. Mass Spec. Ion Phys.* **21**, 349 (1976).

Laboratory investigation of the ionospheric O_2^+ ($X^2\Pi_g$, $v \neq 0$) reaction with NO. Lindinger, W. and Ferguson, E. E. *Planet. Space Sci.* **31**, 1181 (1983).

Rate coefficient at 300 K for the vibrational energy transfer reactions from $N_2(v = 1)$ to $O_2^+(v = 0)$ and $NO^+(v = 0)$. Ferguson, E. E., Adams, N. G., Smith, D., and Alge, E. *J. chem. Phys.* **80**, 6095 (1984).

The reaction $O_2^+ + CH_4 \rightarrow CH_3O_2^+ + H$ studied from 20 to 560 K in a supersonic jet and in a SIFT. Rowe, B. R., Dupeyrat, C., Marquette, J. B., Smith, D., Adams, N. G., and Ferguson, E. E. *J. chem. Phys.* **80**, 241 (1984).

Section 6.5

Physics and chemistry of atmospheric ions (stratosphere and troposphere). Arnold, F., in *Atmospheric chemistry* (ed. Goldberg, E. D.), pp. 273–300, (Dahlem Konferenzen). (Springer-Verlag, Berlin, 1982.)

Mid-latitude lower ionosphere structure and composition measurements during winter. Arnold, F. and Krankowsky, D. *J. atmos. terr. Phys.* **41**, 1127 (1979).

Positive and negative ions in the stratosphere. Arijs, E. *Ann. Geophysicae* **1**, 149 (1983).

Positive ion composition measurements between 33 and 20 km altitude. Arijs, E. Nevejans, D., Ingels, J., and Frederick, P. *Ann. Geophysicae* **1**, 161 (1983).

Modelling of stratospheric ions: a first attempt. Brasseur, G. and Chatel, A. *Ann. Geophysicae* **1**, 173 (1983).

Composition measurements of tropospheric ions. Heitmann, H. and Arnold, F. *Nature, Lond.* **306**, 747 (1983).

Ion measurements can yield valuable data about the neutral trace gas abundances.

Upper stratosphere negative ion composition measurements and inferred trace gas abundances. Arnold, F. and Qui, S. *Planet. Space Sci.* **32**, 169 (1984).

Stratosphere *in situ* measurements of H_2SO_4 and HSO_3 vapours during a volcanically active period. Qui, S. and Arnold, F. *Planet. Space Sci.* **32**, 87 (1984).

Ion chemistry in the stratosphere can lead to significant changes in the composition of neutral species.

Effects of intense stratospheric ionisation events. Reid, G. C., McAffee, J. R., and Crutzen, P. J. *Nature, Lond.* **175**, 489 (1978).

Stratospheric ions may also be of importance in nucleation of aerosol growth.

Ion nucleation—a potential source for stratospheric aerosols. Arnold, F. *Nature, Lond.* **299**, 134 (1982).

Stratospheric sources of CH_3CN and CH_3OH. Murad, E., Swider, W., Moss, R. A., and Toby, S. *Geophys. Res. Lett.* **11**, 147 (1984).

Implications for trace gases and aerosols of large negative ion clusters in the stratosphere. Arnold, F., Viggiano, A. A., and Schlager, H. *Nature, Lond.* **297**, 371 (1982).

Condensation nuclei events at 30 km and possible influences of solar cosmic rays. Hofmann, D. J. and Rosen, J. M. *Nature, Lond.* **302**, 511 (1983).

7 The airglow

7.1 Optical emission from planetary atmospheres

Sources located outside our atmosphere and within it illuminate the night sky. Moon, stars, and surface lights all contribute, but if the light from these sources were eliminated, the sky would not be completely black. A faint glow would remain that has its origins in atmospheric photochemical processes, and to which the name *airglow* is given. Astronauts see this airglow as an envelope that sheaths the night side of the Earth, but Earth-bound observers are also able to perceive it. The visible radiation is rather feeble, and has been equated in intensity to the light of a candle at a distance of 100 m. In fact, the night sky would appear far brighter were it not that the strongest airglow emission features lie in the near infra-red region, just beyond the response of the human eye. By day, the airglow intensities are orders of magnitude larger, but are not detectable by eye because they are completely dominated by atmospherically scattered sunlight. According to the time of day at which observations are made, the airglow is described as *nightglow* or *dayglow*. When the Sun is below the horizon at ground level, but in view from the upper atmosphere, the emission is termed the twilight glow. Several of the lines and bands are identical to those seen in the aurora (Section 6.1.1), and some distinction needs to be drawn between the two phenomena of airglow and aurora. Airglow occurs continuously, is weak, and is observed at all latitudes. In contrast, auroras are much more intense, but irregular in form and occurrence, and are restricted to a region near the geographical poles. The differences arise from the excitation mechanisms. Airglow is driven by photons from the Sun, whereas auroras are excited by the impact of energetic solar particles.

Airglow emissions are a feature of most planetary atmospheres. The light consists of atomic and molecular line, band, and continuum systems of atmospheric constituents, both neutral and ionized. Because many of the emission features are measurable at the Earth's surface, airglow investigations provided a valuable source of information about the composition of our own as well as other planetary atmospheres long before the era of direct investigations by rocket probes and satellites. Other emission features are absorbed by the terrestrial atmosphere, so that their study is only possible with instrumentation on space vehicles. Studies of the airglows of the planets continue to assist interpretations of the photochemistry and dynamics of the atmospheres.

Most of the spectral features of the airglow have their origins in transitions from *electronically excited* species (although very important components of the Earth's airglow are due to *vibrationally* excited OH, as described in Section 7.4.3, and NO). Many of the transitions are formally *forbidden* by the electric dipole selection rules (see Section 3.1). Indeed, there may be no lower level to which an electronically excited species can make an allowed transition, in which case the species will have a long radiative lifetime, and is said to be *metastable*. Metastable species can therefore be transported away from the region in which they are formed, and subsequently give up their energy in collisional quenching processes. Excited species can participate in otherwise endothermic reactions (Section 3.2.2). Significant non-local effects are introduced, particularly into thermospheric chemistry, as a result of the formation and transport of metastable atoms and molecules. Whatever the ultimate fate of the excitation energy, the emitting species responsible for the airglow are far removed from thermal equilibrium with their surroundings so far as their electronic temperature is concerned. One of our interests will be to look at the physical and chemical processes that can lead to specific excitation of a particular energy level (Section 7.2).

For many of the atoms and molecules of the atmosphere, the first allowed optical transition to and from the ground state (the *resonance* transition) lies in the 'vacuum' ultraviolet, and so by definition does not penetrate the Earth's atmosphere. Resonance emissions from helium ($\lambda = 58.4$ nm), nitrogen (120.0 nm), hydrogen (121.6 nm), and oxygen (130.2, 130.4, 130.6 nm) which might otherwise be expected to be strong, are not observed at the ground, although they are major features of the Earth's airglow as seen from space. We have already alluded to the resonant absorption and re-radiation of sunlight by hydrogen and helium in connection with sources of night-time ionization (Section 6.3). Emissions at wavelengths longer than the atmospheric cut-off may sometimes be removed by specific absorption processes. For example, the strongest feature of the terrestrial dayglow is due to a transition from the first excited state of O_2, the $^1\Delta_g$, to the ground state, $^3\Sigma_g^-$, and is known as the *Infra-red Atmospheric Band*. The (0, 0) band of this system lies at $\lambda = 1270$ nm, just into the infra-red region, and in a part of the spectrum generally free of atmospheric absorptions. The exception, of course, is for the lines of the oxygen bands originating in the $v'' = 0$ level of the ground state. Although the $^1\Delta_g - {}^3\Sigma_g^-$ transition is highly forbidden (note that the electric dipole selection rules involving S, Λ, and g/u are all broken!), the great optical path of atmospheric oxygen filters out from the surface most of the $\lambda = 1270$ nm line produced at altitudes of around 50 km (Section 7.4.1). Atmospheric absorption of the (0, 1) band (i.e., $^1\Delta_g$, $v' = 0 \leftarrow {}^3\Sigma_g^-$, $v'' = 1$) at $\lambda = 1580$ nm is almost non-existent because the population of $v'' = 1$ in O_2 is small at atmospheric temperatures. This band can thus be observed from the ground, while rockets, balloons, or aeroplanes must be used to study the (0, 0) emission. Cunning experimentation allows the (0, 0) transition from *other* planetary airglows to be observed with ground-based telescopes, even though at first sight such investigations might appear doubly restricted. However, high-resolution spectroscopy (using interferometric methods) can separate out individual rotational lines in the bands. If Earth and the target planet have a sufficiently high relative velocity, then the planetary airglow lines are Doppler-shifted away from the terrestrial absorption lines. Successful investigations of $O_2(^1\Delta_g)$ in the airglows of Mars and Venus have been achieved by the technique.

Altitude profiles of the emitting species often afford valuable insight into excitation and deactivation mechanisms of specific airglow features. Early studies used the variation in intensity with viewing angle to estimate the profile, but this rather unreliable method has been replaced almost completely by rocket measurements. The total (integrated) overhead intensity is measured as a function of altitude attained by the rocket, and the profile obtained by numerical differentiation of the result. Figures 7.1 and 7.2 illustrate the difficulties inherent in the method. Data for overhead intensities of the night-glow Infra-red Atmospheric Band are given in the first figure (the intensity units are explained in Section 7.3) and the derived $[O_2(^1\Delta_g)]$ profile in the second. The two concentration peaks arise from relatively small changes of slope in Fig. 7.1. Noise on the data could greatly affect the magnitude and position of the peaks, especially in the case of the higher altitude one where the remaining overhead intensity is low. In Section 7.3, we consider what factors determine the absolute intensities and the altitude dependence of airglow emissions.

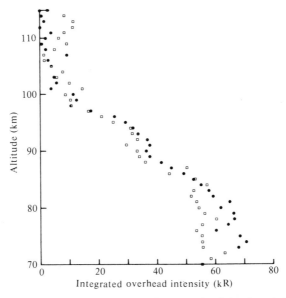

Fig. 7.1.   Infra-red Atmospheric Band ($O_2\,^1\Delta_g \rightarrow {}^3\Sigma_g^-$) in the nightglow at $\lambda =$ 1270 nm. Overhead intensities are measured during the ascent (squares) and descent (circles) of a rocket-borne instrument. [From Evans, W. F. J., Llewellyn, E. J., and Vallance-Jones, A. *J. geophys. Res.* **77**, 4899 (1972).]

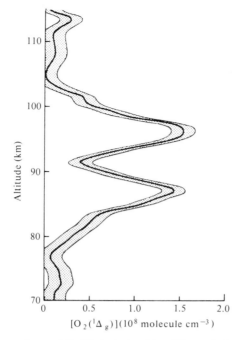

Fig. 7.2. Concentrations of $O_2(^1\Delta_g)$ derived by differentiation of the data of Fig. 7.1. The width of the band represents the estimated probable error. Note how the two marked concentration peaks arise from small inflections in the measured total intensities. (Source as for Fig. 7.1.)

Dozens of features are known in the terrestrial airglow. Our intention is not to catalogue and explain all these features. Instead, we discuss in Section 7.4 a few of the more interesting and important transitions that are observed. First, however, we summarize the types of process that lead to excitation, deliberately using as illustrations some additional features of the airglow.

7.2 Excitation mechanisms

During the day, sunlight can be absorbed directly by atmospheric constituents, to yield electronically excited products. Radiative decay then contributes to the airglow, and the absorption-radiation sequence is *resonance fluorescence* or *resonance scattering*.[a] This is the process we discussed in connection with the residual ionization at night (Section 6.3) caused predominantly by H

---

[a] Properly speaking, the emitted radiation must be at the same wavelength as the absorption for the fluorescence to qualify as a resonant process. However, in molecular systems the concept of resonance is often extended to include emission from and to vibrational levels (of the *same* electronic states) different from those initially populated.

and He atoms. Atoms such as H have particular roles to play in resonance scattering, since the hydrogen resonance line is a major component of the solar spectrum and there is thus a match that efficiently couples energy transfer from Sun to Earth. The upper and lower electronic levels involved are the ^{2}P and ^{2}S states of H, so that the excitation-emission sequence may be written

$$H(^2S) + hv(\text{solar}; \lambda = 121.6 \text{ nm}) \rightarrow H(^2P), \tag{7.1}$$

$$H(^2P) \rightarrow H(^2S) + hv(\text{airglow}; \lambda = 121.6 \text{ nm}). \tag{7.2}$$

Other species, for which there is not a specific solar emission line, can absorb from the solar continuum. Even molecular species such as NO or OH are efficiently excited in this way, the NO 'γ system' ($A^2\Sigma^+ \rightarrow X^2\Pi$) being the strongest dayglow feature in the spectral range 200–300 nm (but observable only by rockets because of atmospheric absorption). The OH resonance band of the OH ($A^2\Sigma^+ \rightarrow X^2\Pi$) transition at $\lambda = 306.4$ nm is another intense day-glow emission observed from rockets, and the $A^2\Sigma^+$ state is mainly populated by resonance absorption. A second process may, however, supplement the direct absorption route. In Section 3.2.1, we pointed out that photodissocia-tion often yields one or more electronically excited fragments. Water vapour dissociation (for example, by Lyman-α radiation) yields some $OH(A^2\Sigma^+)$,

$$H_2O + hv \rightarrow OH(A^2\Sigma^+) + H, \tag{7.3}$$

that contributes a small intensity to the OH UV dayglow. One of the most important excitation reactions of this kind involves ozone photolysis. At $\lambda \leqslant 310$ nm *both* the atomic O and the molecular O_2 fragments are excited (see Section 3.2.1), and contribute to the oxygen airglow as will be discussed in Section 7.4.1. Photoionization is a particular case of photofragmentation, and can lead to excitation in the positive ion. The process is probably responsible for exciting a variety of N_2^+ states such as $A^2\Pi_u$ and $B^2\Sigma_u^+$ that make important contributions to the dayglow throughout the visible and near infra-red regions:

$$N_2(X^1\Sigma_g^+) + hv(\lambda \leqslant 79.6 \text{ nm}) \rightarrow N_2^+(B^2\Sigma_u^+, A^2\Pi_u, X^2\Sigma_g^+) + e. \tag{7.4}$$

Photoelectrons produced in ionization processes such as (7.4) may be created with excess energy, since solar photon energies can be appreciably greater than ionization potentials. At high altitudes, the electrons have a mean kinetic energy of 10 eV. Inelastic collisions of the electrons with molecules generally lead to the dissipation of the excess energy as vibration and rotation. Some-times, however, low-lying excited electronic states may be populated by the electron impact. Several excited states of neutral N_2 seem to be formed by this mechanism

$$N_2(X^1\Sigma_g^+) + e(\text{'hot'}) \rightarrow N_2(C^3\Pi_u, B^3\Pi_g, A^3\Sigma_u^+) + e, \tag{7.5}$$

and transitions of systems such as C → B ('Second positive'), B → A ('First positive') and A → X ('Vegard-Kaplan') are significant components of the dayglow. Note that the C → B, and B → A radiative transitions populate the lower—but still excited—electronic levels by cascading from the higher ones. Radiationless transitions can also populate, in an intramolecular energy transfer step, excited states distinct from those initially excited. Rather complicated processes of this kind occur in the N_2 system: not only does the upper B state populate the lower A in a radiationless crossing, but high vibrational levels of the A state can cross back to lower vibrational levels of the B. We shall meet cases of *inter*molecular energy transfer, in which a separate and chemically distinct species is excited, in Section 7.4.1. Reference back to Fig. 3.1, which depicts the fates of electronically excited species, will show that pathways (i) to (v) thus all make their contribution to some feature or other of the dayglow.

At night, the source of photochemical excitation is removed, yet the existence of the nightglow shows that excited species persist. In general, this result must mean that solar energy has been stored during the day, and that reactions are releasing the energy at night. Neutral atoms (especially oxygen) are a very important energy reservoir at altitudes below ~ 100 km, while ions are very important at higher levels. We shall see in Section 7.4.1 how several atomic and molecular transitions in the nightglow owe their existence to atomic oxygen. Reaction exothermicity is used to populate preferentially certain states of the species, which are thus out of thermal equilibrium with their surroundings. Emission of *chemiluminescence* then contributes to the nightglow. A typical example from the laboratory is the reaction

$$NO + O_3 \rightarrow NO_2^* + O_2, \tag{7.6}$$

in which emission is seen from NO_2^* populated virtually up to the exothermicity (205 kJ mol^{-1}) of the reaction. It is not clear whether reaction (7.6) is of significance in the airglow, but NO_2 emission is implicated in a weak background continuum of the nightglow, and it may be produced in the *two-body*, radiatively stabilized, combination of O with NO:

$$NO + O \rightarrow NO_2 + h\nu. \tag{7.7}$$

Dissociative recombination of positive ions with electrons (Section 6.4.1) liberates the molecular ionization energy in the fragments, and electronic excitation may result. Excited O-atoms are formed by dissociative recombination of O_2^+ (cf. p. 290), and the first electronically excited state of N, the ^2D, is probably populated in the ionosphere by the analogous reaction with NO^+

$$NO^+ + e \rightarrow N(^2D) + O. \tag{7.8}$$

Nightglow excitation mechanisms can obviously operate during the day, but they are *usually* dominated while the Sun is present by processes more directly utilizing solar energy. Some exceptions do exist, a notable one being the chemical excitation of *vibrationally* excited OH radicals described in Section 7.4.3.

7.3 Airglow intensities and altitude profiles

Emission from an excited species A* is described by a rate law

$$I = k_r[A^*] \tag{7.9}$$

where I is here a total emission rate in quanta (or photons) per unit volume per unit time. The rate coefficient, k_r, is equivalent to the transition probability, or Einstein 'A' coefficient for spontaneous emission; a *radiative lifetime*, τ, is often defined by the relation $\tau = \ln 2/k_r$. To a first approximation, k_r is independent of concentration or pressure (although for certain forbidden transitions, collision-induced processes do endow k_r with some pressure dependence).

Radiation is an isotropic process, so that each volume element of the airglow emits equally in all directions. A human observer or an instrument does not integrate all this radiation, but rather perceives a 'brightness' that depends on the flux of photons per unit area per unit time. For this reason, airglow brightnesses are usually measured in *Rayleigh* units (R). One Rayleigh is the brightness of a source emitting 10^6 photons cm^{-2} s^{-1} in all directions.[a] The convenience of the unit lies in the direct relationship between it and the emission rate of eqn (7.9), since (for an optically thin medium) I need only be multiplied by the depth of the emitting layer to give the Rayleigh brightness. For correct scaling of the 10^6 factor in the definition, the depth has to be measured in units of 10^6 cm (10 km), which is, in fact, a typical thickness for airglow emission layers. The name of the unit honours Robert John Strutt, Fourth Lord Rayleigh, who performed much pioneering work on the airglow. He is sometimes called 'the airglow Rayleigh' to distinguish him from his father, the Third Lord Rayleigh, or 'scattering Rayleigh' (cf. p. 74).

Intensities of airglow features are related to atmospheric concentrations of excited species through the transition probability, k_r, of eqn (7.9). Large intensities of strong emitters may not necessarily correspond to higher concentrations than lower intensities of weak emitters. For example, the dayglow intensity of the $N(^2D - {}^4S)$ lines at $\lambda \sim 520$ nm is only 90 R, but since k_r for this highly forbidden transition is $\sim 7.4 \times 10^{-6}$ s^{-1} (lifetime 26 *hours*) the concentration of $N(^2D)$ in our hypothetical layer of 10 km thickness would be 10^7 atom cm^{-3}. Emission from sodium in the dayglow, of the allowed $Na(^2P - {}^2S)$ resonance transition at $\lambda \sim 589$ nm, is more than 300 times as intense (30 kR). However, k_r is $\sim 6.3 \times 10^7$ s^{-1} (lifetime 12 *nanoseconds*) which suggests a concentration of excited sodium atoms of a few hundred per *cubic metre* (4.8×10^{-4} atom cm^{-3}) for a 10 km emitting layer.

Radiative decay inevitably competes with production of emitting excited

---

[a] The light is radiated into a solid angle of 4π steradians, so that the source is emitting $10^6/4\pi$ photons cm^{-2} steradian^{-1} s^{-1}, a quantity called the *surface brightness* (sometimes 'intensity') in optical measurements.

species. Non-radiative intra- and intermolecular processes (cf. Fig. 3.1) provide additional loss mechanisms. For many atmospherically significant airglow emitters, physical quenching [pathway (vi) in the figure] is the most important of these processes. If we write the quencher species as M and the rate coefficient as k_q, then the generalized excitation–de-excitation mechanism becomes

$$\text{Source} \rightarrow A^* \qquad \text{Excitation; Rate} = P \qquad (7.10)$$

$$A^* \rightarrow A + h\nu \qquad \text{Emission; Rate} = k_r[A^*] \qquad (7.11)$$

$$A^* + M \rightarrow A + M \qquad \text{Quenching; Rate} = k_q[A^*][M]. \qquad (7.12)$$

So long as the excitation rate, P, does not change rapidly, then a steady state for $[A^*]$ may be set up, and

$$
\begin{aligned}
I = k_r[A^*] &= k_r P/(k_r + k_q[M]) \\
&= P(1 + k_q[M]/k_r)^{-1}.
\end{aligned}
\qquad (7.13)
$$

This equation should be compared with eqns (3.47) to (3.49) in Section 3.5, where the quantitative justification of steady state treatments is to be found. In the discussion that follows we ignore physical transport in order to high-light the chemical kinetic controls on airglow. Physical quenching determines the intensity of the airglow for a given excitation rate according to the relative magnitudes of k_r and $k_q[M]$. For a species such as $N(^2D)$, with k_r less than 10^{-5} s^{-1}, quenching dominates at all emitting altitudes. Reported values of k_q for quenching of $N(^2D)$ by atmospheric gases are of the order of 10^{-11} cm^3 molecule^{-1} s^{-1}, so that deactivation is dominated by radiation only for $[M] \leqslant 10^6$ molecule cm^{-3}. Total particle concentrations of this magnitude are found at altitudes above ~ 500 km in the Earth's atmosphere. The peak of $N(^2D)$ emission in the dayglow appears to originate at ~ 200 km, where $[M] \sim 10^{10}$ particle cm^{-3}, so that only one in 10^4 of the atoms excited can actually emit. Airglow emission profiles are obviously dependent both on the variation of the excitation rate, P, with altitude, and on the $[M]$ dependence of the function $(1 + k_q[M]/k_r)^{-1}$ found in eqn (7.13). The arguments we have presented in earlier chapters about the compromise between atmospheric penetration of energetic solar ultraviolet and the concentration of reactive species apply equally to airglow excitation, and layer-like airglow regions can be expected. Species requiring high energy for their production and excitation are generally found at elevated altitudes. As an additional feature in the airglow, however, the lower boundary of the emitting layer is often sharpened by increased rates of loss by quenching acting together with decreased rates of excitation.

In circumstances where P changes rapidly, the steady state may not be maintained. Such a situation may arise at twilight or dawn, or during an eclipse, with an excitation process dependent on the presence of sunlight. For the extreme case, where P goes to zero instantaneously, airglow intensity will decay with a first-order rate determined by the composite coefficient

$(k_r + k_q[M])$. More realistic cases can be treated numerically, and allowance made for physical transport. Observations of the time dependence of the airglow intensity during periods of change can clearly provide information about [M], and hence emission altitude, if k_q and k_r are known, or conversely about the rate constants if the emission altitude can be estimated.

7.4 Specific emission sources

7.4.1 Atomic and molecular oxygen

Excited states of O and O_2 make an extremely important series of contributions to the airglow of Earth and other planets. Figures 7.3 and 7.4 introduce this topic by showing some transitions in O and O_2 that can be observed at ground level. *All* the relatively long wavelength ($\lambda \gtrsim 300$ nm) transitions are forbidden by the electric dipole rules, and the excited states are therefore metastable.

The $O(^1S \rightarrow {}^1D)$ transition at $\lambda = 557.7$ nm was the first component of the airglow to be identified, by high resolution spectroscopy, with a specific atomic or molecular event. Because the radiation was already known in the aurora, this well-known line is often called the *auroral green line*, even when it originates in the airglow. Both the green line, and the red doublet at $\lambda = 630.0$ and 636.4 nm, due to the $O(^1D \rightarrow {}^3P_{2,1})$ transition, seem to arise from two different altitude regions of the atmosphere. The high-altitude excitation

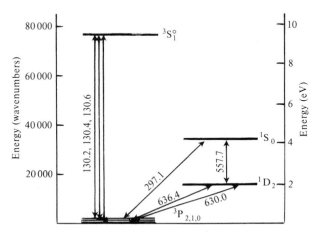

Fig. 7.3. Some low-lying energy levels of atomic oxygen. Atmospherically important optical transitions are shown by the arrows, and the wavelengths of the lines are given in nanometres. (Data from Moore, C. E., *Atomic energy levels*, Vol. 1, NSRDS-NBS35, Washington DC, 1971.)

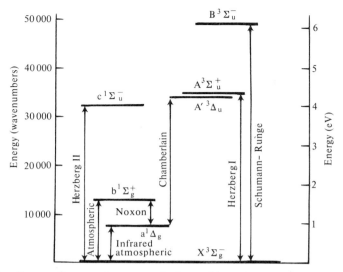

Fig. 7.4. Nomenclature of some optical transitions in molecular oxygen. With the exception of the B ↔ X (Schumann-Runge) system, all these transitions are 'forbidden' by electric dipole selection rules. [Data from Krupenie, P. H. *J. Phys. Chem. Ref. Data* **1**, 423 (1972).]

processes include electron impact and dissociative recombination of O_2^+, which can yield both $O(^1S)$ and $O(^1D)$;

$$O_2^+ + e \rightarrow O(^1S) + O(^3P) + 2.78 \text{ eV} \tag{7.14}$$

$$O_2^+ + e \rightarrow O(^1D) + O(^3P) + 4.99 \text{ eV}. \tag{7.15}$$

This dissociative recombination process is, of course, one of the two most important charge neutralization steps in the F region, the other being reaction (7.8) (cf. Sections 6.2 and 6.4.1). The yield per recombination of $O(^1D)$ is independent of vibrational excitation in O_2^+, while the yield of $O(^1S)$ is energy-dependent, ranging from 0.09 to 0.23. Reactions (7.14) and (7.15), with ions produced by energetic solar particles (see Section 6.3), are likely to be significant auroral sources of $O(^1S)$ and $O(^1D)$. It is probably the vibrational energy dependence of the $[O(^1S)]/[O(^1D)]$ ratio in the recombination step that accounts for the variability of enhancement of the oxygen lines in auroras.

Reactions of neutral species are responsible for the lower altitude (90–100 km) O-atom airglow. The main emitting layer of $O(^1S)$ is in this region, and the excited atoms are formed by day and by night in a process that we shall describe in detail later. A little $O(^1D)$ is populated radiatively in the $O(^1S \rightarrow {}^1D)$ emission of the green line. By day, however, two familiar photo-dissociation steps can generate $O(^1D)$. Optical dissociation of O_2 in the

Schumann-Runge continuum ($\lambda \lesssim 175$ nm; see Section 3.2.1) yields one ^{1}D and one ^{3}P atom

$$O_2 + h\nu(\lambda \lesssim 175 \text{ nm}) \rightarrow O(^1D) + O(^3P). \tag{7.16}$$

Maximum solar energy absorption in the Schumann-Runge system occurs at altitudes of 80–110 km, corresponding quite well with the maximum in $O(^1D)$ daytime emission. At longer wavelengths, and thus potentially at lower altitudes, ozone photolysis is a source of $O(^1D)$, as discussed at length in Section 3.2.1

$$O_3 + h\nu(\lambda < 310 \text{ nm}) \rightarrow O(^1D) + O_2(^1\Delta_g). \tag{7.17}$$

The maximum in $O(^1D)$ production rate from this reaction arises at an altitude of ~ 40 km. However, quenching of $O(^1D)$ by N_2 and O_2 is efficient, and at 40 km the collisional lifetime is less than 1 μs, so that radiation (lifetime 110 s) is very improbable.

The $^1\Delta_g$ *molecular* fragment produced in reaction (7.17) is, by contrast with $O(^1D)$, only very weakly quenched by N_2 or O_2. In the upper stratosphere, the collisional lifetime is of the order of tens to hundreds of seconds. Although the radiative lifetime is also very long (44 min) for the highly forbidden $O_2(^1\Delta_g \rightarrow {}^3\Sigma_g^-)$ Infra-red Atmospheric Band at $\lambda = 1270$ nm (see p. 309 and Fig. 7.4), a sensible fraction of $O_2(^1\Delta_g)$ can emit because of the inefficiency of physical deactivation. In fact, the Infra-red Atmospheric Band is the most intense feature of the dayglow, the (0, 0) transition (observed at altitudes high enough to avoid self-absorption, see p. 309) having an intensity of 20 MR. Consideration of energy balance alone points to ozone photolysis as the source of $O_2(^1\Delta_g)$, since no species other than ozone absorbs enough sunlight at the altitudes where the dayglow emission originates. Peak concentrations of $O_2(^1\Delta_g)$, determined by rocket photometry, are typically 2×10^{10} molecule cm^{-3} at 50–60 km altitude. Concentrations of this magnitude mean that the metastable excited molecules are present at the parts per million level. Laboratory measurements of the quantum yield for singlet oxygen production in reaction (7.17), and of rate coefficients for quenching by atmospheric gases, may be put together with experimentally determined atmospheric ozone concentrations and solar irradiances to predict an $[O_2(^1\Delta_g)]$—altitude profile. The results of such a calculation are displayed as the solid line in Fig. 7.5, while some rocket measurements of $[O_2(^1\Delta_g)]$ are shown as circles on the figure. Agreement of this kind may be taken as strong confirmation of the mechanism suggested for daytime excitation of $O_2(^1\Delta_g)$. After the Sun falls below the horizon, or is eclipsed, concentrations of $O_2(^1\Delta_g)$ fall rapidly. An analysis of the intensity-time dependence in eclipses has been used to derive a composite quenching rate coefficient by atmospheric gases of 3×10^{-19} cm^3 molecule^{-1} s^{-1}. This value is almost exactly identical to that calculated for air from the laboratory quenching data on the individual atmospheric gases. Confidence in the model is thus so great that dayglow measurements of the

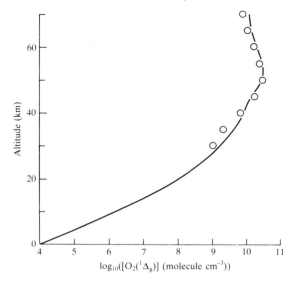

Fig. 7.5. Atmospheric concentrations of $O_2(^1\Delta_g)$ during the day. Experimental values (circles) are derived from rocket measurements of the Infra-red Atmospheric Band intensity in the dayglow, while the calculations (solid line) are based on laboratory kinetic and spectroscopic data. [From Crutzen, P. J., Jones, I. T. N., and Wayne, R. P. *J. geophys. Res.* **76**, 1490 (1971).]

Infra-red Atmospheric Band can be used to derive atmospheric ozone concentrations and profiles. For example, a secondary peak in emission intensity found at a height of ~ 90 km is probably to be associated with a secondary maximum in $[O_3]$ at this altitude. One of the instruments on the SME satellite (Section 4.4.5) now routinely inverts the airglow intensities at $\lambda = 1270$ nm, taking into account quenching rates, to derive ozone concentrations.

The Infra-red Atmospheric Band at $\lambda = 1270$ nm is present in the airglow of both Mars and Venus. As discussed in Section 7.1, Doppler-shifted lines can be observed by telescope. Intensities on Mars are as high as 26 MR in the winter north-polar region. Ozone photolysis, as on Earth, appears to be the only reasonable source of the singlet oxygen. Martian atmospheric ozone concentrations are quite small (see Section 8.3); quenching of $O_2(^1\Delta_g)$ by CO_2, the major component of the atmosphere, is very slow, so that high intensities of the Infra-red Atmospheric Band are still possible. Ozone photolysis cannot be the source of $O_2(^1\Delta_g)$ on Venus, because the column emission rate on the day side (1.8 MR) is close to the rate at which O atoms are produced by CO_2 photolysis. Photodissociation of ozone can account for only 130 kR of the emission. Favoured alternative excitation reactions have included the exothermic processes

$$Cl + O_3 \rightarrow ClO + O_2 + 161 \text{ kJ mol}^{-1} \tag{7.18}$$

$$ClO + O \rightarrow Cl + O_2 + 236 \text{ kJ mol}^{-1}. \tag{7.19}$$

Each reaction is sufficiently exothermic to populate either $O_2(^1\Delta_g)$ or $O_2(^1\Sigma_g^+)$. Together they represent the Cl_x catalytic chain (see Section 4.4.1), but the chain needs to be fed by O_3 as well as by O. Laboratory evidence for the excitation of $O_2(^1\Delta_g)$ in either reaction is very meagre, and the presence of chlorine-containing species on Venus is currently disputed! Interpretation of $O_2(^1\Delta_g)$ airglow intensities on Venus thus remains uncertain.

The second excited singlet of O_2, the $b^1\Sigma_g^+$ state (Fig. 7.4), also produces a strong emission system in the dayglow. Just beyond the visible region at $\lambda = 762$ nm, the forbidden transition $O_2(^1\Sigma_g^+ \rightarrow {}^3\Sigma_g^-)$ is readily observed by rocket photometry. Radiation of this system is called the *Atmospheric Band*. Between 65 and 100 km, the dominant excitation mechanism is resonance absorption of solar radiation

$$O_2 + h\nu(\lambda = 762 \text{ nm}) \rightarrow O_2(^1\Sigma_g^+). \tag{7.20}$$

Above 100 km, and below 65 km, $O_2(^1\Sigma_g^+)$ is mainly populated by energy transfer from $O(^1D)$ to O_2

$$O(^1D) + O_2 \rightarrow O_2(^1\Sigma_g^+) + O(^3P). \tag{7.2.1}$$

Excited oxygen atoms are produced by O_2 photolysis, reaction (7.16), in the higher region, and by O_3 photolysis, reaction (7.17), in the lower. The derived quenching rate constant for air is in good agreement with laboratory data.

The night-time persistence of excited atomic and molecular oxygen, as demonstrated by the nightglow, offers some difficulties in interpretation. Direct photochemical excitation is obviously excluded. So is indirect excitation by photochemically generated species, such as $O(^1D)$, whose lifetime against reaction is short. An obvious energetic reservoir for the oxygen species is ground state atomic oxygen. We shall concentrate in the rest of this section on how O atoms could populate the states observed and on whether the intensities and altitude distributions are compatible with the proposed mechanisms.

Our starting point is the $O(^1S)$ state responsible for the green line of the nightglow. For years, controversy existed over the relative merits of a one-step termolecular reaction

$$O + O + O \rightarrow O(^1S) + O_2, \tag{7.22}$$

suggested initially by Chapman, and a two-step scheme,

$$O + O \xrightarrow{M} O_2^*, \tag{7.23}$$

$$O + O_2^* \rightarrow O(^1S) + O_2, \tag{7.24}$$

due to Barth; O_2^* is some sufficiently energy-rich state of O_2 whose identity we shall shortly discuss. Both schemes predict a production rate of $O(^1S) \propto [O]^3$. Anomalies in the laboratory data have now been resolved, and theoretical and experimental considerations all favour the two-step 'Barth' mechanism.

Recombination of ground state O atoms in reaction (7.23) could, in principle, populate any or all of the singlet, triplet (and quintet) molecular states that correlate with $O(^3P)$. The potential energy diagram of Fig. 3.2 shows some of these states. Excitation of $O(^1S)$ in reaction (7.24) requires 4.19 eV, so that the $v' = 0$ levels of $c^1\Sigma_u^-$ (4.05 eV), $A'^3\Delta_u$ (4.24 eV) and $A^3\Sigma_u^+$ (4.34 eV) are all possible candidates for O_2^*. Recent experimental evidence has drawn particular attention to the $c^1\Sigma_u^-$ state. Nightglow emission ($\lambda \sim 400$–650 nm) from Venus that was detected by the Soviet Venera 9 spacecraft was subsequently identified as resulting from a $c^1\Sigma_u^- \rightarrow X^3\Sigma_g^-$ ('Herzberg II': see Fig. 7.4) progression in O_2, overlapped by weaker bands of the $A'^3\Delta_u \rightarrow a^1\Delta_g$ ('Chamberlain') system. Re-examination of terrestrial airglow spectra in the light of these findings revealed that the Herzberg II bands, due to $O_2(c^1\Sigma_u^-)$, are also present in the Earth's nightglow. Intensities are, however, much weaker on Earth (~ 100 R) than on Venus (2700 R), and the vibrational distribution is totally different, with the terrestrial distribution peaking for $v' = 7$ and the Venus spectrum showing emission only from $v' = 0$. On Mars, the Herzberg II system is even weaker, the intensity being $\leqslant 30$ R. Identification of the state of O_2^* that excites $O(^1S)$ is helped by the Venusian nightglow studies. Green-line emission is very weak (< 5 R) on Venus, so that the precursor state present in the Earth's atmosphere is virtually absent in the upper atmosphere of Venus. But $v' = 0$ of $c^1\Sigma_u^-$, $A'^3\Delta_u$, and $A^3\Sigma_u^+$ are all known to be formed on Venus. The difference between the species is that the $c^1\Sigma_u^-$ state lies 0.14 eV below the $O(^1S)$ excitation energy, while the A' and A states lie above. An activation energy of the order of the endothermicity would make reaction (7.24) slow and the emission weak, since the temperature is only ~ 188 K at those altitudes in the Venusian atmosphere where [O] is highest. Vibrational relaxation of $O_2(c^1\Sigma_u^-)$ is much less pronounced in the Earth's atmosphere than on Venus, with peak intensities from $v' = 7$. Reaction (7.24) becomes exothermic for $v' > 2$ or 3, so that the process can be an efficient source of $O(^1S)$ on Earth.

Laboratory studies have demonstrated explicitly that $O_2(c^1\Sigma_u^-)$ is, indeed, formed in the recombination of $O(^3P)$, so that reaction (7.23) can be written out in full for production of the $c^1\Sigma_u^-$ state

$$O + O + M \rightarrow O_2(c^1\Sigma_u^-) + M. \tag{7.25}$$

Emission of the $c \rightarrow X$, Herzberg II, bands is greatest when carbon dioxide is used as the third-body M, but CO_2 is not essential. Carbon dioxide has a peculiar influence either because it is a particularly effective third-body, or because it promotes vibrational relaxation in the recombining O atoms to funnel most of the excited population to $v' = 0$ of $O_2(c^1\Sigma_u^-)$. Venus, with a CO_2 atmosphere, therefore has a relatively large concentration of the excited molecules, but with $v' = 0$; Earth, on the other hand, has a lower concentration, but with v' up to 7, because M is predominantly N_2 and O_2. Emission of the Herzberg II system from Mars is very weak indeed, yet the atmosphere is, like that of Venus, largely CO_2. However, O_2 is at least 10^3 times as abundant on Mars as on Venus, and quenching of $O_2(c^1\Sigma_u^-)$ by O_2 probably suppresses the Martian Herzberg II bands.

At first sight, recombination of $O(^3P)$ in reaction (7.23) is a plausible night-time source not only of $O_2(c^1\Sigma_u^-)$, but also of the $a^1\Delta_g$ and $b^1\Sigma_g^+$ states that give rise to the Infra-red Atmospheric and Atmospheric Bands of the nightglow. Channelling into the available molecular electronic states might be expected to be determined statistically, according to the electronic degeneracies. Much depends, however, on the energy barriers experienced by the atoms at large internuclear separations as they approach each other on the various potential curves. Nightglow concentrations of $O_2(^1\Delta_g)$, such as those derived for Fig. 7.2, would require 25 per cent of O atom recombination to follow the $O_2(^1\Delta_g)$ curve, which seems an intolerably large proportion. Theoretical considerations, in fact, favour $^5\Pi_g$ as the dominant initial state, with the lower electronic levels being filled by radiationless and collisional processes. Quite apart from the problems with absolute rates of $O_2(^1\Delta_g)$ and $O_2(^1\Sigma_g^+)$ production, the simple recombination mechanism also fails to predict correctly the altitudes and thicknesses of the emitting layers of the various bands, and to reconcile them with the altitude of [O] maximum.

An alternative mechanism for nightglow excitation calls again on the $c^1\Sigma_u^-$ state of O_2 as a precursor species. In this scheme, $O_2(c^1\Sigma_u^-)$ is formed vibrationally excited in the O-atom recombination step (perhaps with the $^5\Pi_g$ as an intermediate state). Quenching of $O_2(c^1\Sigma_u^-)$ by atomic oxygen is rapid, but the products depend on the degree of vibrational excitation. Higher levels ($v' > 2$ or 3) can excite $O(^1S)$ in reaction (7.24); lower levels still have enough energy to produce $O_2(a^1\Delta_g)$ as a quenching product. *Molecular* oxygen quenches $O_2(c^1\Sigma_u^-)$ to $O_2(b^1\Sigma_g^+)$ in this hypothesis. The set of quenching reactions is thus:

$$O_2(c^1\Sigma_u^-, v' > 2\text{-}3) + O \rightarrow O(^1S) + O_2(^3\Sigma_g^-), \tag{7.26}$$

$$O_2(c^1\Sigma_u^-, v' < 2\text{-}3) + O \rightarrow O_2(^1\Delta_g) + O, \tag{7.27}$$

$$O_2(c^1\Sigma_u^-, v') + O_2 \rightarrow O_2(^1\Sigma_g^+) + O_2. \tag{7.28}$$

The intensity of the upper emission layer[a] of the Infra-red Atmospheric Band (Fig. 7.2) is expected to roughly follow $[O(^3P)]$ but is possibly shifted to slightly lower altitudes than the peak at ~ 98 km of the green line due to $O(^1S)$, since the increased vibrational relaxation at higher densities favours reaction (7.27) over (7.26). Maximum production rates for $O_2(^1\Sigma_g^+)$ are determined by

---

[a] An entirely different mechanism, perhaps involving vibrationally excited OH (Section 7.4.3), must be responsible for the lower peak.

the product $[O_2][M]$ as well as $[O]^2$, so that the peak Atmospheric Band intensity should come from yet lower altitudes (~ 94 km), in agreement with the observations. Thus the mechanism, although hypothetical, does consistently explain the nightglow emissions of the three singlet states of O_2.

7.4.2 Atomic sodium

Airglow emissions from atomic sodium and other metals are of interest because the *allowed*, resonance, transitions fall within the visible region and are therefore observable from the Earth's surface. Metallic *ions* have been observed consistently at D- and E-region heights by rocket-borne mass spectrometers, and, as pointed out in Section 6.4.2, ablation of meteors seems to be a major source of the metals. The mesospheric layer of neutral sodium atoms lies roughly between 86 and 100 km, with peak concentrations being found typically at ~ 93 km. Normal peak concentrations are $\sim 3 \times 10^3$ atom cm^{-3}. Increases in concentration observed during meteor showers support the idea of a meteor ablation source for the atoms. Dayglow emission of the sodium yellow $Na(^2P_{1/2, 3/2} \rightarrow {}^2S)$ resonance doublet ($\lambda = 589.6, 589.0$ nm) is excited by direct absorption of solar radiation. Intensities are around 30 kR. As pointed out in Section 7.3, this intensity corresponds to an excited state concentration of only $\sim 5 \times 10^{-4}$ atom cm^{-3}, so that just over one atom in 10^7 is excited. Artificial excitation of the resonance emission provides a method for the measurement of sodium concentration–altitude profiles from the ground. A dye laser is tuned to one of the resonance lines, and a short pulse of light directed into the atmosphere. Sodium atoms present in the atmosphere can resonantly absorb and re-emit the radiation. Examination of the intensity of the scattered light, and the time delay between outgoing and returning pulses, allows the profile to be derived. By analogy with radio-frequency radar sounding, the laser technique is called *lidar*.

Nightglow intensities of the yellow sodium lines are much weaker (~ 100 R) than dayglow ones, but that they are present at all means that some reaction must be capable of exciting $Na(^2P)$. Interferometric measurements on the spectral line shapes show that the $Na(^2P)$ atoms possess a kinetic energy too small to arise from excitation in ionic processes such as dissociative recombination. Chemical reaction, however, implies consumption of atomic sodium, and the elemental sodium in the layer would soon be depleted unless a regenerative step also occurs. An oxidation-reduction cycle, involving the species NaO, seems the only plausible scheme, although the reactions have not yet been substantiated in laboratory experiments. The cycle starts with the oxidation of sodium by O_3,

$$Na + O_3 \rightarrow NaO + O_2, \qquad (7.29)$$

and this reaction is followed by reduction of NaO by atomic oxygen to yield either excited, ^2P, or ground state, ^2S, sodium

$$NaO + O \xrightarrow{\quad \alpha \quad} Na(^2P) + O_2, \tag{7.30}$$

$$NaO + O \xrightarrow{\quad (1-\alpha) \quad} Na(^2S) + O_2. \tag{7.31}$$

Involvement of ozone is inevitable in view of the emission intensity. Given that $[Na] \sim 3 \times 10^3$ atom cm^{-3} at its peak, and with 100 ^2P atoms excited per second (100 R in a 10 km layer), the product of rate coefficient and co-reactant must be $100/(3 \times 10^3) \sim 3 \times 10^{-2}$ s^{-1}. Since the rate coefficient cannot exceed the collision limit ($\sim 3 \times 10^{-10}$ cm^3 molecule^{-1} s^{-1}), the *minimum* concentration of co-reactant is $\sim 10^8$ molecule cm^{-3}. Ozone is almost certainly the only reactive species in sufficient abundance at 90 km that could give an energy-rich product (NaO). Regeneration of atomic sodium requires a reduction process that is at least as fast as the oxidation reaction in the mesosphere, and that is sufficiently exothermic to excite $Na(^2P)$. Even after all reactions other than (7.29) and (7.30) have been excluded, some constraints remain. First, the reactions must be sufficiently fast; and the fraction, α, of reactions of O with NaO that lead to $Na(^2P)$ must be at least as high as 0.3. Theory suggests that these conditions can be met if the reactions are of the electron-jump type.

Cyclic regeneration of free elemental sodium is, of course, demanded if sodium undergoes any reaction, regardless of the emission of light. Reaction (7.30), followed by radiation and by reaction (7.29), provides the necessary sequence so far as NaO is concerned. Other reactive channels include three-body formation of NaO_2:

$$Na + O_2 + M \rightarrow NaO_2 + M. \tag{7.32}$$

Atomic oxygen converts NaO_2 back to NaO, and so this pathway can lead to light emission *via* reaction (7.30). Figure 7.6 shows the major pathways of atmospheric sodium chemistry, and includes ionic species and a possible involvement of neutral NaOH.

An interesting phenomenon associated with meteor train luminosity may have one explanation in the cyclic chain processes (7.29) and (7.30). Most

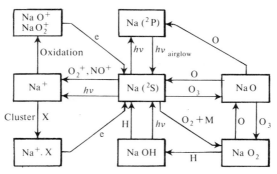

Fig. 7.6. Flow diagram of major processes in the mesospheric chemistry of neutral and ionized atomic sodium.

meteors that produce a visible train on entering the Earth's atmosphere show a very short-lived luminosity, following the impact of the body with the atmosphere. In about one in 125 000 visual meteor events, however, the train lasts for more than ten minutes; enduring luminosity for over an hour has been recorded. Emissions of sodium and magnesium (as well as, perhaps, of the O_2 Atmospheric Band) have been identified in these persistent trains. The afterglows are most strongly emitted from altitudes of ~ 90 km. Oxidation and reduction in a cycle, as proposed for sodium in the airglow, seems an attractive way of explaining long-lasting excitation of additional sodium ablated off a meteor.

7.4.3 Hydroxyl radicals

Vibrational emission phenomena are ordinarily thought of in connection with thermally equilibrated radiative transfer in the atmosphere (Sections 2.2.2 to 2.2.4). However, vibrationally *dis*-equilibrated OH radicals make a strong contribution to the longwave visible and near infra-red airglow, and offer a rare example of vibrational, rather than electronic, airglow emission. The OH airglow has additional interest because, as we shall see later, it provided the first hint of the catalytic chains that are now known to be so important in atmospheric ozone chemistry. Vibrational-rotational transitions among the nine lowest levels, v'', in the ground electronic state of OH give rise to the observed emission system, known as the *Meinel Bands* of OH. Transitions are observed corresponding not only to the allowed $\Delta v'' = 1$ fundamental, but also to the forbidden $\Delta v'' > 1$ overtones. Fundamentals lie in the range ~ 4500 nm ($v'' = 9 \to 8$) to ~ 2800 nm ($v'' = 1 \to 0$), and are intrinsically the strongest features, but are relatively difficult to detect because they lie well into the infra-red. Overtone transitions of the $\Delta v'' = 4, 5 \dots$ series lie at $\lambda < 1000$ nm, and, although much weaker because of their forbidden nature, they are more readily observed without interference from the thermal background of atmosphere or detectors than the fundamentals. Even $\Delta v'' = 9$, for $v'' = 9 \to 0$, is detectable just in the ultraviolet ($\lambda \sim 382$ nm), although the intensity of the band is nearly eight orders of magnitude weaker than the bands of the $\Delta v'' = 1$ series.

Evaluation of possible excitation mechanisms starts from the apparent sharp cut-off in excitation at $v'' = 9$, corresponding to ~ 312 kJ mol^{-1}, with no emission from $v'' = 10$ (~ 337 kJ mol^{-1}) being seen. A reaction is needed for excitation whose exothermicity lies between 312 and 337 kJ mol^{-1}. Atomic hydrogen reacts with ozone to form OH in exactly this way

$$H + O_3 \to OH + O_2 + 322 \text{ kJ mol}^{-1}. \tag{7.33}$$

Laboratory investigations show that this reaction is indeed strongly chemiluminescent, with OH ($v'' \leqslant 9$) being the emitter. The immediate question for atmospheric excitation is why reaction (7.33) does not consume all available H atoms if it is fast enough to give the observed intensities. An explanation came from laboratory studies of the rate of the reaction

$$O + OH \rightarrow O_2 + H, \tag{7.34}$$

which was shown to be fast enough to regenerate H atoms in the atmosphere. Cyclic regeneration of H also explains a further result of the laboratory studies that was initially disturbing. Addition of ozone to atomic oxygen led to weak emission of the OH Meinel bands *even though hydrogen was absent and the reactants supposedly dry*! Realization that *traces* of H or H_2O (source of OH) could participate in a *catalytic* cycle made the result less surprising, and emphasized the need for purity and dryness in reactants. The same realization afforded an explanation of why the apparent rate constant for the reaction

$$O + O_3 \rightarrow 2O_2, \tag{7.35}$$

was much larger when measured in experiments where [O] was high than it was when the atoms were very dilute. Because sufficient precautions had not been taken, traces of H or OH catalysed O_3 decomposition in the cycle (7.33) + (7.34) in the high [O] experiments. From this interpretation of the laboratory results, it is a short step to the postulation of HO_x catalytic cycles in the stratospheric ozone destruction by atomic oxygen (Section 4.4.1).

Further support for the $H–O_3$ mechanism in exciting the OH Meinel Bands of the atmosphere comes from the diurnal intensity variations. Dayglow intensities totalled over all bands are typically 4.5 MR. Reaction (7.33) can give the required excitation rate, so long as cascading between different v'' levels can give several photons for each excitation event. Nightglow intensities are not much less, but a curious phenomenon is observed in both morning dawn and evening twilight periods. At both periods the intensity decreases rapidly and then recovers nearly to the earlier value. Excited OH production is suppressed above ~ 85 km by day, when $[O_3]$ is depleted, but enhanced at lower altitudes, where the supply of O atoms is rate determining. During the twilight periods, concentrations of O, O_3, and H vary in a complicated manner, but models predict a dip in emission intensity consistent with that seen.

This concluding section of the chapter illustrates how airglow studies give an understanding of atmospheric processes not easily available from other types of investigation. Complemented by reliable laboratory experiments on excited species, airglow is thus a prime source of information about the atmosphere of Earth and the other planets.

Bibliography

General introductions to the study of airglow

The radiating atmosphere. McCormack, B. M. (ed.). (D. Reidel, Dordrecht, 1971.)
Spectroscopic emissions. Chapter 7 in *Physics and chemistry of the upper atmosphere.* Rees, M. H. (Cambridge University Press, Cambridge, 1989.)

A review of the photochemistry of selected nightglow emissions from the mesopause. Meriwether, J. W., Jr. *J. geophys. Res.* **94**, 14629 (1989).

The EUV dayglow at high spectral resolution. Morrison, M. D., Bowers, C. W., Feldman, P. D., and Meier, R. R. *J. geophys. Res.* **95**, 4113 (1990).

Upper atmosphere models and research. Ryecroft, M. J., Kasting, G. M., and Rees, D., eds. *Adv. Space Res.* **10**, No. 6 (1990).

Models for aurora and airglow emissions from other planetary atmospheres. Fox, J. L. *Canad. J. Phys.* **64**, 1631 (1986).

The visible airglow experiment—a review. Hays, P. B., Abreu, V. J., Solomon, S. C., and Yee, J. *Planet. Space Sci.* **36**, 21 (1988).

The airglow. McEwan, M. J. and Phillips, L. F., Chapter 5 of *Chemistry of the atmosphere.* (Edward Arnold, London, 1975.)

Physics of the aurora and airglow. Chamberlain, J. W. (Academic Press, New York, 1961.)

The light of the night sky, Roach, F. E. and Gordon, J. L. (D. Reidel, Dordrecht, 1973.)

This review provides a background to the chemistry of the regions of the atmosphere where many of the airglow emissions originate.

Neutral and ion chemistry of the upper atmosphere. Thomas, L., in Handbuch der Physik (ed. Rawer, K.), XLIX/6, 7 Part VI. (Springer-Verlag, Berlin, 1982.)

Sections 7.1 to 7.3

Determination of atmospheric composition and temperature from UV airglow. Meier, R. R. and Anderson, D. E., Jr. *Planet. Space Sci.* **31**, 967 (1983).

The role of metastable species in the thermosphere. Torr, M. R. and Torr, D. G. *Rev. Geophys. & Space Phys.* **20**, 91 (1982).

Low latitude airglow. Meier, R. R. *Rev. Geophys. & Space Phys.* **17**, 485 (1979).

Photographic observations of the Earth's airglow from space. Mende, S. R., Banks, P. M., Nobles, R., Garriot, O. K., and Hoffman, J. *Geophys. Res. Lett.* **10**, 1108 (1983).

Electronic structure of excited states of selected atmospheric systems. Michels, H. H., in *The excited state in chemical physics,* Vol. 2 (ed. McGowan, J. W.). (John Wiley, New York, 1979.)

First optical detection of atomic deuterium in the upper atmosphere from Spacelab 1. Bertaux, J. L., Goutail, F., Dimarellis, E., Kockarts, G., and Van Ransbeek, E. *Nature, Lond.* **309**, 771 (1984).

Section 7.4.1

Atomic oxygen contributions to the airglow

OI emissions. Noxon, J. F., in *Physics and chemistry of upper atmospheres,* pp. 213–18. (D. Reidel, Dordrecht, 1973.)

The OI (6300 Å) airglow. Hays, P. B., Rusch, D. W., Roble, R. G., and Walker, J. C. G. *Rev. Geophys. & Space Phys.* **16**, 225 (1978).

The 'auroral green line'

Rocket photometry and the lower-thermospheric oxygen nightglow. Greer, R. G. H., Murtagh, D. P., McDade, I. C., Llewellyn, E. J., and Witt, G. *Phil. Trans. R. Soc.* **A323**, 579 (1987).

Excitation of 557.7-nm OI line in nightglow. Bates, D. R. *Planet. Space Sci.* **36**, 883 (1988).

The excitation of O(^{1}S) and O$_2$ bands in the nightglow: a brief review and preview. McDade, I. C. and Llewellyn, E. J. *Canad. J. Phys.* **64**, 1626 (1986).

Excitation of O(^{1}S) and emission of 5577 Å radiation in aurora. Rees, M. H. *Planet. Space Sci.* **32**, 373 (1984).

The green light of the night sky. Bates, D. R. *Planet. Space Sci.* **29**, 1061 (1981).

On the mechanism and yields of O(^{1}S) and O(^{1}D) formation in the dissociative recombination of O$_2^+$.

Oxygen green and red line emission and O$_2^+$ dissociative recombination. Bates, D.R., *Planet. Space Sci.* **38**, 889 (1990).

O(^{1}S) from dissociative recombination of O$_2^+$: nonthermal line profile measurements from dynamics explorer. Killeen, T. L. and Hays, P. B. *J. geophys. Res.* **88**, 10163 (1983).

The dissociative recombination of O$_2^+$: the quantum yield of O(^{1}S) and O(^{1}D). Abreu, V. J., Solomon, S. C., Sharp, W. E., and Hays, P. B. *J. geophys Res.* **88**, 4140 (1983).

Molecular oxygen
The review listed first includes a section on airglow emission.

Reactions of singlet molecular oxygen in the gas phase. Wayne, R. P. in *Singlet oxygen*, (ed. Frimer, A. A.). (CRC Press, Florida, 1985).

Singlet oxygen airglow. Wayne, R. P. *J. Photochem.* **25**, 345 (1984).

Metastable oxygen emission bands. Slanger, T. G. *Science* **202**, 751 (1978).

The oxygen nightglow. Bates, D. R. in *Progress in atmospheric physics* Rodrigo, R., Lopez-Moreno, J. J., Lopez-Puertas, M., and Molina, A. (D. Reidel, Dordrecht, 1988.)

Transition probabilities of the bands of the oxygen systems of the nightglow. Bates, D. R. *Planet. Space Sci.* **36**, 869 (1988).

The Infra-red Atmospheric Band

ETON 6: A rocket measurement of the O$_2$ infrared atmospheric (0, 0) band in the nightglow. McDade, I. C., Llewellyn, E. J., Greer, R. G. H., and Murtagh, D. P. *Planet. Space Sci.* **35**, 1541 (1987).

A global study of $^1\Delta$ airglow: day and twilight. Noxon, J. F. *Planet. Space Sci* **30**, 545 (1982).

O$_2$($^1\Delta_g$) in the atmosphere. Llewellyn, E. J., Evans, W. F. J., and Wood, H. C., in *Physics and chemistry of upper atmospheres*, pp. 193–202. (D. Reidel, Dordrecht, 1973.)

Measurement of O$_2$(a$^1\Delta_g$) emission in a total solar eclipse. Bantle, M., Llewellyn, E. J., and Solheim, B. H. *J. atmos. terr. Phys.* **46**, 265 (1984).

Rocket-borne photometric measurements of O$_2$($^1\Delta_g$), green line and OH Meinel bands in the nightglow. Lopez-Moreno, J. J., Vidal, S., Rodrigo, R., and Llewellyn, E. J. *Ann. Geophysicae* **2**, 61 (1984).

A simultaneous measurement of the height profiles of the night airglow OI 5577 A, O$_2$ Herzberg and atmospheric bands. Ogawa, T., Iwagami, N., Nakamura, M., Takano, M., Tanabe, A., Takechi, A., Miyashita, A., and Suzuki, K. *J. Geomagn. Geoelectr.* **39**, 211 (1987).

SME observations of $O_2(^1\Delta_g)$ nightglow: an assessment of the chemical production mechanisms. Howell, C.D., Michelangeli, D.V., Allen, M., Yung, Y.L., and Thomas, R.J. *Planet. Space Sci.* **38**, 529 (1990).

The $c^1\Sigma_u^-$ state of O_2 and the mechanism of excitation of oxygen airglow. The first paper describes the identification of emission from O_2 $c^1\Sigma_u^-$ in the Earth's nightglow. Subsequent papers discuss the possible involvement of O-atom recombination in excitation of various oxygen airglow features, and consider whether an excited state of O_2—conceivably $O_2(c^1\Sigma_u^-)$—might be a precursor in a two step mechanism.

$O_2(c^1\Sigma_u^- \rightarrow X^3\Sigma_g^-)$ emission in the terrestrial nightglow. Slanger, T. G. and Huestis, D. L. *J. geophys. Res.* **86**, 3551 (1981).

Association of atomic oxygen and airglow excitation mechanisms. Wraight, P. C. *Planet. Space Sci.* **30**, 251 (1982).

Auroral population of the $O_2(b^1\Sigma_g^+, v')$ levels. Henriksen, K. and Sivjee, G.G. *Planet. Space Sci.* **38**, 835 (1990).

The excitation of $O_2(b^1\Sigma_g^+)$ in the nightglow. Greer, R. G. H., Llewellyn, E. J., Solheim, B. H., and Witt, G. *Planet. Space Sci.* **29**, 383 (1981).

An assessment of the proposed $O(^1S)$ and $O_2(b^1\Sigma_g^+)$ nightglow excitation parameters. Murtagh, D.P., Witt, G., Stegman, J., McDade, I.C., Llewellyn, E.J., Harris, F., and Greer, R.G.H. *Planet. Space Sci.* **38**, 43 (1990).

Excitation of oxygen emissions in the night airglow of the terrestrial planets. Krasnopolsky, V. A. *Planet. Space Sci.* **29**, 925 (1981).

On the excitation of oxygen emissions in the airglow of the terrestrial planets. Llewellyn, E. J., Solheim, B. H., Witt, G., Stegman, J., and Greer, R. G. H. *J. Photochem.* **12**, 179 (1980).

O_2-triplet emissions in the nightglow. Murtagh, D. P., Witt, G., and Stegman, J. *Canad. J. Phys.* **64**, 1587 (1986).

Excitation of Herzberg I and II bands in the atmospheres of Earth and Venus. Parisot, J.-P. *Ann. Geophys.* **86**, 481 (1986).

Section 7.4.2

Sodium airglow

Night-time Na D emission observed from a polar orbit DMSP satellite. Newman, A.L. *J. geophys. Res.* **93**, 4067 (1988).

Excitation of the Na D-doublet of the airglow. Bates, D. R. and Ojha, P. C. *Nature, Lond.* **286**, 790 (1980).

Laser sounding of atmospheric sodium: interpretation in terms of global atmospheric parameters. Megie, G. and Blamont, J. E. *Planet. Space Sci.* **25**, 1093 (1977).

A meteor ablation model of the sodium and potassium layers. Hunten, D. M. *Geophys. Res. Lett.* **8**, 369 (1981).

Sodium nightglow and gravity waves. Molina, A. *J. atmos. Sci.* **40**, 2444 (1983).

Chemistry of the sodium layer.
The first paper is concerned more particularly with neutral chemistry, and presents suggestions for kinetic parameters based on an 'electron jump' model. The second

paper develops a cluster ion scheme that relates Na^+ concentrations to those of Na as well as discussing the neutral processes and reviewing the atmospheric observations.

Gas phase chemical kinetics of sodium in the upper atmosphere. Kolb, C. W. and Elgin, J. B. *Nature, Lond.* **263**, 488 (1976).

A cluster ion chemistry for the mesospheric sodium layer. Richter, E. S. and Sechrist, C. F., Jr. *J. atmos. terr. Phys.* **41**, 579 (1979).

Meteor trails

The excitation of spectral lines in faint meteor trains. Poole, L. M. G. *J. atmos. terr. Phys.* **41**, 53 (1979).

The source of enduring meteor train luminosity. Baggaley, W. J. *Nature, Lond.* **289**, 530 (1981).

Section 7.4.3

Observations of OH Meinel (7,4) $P(N'' = 14)$ transitions in the night airglow. Pendleton, W., Jr., Espy, P., Baker, D., Steed, A., Fetrow, M., and Henriksen, K. *J. geophys. Res.* **94**, 405 (1989).

Rocket measurements of the altitude distributions of the hydroxyl airglow. Baker, D. J. and Stair, A. T., Jr. *Phys. Scri.* **37**, 611 (1988).

Mesospheric oxygen densities inferred from night-time OH Meinel band emissions. McDade, I. C. and Llewellyn, E. J. *Planet. Space Sci.* **36**, 897 (1988).

Seasonal variability of the OH Meinel bands. Le Texier, H., Solomon, S., and Garcia, R. R. *Planet. Space Sci.* **35**, 977 (1987).

Kinetic parameters related to sources and sinks of vibrationally excited OH in the nightglow. McDade, I. C. and Llewellyn, E. J. *J. geophys. Res.* **92**, 7643 (1987).

Observation and interpretation of hydroxyl airglow emissions. Gattinger, R. L. and Vallance-Jones, A., in *Physics and chemistry of upper atmospheres*, pp. 184–92. (D. Reidel, Dordrecht, 1973.)

Mid-winter hydroxyl night airglow emission intensities in the nothern polar region. Myrabø, H. K. and Deehr, C. S. *Planet. Space Sci.* **32**, 263 (1984).

An oxygen-hydrogen atmospheric model and its application to the OH emission problem. Moreels, G., Megie, G., Vallance-Jones, A., and Gattinger, R. L. *J. atmos. terr. Phys.* **39**, 551 (1977).

Waves in the OH emissive layer. Herse, M. *Science* **225**, 172 (1984).

8
Extraterrestrial atmospheres

8.1 Introduction

Our knowledge about the bodies in the solar system and their atmospheres has expanded phenomenally in the last one or two decades. Increasing sophistication of Earth-based investigations has allowed detailed examination of atmospheres from afar. But the coming of the 'space age' has provided a wealth of information that cannot be obtained at all from Earth's surface. Instruments can be borne aloft beyond our own atmosphere; 'fly-bys', orbiters, and landers have reached the atmospheres of distant planets to give the result of *in situ* measurements. Table 8.1 shows some recent space missions that have included atmospheric observations. This chapter presents some features of atmospheric chemistry as it now appears in the light of recent research. The solar system bodies under consideration include Saturn's satellite Titan, which has a massive atmosphere, as well as the planets. The atmospheres of these bodies seem to divide into two classes: those of the *Inner* and the *Outer* planets. Any gases left over when the inner planets (Venus, Earth, Mars) were formed were lost, and their present atmospheres have been acquired by outgassing from the solid planet followed by chemical modification and evolution (see Chapter 9). Carbon dioxide characterizes the atmospheres of these planets, although Earth is a special case because life has converted CO_2 to O_2 (see Chapters 1 and 9). Outer planets (Jupiter, Saturn, Uranus, Neptune) have retained an atmosphere similar to that with which they were formed because the large escape velocities (cf. Section 2.3.2 and Table 2.1) inhibit loss. Titan's peculiar atmosphere (Section 8.5.1) is probably a consequence of the low escape velocity (Table 2.1) allowing the atmosphere to have evolved more than that of the parent planet, Saturn.

8.2 Venus

8.2.1 Atmospheric composition

Venus is the nearest planet to the Sun to possess an atmosphere, which is about a hundred times as massive as that of the Earth (cf. Table 1.1). Carbon dioxide is the major constituent (96.5 per cent: Table 1.1) with ~ 3.5 per cent of N_2 as the next most abundant species. Incoming solar energies ($\sim 2600 \, \text{Wm}^{-2}$) are nearly twice as great as on Earth, and radiation trapping

Table 8.1 Some recent and proposed missions.

Date		Name	Objective
July–Sept	1976	Viking 1, 2	Mars entry science and lander
Dec	1978	Pioneer Venus	Orbiter and 'Bus' carrying entry probes to Venus
Dec	1978	Venera 11, 12	Soviet probes to Venus
March	1979	Voyager 1	Closest approach to Jupiter
July	1979	Voyager 2	Closest approach to Jupiter
Sept	1979	Pioneer 11	First ever fly-by of Saturn
Nov	1980	Voyager 1	Closest approach to Saturn
Aug	1981	Voyager 2	Closest approach to Saturn
March	1982	Venera 13, 14	Soviet sounders to Venus
July	1985	'VEGA'	Venus fly-by and on to
March	1986	(Venera 15, 16)	Comet Halley (USSR/France)
March	1986	Giotto	Comet Halley mission (ESA)
Jan	1986	Voyager 2	First ever fly-by of Uranus
Aug	1989	Voyager 2	First ever fly-by of Neptune and Triton
–	1991	Magellan	Venus mapping
–	1993	Mars observer	Mars orbiter
–	1995	Galileo	Jupiter, Io long-term study
–	2002	Cassini	Saturn orbiter, Titan probe
–	2002	CRAF	Cometary rendezvous and asteroid fly-by

by the CO_2 leads to high surface temperatures (~ 730 K: cf. Section 2.2). Optical observations of Venus from a distance are hindered by an unbroken layer of cloud, whose tops extend to a height of about 70 km. Our knowledge of the planet has been greatly extended by the Mariner (USA: 1962–74) and Venera (USSR: 1967–present) series of space probes. Our perception of Venus underwent a revolution in December 1978 when several spacecraft reached the planet. Five probes belonged to the Pioneer–Venus (USA) multiprobe mission. Four of the craft reached the surface, two on the night-side and two on the day-side, while the fifth craft was the orbiter, which continues to circle the planet. In the same month, two Soviet spacecraft (Venera 11, 12) reached Venus, and each landed a probe on the surface. The instrumentation included mass spectrometers (for ionic and for neutral species), gas chromatographs, particle size spectrometers, UV spectrometers, nephelometers (cloud density investigations), X-ray spectrometers, thunderstorm detectors, and so on. Seven gas analysers encountered the atmosphere and obtained data on the chemical composition from about 700 km altitude down to the surface. Studies of Venus were continued with the Soviet Venera 13 and 14 lander probes and the Venera 15 and 16 orbiters. Some problems of identification remain. Relatively low resolution mass spectral peaks could have more than one origin: for example, $m/e = 64$ could be due to SO_2 or S_2. Retention times on chromatographic columns are also subject to misinterpretation.

Contamination of mass spectral data, especially by H_2O, is always a difficulty, and one Pioneer–Venus analyser had the problem compounded because the inlet used for sampling became blocked by a cloud droplet between 50 and 28 km during the probe descent. Nevertheless, the results of the different experiments are reasonably consistent, and form the basis of the Venus entry in Table 1.1 (to which reference should be made in subsequent discussion of atmospheric concentrations). The existence of COS (carbonyl sulphide), originally reported to be a major sulphur compound in the lowest part of the atmosphere, is doubtful there, and its mixing ratio at higher altitudes is at most a few parts per million. A serious discrepancy in the data exists for the water vapour abundance, with mixing ratios from 2×10^{-5} (optical spectroscopy) to 5×10^{-3} (gas chromatography) being suggested by different experiments. One of the most significant findings from the Pioneer–Venus and Venera missions concerns the absolute and relative isotopic abundances of the inert gases; we shall discuss this topic in connection with the evolution of planetary atmospheres (Section 9.2.1).

Chlorine-containing compounds are of potential importance in Venusian photochemistry, as will appear later. Hydrogen chloride has been identified (as has HF) by Earth-based infra-red spectrometry. These studies only give concentrations above the Venusian cloud tops ($\gtrsim 70$ km), of course; for HCl, the mixing ratio is $4.2 \pm 0.7 \times 10^{-7}$. Claims have been made for the identification of molecular chlorine in the Venera 11, 12 optical spectroscopy results (mixing ratio $\sim 10^{-7}$). Pioneer–Venus data also support the presence of Cl_2. Ultraviolet absorption was generally ascribed to sulphur species prior to the Pioneer–Venus mission. Models of transparent H_2SO_4 clouds on top of the lower S_8 sulphur clouds appeared consistent with UV reflectance measurements. However, constraints placed on cloud models by the Pioneer–Venus observations seem to rule out sulphur as the UV absorber. Instead, SO_2 appears to be the major source of opacity at short wavelengths, but at $\lambda \gtrsim 320$ nm a second absorber is required. A tenable candidate is Cl_2 if it is present at concentrations of about 1 p.p.m.

8.2.2 Clouds

The clouds provide, from the chemist's point of view, a convenient division between 'lower' and 'upper' atmospheres of Venus. As we discussed in Sections 2.3.1 and 2.3.2, the thermal structure of the Venusian atmosphere allows formal identification of troposphere, stratosphere, mesosphere, and thermosphere. Very roughly, the region below the clouds corresponds to the troposphere, while the stratosphere extends from the cloud tops to ~ 110 km. The clouds also absorb or reflect almost all solar ultraviolet radiation. Photochemical change is thus confined to that part of the atmosphere above the clouds, while below the clouds virtually all chemical change is thermal in nature.

Cloud structure on Venus is quite complex, with three major layers, centred on 51, 54, and 62 km, being identified. In addition, there appear to be lower pre-cloud and haze regions. Three kinds of cloud particles are present. 'Mode 1' are small, ubiquitous aerosols, which extend below the main layers to within 31 km of the surface. 'Mode 2' particles are $\sim 2.0\ \mu$m in diameter, and are seen in all three cloud layers. The refractive index of these particles is 1.44, and they are almost certainly concentrated sulphuric acid (75 per cent H_2SO_4 + 25 per cent H_2O gives the required refractive index). 'Mode 3' particles are larger solid crystals of needle, plate, or dendritic shape, restricted to the middle and lower cloud layers. Sulphuric acid is thus not the only constituent of the Venusian clouds, and, in fact, solid mode 3 particles are about ten times more abundant (mass loading) than mode 2 in the lower two layers.

X-ray fluorescence experiments on Venera 15 and 16 ('Vega': see Section 8.7) suggest that metal chlorides are important components of the clouds, with Fe_2Cl_6 constituting one per cent of the column mass loading above 47 km, and Al_2Cl_6 being the dominant chlorine-bearing species in the middle cloud layer.

Phosphorus compounds also seem to be present in the cloud layers (Venera 15, 16), with P_4O_6 as the main phosphorus-bearing gas. Reaction of P_4O_6 with H_2SO_4 droplets converts the latter into phosphoric acid (H_3PO_4) droplets with release of sulphur dioxide. The H_3PO_4 droplets lose water when they descend to about 25 km, and evaporation of P_4O_{10} should occur. One interpretation of the Venera data thus now suggests a cloud composition of H_2SO_4 (altitude 52–62 km, mass loading 5 mg m^{-3}), Fe_2Cl_6 (47–52 km, 0.2 mg m^{-3}), H_3PO_4 ($\leqslant 52$ km, 5 mg m^{-3}), Al_2Cl_6 (53–58 km, 3 mg m^{-3}), and S_8 (52–62 km, total sulphur polymers ca. 0.1 of H_2SO_4 by mass). Further analysis of the data is needed to see how far the X-ray fluorescence results are supported by the mass spectrometric and gas chromatographic data obtained in the same mission.

8.2.3 Lightning

One of the more unusual experiments on the Venera 11, 12 probes was the 'Groza' ('Thunderstorm') search for low-frequency electromagnetic radiation pulses (*sferics*) which must be emitted by lightning discharges if they are present in the atmosphere of Venus. On both probes, groups of pulses were discovered whose character was like terrestrial thunderstorm radiation. Additional support for Venusian thunderstorms comes from a plasma wave detector on the Pioneer–Venus orbiter. Impulsive signals, of a kind not normally encountered, are occasionally seen when the orbiter penetrates the night-time ionosphere, and these impulses are consistent with an origin from lightning. Re-examination of optical spectrometer data from the earlier Venera 9, 10 missions showed that lightning had been seen, and that an extensive thunderstorm had been encountered with a flashing frequency of

100 flashes per second. Ultraviolet and blue emissions were observed, pointing to the clouds rather than the surface as the location of the lightning, because the observed radiation could not have penetrated the lower atmosphere. Estimates for the mean rate of energy dissipation by lightning in the Venusian atmosphere range from 0.4×10^{-7} J cm^{-2}s^{-1} to as much as fifty times this value.

Lightning discharges may affect cloud chemistry by providing a local source of CO by CO_2 pyrolysis, far removed from the higher altitude CO_2 photolysis region (Section 8.2.5). Possibly more important, NO may be produced. Laboratory investigations of simulated lightning discharges in an oxygen-deficient mixture of CO_2 (95 per cent) and N_2 (5 per cent), representing the composition of the Venusian atmosphere, show that around 4×10^{15} molecule J^{-1} of NO are formed. Nearly 10^{12} molecule cm^{-2}s^{-1} of NO could thus be generated in the Venusian clouds.

Oxides of nitrogen may play an interesting role in the oxidation of SO_2 to SO_3 or H_2SO_4 present in the clouds, and may even be incorporated into the 'mode 3' solid aerosol. Commercial production of sulphuric acid was, until the early years of this century, carried out by the 'Lead Chamber' process in which nitrogen oxides were carrier catalysts for the oxidation of SO_2. The process may be summarized by the sequence

$$NO + NO_2 + 2H_2SO_4 \rightarrow 2NOHSO_4 + H_2O \qquad (8.1)$$

$$2NOHSO_4 + SO_2 + 2H_2O \rightarrow 3H_2SO_4 + 2NO \qquad (8.2)$$

Net $\qquad NO_2 + SO_2 + H_2O \rightarrow H_2SO_4 + NO.$

Nitrosylsulphuric acid (NOHSO$_4$) is an essential part of the scheme (although the detailed mechanism may involve formation of HNO_2 and oxidation of H_2SO_3 in the aqueous phase). Venusian oxidation of SO_2 could follow the same route if sufficient NO$_x$ is present. A further observation from Lead Chamber operation is especially interesting in this context. If not enough H_2O is used, the process can be interrupted before reaction (8.2) occurs, and crystals of NOHSO$_4$ occasionally form midair in the shape of feathery crystals. As a conjecture, then, the mode 3 cloud particles may be NOHSO$_4$. Solid NOHSO$_4$ melts at 347 K, the temperature near the border of the middle and lower cloud levels, where the number of cloud particles rapidly decreases. A dilute solution of NOHSO$_4$ (1.4 per cent) in H_2SO_4 (85 per cent) shows visible and UV absorption compatible with the Venusian spectrophotometric observations (cf. Section 8.2.1). Whether or not NOHSO$_4$ is present on Venus, or plays a part in SO_2 oxidation, will depend on NO$_x$ concentrations. Mixing ratios in the parts per million range seem needed, which demand mean NO$_x$ lifetimes of at least thousands of years if lightning is the source of NO$_x$. Unfortunately, neither gas chromatographic nor mass spectrometric data from the recent *in situ* measurements provide evidence to resolve the question.

8.2.4 Sub-cloud chemistry

Chemistry in the region below the clouds seems to be dominated by the thermal reactions of sulphur- and carbon-containing species. In view of the high temperature and pressure, a starting point in geochemical concepts of the troposphere has been the assumption of complete chemical equilibrium. The surface is often considered to promote rapid equilibration by catalytic acceleration of chemical processes. To a large extent, the equilibrium treatment has to be adopted as a first approximation, because kinetic data are not available for the individual elementary reactions. Rates may, however, be slow enough that vertical transport and mixing dominate chemical reactions in the whole lower atmosphere. Atmospheric composition should then be approximately constant throughout the sub-cloud region, and can be determined by the application of thermochemical equilibrium conditions near the surface.

The *in situ* measurements cast some doubt on both the constancy of composition and the equilibrium assumption itself. Mixing ratios of COS calculated from the equilibrium constant for the reaction

$$3CO + SO_2 \rightleftharpoons 2CO_2 + COS, \tag{8.3}$$

are $\sim 10^{-4}$ near the planet's surface, and reach 10^{-2} by an altitude of 10 km, taking the values of SO_2 and CO concentration given in Table 1.1. Pioneer–Venus measurements set the limit of detectability at a mixing ratio of 10^{-5}, and even Venera 13, 14 data do not exceed 4×10^{-5}. Thus chemical equilibrium does *not* seem to be attained for COS. For S_2 formation, the overall equilibrium can be written

$$4CO + 2SO_2 \rightleftharpoons 4CO_2 + S_2. \tag{8.4}$$

Thermochemical data suggest an S_2 mixing ratio of 1.3×10^{-6} using the concentrations of Table 1.1, but, because of the $[CO]^4$-dependence, a value of 1.8×10^{-7} is obtained for a slightly smaller CO mixing ratio of 1.7×10^{-5} (measured by Venera 12 at 12 km). Venera orbiter spectrometer results suggest a surface mixing ratio for S_2 of 1.6×10^{-7}, which would be consistent with the lower [CO] value. Thus S_2 concentrations *may* support the concept of tropospheric chemical equilibrium *at the surface*. Above the surface, it seems that kinetic constraints must exist. Ascending and descending gas fluxes in the vertical atmospheric motion may possibly have differing compositions, with the ascending gases containing components of crustal origin (e.g. COS, S_2, H_2S, HCl, etc.) and the descending gases containing products of photolysis at the cloud tops (SO_3, Cl_2, O_2, etc.). In this case, the atmospheric composition should be regarded as reflecting a *steady state* condition rather than a thermodynamic equilibrium one. At the surface, chemical equilibrium may be achieved among all gaseous species *except* CO (for which the measured mixing ratio is nearly constant with altitude). This view then allows CO and O_2 to be out of equilibrium with each other and with the COS/SO_2 system in the free troposphere. Surface equilibration requires a substantial amount of sulphur trioxide (SO_3) to be present. The data of Table 1.1 require there to be almost ten times as much SO_3 as SO_2, but none of the *in situ* experiments were sensitive to SO_3, so its presence remains hypothetical.

Elementary reaction steps can be written that explain the interconversions between the different species. Most schemes start by production of SO and thence atomic sulphur:

$$SO_2 + CO \rightarrow SO + CO_2 \tag{8.5}$$

$$SO + CO \rightarrow S + CO_2 \tag{8.6}$$

$$SO + SO_2 \rightarrow S + SO_3. \tag{8.7}$$

Gas-phase S_2 may be formed *via* COS as an intermediate:

$$CO + S + M \rightarrow COS + M \tag{8.8}$$

$$S + COS \rightarrow S_2 + CO \tag{8.9}$$

$$SO + COS \rightarrow S_2 + CO_2. \tag{8.10}$$

Concentrations of species such as COS and CO are determined by the competition between production and loss processes. Observational data for COS from the Pioneer–Venus mission were confined to an upper limit of a few p.p.m. Preliminary analysis of Venera 13, 14 (1982) results has suggested slightly higher concentrations (40 ± 20 p.p.m.) together with some H_2S (80 ± 40 p.p.m.) in the altitude range 27 to 37 km, but these data are still subject to considerable refinement. It is, however, clear that the reduced gases COS and H_2S are about an order of magnitude less abundant than SO_2. Low concentrations of CO also suggest that reactions consuming the gas are efficient relative to those forming it below the clouds. Spectroscopic data from the Venera probes give strong circumstantial evidence for the presence of S_2, S_3, and higher sulphur oligomers, S_x. Reactions such as

$$S_2 + COS \rightarrow CO + S_3, \tag{8.11}$$

and similar processes involving $S_3 \ldots S_x$ could conceivably have quite large rate coefficients, but they are unlikely to be important in the Venusian troposphere because of the small concentrations of COS. Sequential addition of the type

$$S_n + S_m(+M) \rightarrow S_{n+m}(+M) \tag{8.12}$$

is an alternative source of S_x ($x = n + m$), with the stable S_8 probably being the terminal species. Photodissociation of polysulphur may be the process that drives Venusian sulphur chemistry out of thermal equilibrium, and that initiates the reversal of the reducing steps (8.5) to (8.10). Because the $S_{(x-1)}$–S bond strength is relatively weak for $x > 3$, photodissociation thresholds for the steps

$$S_x + h\nu \rightarrow S^{(*)} + S_{x-1} \tag{8.13}$$

can be displaced to longer wavelengths compared to the S_2 limit (285 nm) and into the spectral region where some radiation penetrates the clouds. Excited S atoms may be a product of the dissociation, so that reactions such as

$$S^* + H_2O \rightarrow SH + OH, \tag{8.14}$$

$$S^* + HCl \rightarrow SH + Cl \tag{8.15}$$

could become thermochemically possible. Radicals like SH and OH initiate further chemistry, with SH playing a part in H_2S and sulphane (H_yS_z) formation. Several cyclic processes can be written that utilize the photolysis of S_3 below the cloud tops. A typical sequence is

Cycle 1:

$$S_3 + h\nu \rightarrow S_2 + S^* \tag{8.16}$$

$$S^* + H_2O \rightarrow SH + OH \tag{8.14}$$

$$OH + CO \rightarrow CO_2 + H \tag{8.17}$$

$$H + SH \rightarrow H_2 + S \tag{8.18}$$

$$S + S_2 + M \rightarrow S_3 + M \tag{8.19}$$

$$\text{Net} \qquad CO + H_2O + h\nu \rightarrow CO_2 + H_2.$$

Such a reaction sequence may be a source of H_2 and a sink of H_2O in the lower atmosphere that could balance the reverse net reaction occurring above the clouds. Since the concentrations of the S_x precursors of S^* are temperature dependent, the role of these molecules, especially S_3 ('thiozone'), may bear a remarkable analogy to the role of ozone in the Earth's upper atmosphere.

8.2.5 Stratospheric chemistry

Three inter-related questions about the chemistry of the Venusian stratosphere deserve particular attention. First, what controls the extent of CO_2 photolysis, and the abundances of CO and O_2? Secondly, what part does SO_2 present above the cloud tops play in stratospheric chemistry? And, thirdly, what is the abundance of H_2 in the bulk atmosphere?

Photolysis of CO_2 occurs readily above the cloud tops,

$$CO_2 + h\nu(\lambda \lesssim 204 \text{ nm}) \rightarrow CO + O, \tag{8.20}$$

but the recombination reaction,

$$O + CO + M \rightarrow CO_2 + M, \tag{8.21}$$

is spin forbidden and slow. Atomic oxygen is lost mainly in the direct and indirect (see later) formation of O_2, which we represent by the equation

$$O + O + M \rightarrow O_2 + M. \tag{8.22}$$

Measurements of the Infra-red Atmospheric Band airglow indicate that a substantial fraction of CO_2 photolysis leads to O_2 production, as discussed in Section 7.4.1. Without some oxidation step for CO, the present-day abundance (~ 45 p.p.m. at the cloud tops) could be produced photochemically in about 200 years, and the stratospheric O_2 concentration (~ 1 p.p.m.) would be formed in a few years. The stoicheiometry implied by reactions (8.16) and (8.18) would predict $[CO]/[O_2] = 2$, whereas the measured ratio is at least 45. Some additional reactions are therefore demanded that will oxidize CO back to CO_2 and that will break O–O bonds. Catalytic cycles, such as those important in terrestrial chemistry (Sections 4.4 and 5.3) are likely to be involved.

The simplest possible oxidation of CO starts with the reaction (Section 5.3.5) with OH

$$CO + OH \rightarrow CO_2 + H. \tag{8.17}$$

We shall see (Section 8.3.2) that a cycle based on this reaction is very important on Mars. However, the cycle cannot account for the efficient removal of O_2 on Venus, because there is insufficient H_2O to yield the necessary concentrations of HO_x radicals.

Considerable interest has been focused on the possible roles of ClO_x and SO_x species in catalytic cycles since the discovery of HCl and SO_2 in the upper atmosphere, and chlorine-containing compounds in the lower atmosphere (p. 333). In the Earth's atmosphere, ClO_x only catalyses the conversion of odd oxygen to O_2 (Section 4.4). On Venus, however, a dual role is possible, with two sequences leading to recombination of O with CO. Three-body formation of chloroformyl radicals (ClCO) and then of peroxychloroformyl radicals ($ClCO_3$) is involved in both cycles

Cycle 2:

$$Cl + CO + M \rightarrow ClCO + M \tag{8.23}$$

$$ClCO + O_2 + M \rightarrow ClCO_3 + M \tag{8.24}$$

$$ClCO_3 + O \rightarrow Cl + CO_2 + O_2 \tag{8.25}$$

Net $\qquad CO + O \rightarrow CO_2$

Cycle 3:

$$Cl + CO + M \rightarrow ClCO + M \tag{8.23}$$

$$ClCO + O_2 + M \rightarrow ClCO_3 + M \tag{8.24}$$

$$ClCO_3 + Cl \rightarrow Cl + CO_2 + ClO \tag{8.26}$$

$$O + ClO \rightarrow O_2 + Cl \tag{8.27}$$

Net $\qquad CO + O \rightarrow CO_2$.

We note that these cycles convert CO back to CO_2, and consume atomic oxygen, but they do not break the O–O bond and do not, therefore, lead to the destruction of O_2. Sulphur-containing species seem capable of catalysing the (photo)dissociation of O_2. Here, the initial step is photolysis of SO_2:

$$SO_2 + hv(\lambda \lesssim 219 \text{ nm}) \rightarrow SO + O, \tag{8.28}$$

followed by atomic sulphur formation from SO by a photochemical and a thermal step

$$SO + hv(\lambda \lesssim 235 \text{ nm}) \rightarrow S + O \tag{8.29}$$

$$SO + SO \rightarrow S + SO_2. \tag{8.30}$$

Atomic sulphur is then responsible for the fission of the O–O bond

$$S + O_2 \rightarrow SO + O. \tag{8.31}$$

An important cycle has recently been identified following the discovery of a reaction between ClO and SO. The cycle thus links ClO_x and SO_x species

Cycle 4:

$$SO_2 + hv(\lambda \lesssim 219 \text{ nm}) \rightarrow SO + O \tag{8.28}$$

$$Cl + CO + M \rightarrow ClCO + M \tag{8.23}$$

$$ClCO + O_2 + M \rightarrow ClCO_3 + M \tag{8.24}$$

$$ClCO_3 + Cl \rightarrow Cl + CO_2 + ClO \tag{8.25}$$

$$ClO + SO \rightarrow Cl + SO_2 \tag{8.32}$$

$$\overline{}$$

Net $\qquad\qquad CO + O_2 + hv \rightarrow CO_2 + O.$

A key feature of the scheme is the coupling of reactions (8.28) and (8.32), which has the effect of using SO_2 photosensitization to break a Cl–O bond. Photolysis of SO_2 thus synergizes the ClO_x catalysis. The mechanism bears an interesting analogy with the tropospheric oxidation cycle on Earth (Section 5.3):

Cycle 5:

$$CO + OH \rightarrow CO_2 + H \tag{8.17}$$

$$H + O_2 + M \rightarrow HO_2 + M \tag{8.33}$$

$$HO_2 + NO \rightarrow NO_2 + OH \tag{8.34}$$

$$NO_2 + hv \rightarrow NO + O \tag{8.35}$$

$$\overline{}$$

Net $\qquad\qquad CO + O_2 + hv \rightarrow CO_2 + O$

with SO_2 playing the part of NO_2, and ClO_x playing the part of HO_x.

Cycle 4 may, indeed, be supplemented by cycles such as 5 on Venus if sufficient NO_x is present in the atmosphere. Although some NO is undoubtedly present above the clouds, because NO emission bands are seen in the ultraviolet airglow, it is not clear whether the lightning source (Section 8.2.3) in reality supplies enough NO to make cycle 5 important. Similar arguments apply to HO_x catalytic schemes that demand high concentrations of H_2. Mass spectrometric measurements of H_2 mixing ratios are the subject of some controversy, since the $m/e = 2$ ion may be either H_2^+ or D^+ (and the $m/e = 19$ ion can likewise be either H_3O^+ or HDO^+). Indeed, low H_2 favours the role of ClO_x chemistry on Venus. The terrestrial sink reaction (Sections 4.4.2 and 4.4.3),

$$Cl + CH_4 \rightarrow HCl + CH_3, \tag{8.36}$$

is replaced on Venus by the analogous process

$$Cl + H_2 \rightarrow HCl + H, \tag{8.37}$$

and excess H_2 will suppress catalytic cycles involving ClO_x. Important processes reverse the formation of HCl, of which the dominant one is photochemical

$$HCl + hv \rightarrow H + Cl. \tag{8.38}$$

In addition, thermal processes destroy HCl:

$$H + HCl \rightarrow H_2 + Cl, \tag{8.39}$$

$$O + HCl \rightarrow OH + Cl, \tag{8.40}$$

$$OH + HCl \rightarrow H_2O + Cl, \tag{8.41}$$

with the consequent formation of H_2 and H_2O. Molecular and atomic hydrogen are transferred to the Venusian exosphere, and so the rate of escape from the atmosphere is determined by the rate of HCl photolysis. Water formation in reaction (8.41) is a permanent sink in the Venusian stratosphere, since H_2O photolysis is prevented by the absorption of short-wave radiation, and H_2O is scavenged by SO_3 to form H_2SO_4 (see below). The net chemical change in the stratosphere is thus photochemical conversion of HCl and CO_2 to form H_2O, CO, and Cl_2, and the build-up and downward flux of free chlorine is the only way to conserve the atmospheric oxidation state. Reversal of photochemical change by thermochemical equilibrium chemistry must complete the cycle in the lower atmosphere. Imperfect balance in the cycle leads to a permanent and irreversible leakage of hydrogen by escape (cf. Section 9.3).

Figures 8.1 and 8.2 summarize the results of a one-dimensional model calculation based on the scheme of cycle 4, with no source of H_2 from below the clouds. Note that the mixing ratio for H_2 reaches 10^{-7} by 100 km, even though it is 10^{-13} at the lower boundary. As can be seen from Fig. 8.1, the model matches the measured concentration and apparent scale height for SO_2 up to ~ 70 km, and the CO measurements up to ~ 100 km. The limit for O_2

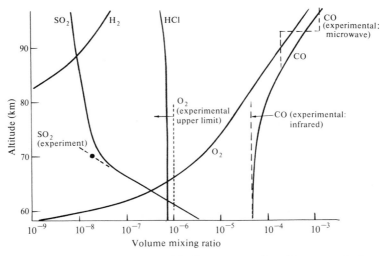

Fig. 8.1. Venusian abundances of bulk gases predicted by a model based on extended chlorine chemistry. Some experimental data are provided for comparison: the upper limit for O_2 is derived from measurements from Earth. [Data of Yung, Y. L. and DeMore, W. B. *Icarus* **51**, 199 (1982).]

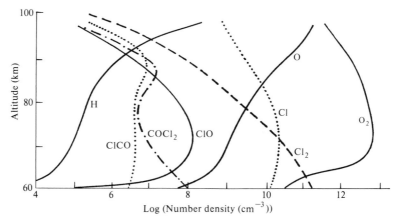

Fig. 8.2. Concentrations of minor and intermediate species derived by the model calculations of the Venusian atmosphere. (Data from same source as for Fig. 8.1.)

concentration (derived from Earth-based optical measurements) is also consistent with the model for heights less than ~ 70 km. Figure 8.2 shows that by far the most abundant intermediate species in the lower stratosphere is Cl, with O dominating only above about 90 km. One particularly significant conclusion of the model is that rates of production and loss of CO are about equal at ~ 80 km (i.e., there is a photochemical stationary state there). Without such a steady state, it is difficult to explain observations that mesospheric CO concentrations drop at night. In other models, that adopt 'high' H_2 and NO concentrations, the bulk of CO is destroyed around 70 km, and there is no significant chemical sink in the mesosphere between 80 and 90 km, and there should then be no diurnal variation in CO.

One further feature of the SO_2 and SO photolysis reactions, (8.28) and (8.29), is that SO_2 effectively sensitizes its own conversion to H_2SO_4. Atomic oxygen reacts with SO_2 to form SO_3 in the three-body process

$$O + SO_2 + M \rightarrow SO_3 + M. \tag{8.42}$$

The oxidation of SO_2 may also be catalysed by Cl atoms in the sequence

$$Cl + SO_2 + M \rightarrow ClSO_2 + M \tag{8.43}$$

$$ClSO_2 + O_2 + M \rightarrow ClSO_4 + M \tag{8.44}$$

$$ClSO_4 + Cl \rightarrow ClO + ClSO_3{}^* \tag{8.45}$$

$$ClSO_3{}^* \rightarrow Cl + SO_3. \tag{8.46}$$

In both direct and catalysed oxidations, the conversion of SO_2 to SO_3 provides an effective sink for O_2 on Venus and H_2SO_4 is produced in the reaction of SO_3 with water

$$SO_3 + H_2O + M \rightarrow H_2SO_4 + M. \tag{8.47}$$

The photochemical aerosols produced by reactions (8.42) to (8.47) constitute the cloud layer that separates troposphere from stratosphere in the Venusian atmosphere. The consequences of cloud formation are profound, because scattering increases planetary albedo and absorption prevents most ultraviolet radiation from reaching the lower atmosphere. At the same time, the chemical reactions consume both O_2 and H_2O, and the H_2SO_4 aerosol further desiccates the upper atmosphere.

8.3 Mars

8.3.1 Atmospheric structure and composition

Spacecraft exploration is in great part responsible for the dramatic improvement brought in the last fifteen years to our knowledge of the atmosphere of Mars. Starting with Mariners 4 to 9 (1965–72), and Mars 2 to 6 (1971–4) the programme culminated with the spectacularly successful Viking landers and orbiters, which reached Mars in the summer of 1976. Detailed information on the composition and meteorology of the Martian atmosphere came from *in situ* and remote-sensing measurements taken over more than a Martian year.

In contrast to the atmosphere of Venus, the Martian atmosphere is characterized by the variability of its physical parameters with the season, time of day, latitude and longitude, the time in the solar cycle, and with meteorological events such as dust storms and clouds, and other dynamical phenomena.

Mars has a rather tenuous atmosphere composed mainly of CO_2. Exerting a surface pressure of only 6–8 mbar (4.5–6.0 Torr), the atmosphere's total mass is less than one per cent of that of the Earth. Surface temperatures calculated from radiative transfer models (Section 2.2, Table 2.1) are ~ 217 K, in reasonable agreement with direct measurement. During the winter months, a thin sheet of cloud (the *polar hood*, probably composed of ice crystals) gradually spreads from the polar regions to middle latitudes. During the rest of the year, the atmosphere is generally cloud-free, but the surface is sometimes obscured for several weeks by dust raised by winds from the surface. Layered deposits cover the poles, and the area of the polar caps expands greatly during the local winter. Surface temperatures fall as low as 125 K at these periods: condensation from a CO_2 vapour pressure of 6.8 mbar becomes possible at 148 K, and the winter caps are probably a mixture of water ice and solid CO_2. Residual summer polar caps must consist largely of water-ice, since the cap temperatures are then 200–205 K; the thickness is estimated as between 1 and 1000 m. A permafrost layer covers most of the planet in winter, but, by recession towards the poles in spring and summer, this reservoir releases substantial quantities of water vapour to the atmosphere. Some further aspects of the effects of CO_2 condensation and the small heat capacity of the atmosphere on pressure variability, temperature structure, and circulation are discussed at appropriate points in Chapter 2.

Carbon dioxide was first identified in 1947 as a constituent of the Martian atmosphere by optical spectroscopy from Earth, and H_2O was discovered 16 years later. Subsequent experiments identified CO and O_2, and the UV spectrometers on Mariners 6 and 7 showed the presence of H, O, and O_3. Finally, analysis of Viking orbiter and lander mass spectrometric data revealed that N_2, Ar, Kr, Xe, and Ne were bulk atmospheric components; O_2, O, NO, and CO were confirmed in the upper atmosphere. Ion composition, density, and temperatures were measured during the entry phase. Isotopic ratios were determined for a variety of elements, and will be discussed further in Chapter 9 since they bear largely on models for the evolution of the planet and its atmosphere.

Bulk atmospheric composition data, as determined by the Viking missions, are summarized in the entry for Mars in Table 1.1. Water-detector instruments have revealed considerable global and seasonal variability, so that the mixing ratio for H_2O of 3×10^{-4} is only a 'typical' value. Ozone mixing ratios vary from 0.04 to 0.2×10^{-6}. In the upper atmosphere (i.e. above ~ 100 km), [NO] $\sim 10^{-4}$ [CO_2].

Mariner airglow data illustrate one of the most interesting facets of Martian atmospheric chemistry. Part of the ultraviolet spectrum obtained from Mariner 9 is reproduced in Fig. 8.3. Amongst the transitions seen are the

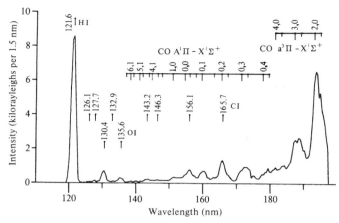

Fig. 8.3. Mars airglow spectrum, 110–190 nm. To obtain this spectrum, 120 individual limb observations, at 1.5 nm resolution, were averaged. [Mariner 9 data, from Barth, C. A., Stewart, A. J., Hord, C. W., and Lane, A. L. *Icarus* **17**, 457 (1972).]

resonance lines of H ($\lambda = 121.6$ nm), O ($\lambda \sim 130.4$ nm), and molecular band systems of CO, while, at longer wavelengths, systems due to CO^+ and CO_2^+ also appear. Apparent emission rates may be plotted as a function of altitude (Fig. 8.4), and the results show large differences in behaviour for the several airglow features. With the exception of H and O lines (and a portion of the $CO_2^+ A^2\Pi \rightarrow X^2\Pi$ transition, not included in our figure), the scale heights are small (~ 18 km) and identical. With the same exceptions, which are excited by resonance scattering, the emissions are probably excited by the action of solar photons and photoelectrons on CO_2, and thus have the same scale height as CO_2. These two conclusions mean that the upper atmosphere is essentially undissociated and cold. Since atomic oxygen emissions are observed above 250 km altitude, *some* dissociation of the CO_2 has occurred. Because part of the CO_2^+ (A $\rightarrow$ X) emission is attributed to fluorescent scattering, CO_2^+ is a constituent (but not necessarily a major one) of the Martian ionosphere.

The large scale heights, in the upper atmosphere, of oxygen (at mixing ratios of 5 to 10×10^{-3}) and hydrogen (at an almost *constant* concentration of 3×10^4 atom cm^{-3}) indicate clearly that these atomic species are escaping the gravitational field of the planet (Section 2.3.2). The base of the exosphere (*exobase*) lies at ~ 230 km on Mars. Although temperatures are relatively low (~ 320 K), thermal escape of hydrogen is possible because of the small value of g, and hence escape velocity, on Mars (~ 5 km s^{-1}: Table 2.1). Escape fluxes of about 1.2×10^8 atom $cm^{-2}s^{-1}$ of hydrogen are predicted for the measured concentration of 3×10^4 atom cm^{-3}. That any hydrogen remains in the exosphere therefore means that there is an equivalent source, presumably

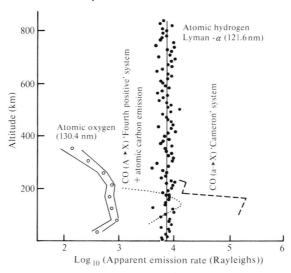

Fig. 8.4. Altitude profiles for dayglow emission features determined from the Mariner 9 data. The line through the HI (Lyman-α) data is a least-squares fit, and the envelope on the OI points indicates error limits. (Source as for Fig. 8.3, and McEwan, M. J. and Phillips, L. F. *Chemistry of the atmosphere*, Edward Arnold, London, 1975).

dissociation of a hydrogen-bearing molecule such as H_2O, which is currently operating in the Martian atmosphere. We shall see in Section 8.3.3 that the H atoms are liberated through the intermediacy of ionic processes. Furthermore, O-atom escape also involves ions, and the rates of oxygen and hydrogen escape processes are self-regulating to be equivalent to loss of H_2O. Before we turn to a consideration of the Martian ionosphere, however, we examine those aspects of the 'neutral' photochemistry that govern the concentrations of minor constituents such as CO and H- and O-bearing molecules.

8.3.2 Carbon dioxide photochemistry

Mars, like Venus, has an atmosphere whose bulk chemistry is dominated by carbon dioxide photolysis

$$CO_2 + h\nu(\lambda \lesssim 204 \text{ nm}) \rightarrow CO + O. \qquad (8.20)$$

As on Venus, it is again necessary to find a catalytic route for the recombination of O and CO, since the direct process (8.21) cannot possibly maintain CO and O_2 at low enough concentrations (Section 8.2.5). Present-day CO concentrations could be produced in two to three years, and the entire atmosphere be modified in about 2000 years. Water vapour on Mars is at least ten times more abundant than on Venus, and there is evidence for condensed-phase water on the surface. HO_x chemistry can thus provide a recombination mechanism that is fast enough to compete with CO_2 photolysis. Cycles that can be readily identified include

Cycle 6:

$$H + O_2 + M \rightarrow HO_2 + M \qquad (8.33)$$

$$O + HO_2 \rightarrow O_2 + OH \qquad (8.48)$$

$$CO + OH \rightarrow CO_2 + H \qquad (8.17)$$

Net $\qquad CO + O \rightarrow CO_2$

and

Cycle 7:

$$O + O_2 + M \rightarrow O_3 + M \qquad (8.49)$$

$$H + O_3 \rightarrow O_2 + OH \qquad (8.50)$$

$$CO + OH \rightarrow CO_2 + H \qquad (8.17)$$

Net $\qquad CO + O \rightarrow CO_2.$

Odd-hydrogen compounds are supplied by photochemical decomposition of H_2O either directly in photolysis,

$$H_2O + h\nu \rightarrow OH + H, \qquad (8.51)$$

or by reaction with $O(^1D)$ derived from CO_2, O_2, and O_3 photolysis (cf. Sections 3.2, 4.4.3, and 5.3.1):

$$O(^1D) + H_2O \rightarrow OH + OH. \qquad (8.52)$$

Water may be re-formed by the reaction

$$OH + HO_2 \rightarrow H_2O + O_2, \qquad (8.53)$$

but there is a small net sink for H_2O at low altitudes associated with the formation of molecular hydrogen

$$H + HO_2 \rightarrow H_2 + O_2. \qquad (8.54)$$

It is the supply of H_2 from this process that ultimately limits the rate at which hydrogen escapes from the exosphere. Since the reaction is fed by the leak of H_2O that is permanently destroyed, escape of two H atoms is necessarily accompanied by the escape of one O atom, a point to which we shall return in the next section.

Molecular oxygen is formed mainly in the reaction

$$O + OH \rightarrow O_2 + H, \tag{8.55}$$

the catalytic cycles having no *net* effect on O_2 concentrations. Ozone is created largely by the $O + O_2$ combination reaction (8.49), of cycle 7; but its destruction is dominated by photolysis rather than by removal in reaction (8.50). Concentrations of O_3 therefore reflect in a direct fashion the concentrations of O, which are themselves highly dependent on the odd hydrogen reactions of cycles 6 and 7. Spatial variability of O_3 is very large on Mars (it approaches a factor of 30), and it seems reasonable that this variability should reflect large inhomogeneities in the spatial distribution of odd hydrogen, with $[O_3]$ large where $[HO_x]$ is small. Part of the variation can be ascribed to condensation of H_2O in low-temperature regions, but additional factors such as changing vertical and horizontal transport of O and O_3 are probably also involved.

Atmospheric concentrations of O_3 and reaction intermediates (e.g. H, OH, HO_2, and O) can be calculated using appropriate models. Figure 8.5 shows the results of a one-dimensional model, while Fig. 8.6 shows the concentration profiles of bulk constituents on which it is based.

Progress in heterogeneous chemistry associated with interpretations of the Earth's Antarctic ozone hole (Section 4.7.3) has prompted suggestions that heterogeneous processes might also play a part in the chemistry of the Martian atmosphere. The widespread presence of dust means that a molecule in the Martian atmosphere collides, on average, with a silicate dust particle every thousand seconds. Some conversion of CO to CO_2 could thus plausibly occur

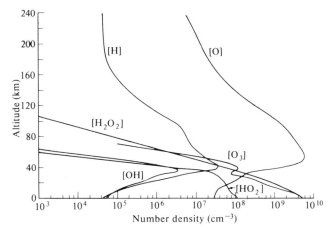

Fig. 8.5. Martian concentration profiles for H, OH, HO_2, H_2O_2, O, and O_3 predicted by a one-dimensional model. The averaged model atmosphere used in these calculations is described by Fig. 8.6. [From McElroy, M. B., Kong, T. Y., and Yung, Y. L. *J. geophys. Res.* **82**, 4379 (1977).]

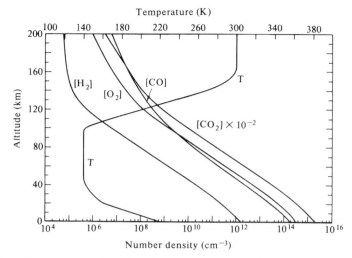

Fig. 8.6. Time-averaged model of temperature and of bulk constituents of the Martian atmosphere. (Source as for Fig. 8.5.)

on the surfaces of dust grains. Similarly, water-ice particles could denitrify the Martian atmosphere in the same way that they do in the Earth's polar stratosphere. In that case, large amounts of the oxides of nitrogen might be sequestered on the Martian surface. For the time being, such ideas remain purely speculative, but they do illustrate the growing awareness of the chemical, as well as physical, effects of suspended particles in atmospheres (cf. Sections 1.3, 4.7.4, 4.7.7, and 5.3.9).

8.3.3 Ionospheric chemistry

The ionosphere of Mars merits special attention because of the part it may play in determining the bulk composition of the atmosphere. Ionic processes can release sufficient kinetic energy for product translational velocities to exceed the rather small Martian escape velocity (~ 5 km s^{-1}). It is therefore important to understand what factors control the ion composition and the rates of ionization in the Martian atmosphere.

Figure 8.7 is a concentration–altitude profile obtained by the Viking 1 lander for ions in the Martian atmosphere: it represents the first *in situ* measurement of ions from the ionosphere of a planet other than our own. Two interesting features of the profile are immediately apparent. First, the dominant ion throughout the atmosphere is O_2^+; and CO_2^+ follows a similar-shaped profile, but is about ten times less abundant. Secondly, the total ion density shows a single peak in concentration at an altitude of about 130 km, in

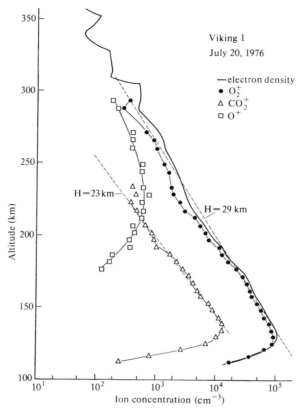

Fig. 8.7.   Ion composition profiles on Mars as measured by the Viking 1 lander. [From Hanson, W. B., Sanatini, S., and Zuccaro, D. R. *J. geophys. Res.* **82**, 4351 (1977).]

contrast with the multi-layered structure of the Earth's ionosphere (Section 6.1.3). Explanations of both features turn out to be related.

Primary ion formation in the atmosphere of Mars (as in that of Venus) is achieved by photoionization of the main atmospheric component, CO_2

$$CO_2 + h\nu(\lambda \lesssim 93 \text{ nm}) \rightarrow CO_2^+ + e. \tag{8.56}$$

However, the CO_2^+ ion reacts very rapidly with neutral atomic oxygen to yield O^+ and O_2^+,

$$CO_2^+ + O \rightarrow O^+ + CO_2, \tag{8.57}$$

$$CO_2^+ + O \rightarrow O_2^+ + CO, \tag{8.58}$$

in comparable amounts. An even faster reaction between O^+ and CO_2

$$O^+ + CO_2 \rightarrow O_2^+ + CO, \tag{8.59}$$

prevents the development of a significant layer of O^+ below ~ 200 km. The terminal ion O_2^+ is thus formed with a maximum rate at the altitude of maximum (day-side) ionization of CO_2, where the optical depth is unity for solar ultraviolet radiation in the approximate wavelength range 20–90 nm. Such behaviour is the analogue of an F_1 mechanism in terms of the Earth's ionosphere (Section 6.4.1), and there is no Martian F_2 peak comparable to Earth's. It will be recalled that the F_2 peak arises because of the bottle-neck in ion–electron recombination brought about by the step forming molecular ions

$$O^+ + N_2 \rightarrow NO^+ + N. \tag{8.60}$$

Curiously, the rate coefficient for the reaction with CO_2, (8.59), is 1000 times larger than that for reaction (8.60), so that conversion of atomic to molecular ions is not a rate-determining step in the Martian ionosphere at the altitudes of maximum ionization. Calculations of $[CO_2^+]$, $[O_2^+]$ (and $[O^+]$) profiles match the experiments well, and the ratio $[O_2^+]/[CO_2^+]$ is correctly predicted, as can be seen in Fig. 8.8.

Ion–electron recombination,

$$AB^+ + e \rightarrow A + B, \tag{8.61}$$

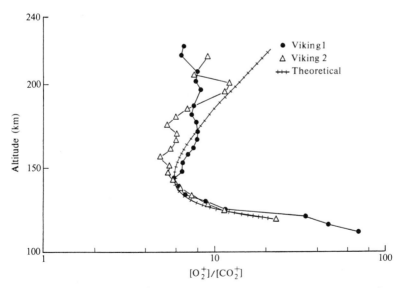

Fig. 8.8. Ratio of O_2^+ to CO_2^+ ion concentrations in the Martian atmosphere as calculated from the measurements of the two Viking landers. Agreement with the theoretical predictions of a model is excellent. (Source as for Fig. 8.7).

Table 8.2   Escape from Mars.

	Species			
	C	N	O	CO
Escape energy $(kJ\ mol^{-1})$	144	168	192	336
Reaction exothermicity in each fragment $(kJ\ mol^{-1})$				
$CO_2^+ + e$	–	–	**509**	291
$O_2^+ + e$	–	–	239	–
$N_2^+ + e$	–	**279**	–	–
$CO^+ + e$	159	–	119	–
$NO^+ + e$	–	138	121	–

Note—Reaction exothermicities are calculated for electronic ground-state products *except* for O_2^+, where the lowest energy channel is assumed to be $O(^3P) + O(^1D)$. If excited products are formed in other recombinations, less energy will be available as translation: e.g., if N_2^+ yields $N(^4S) + N(^2D)$ the reaction exothermicity is $164\ kJ\ mol^{-1}$ for each atom.

can release, as translation apart of fragments A and B, an energy up to the ionization potential of AB. Table 8.2 shows the energy that can be released in the recombination products of several possible Martian ions, together with the energy required for the fragments to escape the gravitational field of the planet (cf. Section 2.3.2). Products for which energies are listed in bold-face type can therefore escape. Most important of these in the present-day atmosphere must be atomic oxygen from recombination of O_2^+, the dominant ion,

$$O_2^+ + e \rightarrow O^*(^1D, {}^1S) + O. \tag{8.62}$$

This source of 'hot' O atoms is, in fact, the complement to the thermal escape of H atoms that together make up the loss of H_2O from the planet described in the previous section. Escape rates for oxygen are set by the rate at which CO_2 is photoionized in the exosphere. Hydrogen escape is limited by the supply of H_2 to the exosphere from below, atoms being released in the exosphere by the ionic reactions

$$CO_2^+ + H_2 \rightarrow CO_2H^+ + H, \tag{8.63}$$

$$CO_2H^+ + e \rightarrow CO_2 + H. \tag{8.64}$$

As we saw earlier, the abundance of H_2 is set by the reactions

$$CO + OH \rightarrow CO_2 + H, \tag{8.17}$$

$$H + O_2 + M \rightarrow HO_2 + M, \tag{8.33}$$

$$H + HO_2 \rightarrow H_2 + O_2, \tag{8.54}$$

and it is therefore sensitive to the net oxidation state of the atmosphere. Consequently, the H atom escape rate is set ultimately by the rate at which oxygen is lost, and the processes are self-regulating to achieve a 2 : 1 stoicheiometry for H : O loss. At the present epoch, water is being 'processed' by the atmosphere and transferred to space at a rate of 6×10^7 molecule $cm^{-2}s^{-1}$. The water vapour is likely to be derived from condensed-phase water on the planet's surface. If the present escape rate had applied over the life of the solar system ($\sim 4.6 \times 10^9$ yr), the water lost would have coated the surface of Mars with ice to an average depth of 2.5 m. In reality, solar ultraviolet fluxes are likely to have been greater in the early history of the solar system (cf. p. 400, Section 9.4), and the greater resultant escape rates increase the estimates of surface depth lost by a factor of up to eight (i.e. 20 m). Some geologic evidence can be interpreted to suggest much greater initial inventories of water (up to 500 m), so that additional mechanisms for hydrogen loss may have operated in earlier periods of the planetary history. An apparent substantial D/H enrichment (Section 9.3) requires such enhanced loss rates in the past.

Escape of species for which there is *not* a surface source raises the interesting question of whether the atmosphere is undergoing continual change in its composition. Viking isotopic measurements of the $^{15}N/^{14}N$ ratio show an enrichment of about 1.6 compared with the terrestrial abundance (the ratio is 3.66×10^{-3} on Earth, and 5.94×10^{-3} on Mars). Such enrichment could be brought about on Mars, from N_2 having initially the terrestrial composition, by preferential escape of the lighter ^{14}N isotope over geological time. Translationally 'hot' nitrogen, able to escape, can be formed by N_2^+ recombination (Table 8.2), or by impact dissociation of N_2 by energetic photoelectrons. The preference for loss of ^{14}N over ^{15}N is only small, so that large total amounts of nitrogen must have been lost during the planet's history. Initial partial pressures of N_2 must have been *at least* 1.3 mbar (they are now 0.17 mbar). Some calculations that allow for sinks of nitrogen such as fixation by lightning suggest that the early Martian atmosphere may have contained as much as 100 mbar of N_2 and 10 bar of CO_2. On the assumption that the Martian H/N ratio was the same as on Earth, these amounts of N_2 imply a surface layer of H_2O as much as 500 m deep, in accordance with the estimates based on geological evidence. Mars therefore provides an example of an evolving atmosphere. Loss of nitrogen and other constituents of the Martian atmosphere is evidently a result of the small gravitational acceleration of this particular planet. A more general discussion of atmospheric evolution is presented in Chapter 9.

8.4 Jupiter and Saturn

Beyond the asteroid belt lie the giant planets Jupiter, Saturn, Uranus, and Neptune, with their extensive satellite systems, and Pluto, which is more like a satellite than any of its giant companions. Jupiter and Saturn have low mean

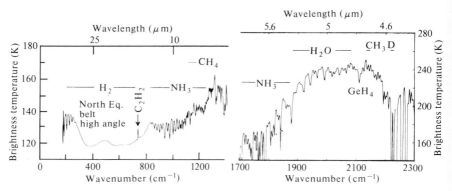

Fig. 8.9. Thermal emission spectrum from Jupiter recorded by the Infra-red Interferometric Spectrometer (IRIS) carried by Voyager 1. Strong spectral features are seen for the gases H_2, C_2H_2, NH_3, CH_4, H_2O, GeH_4, and CH_3D, while between 1100 and 1200 cm^{-1} there is significant absorption by PH_3 (Q branch at 1122 cm^{-1}). [From Hanel, R., Conrath, B., Flasar, M., Kunde, V., Lowman, P., Maguire, W., Pearl, J., Pirraglia, J., Samuelson, R., Gautier, D., Gierasch, P., Kumar, S., and Ponnamperuma, C. *Science* **204**, 972 (1979).]

densities, about one-quarter and one-eighth that of Earth respectively, suggesting that they are composed almost entirely of light elements. Our knowledge of the Jovian and Saturnian systems has been advanced enormously first by the Pioneer 10 and 11, and then by the Voyager 1 and 2 encounters. Voyager 2 has provided data of a similar quality about Uranus, Neptune and some of their satellites as well. It has been said that in the few hours of the Voyager encounters, more was learnt about the planets and satellites than had been found out in the rest of human history! Atmospheric composition was investigated by several instruments on the Voyager spacecraft. Experiments studying infra-red radiation (IRIS), ultraviolet spectroscopy (UVS), photopolarimetry (PPS: aerosols), and radioscience (RSS: ions) were of particular value. Figure 8.9 indicates the type of spectrum obtained in the IRIS experiments (Voyager 1 at Jupiter): from the data, the identity and concentration of infra-red emitters can be deduced. Table 8.3 describes the chemical compositions of the Jovian and Saturnian atmospheres as they now appear after the Voyager missions. As expected, the bulk of the atmosphere is made up of hydrogen and helium. The fractional abundance of He is, however, markedly smaller than the solar ratio (molar ratio 0.16), especially in the atmosphere of Saturn. This result suggests that helium has undergone gravitational separation from hydrogen within the interiors of the planets. The CH_4/H_2 ratio in Jupiter's atmosphere is roughly 2.1 times the solar equivalent value (although that may not be the case for the bulk planetary composition). Analysis of NH_3 data also leads to the conclusion that the

Table 8.3   Composition of the atmospheres of Jupiter and Saturn.

Constituent	Volume mixing ratio	
	Jupiter	Saturn
H_2	0.90	0.96
He	0.10	0.04
CH_4	1.75×10^{-3}	1.76×10^{-3}
NH_3	1.8×10^{-4b}	–
H_2O	$1–30 \times 10^{-6b}$	–
C_2H_6	5×10^{-6}	4.8×10^{-6}
PH_3	6×10^{-7}	2×10^{-7}
CH_3D	3.5×10^{-7}	2.3×10^{-7}
C_2H_2	2×10^{-8}	1.1×10^{-7}
CO	$1–10 \times 10^{-9}$	–
HCN	2×10^{-9c}	–
GeH_4	7×10^{-10}	–
C_2H_4	7×10^{-9a}	–
CH_3C_2H	2.5×10^{-9a}	no estimate[a]
C_3H_3	–	no estimate[a]
C_6H_6	2×10^{-9a}	–

[a] Tentative identification.
[b] Value at 1–4 bar.
[c] Tropopause and below.
Data summarized by Strobel, D. F. *Int. Rev. phys. Chem.* **3**, 145 (1983) and Atreya, S. K., *Atmospheres and ionospheres of the outer planets and their satellites*, Springer-Verlag, Heidelberg (1986), who give references to the original publications.

nitrogen/hydrogen ratio is enhanced by a factor of two over the solar ratio, while oxygen and germanium seem to be *depleted* by factors of $\sim 50–150$ and ~ 10 respectively. Taken at face value, these elemental ratios seem to argue against formation of the planetary system from a primitive homogeneous nebula in favour of an accretion hypothesis (see Chapter 9).

Deep in the atmosphere, thermal chemistry yields compounds of the elements consistent with thermochemical equilibrium at the temperatures and pressures encountered (although the detection of PH_3 and GeH_4 implies the occurrence of disequilibrating processes). Photochemistry in the atmospheres of Jupiter and Saturn can convert CH_4 to heavier hydrocarbons, and NH_3 to N_2H_4, as we shall see shortly. Our current understanding is that the photochemical products are transported downwards to the hot, dense interiors where thermal decomposition and subsequent reaction with H_2 recycle CH_4 and NH_3. Some of the possible photochemically-generated species are condensable at the low temperatures encountered in the atmospheres, so that a knowledge of the vertical temperature structure of the atmosphere is needed

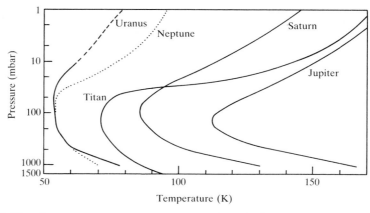

Fig. 8.10. Temperature–pressure structure of the Jovian planets, and of Saturn's satellite, Titan [Data from Strobel, D. F., *Int. Rev. Phys. Chem.* **3**, 145 (1983); Hanel, R. *et al.*, *Science* **233**, 70 (1986); and Conrath. B. *et al.*, *Science* **246**, 1454 (1989)].

for proper interpretation of the photochemistry. Figure 8.10 gives temperature–altitude profiles for the outer planets and for Titan, over the regions of photochemical interest. On Jupiter and Saturn, the temperature profile starts at the transition level from an adiabatic lapse rate (convectively controlled) to an increasing temperature above (radiatively controlled). The temperature minima act as 'cold traps' which limit mixing ratios of condensable gases in the upper atmospheres. Thus NH_3 is limited to a mixing ratio of less than 10^{-7} above the Jovian minimum of ~ 110 K, while the cooler 'tropopause' of Saturn (~ 85 K) prevents detectable concentrations of NH_3 reaching the upper atmosphere. An atmospheric 'parcel' ascending from great depths will lose constituents as they condense out as various liquid or solid phases. Condensates remain as aerosols at the appropriate levels. For example, on Jupiter, dense water clouds form at ~ 270 K, while near the 200 K level, H_2S is thought to react with NH_3 to form a cloud of solid NH_4SH particles. White crystals of ammonia precipitate out at ~ 154 K to produce the visible upper cloud layer.

Above the clouds, photochemical transformations can take place. Although the chemistry of the Jovian and Saturnian atmospheres is very different from anything we have encountered so far, one well-established principle still applies, and that is that the processes requiring the highest energy occur highest in the atmosphere, because short-wavelength radiation is generally attenuated more rapidly than longer wavelength radiation. Figure 8.11 provides a pictorial summary of some of the most important chemical steps. Atomic hydrogen is formed photochemically from molecular hydrogen, and

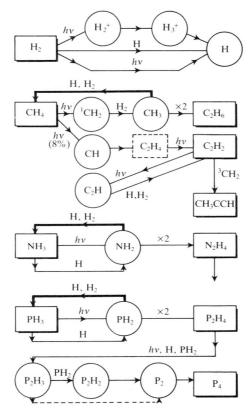

Fig. 8.11. Most important steps in the chemistry of the atmospheres of Jupiter and Saturn.

the chemistry of the atmospheres of Jupiter and Saturn is greatly influenced by the reactions of other species with both H and H_2. Hydrides such as CH_4, NH_3, and PH_3 also undergo photolysis to produce the intermediates CH_2, CH, NH_2, and PH_2 that participate in further reactions to yield some of the compounds observed experimentally. Note that the back reactions of CH_3 (derived from CH_2), NH_2, and PH_2 with both H and H_2 regenerate the starting reactants, and dominate over the formation of the more complex products.

We shall now examine the individual steps in more detail. Hydrogen is the dominant constituent of the atmospheres under discussion. It absorbs to dissociate only at $\lambda < 100$ nm, with a dissociation continuum having its onset at $\lambda = 84.5$ nm, and an ionization continuum starting at $\lambda = 80.4$ nm. Ionization of H_2 leads to the formation of at least two H atoms *via* the reactions

$$H_2^+ + H_2 \rightarrow H_3^+ + H, \tag{8.65}$$

$$H_3^+ + e \rightarrow H_2 + H, \tag{8.66}$$

$$H_3^+ + e \rightarrow 3H, \tag{8.67}$$

so that the effect for neutral chemistry of either ionization or dissociation is H-atom formation. Since the three-body recombination of H atoms is exceedingly slow at ionospheric pressures, there is a net downward flux of H atoms from the ionosphere to lower altitudes, where methane and ammonia photochemistry are important. Methane photolysis requires photons with $\lambda \lesssim 145$ nm, and occurs preferentially high in the stratosphere. Ammonia photolysis is driven by photons with $\lambda \lesssim 160$ nm, and occurs primarily near the tropopause, where condensation and photolysis have not severely depleted the species.

One of the most interesting aspects of the chemistry of the outer solar system is the synthesis of organic compounds. Since methane is a relatively abundant source molecule, it is necessary to investigate how it might be converted to more complex molecules. Intense solar Lyman-α ($\lambda = 121.6$ nm) dissociates CH_4 *via* several primary paths

$$CH_4 + h\nu \rightarrow {}^1CH_2 + H_2 \tag{8.68}$$

$$\rightarrow {}^{1,3}CH_2 + 2H \tag{8.69}$$

$$\rightarrow CH + H + H_2 \tag{8.70}$$

Reaction (8.70) is responsible for about 8 per cent of the photolysis, while the other two paths together make up 92 per cent and are of comparable importance. The methylene (CH_2) radical formed in reaction (8.69) may be in the triplet ground ($^3CH_2$) or singlet excited ($^1CH_2$) states, and the reactivity depends on which state is involved. Important secondary reactions that lead to observed products include

$$CH_2 + H_2 \rightarrow CH_3 + H, \tag{8.71}$$

$$CH_3 + CH_3 \xrightarrow{(M)} C_2H_6. \tag{8.72}$$

Ethene (ethylene) is formed by the reaction

$$CH + CH_4 \rightarrow C_2H_4 + H, \tag{8.73}$$

but is rapidly photolysed to ethyne (acetylene),

$$C_2H_4 + h\nu \rightarrow C_2H_2 + H_2, \tag{8.74}$$

which is itself photochemically rather stable, because its dissociation products (C_2H and C_2) react with H_2 to regenerate C_2H_2. As a result, concentrations of C_2H_2 in the upper atmospheres are much greater than those of C_2H_4. Higher hydrocarbons, and possibly even polymeric materials, can be formed by reactions of C_2H_2, as for example with $^3CH_2$ to yield an observed product

$$^3CH_2 + C_2H_2 \xrightarrow{\text{(M)}} CH_3C_2H \tag{8.75}$$

propyne (methylacetylene).

While the reactions just discussed can account for the chemical identity of the molecules observed in the atmospheres of Jupiter and Saturn, it is not yet evident that photochemical models can correctly predict the concentrations found. In particular, current models predict much lower values (near unity) for $[C_2H_6]$ relative to $[C_2H_2]$ compared to the ratio actually found (from 40 to 200). It may be that the kinetic data employed have been incorrectly extrapolated to the pressures and temperatures relevant to the planetary atmospheres. But another explanation is that C_2H_2 undergoes photopolymerization reactions to produce more complex hydrocarbons. Such reactions would offer a sink for C_2H_2, and reduce its atmospheric concentration. In addition, the hydrocarbon molecules formed could produce an aerosol haze, thus accounting for the rather low albedo of Jupiter and Saturn. Photochemical hazes of this kind are certainly important on Titan (Section 8.5), where the ultraviolet albedo is yet lower.

Ammonia photochemistry in the Jovian and Saturnian atmospheres is rather simple, since the primary step for the available wavelengths (and in the presence of H_2) can be written

$$NH_3 + h\nu \to NH_2 + H. \tag{8.76}$$

Hydrazine formation or regeneration of ammonia follow

$$NH_2 + NH_2 \xrightarrow{\text{(M)}} N_2H_4, \tag{8.77}$$

$$NH_2 + H \xrightarrow{\text{(M)}} NH_3, \tag{8.78}$$

with condensation of the N_2H_4 as a haze. Entirely analogous reactions describe the main part of phosphine photochemistry

$$PH_3 + h\nu \to PH_2 + H, \tag{8.79}$$

$$PH_2 + PH_2 \xrightarrow{\text{(M)}} P_2H_4, \tag{8.80}$$

$$PH_2 + H \xrightarrow{\text{(M)}} PH_3, \tag{8.81}$$

and solid P_2H_4 is probably formed as a condensation product. Coupling between NH_3 and PH_3 chemistry is brought about by the reaction

$$H + PH_3 \to PH_2 + H_2, \tag{8.82}$$

since the hydrogen liberated by NH_3 photolysis, (8.76), accelerates the conversion of PH_3 to P_2H_4. Both PH_3 and NH_3 concentrations decrease rapidly above the tropopause, and measurements utilizing ultraviolet absorption therefore detect orders of magnitude less NH_3 than those using infra-red techniques that penetrate deeper into the atmosphere.

Both NH_3 and PH_3 photochemistry have minor pathways that we shall exemplify in the case of PH_3. P_2H_4 is susceptible to photolysis, or attack by H and PH_2, to yield P_2H_3 and then P_2H_2,

$$P_2H_4 + h\nu(\text{or } H, PH_2) \rightarrow P_2H_3 + H(\text{or } H_2, PH_3), \qquad (8.83)$$

$$P_2H_3 + PH_2(\text{or } H, P_2H_3) \rightarrow P_2H_2 + PH_3(\text{or } H_2, P_2H_4). \qquad (8.84)$$

Elemental phosphorus is an end-product

$$P_2H_2 \rightarrow P_2 + H_2, \qquad (8.85)$$

$$2P_2H_3 \rightarrow P_4H_6 \rightarrow 2PH_3 + P_2, \qquad (8.86)$$

$$P_2 + P_2 \xrightarrow{\text{(M)}} P_4. \qquad (8.87)$$

Likewise, N_2 is the end-product of the N_2H_4 sequence. Were it not for the condensation of N_2H_4, and thus its transport to the troposphere and protection from chemical and photon attack, N_2 would be the ultimate NH_3 photolysis product.

Formation of P_4 in reaction (8.87) brings us inevitably to a discussion of the red and yellow colourings seen on Jupiter and Saturn. Colours of a wide variety are found on the bodies of the outer solar system, but their sources are not known. Red phosphorus is one obvious candidate, but there has been some controversy over the actual colour of the phosphorus formed during photolysis of PH_3. Spectroscopy authenticates the allotrope as 'red', but visually the colours range from yellow to violet. Sulphur species offer another obvious inorganic contributor to the observed colourings. Coloration by complex *organic* molecules is another possibility that carries additional interest because of the implications for the origin of life in reducing atmospheres. Simple hydrocarbons such as C_2H_6 and C_2H_2 do not accumulate in sufficient concentrations on Jupiter and Saturn to condense, nor do they absorb in the visible region. What is needed is complex molecules with conjugated bonds (e.g. polyacetylenes) and nitriles. In the context of complex molecule formation, the observed presence of HCN may be significant. Convection from the hot, dense interior atmosphere cannot be the source of the molecule, nor can local heating due to lightning discharges produce as much as is found. One promising pathway to HCN generation seems to be the ultraviolet photolysis of the cyclic isomer of C_2H_5N, aziridine (ethyleneimine), itself formed indirectly from NH_2 radicals and C_2H_2. Laboratory studies show that aziridine is photolysed by vacuum ultraviolet light to yield HCN so that the pathway is plausible, although not substantiated, for the planetary atmospheres.

Although the atmospheres of Jupiter and Saturn are usually thought of as totally reducing, the solar oxygen abundance requires substantial amounts of H_2O and other oxygen species. Most H_2O condenses out below the 'photochemical' regions, but the discovery of CO on Jupiter proves that photochemical processes involving oxygen species do occur. At one time, it was thought that an extraplanetary source of H_2O might be needed to explain the CO. However, measurements of the linewidths of the CO spectra have now demonstrated that the CO is formed by the reaction of methane with water in the planetary interior. Oxygen in any form is virtually bound to be converted ultimately to CO. Once formed, CO cannot be photolysed, and it is not attacked chemically in the lower stratosphere because condensation of water removes the OH source that would oxidize it to CO_2, while CO_2 photolysis itself rapidly re-forms carbon monoxide.

8.5 Titan and Io

8.5.1 Titan

Saturn's largest satellite, Titan, is the only satellite in the solar system to possess a *massive* atmosphere. As early as 1908, visual observations had suggested the existence of an atmosphere, and by 1944 absorption bands of methane had been discovered. With the Voyager fly-bys have come our first definitive data about the atmospheric composition and structure. Ultraviolet spectroscopy (UVS) experiments detected emission lines from molecular and atomic nitrogen in the upper atmosphere (Fig. 8.12), characteristic of electron-excited N_2. Together with measurements of temperature and scale height, the UVS results show that the atmosphere is predominantly N_2, with methane a minor constituent. Temperature and pressure at the surface were found to be 94.5 ± 0.4 K and 1.5 bars, so that the pressure is roughly 50 per cent greater than that on Earth. Titan is the only body in the solar system besides Earth that has an atmosphere composed largely of nitrogen. Voyager's infra-red (IRIS) experiments detected a suite of hydrocarbons and nitrogen compounds in addition to CH_4, and our present knowledge of Titan's atmospheric composition is given in Table 8.4. Television pictures returned by the Voyager cameras showed that the satellite is covered by coloured clouds. The clouds must be aerosols derived from the gaseous organic compounds, and are thus Titan's equivalent of photochemical smog (cf. Section 5.5.7)! Above the coloured cloud layers, which extend from the surface to an altitude of ~ 200 km, lies a thinner haze layer of aerosol particles. Absorption of solar radiation by the various aerosols leads to a net heating above an altitude of ~ 40 km, and thus to a temperature inversion in Titan's 'stratosphere'. The curve for Titan in Fig. 8.10 shows that the minimum temperature, at the tropopause, is about 70 K. Titan's atmosphere is reducing in bulk chemical composition, the oxidation state lying between that of Jupiter and Saturn on the one hand, and the terrestrial planets on the other. For many of the organic compounds listed in Table 8.4, the measured abundances exceed the saturated vapour pressures at the tropopause, so that the lower atmosphere cannot be the source of complex molecules found in the stratosphere, even though certain species (e.g. HCN) might be abundant on the surface and in the interior of the satellite.

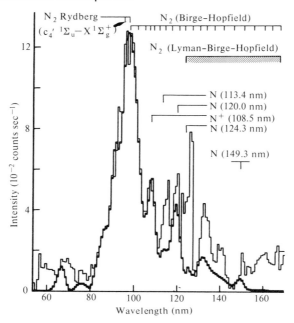

Fig. 8.12. Titan's emission spectrum in the extreme ultraviolet (Voyager 1). The spectrum is obtained for the averaged daytime disc, and has the strong feature due to atomic hydrogen Lyman-α ($\lambda = 121.6$ nm) removed by computer. The heavy overplotted spectrum is a synthetic model spectrum composed of N_2, N, and N^+ emissions excited by electron impact on N_2. [From Strobel, D. F., *J. geophys. Res.* **87**, 1361 (1982).]

The observed compounds must, therefore, be derived from volatile parent molecules. Nitrogen does not condense at the temperatures of the tropopause, although CH_4 clouds may exist. The surface temperature is probably just above the triple point of CH_4 (90.7 K), so that liquid methane pools or oceans may cover the satellite's surface. Negligible amounts of NH_3 are present in the stratosphere because of the 70 K 'cold trap' temperature. Any N_2H_4 formed by ammonia photolysis [reactions (8.76) and (8.77)] near the surface will also be trapped out. We shall see in Chapter 9 that ammonia photochemistry is one possible source of Titan's present-day nitrogen. In that case, trapping of NH_3 must not have been significant, and the atmosphere must have been at least 50 K warmer than it is now, during about 4 per cent of its evolutionary history.

Escape of atomic and molecular hydrogen from Titan's atmosphere is ensured by the combination of very low escape velocity (2.1 km s^{-1}: Table 2.1), extended atmosphere, and 'warm' thermosphere (~ 186 K). Thermal escape of H and H_2 can cope with the rate of hydrogen formation in direct and

Table 8.4   Composition of Titan's atmosphere.

Constituent	Volume mixing ratio		
	Near surface	Stratosphere (40–100 km)	Thermosphere (3900 km)
N_2		> 0.97	
CH_4	$< 3 \times 10^{-2}$	$1–3 \times 10^{-2}$	$8 \pm 3 \times 10^{-2}$
CH_3D		detected	
H_2	$2 \pm 1 \times 10^{-3}$	2.0×10^{-3}	
CO	$10 \pm 5 \times 10^{-5}$	6×10^{-5}	
CO_2		$7–30 \times 10^{-10}$	
H_2O		$< 1 \times 10^{-9}$	
Ar	< 0.16		$< 6 \times 10^{-2}$
Ne	$< 2 \times 10^{-3}$		$< 1 \times 10^{-2}$
C_2H_6		2×10^{-5}	
C_2H_4		4×10^{-7}	
C_2H_2		2×10^{-6}	$\sim 1.5 \times 10^{-3}$
C_3H_8		$2–4 \times 10^{-6}$	
CH_3CCH		3×10^{-8}	
CHCCCH		$1–10 \times 10^{-8}$	
HCN		2×10^{-7}	$< 5 \times 10^{-4}$
C_2N_2		$1–10 \times 10^{-8}$	
HCCCN		$1–10 \times 10^{-8}$	

Data summarized by Strobel, D. F. *Int. Rev. Phys. Chem.* **3**, 145 (1983), and Yung, Y. K., Allen, M., and Pinto, J. P., *Astrophys. J. Supp. Ser.* **55**, 465 (1984), who give references to the original publications.

catalysed photolysis of hydrocarbons, principally CH_4. Build-up of heavier hydrocarbons, such as C_2H_6 and C_3H_8, at the expense of CH_4, therefore takes place, and H_2 is a minor constituent of Titan's atmosphere. Recycling back to CH_4 is impossible, so that there is a one-way evolution towards more complex organic species that condense at the tropopause and are ultimately deposited on the surface. Hydrogen lost from the satellite has formed a dough-nut shaped torus around Saturn, through which Titan is continually sweeping.

The continuing production of complex species leads to the formation of the photochemical aerosol that in turn absorbs sunlight to give a thermal inversion and yield dynamical stability of the atmosphere. There are obvious parallels with the ozone layer in Earth's atmosphere.

Absence of H_2, and the presence of abundant N_2, modify Titan's atmospheric photochemistry considerably from that of Jupiter or Saturn. On the planets, radicals such as CH_3 or NH_2 derived from CH_4 or NH_3 react most frequently by abstraction of hydrogen from H_2 or by three-body association with H-atoms to return to the starting compound in a 'do nothing' cycle. Such

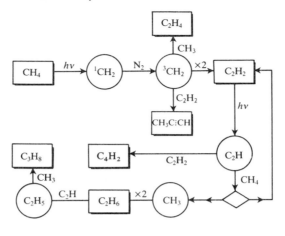

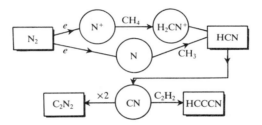

Fig. 8.13. Important steps in the chemistry of Titan's atmosphere. This diagram should be compared with Fig. 8.11 for the parent planet.

reactions cannot occur on Titan, so that the less hydrogen-rich hydrocarbons are favoured. Mixing ratios of C_2H_6 are four times greater on Titan than on Saturn, those of C_2H_2 are 27 times larger, and C_2H_4 is detected on the satellite but not on the planet. The extensive airglow of N_2, N, and N^+ emission features means that N_2 is excited, dissociated, and ionized by electron impact processes in the thermosphere. Reactions of the energetic nitrogen species must therefore be included in any consideration of chemistry in Titan's atmosphere. Figure 8.13 provides an overview of some of the most important steps leading to identified products, and should be compared with Fig. 8.11 for Jovian chemistry. The three key differences are the absence of back-reactions involving H and H_2, the presence of processes involving N and N^+, and the quenching of $^1CH_2$ to $^3CH_2$ by N_2 and the consequent formation of C_2H_4, C_2H_2, and C_3H_4 from the triplet.

Because N_2 is the bulk constituent of Titan's atmosphere, this latter quenching means that $^3CH_2$ is the net product of CH_4 photolysis on Titan (with only minor pathways to CH and possibly CH_3 [cf. reactions (8.68)–(8.70), p. 358]. These triplet radicals react primarily to form C_2H_2, so that the first stages in Titan's hydrocarbon photochemistry are

$$CH_4 + hv \xrightarrow{\text{N}_2} {}^3CH_2 + H_2(\text{or } 2H), \tag{8.88}$$

$$^3CH_2 + {}^3CH_2 \rightarrow C_2H_2 + H_2(\text{or } 2H), \tag{8.89}$$

followed by photolysis of C_2H_2 to yield C_2H radicals, which catalyse dissociation of CH_4

$$C_2H_2 + hv \rightarrow C_2H + H \tag{8.90}$$

$$C_2H + CH_4 \rightarrow C_2H_2 + CH_3 \tag{8.91}$$

Net $\qquad CH_4 \xrightarrow{\;hv\;} CH_3 + H.$

Reactions of the methyl (CH_3), methylene (CH_2), and ethynyl (C_2H) radicals can then yield many of the compounds observed in the atmosphere.

$$^3CH_2 + CH_3 \rightarrow C_2H_4 + H, \tag{8.92}$$

$$CH_3 + CH_3 \xrightarrow{\text{(M)}} C_2H_6, \tag{8.72}$$

$$C_2H + C_2H_6 \rightarrow C_2H_2 + C_2H_5, \tag{8.93}$$

$$C_2H_5 + CH_3 \xrightarrow{\text{(M)}} C_3H_8, \tag{8.94}$$

$$C_2H + C_2H_2 \rightarrow H + HC\!:\!CC\!:\!CH \text{ (diacetylene)}, \tag{8.95}$$

$$^3CH_2 + C_2H_2 \xrightarrow{\text{(M)}} CH_2\!:\!C\!:\!CH_2 \text{ (allene)}, \tag{8.96}$$

$$CH_2\!:\!C\!:\!CH_2 \rightarrow CH_3C\!:\!CH \text{ (methylacetylene)}. \tag{8.97}$$

Allene ($CH_2\!:\!C\!:\!CH_2$) is the isomer of C_3H_4 favoured in reaction (8.96), but is not found on Titan: isomerization can yield the methylacetylene ($CH_3C\!:\!CH$) observed. Polyacetylenes can be formed by successive reactions analogous to eqn (8.95), but their growth may be inhibited by H atoms even at the low [H] found on Titan. It is thus not clear at present whether such molecules are responsible for aerosol formation.

Hydrogen cyanide formation is more easily explained for Titan's atmosphere than it is for Jupiter's (or Saturn's (Section 8.4)) because the nitrogen source is N_2 itself. Electron impact and solar radiation form N^+ ions and N atoms from N_2. Routes exist from both these species to HCN, as exemplified by the reactions

$$N^+ + CH_4 \rightarrow H_2CN^+ + 2H, \tag{8.98}$$

$$H_2CN^+ + e \rightarrow HCN + H, \tag{8.99}$$

and

$$N + CH_3 \rightarrow HCN + H_2. \tag{8.100}$$

Photolysis of HCN itself generates the CN radical

$$\text{HCN} + hv \rightarrow \text{H} + \text{CN}, \qquad (8.101)$$

that can react with known atmospheric constituents to form products that have been detected

$$\text{CN} + \text{C}_2\text{H}_2 \rightarrow \text{HC}\!:\!\text{CCN} + \text{H}, \qquad (8.102)$$

$$\text{CN} + \text{HCN} \rightarrow \text{C}_2\text{N}_2 + \text{H}. \qquad (8.103)$$

Synthesis of relatively complex organic molecules by ion–molecule and by radical–radical reactions bears an interesting analogy with the chemistry of dense interstellar clouds, which we shall discuss briefly in Section 9.1.2. Because the classes of reaction involved often have near-zero activation energies, and the ionic processes are characterized by long-range attractive forces, they continue to drive chemical change at temperatures and particle densities much lower than those favouring activated reactions of neutral species. Further reading on the subject is described in the Bibliography associated with Section 9.1.2.

Recent discoveries of CO_2 and CO on Titan have opened up new vistas in the chemistry of reducing atmospheres. Continual input of H_2O to the satellite is demanded in order to maintain any CO_2 in the atmosphere at all. Without water, photolysis would convert CO_2 to CO and there would be no route for reversing the process, as we have discussed in connection with CO_2 chemistry on Venus and Mars (Sections 8.2 and 8.3). Hydroxyl radicals can be formed photolytically when water is present, and so permit the usual oxidation step

$$\text{CO} + \text{OH} \rightarrow \text{CO}_2 + \text{H}. \qquad (8.17)$$

Meteorites, perhaps supplemented by sputtering from Saturn's rings and icy satellites, can account for the requisite water supply. Input of H_2O can also be responsible, *via* OH radicals, for the partial oxidation of CH_4 to CO itself. However, the evidence so far available does not exclude the possibility of Titan having been formed with an atmosphere containing CO (and, in fact, N_2: see Chapter 9). Carbon monoxide in the contemporary atmosphere appears to be roughly in a steady state between formation and oxidation to CO_2. Carbon dioxide is destroyed by photolysis (and perhaps by reaction with CH_2), but at a rate insufficient to overcome its production. Its vapour pressure therefore builds up to reach saturation near the tropopause. Solid CO_2 condenses and is precipitated to the surface. The measured CO_2 abundance is consistent with saturation at 75 K and a total pressure of ~ 100 mbar, so that CO_2 seems to be in a steady state, controlled by condensation and sublimation. If the large influx of water postulated for Titan is borne out by further observation, H_2O vapour would have to be considered as a potentially important species on Saturn as well. Spectroscopy shows that CO on Jupiter is predominantly present in the troposphere (p. 361), and is formed by reactions deep in the interior. For Saturn, the altitude distribution of CO is not known, but an influx of water would make a stratospheric source of CO likely. The chemistry of H_2O, CO, and CO_2 may ultimately prove to be as significant to reducing atmospheres as that of the 'trace constituent' reduced compounds (e.g. CH_4, H_2) is to oxidizing ones.

8.5.2 Io

A massive atmosphere like Titan's is not found on any of the other satellites of the gas giant planets. However, Jupiter's satellite Io is able to retain a tenuous atmosphere which is of interest because its composition is quite unlike that of any other body in the solar system. One of the first indications that Io might have an atmosphere was provided by a brightening of the satellite in the period immediately after its emergence from an eclipse, suggesting that condensation of an atmospheric gas had occurred during the period of darkness. Experiments on Pioneer 10 established the existence of an ionosphere with relatively high electron densities, and thus demonstrated that the atmosphere was present, although surface pressures were perhaps only 10^{-7} of those on Earth. Atomic emission lines from sodium (and weaker ones from potassium) show that alkali metals are present in a torus of atoms that have escaped from Io, but are still orbiting Jupiter. The highlight of the visits of the Voyager craft to Io was the discovery of active volcanoes on the satellite. The infra-red absorption experiment on Voyager identified SO_2 in the atmosphere above one of the volcanic hot spots, and the fly-bys showed that sulphur and oxygen (as ions) made up a giant plasma torus encircling Jupiter at Io's orbital distance. The International Ultraviolet Explorer (IUE) has also detected optical emissions from *neutral* sulphur and oxygen atoms near Io. These observations, coupled with a further inference that SO_2 and S are present in condensed phases at the satellite's surface, suggest an atmosphere in which SO_2 is a major constituent, with minor amounts of the photochemically derived products S and O, and some Na and K.

The primary pathways for photodissociation of SO_2 are

$$SO_2 + hv(\lambda < 221\,\text{nm}) \rightarrow SO + O \qquad (8.104)$$

$$SO_2 + hv(\lambda < 207\,\text{nm}) \rightarrow S + O_2 \qquad (8.105)$$

and two important secondary reactions can also act as sources of S and O atoms

$$SO + SO \rightarrow SO_2 + S \qquad (8.106)$$

$$S + O_2 \rightarrow SO + O. \qquad (8.107)$$

Atomic oxygen can readily escape from Io, but the relatively slower thermal escape of atomic sulphur may lead to net accumulations of sulphur in the atmosphere and at the surface.

Ionospheric species are produced from photoionization and electron impact ionization of SO_2, S, and O, with a balancing charge neutralization *via* dissociative recombination (cf. reaction (6.8) and the associated discussion in Section 6.2) of the molecular ion

$$SO_2^+ + e \rightarrow SO + O. \qquad (8.108)$$

Attempts have been made at modelling the neutral and ionized atmospheres of Io, but more data are needed for resolution of several problems. The Galileo spacecraft is scheduled to make a fly-by of Io at an altitude of 1000 km (as well as remaining in the Jovian system for several years), and the experiments it carries should complement ground-based observations in improving our understanding of the atmosphere of this interesting satellite.

8.6 Uranus, Neptune, Triton, and Pluto

Beyond Saturn lies a cold and dim part of the solar system hardly conducive to an active chemistry. However, some of the bodies at these great distances possess atmospheres with interesting features that are only now slowly becoming apparent. The successful encounters of Voyager 2 with the Uranian and Neptunian systems in 1986 and 1989 have brought a wealth of observational data that will take years for detailed analysis. Even the preliminary results have vastly enhanced the knowledge that has been built up from ground-based observations. Uranus and Neptune are much smaller than Jupiter and Saturn, with masses only a few percent of the mass of Jupiter, but they are denser, suggesting a rocky or icy core and an overall departure from solar composition. However, the composition of the atmospheres of Uranus and Neptune is closer to that presumed for the Sun than that of Jupiter or Saturn. The atmospheres are mainly hydrogen, and contain helium with a mole fraction abundance of about 0.15, almost exactly that of the Sun. Table 8.3 shows that helium is depleted with respect to this abundance on Jupiter, and, especially, on Saturn. Gravitational separation of helium towards the centre of the planet, which probably accounts for depletions higher up on the gas giants, cannot, of course, arise on planets with solid cores. There is certainly plenty of dynamic activity within the atmospheres. Neptune must be one of the windiest places in the solar system, with differential wind speeds approaching the speed of sound!

Quadrupole and pressure-induced bands of H_2 measured from Earth had demonstrated the presence of H_2 in the atmospheres before the Voyager encounters. Airglow and auroral emissions from atomic and molecular hydrogen in the vacuum ultraviolet region confirm the presence of H and H_2 to great altitudes. Excitation is effected either by impact of low-energy electrons or by Rayleigh and resonance scattering of sunlight (cf. Sections 2.6 and 7.2). Spectra of the planets are crossed by strong absorption bands of CH_4 in their atmospheres, and the visible component of the absorption lends a greenish colour to the planets. The Voyager instruments detected clouds of methane in the atmospheres of both Uranus and Neptune. Below the cloud decks, CH_4 could possess abundances of two per cent (Uranus) and not less than one per cent (Neptune). These abundances are 10 to 20 times the solar values, as might be expected if the planets accreted from ice-rich material. Above the clouds, there seems to be a significant density in methane loading between the two planets. On Uranus, the mixing ratio is less than 10^{-7}, while on Neptune the initial estimate of the mixing ratio is 3×10^{-5}, at least 300 times more abundant. This result may reflect the greater mixing of material from below the clouds in the case of Neptune, although it could also show a real difference in atmospheric composition. Acetylene, C_2H_2, is found on both planets. Ultraviolet absorption in the atmosphere of Uranus can be matched with $[C_2H_2] : [CH_4] \cong 0.3$ (although other measurements are consistent with a mixing ratio of C_2H_2 as high as 2×10^{-7}). In Neptune's atmosphere, the C_2H_2 mixing ratio is 2×10^{-7}, and C_2H_6 is also present with a mixing ratio of 3×10^{-5}.

The production of C_2H_2 and C_2H_6 in the atmospheres of Uranus and Neptune is likely to be a result of photochemical conversions of CH_4 similar to those that operate in the atmospheres of Jupiter and Saturn (cf. Fig. 8.11 and reactions (8.68) to (8.74)). Photolysis of C_2H_6, and any C_2H_4 formed, will provide a further source of C_2H_2. Synthesis of higher hydrocarbons is likely, and aerosols may be formed. Changes in the brightness of Neptune are in antiphase with changes in solar activity (as they are on Titan), suggesting that sunlight is connected with the formation of aerosol particles that decrease the planetary albedo. Hazes and clouds are probably present in both atmospheres, although the densities and height distributions are very different. Uranus appears much clearer to great depths than does Neptune. Indeed, Neptune's atmosphere is so laden with aerosol particles that heating occurs in the atmosphere and a temperature inversion arises as in Titan's atmosphere. Figure 8.10 shows the temperature–pressure structure of the atmospheres of these planets; the 'tropopause' is evident for Neptune.

The outer solar system contains numerous smaller solid bodies in addition to the two large planets. All the bodies appear to be composed mainly of ices and other compounds, and all are probably aggregates of the solids that condensed from the solar nebula far from the heat of the proto-Sun. Pluto is the most familiar of these bodies. Often regarded as a planet, it has an unusual orbit, and may once have been the satellite of one of the outer planets. It has its own satellite, Charon, that may or may not have been captured. Triton is another peculiar body that may have been captured. It is the largest satellite of the most distant planet, Neptune (Pluto being at present closer to the Sun), and within 10–100 million years it will be ruptured as it approaches its parent.

Both Triton and Pluto, at least, have tenuous atmospheres according to ground-based observations. Before the Voyager fly-by of Triton, the nature of this interesting satellite's atmosphere could only be inferred from the presence of CH_4 and N_2 surface frosts. Ultraviolet spectroscopy experiments on Voyager 2 show strong airglow features from neutral molecular nitrogen and ionized atomic nitrogen, similar in some ways to the airglow of Titan (cf. Section 8.5.1 and Fig. 8.12). These results show that the atmosphere is predominantly made up of N_2, but CH_4 is also present in the lower atmosphere with a mixing ratio of about 10^{-4}. A very intense emission feature from H Lyman-α in the airglow spectrum shows that atomic hydrogen is present in the atmosphere, and it must be derived from photolysis of CH_4, as on Titan, Uranus, and Neptune. The surface pressure is about 1.6×10^{-5} that on Earth, so the atmosphere really is tenuous. Triton has the lowest observed surface temperature of any body in the solar system (38 ± 4 K), which is probably a consequence of the very high surface reflectivity. All the surface features on the satellite seem to be overlain with a relatively thin layer of CH_4 and N_2 ices and their derivatives. There are reddish colorations and streaks on the surface that might be due to organic polymers produced from CH_4 photochemically and by charged-particle bombardment. Clouds and hazes were also detected in the atmosphere, suggesting both condensation and photochemical production of aerosol particles. One of the most intriguing demonstrations of coupling between the surface and the atmosphere was provided by the observation of geyser activity. Plumes were discovered that rise vertically to an altitude of about 8 km. A dense cloud of material forms which serves as a source of a wind-driven trail of material over 100 km long. Geyser plumes like this may arise from the penetration of solar radiation into the solid nitrogen surface, and subsequent explosive release of vaporized nitrogen which would carry ice-entrained dark material up into the atmosphere.

The observed surface temperature and pressure are consistent with saturation equilibrium of N_2 in the atmosphere. For CH_4, however, the abundance seems to be far below saturation, and at 40 km the mixing ratio is only 2.5×10^{-6}, some 30 times smaller than expected if the concentration were controlled by temperatures at the surface and in the lowest part of the atmosphere (say up to 10 km). The mixing ratios for CH_4 imply a decrease in concentration with altitude much more rapid than predicted by a hydrostatic distribution (Section 2.1). A possible explanation is that methane is removed photochemically, and that vertical mixing is weak enough that the concentration gradients can be maintained. If this interpretation is correct, then photochemistry will occur predominantly in the first 20–30 km of the atmosphere. Photolysis of CH_4 by solar and interstellar Lyman-α ($\lambda = 121.6$ nm) radiation yields CH_2, CH, H_2, and H (cf. reactions (8.68) to (8.70)). Because nitrogen is present, Titan's atmosphere may provide a better analogy with Triton's than does Jupiter's. As illustrated in Fig. 8.13, C_2 hydrocarbons such as C_2H_6, C_2H_4, and C_2H_2 may be formed initially, but because the atmosphere of Triton is so cold, the heavier hydrocarbons condense without undergoing photolysis or further reaction, and hydrogen escapes from the satellite. According to one calculation, the flux of C_2 hydrocarbons to the surface would have been sufficient to create a layer on Triton 6 m thick over the life-span of the solar system. The atomic and molecular hydrogen produced ultimately escapes from Triton to enter Neptune's magnetosphere.

For Pluto, the experimental information comes mainly from ground-based observations and the Infra-red Astronomical Satellite (IRAS). The near infrared spectrum of Pluto is dominated by bands similar to those produced by solid (i.e. surface) methane, although there are also features that can be ascribed to gas-phase absorption. There are variations of absorption with the planet's orbital phase which imply a system with surface frost as well as atmospheric absorption. Although it is still cold, Pluto is at a substantially higher temperature (ca. 55 K near the surface) than Triton, and higher mixing ratios of CH_4 are expected. A reasonable model puts the total gas phase abundance as equivalent to a column 540 cm thick if reduced to 273 K and one (Earth) atmosphere (STP). Pluto is small (radius ca. 1500 km), and the escape velocity is only 0.95 km s^{-1}, less than that of Titan. It has been suggested that the rate of escape of methane from Pluto could be so great that the whole mass of the planet could have been lost over its life unless some heavier gas (e.g. nitrogen or argon) were present to limit the escape rate by providing a diffusional barrier. Allowing for energy transport in the atmosphere, however, greatly reduces the predicted escape rate, and it seems that the loss of CH_4 over the age of the solar system could be as little as 0.5 per cent of the planetary mass. There is thus no need to invoke a heavier secondary gas to maintain the CH_4 on Pluto. On Pluto's smaller-mass companion, Charon, methane is absent because of hydrodynamic escape. It seems likely that photochemistry similar to that on Titan or Triton will occur. Solar fluxes on both Pluto and Triton are roughly ten times lower than at Titan, so that solar photochemistry is correspondingly slower. In these circumstances, chemistry initiated by charged particles and cosmic rays is likely to become relatively more important, but the secondary steps are likely to be similar to those already encountered. Pluto has a red tint to its surface, suggesting that methane is being converted to higher hydrocarbons as it is in Triton's atmosphere.

Voyager 2 took advantage of an opportunity that arises only once in 176 years when a unique alignment of the planets made feasible a visit to them. The spacecraft and its instrumentation made a remarkable journey that lasted more than 12 years from the launch and successfully returned information, much dramatic and unexpected, about the planets and their atmospheres. The Voyager and Pioneer spacecraft have now passed on beyond Neptune and Triton to become the most distant identifiable objects in the solar system with the exception of the comets, and it is to these latter remarkable bodies that we turn finally in this chapter.

8.7 Comets

Comets are accompanied by extended tails of gases that might be thought of as atmospheres. In the classical model, a comet nucleus consists of dust and ice in the manner of a very loosely packed snowball. When a comet's orbit brings it near enough to the Sun, heating of the outer layers vaporizes material that is released to space. The familiar coma and tail of a comet are formed by the sublimation of ices; the vaporized gases entrain dust and ice particles. Photochemical reactions can alter the composition of the gases just as in any other atmosphere, but almost all of the gases released from the nucleus are lost to space. Cometary science changed dramatically in the mid-1980s with spacecraft encounters with cometary bodies, especially with comet Halley in March 1986. Six spacecraft from four agencies approached the comet as a 'Halley Armada'. Giotto was the mission of the European Space Agency, and two craft of the Venera series (cf. Section 8.2) were sent *via* Venus as a Franco–Soviet venture. 'Venera' was renamed 'Vega' for the encounters, which is a pun: the word is a contraction of 'Venera–Halley, but Russian has no 'H', and 'G' has to be used instead, turning the name into a bright star! Giotto and Vega carried mass spectrometers for both neutral and ionized species, infra-red spectrometers to study the gases, and particulate impact analysers to investigate the chemical and isotopic composition of the solid dust particles. The *in situ* observations were backed up by a full-scale remote ground- and satellite-based campaign to provide the most intensive study of a comet ever attempted.

Molecules found in the coma of comet Halley that are likely to be parent species from the nucleus include CO, CO_2, CH_4, NH_3, N_2, $HCHO$, HCN, and saturated and unsaturated hydrocarbons, in addition to water vapour. Carbon monoxide is the most abundant species after water, perhaps making up 20 per cent of the gases leaving the nucleus. The parent molecules are then the source of daughter atoms, radicals, and ions that have been known for some time from their optical emission spectra. Estimated rates of gas sublimation suggest that the comet loses about one metre of its thickness every time it orbits the Sun, so that the life of the comet is limited to less than 10 000 revolutions. The dust component is weighted towards large abundances of very small particles, some of which have a mass as little as 10^{-17} g (which might correspond to 100 atoms). Several types of dust particle were discovered. They include particles rich in the lighter elements, H, C, N, and O (and named *CHON particles*), and others that are rich in heavier elements such as Mg (and are probably dominated by silicates). The weight ratio of gas : dust is about 0.9, so that for every 100 atoms of Mg in the dust, there are 312 atoms of carbon in the gas. A wide variety of organic molecules has been inferred to be present in the CHON particles, including saturated and unsaturated hydrocarbons, compounds possessing –CN and –NH– groups, and ring compounds such as hydrocarbons, pyrrole, pyridine, purine, adenine, and xanthine. The dust grains serve as an extended source of gas in the inner coma. For example, a large source of CO extends up to about 15 000 km from the nucleus, and constitutes more than half the total CO flux. Strange jets contain CN radicals up to 50 000 km from the nucleus, and may be related to a source involving CHON particles. The CHON particles have an abundance of carbon almost identical to the cosmic abundance, and this, and some other observations of atomic and isotopic abundances, suggest that dust from Halley's comet is the most unaltered material from the early solar system yet analysed in man's experiments.

The nucleus of comet Halley has an extremely low albedo (ca. 0.04), making it one of the darkest objects in the solar system. Bright dust jets come from relatively small active sites in the sunlit hemisphere. The surface darkness may be a consequence of residual complex and non-volatile organic species left near the surface. Measurements by the ion mass spectrometers show the presence of the dissociation products of methane. If a low-temperature condensate such as methane is present in the near-surface layers of the nucleus, then those layers must have remained unaltered until they were exposed to the Sun. Other compounds besides CH_4, such as CO and CO_2, are also present near the surface of comet Halley, and these volatile materials cause pronounced activity of the comet when it is still up to five times more distant from the Sun than the Earth is. This evidence, and the measured atomic abundances in the dusts and gases, make it likely that the comet is made up of pristine matter from the solar nebula, and that it must have formed in the outermost reaches of the solar system. The comets are thus among the most primitive objects that remain in the solar system, and their study provides one line of evidence about the nature of the system soon after its formation. Much more detailed information is expected from a full analysis of the data from Giotto and Vega, and there are plans for a Cometary Rendezvous and Fly-by (CRAF) mission, and even for a mission that will obtain a sample of a cometary nucleus and return it to Earth. However, knowledge already gained over the last fifteen years from the spacecraft investigations of the planets, their satellites, and the comets has enabled us to construct informed views about the past and probable future of Earth and other planets and their atmospheres. These extrapolations are the subject of our next chapter.

Bibliography

The book given as the first reference describes the Solar System as it now appears after more than a decade of planetary exploration. Atmospheres are discussed in this context, and the book makes an excellent introduction to the study of the planets and their atmospheres.

The new solar system. Beatty, J. K., O'Leary, B., and Chaikin, A. (eds.), 2nd edn. Cambridge University Press, Cambridge, 1982.

The photochemistry of atmospheres Levine, J. S. (ed.). (Academic Press, Orlando, 1985).

Atmospheric compositions: key similarities and differences. Pepin, R. O., in *Origin and evolution of planetary and satellite atmospheres* Atreya, S. K., Pollack, J. B., and Matthews, M. S., (eds.). (University of Arizona Press, Tucson, 1989).

Planetary atmospheres. Hunten, D. M., Pepin, R. O., and Owen, T. in *Meteorites and the early solar system* Kerridge, J. F., and Matthews, M. S. (eds.) (University of Arizona Press, Tucson, 1989).

Planetary exploration. *Phil. Trans. R. Soc.* (Lond.) A **303** (1981), special issue.

The atmospheres of the planets. Mason, B. J., *Meteorol. Mag.* **107**, 67 (1978).

Chemical processes in the solar system: a kinetic perspective. McElroy, M. B., *Int. Rev. Sci. Phys. Chem.* Vol. 9 (*Chemical kinetics*), 127 (1975).

Section 8.2

Venus

These references cover most aspects of composition, structure, meteorology, and chemical transformation in the Venusian atmosphere as seen after the Pioneer–Venus and Venera missions.

The atmospheres of Venus, Earth, and Mars: A critical comparison. Prinn, R. G. and Fegley, B., Jr. *Ann. Rev. Planet. Space Sci.* **15**, 171 (1987).

Photochemistry of the atmospheres of Mars and Venus. Krasnopolsky, V. A. (Springer-Verlag, Berlin, 1986).

The photochemistry of the atmosphere of Venus. Prinn, R. G. in *The photochemistry of atmospheres.* Levine, J. S. (ed.). (Academic Press, Orlando, 1985).

The volcanoes and clouds of Venus. Prinn, R. G. *Scient. Am.* **252** (3), 36 (March 1985).

The atmosphere of Venus. Schubert, G. and Covey, C. *Scient. Am.* **245**, 44 (July 1981).

Magellan arrives at Venus. Grimm, R. E., guest editor. Collected papers in special issue (August 1990) of *Geophys. Res. Lett.* **17** (1990).

The Venus atmosphere. Keating, G. M. (ed.). *Adv. Space Res.* **10**, No. 5 (1990).

Venus. (ed. Hunten, D. M., Colin, L., Donahue, T., and Moroz, V. I.). (University of Arizona Press, Tucson, 1983.)

The atmosphere of Venus. Moroz. V. I. *Space Sci. Rev.* **29**, 3 (1981).

Pioneer Venus (special issues). *J. geophys. Res.* **85**, No. A13 (1980); *Science* **205**, 41 *et seq.* (1979); *Science* **203**, 743 *et seq.* (1979).

The deep atmosphere of Venus revealed by high-resolution night-side spectra. Bézard, B, de Bergh, C., Crisp, D., and Maillard, J.-P. *Nature* **345**, 508 (1990).

Detailed discussions of some aspects of composition.

Measurements of the Venus lower atmosphere composition: a critical comparison. Hoffman. J. H., Oyama, V. I., and von Zahn, U. *J. geophys. Res.* **85**, 7871 (1980).

Chemical composition of the atmosphere of Venus. Krasnopolsky, V. A. and Parshev, V. A. *Nature, Lond.* **292**, 610 (1981).

Section 8.2.2

Clouds are discussed at length in the more general references given above. These papers deal with additional points, including high altitude hazes, and clouds on the night side of the planet

Vega mission results and the chemical composition of Venusian clouds. Krasnopolsky, V. A. *Icarus* **80**, 210 (1989).

Venus: mesospheric hazes of ice, dust, and acid aerosols. Turco, R. P., Toon, O. B., Whitten, R. C., and Keesee, R. G. *Icarus* **53**, 18 (1983).

Cloud structure on the dark side of Venus. Allen, D. A. and Crawford, J. W. *Nature, Lond.* **307**, 222 (1984).

A re-examination of the evidence for large, solid particles in the clouds of Venus. Knollenberg, R. G. *Icarus* **57**, 161 (1984).

Large, solid particles in the clouds of Venus: do they exist? Toon, O. B., Ragent, B., Colburn, D., Blamont, J., and Cot, C. *Icarus* **57**, 143 (1984).

Section 8.2.3

Lightning may be of importance in Venusian atmospheric chemistry. The first reference here discusses the general phenomenon of lightning, and the others explore aspects of lightning relevant to Venusian observations.

Lightning generation in planetary atmospheres. Levin, Z., Borucki, W. J., and Toon, O. B. *Icarus* **56**, 80 (1983).

Lightning measurements from the Pioneer Venus orbiter. Scarf, F. L. and Russell, C. T. *Geophys. Res. Lett.* **10**, 1192 (1983).

Comparison of Venusian Lightning Observations. Borucki, W. J. *Icarus* **52**, 354 (1982).

Lightnings and nitric oxide on Venus. Krasnopolsky, V. A. *Planet. Space Sci.* **31**, 1363 (1983).

Production of nitric oxide by lightning on Venus. Levine, J. S., Gregory, G. L., Harvey, G. A., Howell, W. E., Borucki, W. J., and Orville, R. E. *Geophys. Res. Lett.* **9**, 893 (1982).

Section 8.2.4

Sulphur chemistry

International Ultraviolet Explorer observation of Venus SO_2 and SO. Na, C.Y., Esposito, L.W., and Skinner, T.E. *J. geophys. Res.* **95**, 7485 (1990).

Sulfur trioxide in the lower atmosphere of Venus? Craig, R. A., Reynolds, R. T., Ragent, B., Carle, G. C., Woeller, F., and Pollack, J. B. *Icarus* **53**, 1 (1983).

The clouds of Venus: sulfuric acid by the lead chamber process? Sill, G. T. *Icarus* **53**, 10 (1983).

On the possible roles of gaseous sulfur and sulfanes in the atmosphere of Venus. Prinn, R. G. *Geophys. Res. Lett.* **6**, 807 (1979).

Atmosphere–surface interactions on Venus and implications for atmospheric evolution. Khodakovsky, I. L. *Planet. Space Sci.* **30**, 803 (1982).

Sulfur dioxide: episodic injection shows evidence for active Venus volcanism. Esposito, L. W. *Science* **223**, 1072 (1984).

Section 8.2.5

Catalytic Processes in the atmospheres of Earth and Venus. DeMore, W. B. and Yung, Y. L. *Science* **217**, 1209 (1982).

Photochemistry of the stratosphere of Venus: implications for atmospheric evolution. Yung, Y. L. and DeMore, W. B. *Icarus* **51**, 199 (1982).

Venusian airglow and oxygen spectroscopy.

Hot hydrogen and oxygen atoms in the upper atmosphere of Venus and Mars. Nagy, A. F., Kim, J., and Cravens, T. E. *Ann. Geophys.* **8**, 251 (1990).

Identification of the UV nightglow from Venus. Feldman, P. D., Moos, H. W., Clarke, J. T. and Lane, A. L. *Nature, Lond.* **279**, 221 (1979).

Ultraviolet night airglow of Venus. Stewart, A. I. and Barth, C. A. *Science* **205**, 59 (1979).

Spectroscopy of molecular oxygen in the atmospheres of Venus and Mars. Trauger, J. T. and Lunine, J. I. *Icarus* **55**, 272 (1983).

The upper atmosphere and ionosphere of Venus.

The dayside Venus ionosphere. 1. Pioneer-Venus retarding potential analyser experimental observations. Miller, K. L., Knudsen, W. C., and Spenner, K. *Icarus* **57**, 386 (1984).

A two-dimensional model of the ionosphere of Venus. Cravens, T. E., Crawford, S. L., Nagy, A. F., and Gombosi, T. I. *J. geophys. Res.* **88**, 5595 (1983).

The chemistry of metastable species in the Venusian ionosphere. Fox, J. L. *Icarus* **51**, 248 (1982).

Variations in ion and neutral composition of Venus: Evidence of solar control of the formation of predawn bulges in H^+ and He. Taylor, H. A., Jr., Mayr, H., Brinton, H., Niemann, H., Hartle, R., and Daniell, R. E., Jr. *Icarus* **52**, 211 (1982).

Day and night models of the Venus thermosphere. Massie, S. T. Hunten, D. M., and Sowell, D. L. *J. geophys. Res.* **88**, 3955 (1983).

Global empirical model of the Venus thermosphere. Hedin, A. E., Niemann, H. B., Krasprzak, W. T., and Seiff, A. *J. geophys. Res.* **88**, 73 (1983).

A two-dimensional model of the nightside ionosphere of Venus. Bougher, S. W. and Cravens, T. E. *J. geophys. Res.* **89**, 3837 (1984).

Section 8.3

These references, and especially those to the special reports on the Viking missions, cover most of the topics discussed in the section.

Photochemistry of the atmospheres of Mars and Venus. Krasnopolsky, V. A. (Springer-Verlag, Berlin, 1986).

The photochemistry of the atmosphere of Mars. Barth, C. A. in *The photochemistry of atmospheres* Levine, J. S. (ed.). (Academic Press, Orlando, 1985).

The atmosphere of Mars. Leovy, C. B. *Scient. Am.* **237**, 34 (July 1977).

Vikings at Mars: (special reports), *J. geophys. Res.* **82**, No. 28 (30 Sept. 1977).

Photochemistry and evolution of Mars' atmosphere, a Viking perspective. McElroy, M. B., Kong, T. Y., and Yung, Y. L. *J. geophys. Res.*, **82**, 4379 (1977).

Stability of the Martian atmosphere: possible role of heterogeneous chemistry. Atreya, S. K. and Blamont, J. E. *Geophys. Res. Lett.* **17**, 287 (1990).

The composition of the Martian atmosphere. Owen, T. *Adv. Space Res.* **2**, 75 (1982).

Ozone on Mars

Mariner 9 Ultraviolet Spectrometer Experiment: Seasonal variation of ozone on Mars. Barth, C. A., Hord, C. W., Stewart, A. I., Lane, A. L., Dick, M. L., and Anderson, G. P. *Science* **179**, 795 (1973).

The global distribution of O_3 on Mars. Kong, T. Y. and McElroy, M. B. *Planet. Space Sci.* **25**, 839 (1977).

Airglow and ultraviolet spectroscopy

Hot hydrogen and oxygen atoms in the upper atmosphere of Venus and Mars. Nagy, A. F., Kim, J., and Cravens, T. E. *Ann. Geophys.* **8**, 251 (1990).

Mariner 9 Ultraviolet Spectrometer Experiment: Mars airglow spectroscopy and variations in Lyman alpha. Barth, C. A., Stewart, A. I., Hord, C. W., and Lane, A. L. *Icarus* **17**, 457 (1972).

Ultraviolet spectroscopy of the inner solar system from Mariner 10. Broadfoot, A. L. *Rev. Geophys. & Space Phys.* **14**, 625 (1976).

The ionosphere
This reference is also relevant to the Venusian ionosphere.

Ionospheres of the terrestrial planets. Schunk, R. W. and Nagy, A. F. *Rev. Geophys. & Space Phys.* **18**, 813 (1980).

Sections 8.4 to 8.6

The outer solar system

The composition of outer planet atmospheres. Gautier, D. and Owen, T. in *Origin and evolution of planetary and satellite atmospheres.* Atreya, S. K., Pollack, J. B., and Matthews, M. S. (eds.). (University of Arizona Press, Tucson, 1989).

Atmospheres and ionospheres of the outer planets and their satellites. Atreya, S. K. (Springer-Verlag, Berlin, 1986).

Saturn. Gehrels, T., and Matthews, M. S. (eds.). (University of Arizona Press, Tucson, 1984).

The atmospheres of the outer planets. Hunt, G. E., *Annu. Rev. Earth & Planet. Sci.* **11**, 415 (1983).

The atmospheres of the outer planets and satellites. Trafton, L. *Rev. Geophys. & Space Phys.* **19**, 43 (1981).

Jupiter and Saturn. Ingersoll, A. P. *Scient. Am.* **254**, 66 (Dec 1981).

Section 8.4

Reports on the Pioneer 11 and Voyager mission findings

Pioneer 11 at Saturn (special reports). *Science* **207**, 400 (1980); *J. geophys. Res.* **85**, 5651 (1980).

Voyagers at Jupiter and Saturn (special reports): *Jupiter*: Voyager 1, *Science* **204**, 945 *et seq.* (1979): Voyager 2, *Science* **206**, 925 *et seq.* (1979); *Saturn*: Voyager 1, *Science* **212**, 159 *et seq.* (1981): Voyager 2, *Science* **215**, 499 *et seq.* (1981); Voyagers 1, 2, *J. geophys. Res.* **88**, 8625 *et seq.* (1983).

A new analysis of the Jovian 5 micrometre Voyager/IRIS spectra. Lellouch, E., Drossart, P., and Encrenaz, T. *Icarus* **77**, 457; **80**, 224 (1989).

Photochemistry in the atmospheres of the outer solar system

Photochemistry and clouds of Jupiter, Saturn, and Uranus. Atreya, S. K., and Romani, P. N. in *Recent advances in planetary meteorology.* Hunt, G. E. (ed.). (Cambridge University Press, Cambridge, 1985).

The photochemistry of the atmospheres of the outer planets and their satellites. Strobel, D. F. in *The photochemistry of atmospheres*. Levine, J. S. (ed.). (Academic Press, Orlando, 1985).

Composition and chemistry of Saturn's atmosphere. Prinn, R. G., Larson, H. P., Caldwell, J. J., and Gautier, D. in *Saturn*. Gehrels, T. and Matthews, M. S. (eds.). (University of Arizona Press, Tucson, 1984).

Aeronomy of the major planets: photochemistry of ammonia and hydrocarbons. Strobel, D. F. *Rev. Geophys. & Space Phys.* **13**, 372 (1975).

Photochemistry of the reducing atmospheres of Jupiter, Saturn and Titan. Strobel, D. F. *Int. Rev. phys. Chem.* **3**, 145 (1983).

Jovian H_2 dayglow emission. McGrath, M.A., Ballester, G.E., and Moos, H.W. *J. geophys. Res.* **95**, 10365 (1990).

Phosphine photochemistry, and the possibility that there is an interaction with ammonia photochemistry; organophosphorus compounds; arsine chemistry.

Photochemistry of phosphine and Jupiter's great red spot. Noy, N., Pudolak, M., and Bar-Nun, A. *J. geophys. Res.* **86**, 11985 (1981).

Phosphine photochemistry in the atmosphere of Saturn. Kaye, J. A., and Strobel, D. F. *Icarus* **59**, 314 (1984).

Gas phase synthesis of organophosphorus compounds and the atmospheres of the giant planets. Bossard, A. R., Kagama, R., and Rawlin, F. *Icarus* **67**, 305 (1986).

The chemistry of arsine (AsH_3) in the deep atmospheres of Saturn and Jupiter. Fegley, M. B. *Bull. Amer. Astron. Soc.* **20**, 879 (1988).

The abundance of AsH_3 in Jupiter. Noll, K. S., Larson, H. P., and Geballe, T. R. *Icarus* **83**, 494 (1990).

Some minor components of the Jovian atmosphere.

The abundances of ethane and acetylene in the atmospheres of Jupiter and Saturn. Noll, K. S., Knacke, R. F., Tokunaga, A. T., Lacy, J. H., Beck, S., and Serabyn, E. *Icarus* **65**, 257 (1986).

Hydrogen cyanide is detected, and the problem here is with the route to its formation.

HCN formation on Jupiter: The coupled photochemistry of ammonia and acetylene. Kaye, J. A. and Strobel, D. F. *Icarus* **54**, 417 (1983).

The discovery of CO prompted speculations that the source of oxygen might be extraplanetary, although that idea is now discounted. The amounts of oxygen-containing species in the atmospheres are important in determining the existence of oxidized carbon compounds; the references to CO and CO_2 on Titan (see Section 8.5.1 below) are also relevant to a discussion of this point.

The origin and vertical distribution of carbon monoxide on Jupiter. Noll, K. S., Knacke, R. F., Geballe, T. R., and Tokunaga, A. T. *Astrophys. J.* **324**, 1210 (1988).

Chemical constraints on the water and total oxygen abundances in the deep atmosphere of Jupiter. Fegley, M. B., Jr. and Prinn, R. *Astrophys. J.* **324**, 621 (1988).

Detection of carbon monoxide in Saturn. Noll, K. S., Knacke, R. F., Geballe, T. R., and Tokunaga, A. T. *Astrophys. J.* **309**, L91 (1986).

Lightning probably occurs in the atmospheres of the outer planets. See also the first reference for Section 8.2.3.

Moist convection and the abundances of lightning-produced CO, C_2H_2, and HCN on Jupiter. Podolak, M. and Bar-Nun, A. *Icarus* **75**, 566 (1988).

Lightning activity on Jupiter. Borucki, W. J., Bar-Nun, A., Scarf, F. L., Cook, A. F., and Hunt, G. E. *Icarus* **52**, 492 (1982).

The heat of conversion from $p-H_2$ to $o-H_2$ could drive high zonal flows in the atmospheric 'wind' systems. Voyager infrared measurements appear to support the idea of $o-p$ disequilibrium.

Global variation of the *para*-hydrogen fraction in Jupiter's atmosphere and implications for dynamics on the outer planets. Conrath, B. J. and Gierasch, P. J. *Icarus* **57**, 184 (1984).

Dynamical consequences of orthohydrogen-parahydrogen disequilibrum on Jupiter and Saturn. Gierasch, P. J. *Science* **219**, 847 (1983).

Evidence for disequilibrium of ortho and para hydrogen on Jupiter from Voyager IRIS measurements. Conrath, B. J. and Gierasch, P. J. *Nature, Lond.* **306**, 571 (1983).

Satellites of Jupiter and Saturn.
These satellites are an interesting collection of bodies, and several may possess a very tenuous atmosphere, perhaps sweeping through a torus around the parent planet.

The satellites of Jupiter and Saturn. Morrison. D. *Annu. Rev. Astron. Astrophys.* **20**, 469 (1982).

Section 8.5.1

Titan. Owen, T. *Scient. Am.* **246**, 76 (Feb 1982).

Present state and chemical evolution of the atmospheres of Titan, Triton, and Pluto. Lunine, J. I., Atreya, S. K., and Pollack, J. B. in *Origin and evolution of planetary and satellite atmospheres.* Atreya, S. K., Pollack, J. B., and Matthews, M. S. (eds.). (University of Arizona Press, Tucson, 1989).

Voyagers at Titan (special reports). *J. geophys. Res.* **87**, 1351 *et seq.* (1982); *J. geophys. Res.* **88**, 8625 *et seq.* (1983).

Titan's atmosphere from Voyager infrared observations. 1. The gas composition of Titan's equatorial region. Coustenis, A., Bezard, B., and Gautier, D. *Icarus* **80**, 54 (1989).

The composition and origin of Titan's atmosphere. Owen, T., *Planet. Space Sci.* **30**, 833 (1982).

Chemistry and evolution of Titan's atmosphere. Strobel, D. F. *Planet. Space Sci.* **30**, 839 (1982).

Electron impact excitation, and possible N-containing ion chemistry, in Titan's atmosphere.

EUV emission from Titan's upper atmosphere: Voyager 1 encounter. Strobel, D. F. and Shemansky, D. E. *J. geophys. Res.* **87**, 1361 (1982).

Galactic cosmic rays and N_2 dissociation on Titan. Capone, L. A., Dubach, J., Prasad, S. S., and Whitten, R. C. *Icarus* **55**, 73 (1983).

The discovery of the oxidized species CO and CO_2 in the reducing atmosphere of Titan has opened the way to a greater understanding of the detailed chemistry. Continual input is required of an O-source, probably H_2O.

CO_2 on Titan. Samuelson, R. E., Maguire, W. C., Hanel, R. A., Kunde, V. G., Jennings, D. E., Yung, Y. L., and Aikin, A. C. *J. geophys. Res.* **88**, 8709 (1983).

Abundance of carbon monoxide in the stratosphere of Titan from millimeter heterodyne observations. Marten, A., Gautier, D., Tanguy, L., Lecacheux, A., Rekolen, C., and Paubert, G. *Icarus* **76**, 558 (1988).

Titan: discovery of carbon monoxide in its atmosphere. Lutz, B. L., deBergh, C., and Owen, T. *Science* **220**, 1374 (1983).

The upper atmosphere and torus of Titan.

Distribution of molecular hydrogen in atmosphere of Titan. Bertaux, J. L., and Kockarts, G. *J. geophys. Res.* **88**, 8716 (1983).

Titan: upper atmosphere and torus. Strobel, D. F., in *Saturn* (ed. Gehrels, T.) Chapter IV. (University of Arizona Press, 1983.)

Titan's gas and plasma torus. Eviatur, A., and Podolak, M. *J. geophys. Res.* **88**, 833 (1983).

An estimate of the H_2 density in the atomic hydrogen cloud of Titan. Ip, W.-H. *J. geophys. Res.* **89**, 2377 (1984).

Hazes and aerosols in Titan's atmosphere may be the analogue of photochemical smog on Earth.

Vertical distribution of scattering hazes in Titan's upper atmosphere. Rages, K. and Pollack, J. B. *Icarus* **55**, 50 (1983).

Size estimates of Titan's aerosols based on Voyager high-phase-angle images. Rages, K., Pollack, J. B., and Smith, P. H. *J. geophys. Res.* **88**, 8721 (1983).

Inhomogeneous models of Titan's aerosol distribution. Podolak, M., Bar-Nun, A., Noy, N., and Giver, L. P. *Icarus* **57**, 72 (1984).

Titan: far infrared and microwave remote sensing of methane clouds and organic hazes. Thompson, W. R. and Sagan, C. *Icarus* **60**, 236 (1984).

Production and condensation of organic gases in the atmosphere of Titan. Sagan, C. and Thompson, W. R. *Icarus* **59**, 133 (1984).

An update of nitrile chemistry on Titan. Yung, Y. L. *Icarus* **72**, 468 (1987).

There has been speculation about the possibility of atmospheric chemistry maintaining enough C_2H_6 to supply a condensed 'ocean' on the satellite. The first report shows that CH_4 does not form oceans or global cloud, although there may be a methane ice haze high in the troposphere.

Is Titan wet or dry? Eshelman, V. R., Lindel, G. F., and Tyler, G. L. *Science* **221**, 53 (1983).

Oceans on Titan? Flassar, F. M. *Science* **221**, 55 (1983).

Ethane ocean on Titan. Lunine, J. I., Stevenson, D. J., and Yung, Y. L. *Science* **222**, 1229 (1983).

Section 8.5.2

Io. Johnson, T. V. and Soderblum, L. A. *Sci. Amer.* **249**, 60 (Dec. 1983).

Io's tenuous atmosphere. Johnson, T. V., and Matson, D. L. in *Origin and evolution of planetary and satellite atmospheres.* Atreya, S. K., Pollack, J. B., and Matthews, M. S. (eds.). (University of Arizona Press, Tucson, 1989.)

The 2.5–5.0 micrometre spectra of Io: evidence for H_2S and H_2O frozen into SO_2. Salama, F., Allamandola, L. J., Witterborn, F. C., Cruikshank, D. P., Sandford, S. A., and Bregman, J. D. *Icarus* **83**, 66 (1990).

The global distribution, abundance, and stability of SO_2 on Io. McEwen, A. S., Johnson, T. V., Matson, D. L., and Soderblom, L. A. *Icarus* **75**, 450 (1988).

Sulfur dioxide on Io: spatial distribution and physical state. Howell, H. R., Cruikshank, D. P., and Fanale, F. P. *Icarus* **57**, 83 (1984).

Ejection of sodium from sodium sulfide by the sputtering of the surface of Io. Chrisey, D. B., Johnson, R. E., Boring, J. W., and Phipps, J. A. *Icarus* **75**, 233 (1988).

Io meteorology: how atmospheric pressure is controlled locally by volcanoes and surface frosts. Ingersoll, A. P. *Icarus* **81**, 298 (1989).

Escape of sulfur and oxygen from Io. Cheng, A. F. *J. geophys. Res.* **89**, 3939 (1984).

The SO_2 atmosphere of Io: ion chemistry, atmospheric escape, and models corresponding to the Pioneer 10 radio occultation measurements. Kumar, S. *Icarus* **61**, 101 (1985).

No sulfur flows on Io. Young, A. T. *Icarus* **58**, 197 (1984).

Io's atmosphere from microwave detection of SO_2. Lellouch, E., Belton, M., de Pater, I., Gulkis, S., and Encrenaz, T. *Nature* **346**, 639 (1990).

Section 8.6

Even in the outermost reaches of the System, bodies such as Triton and Pluto may possess atmospheres.

Voyagers at Uranus and Neptune (special reports): *Uranus: Science* **233**, 39 *et seq.* (1986); *J. geophys. Res.* **92**, 14873–15375 (1987); *Neptune: Science* **246**, 1417 (1989).

Uranus deep atmosphere revealed. de Pater, I., Romani, P. N., and Atreya, S. K. *Icarus* **82**, 288 (1989).

Uranus. Ingersoll, A. P. *Sci. Amer.* **256**(1), 20 (Jan 1987).

Nature of the stratospheric haze on Uranus: evidence for condensed hydrocarbons. Pollack, J. B., Rages, K., Pope, S. K., Tomasko, M. G., Romani, P. N., and Atreya, S. K. *J. geophys. Res.* **92**, 15037 (1987).

Chemical models of the deep atmosphere of Uranus. Fegley, M. B. Jr. and Prinn, R. *Astrophys. J.* **307**, 852 (1986).

Methane abundance in the atmosphere of Uranus. Teifel, V. G. *Icarus* **53**, 389 (1983).

Methane photochemistry and haze production on Neptune. Romani, P. N., and Atreya, S. K. *Icarus* **74**, 424 (1988).

Neptune's atmosphere revealed. Romani, P. N., de Pater, I., and Atreya, S. K. *Geophys. Res. Letts.* **16**, 933 (1989).

Stratospheric hazes from CH_4 photochemistry on Neptune. Romani, P. N., and Atreya, S. K. *Geophys. Res. Letts.* **16**, 941 (1989).

Primordial matter in the outer solar system: a study of its chemical composition from remote spectroscopic analysis. Encrenaz, T. *Space Sci. Revs.* **38**, 35 (1984).

Formation of the satellites of the outer solar system: sources of their atmospheres. Coradini, A., Cerroni, P., Magni, G., and Federico, C. in *Origin and evolution of planetary and satellite atmospheres*. Atreya, S. K., Pollack, J. B., and Matthews, M. S. (eds.). (University of Arizona Press, Tucson, 1989).

Triton: do we see to the surface? Cruikshank, D. I., Brown, R. H., Giver, L. P., and Tokumaja, A. T. *Science* **245**, 283 (1989).

Triton's streaks as windblown dust. Sagan, C. and Chyba, C. *Nature* **346**, 546 (1990).

Triton stratospheric molecules and organic sediments. Thompson, W. R., Singh, S. K., Khare, B. N., and Sagan, C. *Geophys. Res. Letts.* **16**, 981 (1989).

Volatiles on Triton: the infrared spectroscopic evidence, 2.0 to 2.5 micrometers. Cruikshank, D. P., Brown, R. H., Tokunaga, A. T., Smith, R. G., and Piscitelli, J. R. *Icarus* **74**, 413 (1988).

Nitrogen on Triton. Cruikshank, D. P., Brown, R. H., and Clark, R. N. *Icarus* **58**, 293 (1984).

Triton: A satellite with an atmosphere. Cruickshank. D. P. and Silvaggio, P. M. *Astrophys. J.* **233**, 1016 (1979).

Frigid oceans for Triton and Titan. Kerr, R. A., *Science* **221**, 448 (1983).

Pluto. Binzel, R. *Scient. Am.* **262**, 26 (June 1990).

Pluto's atmosphere. Elliott, J. L., Dunham, E. W., Bosh, A. S., Slivan, S. M., Young, L. A., Wasserman, L. H., and Millis, R. L. *Icarus* **77**, 148 (1989).

Pluto's atmosphere near perihelion. Trafton, L. M. *Geophys. Res. Letts.* **16**, 1213 (1989).

Upper limits on possible photochemical hazes on Pluto. Stansbury, J. A. *Geophys. Res. Letts.* **16**, 1221 (1989).

Methane absorption variations in the spectrum of Pluto. Buie, M. W., and Fink, U. *Icarus* **70**, 483 (1987).

Observations of Pluto and Charon by IRAS. Sykes, M. V., Cutri, R. M., Lebofsky, L. A., and Binzel, R. P. *Science* **237**, 1336 (1987).

Constraints on bulk composition, seasonal variation, and global dynamics of Pluto's atmosphere. Stern, S. A. and Trafton, L. *Icarus* **57**, 231 (1984).

The surface and atmosphere of Pluto. Cruickshank, D. P. and Silvaggio, P. M., *Icarus* **41**, 96 (1980).

Does Pluto have a substantial atmosphere? Trafton, L. *Icarus* **44**, 53 (1980).

Stability of Pluto's atmosphere. Hunten, D. M. and Watson, A. J. *Icarus* **51**, 665 (1982).

See also references under Sections 8.4 and 8.5

Section 8.7

The origin of comets. Bailey, M.E., Clube, S.V.M., and Napier, W.M. (Pergamon Press, Oxford, 1990).

Encounters with comet Halley (special reports). *Nature* **321**, supplement to no. 6067 (1986).

Giotto encounter with comet Halley (special reports). *Astron. Astrophys.* **187**, nos. 1–2 (1987).

The composition of comets. Jessberger, E. K., Kissel, J., and Rahe, J. in *Origin and evolution of planetary and satellite atmospheres.* Atreya, S. K., Pollack, J. B., and Matthews, M. S. (eds.). (University of Arizona Press, Tucson, 1989.)

Solar nebula origin for volatile gases in Halley's comet. Engel, S., Lunine, J.I., and Lewis, J.S. *Icarus* **85**, 380 (1990).

Did comets form from unaltered interstellar dust and ices? The evidence from infra-red spectroscopy. Tokunaga, A.T. and Brooks, T.Y., *Icarus* **86**, 208 (1990).

Comets and their composition. Spinrad, H. *Ann. Rev. Astron. Astrophys.* **25**, 231 (1987).

Composition measurements and the history of cometary matter. Geiis, J. *Astron. Astrophys.* **187**, 859 (1987).

The organic component in dust from comet Halley as measured by the PUMA mass spectrometer on board Vega 1. Kissel, J., and Krueger, F. R. *Nature* **326**, 755 (1987).

On the temperature and gas composition in the region of comet formation. Bar-Nun, A. and Kleinfeld, I. *Icarus* **80**, 243 (1989).

Observations of cometary nuclei. A'Hearn, M. F. *Ann. Rev. Earth Planet. Sci.* **16**, 273 (1988).

A postencounter view of comets. Mendis, D. A. *Ann Rev. Astron. Astrophys.* **26**, 11 (1988).

Sources of cometary radicals and their jets: gases or grains. Combi, M. R. *Icarus* **71**, 178 (1987).

Abundances in comet Halley at the time of the spacecraft encounters. Wyckoff, S., Tegler, S., Wehinger, P. A., Spinrad, H., and Belton, M. J. S. *Astrophys. J.* **325**, 927 (1988).

Gaseous CN, C_2 and C_3 jets in the inner coma of comet P/Halley observed from the Vega 2 spacecraft. Clairemidi, J., Moreels, G., and Krasnopolsky, V.A., *Icarus* **86**, 115 (1990).

Comet Halley $O(^1D)$ and H_2O production rates. Magee-Sauer, K., Scherby, F., Roesler, F. L., and Harlander, J. *Icarus* **84**, 154 (1990).

Water clusters in the coma of comet Halley and their effect on gas density, temperature, and velocity. Crifo, J. F. *Icarus* **84**, 414 (1990).

Giotto IMS measurements of the production rate of hydrogen cyanide in the coma of comet Halley. Ip, W.-H., Balsiger, H., Geiss, J., Goldstein, B. E., Kettmann, G., Lazarus, A. J., Meier, A., Rosenbauer, H., Schwenn, R., and Shelley, E. *Ann. Geophys.* **8**, 319 (1990).

Ultraviolet spectroscopy and the composition of cometary ice. Feldman, P. D. *Science* **219**, 347 (1983).

Chemistry in comets. Lüst, R. *Topics curr. Chem.* **99**, 73 (1981).

9 Evolution and change in atmospheres and climates

9.1 Sources of atmospheric constituents

9.1.1 Origin and development of atmospheres

Atmospheric compositions can offer valuable clues about the earliest processes in the formation of the solar system and its planets. Geological records are available for Earth, and in a very superficial way for Mars, and they inform us about the development of the planetary crust and of crust–hydrosphere–atmosphere interactions. Planetary probes have returned data that complement chemical and isotopic analyses of meteorites and lunar samples to provide models of the primitive solar nebula and planetary accretion. Some planets (e.g. Jupiter) appear to have retained the *primordial* atmosphere with which they were created. Others (e.g. Mars or the satellite Titan) seem to have atmospheres that have undergone much change. Evolution of our own atmosphere is of more than parochial interest, because the development of an environment suitable for the creation and support of life seems unique in the solar system. With life came the growth of our oxygen-rich atmosphere as it is today (Chapter 1). We have recently become increasingly aware that our atmosphere is susceptible to change both because of extraterrestrial influences, such as changing solar intensity or collision with asteroids or comets, and because of man's activities. Climates, depending as they do on atmospheric composition and temperature, can alter as atmospheres evolve. In the case of our own planet, modifications in climate may be partially controlled by man's deliberate or accidental intervention.

Radioactive dating of meteorites and lunar samples gives ample evidence that the solar system is 4.6 Gyr (4.6×10^9yr) old. Ever since the 'Big Bang' that created the Universe 10–20 Gyr ago, the Universe has been expanding and evolving. During the course of this evolution, nuclear fusion has transformed hydrogen and helium, present from the beginning, into heavier elements such as C, N, O, Mg, Si, and Fe which make up a planet like Earth. Supernova explosions—the death throes of the massive stars—produce elements heavier than iron, and scatter and disperse the heavier elements through the galaxies as tiny dust grains of 10–1000 nm diameter. Spectroscopic investigation suggests that the particles are mainly graphite (C), H_2O ice, and iron and magnesium silicates. Most of the mass of any galaxy still remains as hydrogen and helium, the heavier elements constituting only a very small fraction (about one per cent).

Dust and gas in the Universe are concentrated in the arms of spiral galaxies in which new stars are formed when local regions become sufficiently dense that attractive gravitational forces dominate internal gas pressures. Collapse of the material starts, and a flattened spinning disk—a *solar nebula*—is produced. Compression of the hydrogen gas at the centre of the nebula is ultimately sufficient for fusion processes to be sustained, and a star, in our case the Sun, is formed.

While the concept of a rotating disc-like nebula is common to most current models of the origin of the solar system, there is considerable divergence of opinion about the next stage in the evolution of the solid bodies. One picture maintains that the dust grains accreted into metre-sized *planetesimals* that were the building blocks of which the inner planets were made. Such planetesimals were largely silicate materials in the inner solar system, since ices would have evaporated at the high local temperatures. Collisional processes leading to planet formation were favoured for large objects, so that planets grew at the expense of the multitude of planetesimals by sweeping them up. Many planetesimals are still located in the asteroid belt where Jupiter's gravitational field just cancels the forces of accretion. Material equilibrating and accreting at high temperatures will contain proportionately more iron than material equilibrating at lower temperatures. Taking into account the effects of compression, the densities of Mercury, Venus, Earth, and Mars follow the decrease in density expected from their distances from the Sun. Grains of H_2O, NH_3, and CH_4 ices were available as well as silicates in the outer solar system. Jupiter and Saturn may have grown sufficiently rapidly that they attracted large quantities of hydrogen and helium from the nebula. Further out, Neptune and Uranus were less efficient at capturing the nebular gases. Present-day compositions of the outer planets in terms of H_2 and He are thought to be 80, 70, 15, and 10 per cent for Jupiter, Saturn, Uranus, and Neptune. Astronomical observations of stars that are now young suggest that our early Sun passed through a phase of vigorous solar activity in which a violent solar wind carried off a sizeable amount of the Sun as well as sweeping away remaining gases that had not yet condensed, or accumulated on a solid body.

Events following the accumulation of dust and gas to form larger bodies are entirely different for the inner and the outer planets. The gaseous planets must have undergone an initial period of hot contraction, after which there has probably been little further change. Impact of infalling asteroidal, meteoric, and cometary bodies on the unprotected surface, together with decay of short-lived radioactive elements, heated the inner planets so much that they must have been molten to a considerable depth. Release of gases from the interior (*outgassing*) may have been significant since the intense heating could have led to the dissociation of minerals containing bound H_2O and CO_2, as well as to degassing of physically trapped gases. Atmospheric outgassing is thus tied to the thermal evolution of a planet, and the present composition of the planetary atmosphere provides an indicator of past history. Venus, Earth, Mars and Titan have atmospheres very different from the composition of the Sun or of the primitive solar nebula, as a result of loss of the primary atmosphere and the evolution of a second one.

9.1.2 Interstellar clouds and their chemistry

Stars and their planets account for about 90 per cent of the mass of our own galaxy, the remainder being scattered rather unevenly throughout space. Much of the interstellar medium is very tenuous, possessing between one and a hundred atoms or molecules for every 1000 cm³. In some places much more matter has accumulated to form *interstellar clouds*, of which *diffuse* and *dense* categories are recognized. Particle densities are typically $10^2 cm^{-3}$ in the diffuse clouds, and may be as high as $10^6 cm^{-3}$ in the dense ones, while the temperature may be from 100 K to perhaps as low as 10 K. While the fascinating subject of interstellar cloud chemistry lies largely outside the scope of our present enquiry, there are analogies with processes occurring in planetary atmospheres that should not go unnoticed. Several references are given in the Bibliography that provide a more extensive introduction to the subject.

For nearly fifty years, it has been realized that molecular as well as atomic species were present in the interstellar clouds, visible and ultraviolet absorption spectroscopy identifying several diatomic species. Chemical change is thus occurring in the clouds. Less than thirty years ago, however, it was thought out of the question that complex polyatomic molecules, especially highly evolved organic ones, could ever exist in such cold and relatively thin regions of space. Radio and millimetre wave astronomy has now completely altered our picture of interstellar chemistry. More than 80 molecular species, including at least seven positive ions, had been identified by 1989 as a result of increasingly sophisticated observational techniques and improved spectral data derived from laboratory experiments. The list includes many large polyatomic species as well as diatomics, and new molecules are being added constantly as observations proceed. Polyatomic species detected include HCN, $HCHO$, C_2H_2, C_3O, $HNCO$, $HCOOH$, NH_2CN, CH_3CN, CH_3NH_2, CH_3CHO, C_2H_5OH, and the cyanopolyacetylenes, $H(C \equiv C)_n CN$, with n up to 5. The 'molecular' dense clouds are the most massive objects in the galaxy, possessing up to 10^5 times the mass of our Sun, and they are very active regions in which new stars are being formed continuously. Chemical compounds evolving in the clouds could therefore be present in the gaseous envelopes surrounding any planetary objects formed together with the stars. The compounds may also become incorporated into the frozen volatile materials that probably partially make up the comets. The presence in space of the building blocks of conventional organic synthesis has obvious and interesting implications. Our concern here is primarily with how the molecular species come to be formed in the interstellar clouds, and how they survive there.

Survival of complex molecules is, in fact, intimately bound up with the density of the clouds in which they are found. Starlight provides a galactic ultraviolet flux that would photodissociate and photoionize all polyatomic molecules were they not protected in some way. For cloud densities greater than a few times 10^2cm^{-3}, most of the hydrogen in the interstellar medium is in molecular form (possibly as a result of heterogeneous recombination of H atoms on interstellar grain surfaces). Accumulation of H_2, and perhaps dust grains, ensures that the dense clouds are opaque to the galactic ultraviolet radiation, so that the clouds and the molecules within them become self-shielded from photolysis. It is, indeed, the opacity of the clouds to visible and shorter wavelength radiation that prevented the detection of complex interstellar molecules by conventional optical techniques. Radio-frequency and microwave radiation is not strongly attenuated by H_2 or dust grains, and so can be used to probe the composition of the dense clouds.

Ion-molecule reactions appear to be responsible for the synthesis of a large number of the polyatomic molecules found in dense interstellar clouds, so that ionic processes are seen to play a very significant part in the chemistry of interstellar clouds just as they do in planetary atmospheres (Chapter 6). As explained in Section 3.4.1, collisions between ions and molecules are several times more rapid than collisions between neutral partners. Dipolar fields induced by the ion in the neutral collision partner lead to much longer range attractive forces than are provided by the van der Waals attraction between two neutral species. Not only is the collision frequency enhanced, but spiralling and multiple impact encounters facilitate formation of long-lived intermediate complexes that may be involved in bimolecular and associative ion–molecule reactions. The energy gained in bringing ion and neutral together is often enough to overcome any kinetic activation barrier to reaction, so that most ion–molecule reactions have little if any activation energy. At the temperatures of the interstellar clouds, most reactions between neutral partners would be at a standstill. A few radical–radical reactions of low activation energy form the only exceptions.

We have reviewed important types of ion chemistry in Section 6.2 and Table 6.1, and the principles expounded apply to interstellar as well as to ionospheric chemistry. One difference is that *radiative* association, recombination, and attachment processes were discounted as being of negligible importance in planetary atmospheres, because their rates are so small in comparison with other available reactive pathways. In the interstellar clouds, the temperatures, concentrations, and time-scales may actually make the radiative processes more important than non-radiative ones.

Ionization itself is provided in the dense clouds by interaction of galactic cosmic rays with the most abundant species, hydrogen and helium. It is not our purpose here to give a detailed account of the route followed in the formation of each complex molecule from the primary ions, but it is worthwhile to try to identify some of the more important steps. For every H_2

molecule ionized, an H_3^+ ion is produced, whose function in more complex ion chemistry is to produce OH, H_2O protonated species, and NH_2^+ and NH_3^+. Carbon in the clouds is mainly in the form of atoms and CO; C^+ ions are, however, one of the most abundant ionic species present, because collision of He^+ with CO generates the atomic ions. The C^+ ion is a cornerstone in the formation of organic species. For example, reactions with NH_2 or NH_3 yield HCN^+ and H_2CN^+ ions that lead to HCN in a manner already discussed in connection with Titan's atmosphere (Section 8.5.1). Synthesis of C–C bonds is achieved by the attack of C^+ on carbon-containing neutral radicals to yield (directly or indirectly) first C_2H^+ then $C_2H_2^+$ and finally the ethynyl radical, C_2H, invoked in much of the neutral chemistry of Jupiter, Saturn, and Titan (Sections 8.4, 8.5.1). From that point on, it is obviously possible to postulate routes to most of the polyatomic species observed in the dense clouds. Some of the longest carbon chains are to be found in the cyanopolyacetylenes that can be synthesized from CN, C_2H, and C_2H_2, species all actually detected in the interstellar clouds. Tricarbon monoxide, C_3O, is the first interstellar carbon chain molecule to be detected that contains oxygen.

Abundances of the rare stable isotopes of several elements (e.g., D, ^{13}C, ^{15}N) relative to the more common isotope appear in certain interstellar molecules to be enriched over the abundances typical of the solar system. It is important to know if the isotope ratios in those molecules reflect the ratios in the entire cloud, or if some isotope-selective chemistry has favoured formation of the enriched molecules. Laboratory studies of isotope exchange in ion-neutral reactions have shown that chemical fractionation can indeed explain the apparent isotopic enrichment in some interstellar molecules. We shall meet the subject of isotopic enrichment again in Section 9.3, but for the time being we should heed the lesson learned from the interstellar studies that isotope ratios measured from a particular molecule may not be representative of the overall abundance of an element's isotope.

9.2 Noble gases and nitrogen in planetary atmospheres

9.2.1 Inner planets

Noble gases in planetary atmospheres provide valuable pointers concerning atmospheric origins and evolution. Except for the lightest noble gas, helium, they cannot readily escape to space from any of the inner planets, Venus, Earth, or Mars. Chemical inertness prevents loss of the noble gases (with the possible exception of xenon) to surface rocks. Abundances of the gases thus correspond to the cumulative quantity of gas present in and released to the atmosphere over the entire history of the planet. Two types of noble gas can be distinguished: primordial and radiogenic. Primordial isotopes such as ^{20}Ne, ^{36}Ar, ^{38}Ar, ^{84}Kr, and ^{132}Xe were present in the solar system from the

Table 9.1   Abundances of noble gases[a].

Gas	^{20}Ne	^{36}Ar	^{36}Ar/^{12}C	^{36}Ar/^{38}Ar	^{40}Ar/^{36}Ar[b]	^{20}Ne/^{22}Ne
	← kg kg^{-1} →		←		number ratios	→
Object						
Sun	2.2×10^{-3}	9.0×10^{-5}	2.3×10^{-2}	5.6	< 1.0	13.7
CCI[c]	2.9×10^{-10}	1.3×10^{-9}	3.4×10^{-8}	5.3	–	8.9
Venus	2.9×10^{-10}	2.5×10^{-9}	9.7×10^{-5}	5.6	1.0	11.8
Earth	1.0×10^{-11}	3.5×10^{-11}	2.3×10^{-6}	5.3	296	9.8
Mars	4.4×10^{-14}	2.2×10^{-13}	1.9×10^{-5}	4.1	2840	10.1

[a] Compiled from data given by Pepin, R. O. in *Origin and evolution of planetary and satellite atmospheres*, Atreya, S. K., Pollack, J. B., and Matthews, M. S. (eds.). (University of Arizona Press, Tucson, 1989.) Pepin quotes errors and indicates the uncertainties in measurements and interpretations.

[b] Argon isotope ratios from Pollack, J. B. and Black, D. C. *Icarus* **51**, 169 (1982).

[c] CI carbonaceous chondrites, a class of meteorite.

time of its creation. Radiogenic isotopes, however, have built up from the decay of radioactive nucleides: ^{40}Ar from the decay of ^{40}K, and ^{4}He from the decay of ^{232}Th, ^{235}U, and ^{238}U.

Before the Viking and Pioneer–Venus missions (see Chapter 8), the only information we had about the isotopic composition of planetary atmospheres was for Earth. Table 9.1 shows that radiogenic argon is nearly 300 times more abundant than the primordial isotope. In comparison with solar abundances, on a mass per unit mass basis, the Earth's atmosphere is depleted of ^{36}Ar by a factor of more than two million. For ^{20}Ne, the depletion is 220 million, but for carbon, mainly bound up in involatile compounds, the depletion is only 260. Evidence of this kind is taken as clear proof that Earth has lost almost all its primordial atmosphere, if such an atmosphere existed at all, and that the present atmosphere has been acquired later. In 1975, it appeared that the depletion of noble gases would be even more marked for the hotter planet Venus, but that Mars, having formed in a cooler part of the solar system, might have retained more of its primordial components. Viking (1976) dispelled that idea for Mars, as can be seen in Table 9.1. Natural isotopes of argon and neon are even more deficient on Mars than on Earth, and the radiogenic isotopes relatively more important. Pioneer–Venus (1978) showed that the Mars results were not a freak, but that there was a real tendency for there to be greater abundances of the noble gases in the atmospheres of the planets closer to the Sun. Table 9.1 shows that the relative abundance of ^{20}Ne to ^{36}Ar is comparable on the three planets (in the range 0.1 to 0.3), but much less than the solar ratio (about 25). The patterns of abundances of the noble gases are seen in Fig. 9.1, which also illustrates the decreasing primordial gas residue on the planets at greater distances from the Sun. The depletions

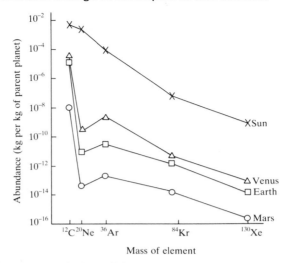

Fig. 9.1. Abundances of primordial gases observed in the atmospheres of Venus, Earth, and Mars, and solar abundances. [Data from same source as Table 9.1].

relative to solar abundances are greatest for the lightest elements. Carbon is much less depleted on the planets than are the noble gases. The planetary patterns for the noble gases clearly resemble each other much more closely than they do the Sun's pattern.

Having established that the atmospheres of Venus, Earth, and Mars are not primordial remnants, it is necessary to examine ways in which the planets could have obtained secondary atmospheres. Several hypotheses exist, which can be broadly classified as *solar nebula, solar wind, comet–asteroid,* and *accretion.* The first two of these hypotheses argue for gravitational capture and retention by the planets after their formation either of gases of the primordial solar nebula or of the solar wind that has flowed over them during their lifetimes. The differences in ratios of noble gases between the Sun and the planets that we have discussed in the last paragraph are evidence against these mechanisms. Substantial numbers of small bodies have impacted with the planets of the inner solar system over the planetary lifetimes, and the comet–asteroid hypothesis proposes that atmospheres were brought to the planets as a result of such impacts. However, Venus and Earth have a roughly equal chance of encountering comets and asteroids, and yet Earth has nearly two orders of magnitude less ^{36}Ar on a mass for mass basis than Venus, thus suggesting that the comet–asteroid hypothesis cannot account for a substantial proportion of the present-day atmospheres. The remaining hypothesis, which is that volatile materials were incorporated into the planet as it accreted, thus seems the most probable. If the planetesimals that formed the planets contained small amounts (perhaps a fraction 10^{-4} by mass) of volatile materials then gases could be released from within the planet as it heated up. One explanation for the large excess of non-radiogenic noble gases on Venus compared to Earth could be that the planetesimals that formed Venus were exposed to an intense solar wind that was absorbed before it reached the part of the solar system where Earth (or Mars) formed. In this context, it is interesting that the pattern of noble gas abundances for Venus (Fig. 9.1) shows a hint of a solar modification of the Earth's pattern; too much emphasis should not be given to this observation, however, since it hinges on a disputed abundance for ^{84}Kr.

The gaseous components of present-day meteorites are of interest because they may reflect the composition of the primitive materials out of which the planets accreted, as well as providing an indicator of the materials present in the solar system that are available for impact degassing (as required by the comet–asteroid hypothesis). For this reason, gas composition data for one important class of meteorite is presented in Table 9.1. The striking similarity in the ^{20}Ne/^{36}Ar ratios for the CI chondrites and the planetary atmospheres has prompted speculation that there may have been a single type of parent gas reservoir, with the present small spread of abundances determined by evolutionary processes. One model based on this idea envisages late-accreting volatile-rich planetesimals to form a 'veneer' on the planetary surface. However, the measured isotopic ratios pose severe constraints on a single type of volatile mass distribution. The ratio of ^{20}Ne/^{22}Ne is significantly higher in the Venusian than the terrestrial atmosphere (and than the ratio for CI chondrites), as shown in Table 9.1, and the ratio of ^{36}Ar/^{38}Ar on Mars is anomalously low compared with that found on other bodies of the solar system. It seems, then, that the similar elemental but disparate isotopic compositions cannot be a result of accretion of planetesimals with constant inventories of volatile species, at least if the compositions were to resemble those of present-day meteorites. The explanation of the similarities, as well as the differences, of atmospheric composition, must therefore be a coincidental result of fractionation and mixing processes that operated, both before and after accretion, on source gases of different origins. Escape of early solar-composition atmospheres from planetesimals and planets is currently thought to be the most likely mechanism that could have achieved the requisite fractionation. Whatever the detailed processes turn out to be, it is clear that the new atmospheric measurements obtained by the planetary missions have provided the basis for reasonable speculation about the origin of the planets and their atmospheres.

9.2.2 Titan

Considerable excitement followed the discovery of N_2 as the major component of Titan's atmosphere (Section 8.5.1) because of the possibility that Titan might resemble Earth in an early, or at least frozen, state of development.

Insight into our own evolution might therefore be achieved by an understanding of the development of Titan's atmosphere to its present form. The first question is how did Titan, 50 times less massive than Earth, acquire so much gas from the preplanetary nebula that it now has an atmosphere one and a half times as dense as Earth's? Capture from the solar nebula (or an unfractionated proto-Saturn nebula) is excluded by the mean molecular weight (i.e. R.M.M.) of the atmosphere, which is slightly larger than 28. This value is not consistent with large amounts of neon in the atmosphere, although cosmic abundances would imply an atmospheric composition of 67 per cent Ne and 33 per cent N_2, with a mean molecular weight of 22.6. Evidently the atmosphere represents a devolatilization of the ices and rocky material that accreted to form the satellite. Two possibilities can now be considered: either NH_3 may be the parent molecule that has undergone photolysis to yield the observed N_2, or N_2 could have been trapped in molecular form. We shall examine these alternatives.

Current models of the Saturnian and solar nebulae do not offer much guidance, some predicting predominantly NH_3 and others predicting predominantly N_2 in the Saturnian nebula during Titan's formation. Ammonia photolysis can form hydrazine (Section 8.4):

$$NH_3 + hv \rightarrow NH_2 + H, \tag{9.1}$$

$$NH_2 + NH_2 \xrightarrow{\text{M}} N_2H_4. \tag{9.2}$$

Hydrazine itself yields N_2 as a photolysis product:

$$N_2H_4 + hv \rightarrow N_2H_3 + H, \tag{9.3}$$

$$N_2H_3 + N_2H_3 \rightarrow N_2 + H_2 + N_2H_4, \tag{9.4}$$

but these processes have to compete with physical removal of N_2H_4 by condensation. As we saw in Section 8.5.1, the present thermal structure of Titan would suggest that NH_3 photolysis generates only N_2H_4. An obvious requirement, if N_2 is to be formed from NH_3, is that during some period of Titan's evolution it was substantially warmer than it is now. Any warm period must have been long enough to yield Titan's N_2, but must obviously have been shorter than the age of the satellite. A surface temperature of 150 K or greater for 1.6×10^8 years suffices and the restriction on time-scale is not contravened. If, as is thought likely, solar ultraviolet fluxes were once much higher than they are now, the increased photolysis rate could reduce the time needed for NH_3 conversion to below 100 000 years.

The alternative hypothesis, that nitrogen was always present as a satellite material in the form of N_2, also turns out to be viable, and in some ways is more attractive since it does not demand an unproven thermal history. Molecular nitrogen cannot be condensed as pure solid N_2, but if Titan were formed at a nebular temperature of ~ 60 K, N_2 would condense out preferentially as a solid clathrate hydrate, $N_2.7H_2O$. Ammonia, methane, and other gases were also probably trapped as water clathrate ices. At $T \sim 60$ K, the vapour pressure of neon is > 40 bars, so that the formation of neon hydrate is out of the question, and the absence of neon in Titan's atmosphere is easily explained. Argon, on the other hand, is a probable constituent of the atmosphere, because it is the only species likely to raise the mean molecular weight to ~ 28.6. Mixing ratios in the upper atmosphere are less than 6 per cent, but could be up to 25 per cent near the surface, where a high vapour pressure of CH_4 (R.M.M. $= 16$) would need to be compensated by relatively large amounts of Ar to maintain the mean R.M.M.

The actual amount of argon present provides a test of the origin of the atmospheric nitrogen. Nitrogen from nebular gases captured as clathrate must be accompanied by argon. Clathrate thermodynamics suggest that $[Ar]/[N_2]$ should be $1-10$ per cent if all present-day N_2 is derived from clathrate. On the other hand, if most of the N_2 is a product of photochemical or chemical processing of NH_3, then the argon released from the clathrate is relatively much less important, perhaps making up as little as 10^{-3} per cent. Direct measurements of argon in Titan's atmosphere may thus provide a diagnostic test of the two models of the origin and early evolution of Titan's atmosphere.

9.3 Isotopic enrichment

Studies of noble gases and their isotopes in planetary atmospheres are particularly rewarding because of the chemical inertness of the elements over geologic time. Abundances of reactive elements also provide useful information, especially if comparisons are made between natural isotopes of the same element, since chemical losses will have been the same. We have seen an example of this kind of study in discussing the loss of nitrogen from Mars (Section 8.3.3). Escape of the ^{14}N isotope is slightly faster than that of ^{15}N because of the lower mass. With the atmosphere as the reservoir of nitrogen, the N_2 remaining will have become slowly richer in ^{15}N over the life of the planet. In comparison with nitrogen on Earth, where neither isotope escapes, the Martian ^{15}N is enriched by a factor of 1.6. It follows that Mars once had a much larger nitrogen atmosphere than it has now (by a factor of ten or more). Continuous degassing of nitrogen from the planet's interior would tend to sustain the original isotopic ratio, so that the observed enrichment favours an evolutionary model in which Mars acquired its nitrogen atmosphere early in its history, with relatively little degassing in later epochs. By way of contrast to the nitrogen isotopes, the Martian $^{16}O/^{18}O$ ratio is almost exactly the same as that for Earth, and ^{18}O has been enriched by less than 5 per cent. Yet Mars is losing O atoms at present at the rate of 6×10^7 atoms s^{-1} for every square centimetre of surface (cf. Sections 8.3.2, 8.3.3). Lack of ^{18}O enrichment implies a source of 'new' oxygen, in a reservoir holding at least 4.5×10^{25} atoms cm^{-2}, presumably in the form of H_2O, since the escape of hydrogen and oxygen from Mars is constrained to have a 2 : 1 stoicheiometry. There is, however, apparently some enrichment of D over H. Preliminary analysis of infra-red measurements of HDO concentrations suggests that D/H is enhanced by a factor of about 6 over the terrestrial value. If D/H for juvenile water on Mars is the same as that for Earth, the observed enhancement must be explained by a divergent history of atmospheric evolution on the two planets. Several steps in the escape of hydrogen favour loss of H over loss of D, but, even so, the observed enrichments can only be attained if some of the (D-enriched) atmospheric water can exchange back with the condensed phase. Model calculations, based on the assumption that D/H in primordial Martian H_2O is the same as the terrestrial value, suggest that a layer about 0.2 m thick must be exchangeable if the escape rate has remained constant over geological time. This quantity is almost two orders of magnitude less than geological inventories of subsurface water, so that it is not unreasonable. Presumably, the much weaker fractionation between ^{18}O and ^{16}O compared with that between D and H has prevented a measurable enhancement of the heavier oxygen isotope even in the presence of exchange with a modest surface reservoir.

Measurement of the extent of deuterium enrichment in atmospheres has, in fact, been regarded as an important tool in determining how much hydrogen has escaped, and thus how much water the planet or satellite originally possessed. To draw the proper inferences, it is obviously necessary to know (at least) the D/H ratio in juvenile water and the relative efficiency of loss of D and H by all plausible mechanisms. Some problems arise even with the 'starting' D/H ratio. Interstellar measurements give present-day D/H = 1×10^{-5}; nuclear burning in stars reduces the relative abundance of deuterium, perhaps by a factor of two over the history of the solar system (4.6 Gyr). A likely value of D/H at the birth of the solar system is thus 2×10^{-5}. Measurements of CH_3D (but not of HD, which may be perturbed) give almost exactly this ratio for both Jupiter and Saturn, in accordance with the expectation that there would have been no fractionation on these gas giants. In Titan's atmosphere, in ocean water on Earth, and in hydrated minerals in meteorites, D/H is much higher, about 1.6×10^{-4}. For Halley's comet, values up to 5.4×10^{-4} have been reported, while in organic molecules in carbonaceous chondrites the ratio can be as large as 2×10^{-3}. One interpretation of the observations is that ices in the solar system condensed from the solar nebula so as to favour partitioning of deuterium into heavier molecules such as water, methane, and ammonia. Solar system bodies that had a relatively large ice content in their make-up (Earth, Titan, meteorites, comets) will carry the signature of an elevated deuterium abundance. Uranus and Neptune, with a much higher ice/gas ratio than Jupiter or Saturn, appear to show some enhancement in deuterium, with D/H perhaps about 7×10^{-5} in the atmosphere of Uranus. Mars and, as we shall see shortly, Venus show further enhancements in D because of preferential escape of H over geological time. The highest of all values of D/H are found in the molecules of the interstellar clouds (Section

9.1.2): ratios of several per cent, three orders of magnitude greater than the cosmic value, have been reported. It has been suggested that comets might become particularly deuterium rich by picking up large quantities of dust grains from interstellar clouds.

The question now arises how the ices became enriched in deuterium. Equilibrium at low enough temperatures is thermodynamically capable of giving the required partitioning into water, methane, and ammonia. However, the time taken to reach this equilibrium is too long compared with the lifetime of the nebula, even if the processes are catalysed by dust grains. An alternative explanation is based on the *kinetic isotope effect*. It is known from laboratory experiments that different isotopes participate in reactions at slightly different rates. The effects are particularly marked for H and D. Mechanisms that can produce the effect include differences in zero-point energy (which make activation energies smaller for molecular reactants containing hydrogen rather than deuterium), and greater quantum mechanical tunnelling for hydrogen. Thus reactions involving breaking of H-bonds are faster than those involving the D-substituted analogues, especially at low temperatures. For example, the forward reaction of an O–H molecule

$$D + OH \rightarrow OD + H \tag{9.5}$$

is expected to be more than seven times faster at temperatures below 300 K than its reverse, which is the O–D analogue. Ultraviolet radiation present throughout the galaxy (the *interstellar radiation field*) can photolyse water, so that we can thus envisage a sequence of steps, initiated photochemically, that leads to preferential formation of a deuterated product

$$H_2O + h\nu \rightarrow OH + H \tag{9.6}$$

$$OH + HD \rightarrow H_2O + D \tag{9.7}$$

$$D + OH \rightarrow OD + H \tag{9.5}$$

$$OD + H_2 \rightarrow HDO + H. \tag{9.8}$$

A similar photochemically initiated series of steps can convert CH_4 to CH_3D preferentially, the key processes being

$$CH_3 + D \rightarrow CH_2D + H \tag{9.9}$$

$$CH_2D + H + M \rightarrow CH_3D + M. \tag{9.10}$$

Once again, the forward process in reaction (9.9) has a greater rate constant than its reverse because the zero-point energy of the CH_2–H bond contributes more to overcoming the activation barrier than does that of the CH_2–D bond. As a consequence, CH_3D is favoured kinetically over CH_4.

The explanation for fractionation into the D-rich species is not, ultimately, very different from that for equilibrium fractionation (since both phenomena go back to the vibrational zero-point energies of D- and H- substituted molecules). However, the formation of radicals photochemically allows the occurrence of radical–radical chemistry, which is relatively fast at low temperatures. Ion–molecule reactions are also fast at low temperatures (Section 3.4.1), and it is kinetic isotope effects acting on such reactions that are believed to be responsible for isotopic fractionation in interstellar molecules.

Chemical enrichment of the kind just outlined may be responsible for the much higher D/H ratio on Titan compared with that on its parent, Saturn. Preferential escape of H cannot be responsible because the low gravity allows both H and D to escape easily. However, the reactions

$$C_2H + CH_4 \rightarrow C_2H_2 + CH_3 \tag{9.11}$$

$$C_2H + CH_3D \rightarrow C_2H_2 + CH_2D \tag{9.12}$$

are the main atmospheric sinks for CH_4 and CH_3D, so that the expected larger rate of reaction (9.11) compared with (9.12) would lead to an enrichment of the deuterated compound over geological time.

Discussion of the loss of water from Mars leads us to the question of whether Venus once possessed large quantities of water. Venus contains quantities of carbon and nitrogen similar to Earth, but hydrogen is deficient. Water abundance on Venus is about 42 kg m^{-2} compared with 2.7×10^6 kg m^{-2} on Earth. There is certainly no liquid water on the surface of Venus today, and the mixing ratio for water in the atmosphere is probably not more than 2×10^{-4}. Either Venus had a low H_2O content from the outset, perhaps because it was formed in a warmer region of the solar nebula, or, alternatively, whatever water was originally present has since disappeared. In the second case, the 'runaway greenhouse effect' (Section 2.2.5) would have demanded that all water was present as vapour, and the escape of hydrogen to space (and the escape and chemical reactions of oxygen) would have played a major role in the evolution of Venus.

Most mechanisms identified as potentially important for escape of hydrogen from Venus discriminate strongly against loss of deuterium, because of the large escape velocity (10.3 km s^{-1}; Table 2.1). Enrichment of deuterium might therefore be expected if Venus had originally possessed a water-rich atmosphere. Several pieces of evidence support deuterium enrichment, although they are not unequivocal. The ion mass spectrometer on Pioneer–Venus detected a signal at $m/e = 2$ from the upper atmosphere that can be attributed to D^+ (although the ion could be H_2^+). Interpretation of the intensity data would require D/H in the bulk atmosphere of $\sim 10^{-2}$. Mass peaks at $m/e = 18.01$ and 19.01 obtained in the lower atmosphere (below 63 km) with the large-probe neutral mass spectrometer may be caused by H_2O and HDO (although the $m/e = 19$ ion could be H_3O^+). If HDO is the source of the heavier ion, then D/H on Venus is $(1.6 \pm 0.2) \times 10^{-2}$, in agreement with the upper atmospheric ion data. On Earth, D/H $\sim 1.5 \times 10^{-4}$ overall (and perhaps twice that value in the upper atmosphere, according to Spacelab 1 observations), so that the deuterium enrichment on Venus is 50–100, implying large quantities of water in the early history of the planet. We shall see below that this enrichment factor is the *maximum* that could arise.

Molecular hydrogen would have been the dominant gas in the early upper atmosphere of Venus. Degassing might be expected to release materials with oxidation states similar to those for terrestrial volcanic gases ($[CO]/[CO_2] \sim 10^{-2}$; $[H_2]/[H_2O] \sim 10^{-2}$). However, at high Venusian temperatures, the gas phase equilibrium

$$CO + H_2O \rightleftharpoons CO_2 + H_2, \tag{9.13}$$

and reactions such as

$$2FeO + H_2O \rightarrow Fe_2O_3 + H_2, \tag{9.14}$$

at the planetary surface would have increased the H_2 content relative to H_2O. Supersonic hydrodynamic outflow, powered by solar ultraviolet heating, would have resulted in the loss of H_2 to space. Interestingly, this flow would have entrained HD, so sweeping deuterium away, *until* the mixing ratio of H_2 dropped below $\sim 2 \times 10^{-2}$. Only after this limit was passed would deuterium enrichment begin, regardless of how much water was originally present. Hydrogen, in the form of water, is now present at a mixing ratio of $\sim 2 \times 10^{-4}$, according to the Venera spectrophotometer data for 54 km altitude. Deuterium enrichment is thus limited to a factor of ~ 100, in accordance with the apparent measured value. The escape rate calculated for loss of H_2 would have exhausted the equivalent of Earth's oceans in about 280 million years.

As the Venusian atmosphere progressed towards its contemporary water vapour content, additional hydrogen loss processes probably began to operate. Translationally 'hot' hydrogen atoms can escape if their velocities exceed 10.3 km s^{-1}. Ion reactions are an obvious source of excess translational energy (cf. Sections 2.3.2 and 8.3.3). Recombination reactions such as

$$OH^+ + e \rightarrow O + H^*, \tag{9.15}$$

can liberate kinetic energy directly, but a more important process on Venus may be elastic collision between 'hot' O^* and ambient H

$$O^* + H \rightarrow O + H^*. \tag{9.16}$$

The atomic oxygen itself can be generated by recombination

$$O_2^+ + e \rightarrow O(^1D) + O(^3P), \tag{9.17}$$

with 239 kJ mol^{-1} excess translational energy (Table 8.2), corresponding to a velocity of ~ 5.5 km s^{-1}. Approximately 15 per cent of collisions between H possessing thermal velocities (at 300 K) and O^*, in reaction (9.8), will produce H^* with speeds in excess of the 10.3 km s^{-1} escape velocity. Mariner 5 Lyman-α (H resonance) airglow observations showed that there is an H atom component with an effective temperature of 1000 K in addition to the atoms that are thermally equilibrated at 300 K. Escape *via* this collisional mechanism could have reduced the hydrogen content from 2 per cent to the contemporary 0.02 per cent in about 4.2 Gyr, so that reactions (9.8) and (9.9) alone could account for the observed deuterium enrichment. Probably both hydrodynamic and ionic-collisional mechanisms operated at the higher hydrogen abundances, with the hydrodynamic loss becoming less important as the water vapour content approached its present level. Whatever the detailed mechanism, the deuterium enhancement suggests that Venus was once much moister than it is now. The contemporary D/H ratio does *not* provide evidence for loss of several oceans' worth of water, although detailed models of escape of hydrogen suggest that large quantities of water might once have been present. Massive loss of hydrogen from water brings with it the problem of disposal of the oxygen. It may be that the oxygen escaped to space along with the hydrogen; alternatively, oxidation of surface material would provide a plausible sink if the surface were molten. Another problem concerns the present-day escape of hydrogen from the atmosphere of Venus. Calculations put the time taken to exhaust hydrogen from the atmosphere at the contemporary escape rate at between 500 to 1500 Myr.

The longer time is perhaps compatible with a gradual depletion of water over the life of the planet, but if the shorter time is correct, then the implication is that water is being replenished as fast as it escapes. Such replenishment could be provided by outgassing from the planetary interior (and possibly by cometary impacts). Mixing ratios for water vapour drop by a factor of about 5 between 10 km altitude and the surface, suggesting that there is a large flux of water *from* the atmosphere into the surface, which could nearly balance a relatively large flux of juvenile water from the interior. Substantial oxidation of the surface would be expected with large water fluxes through it, and some results (e.g. from Venera 13) indicate the presence of Fe(III) minerals that are consistent with a relatively highly oxidized surface.

9.4 Evolution of Earth's atmosphere

Access to the detailed geological record for Earth constrains speculation about our ancient atmosphere (*palaeoatmosphere*) and its development. Isotopic abundances of, for example, the noble gases (cf. Section 9.2.1) argue strongly for our atmosphere being of secondary origin, rather than being a primordial remnant of the solar nebula. Gases trapped in the interior of the Earth outgassed to form the atmosphere, but the composition of the pre-biological palaeoatmosphere did not reflect the composition of the outgassed volatiles. Almost all the H_2O condensed to form the oceans, and the bulk of the CO_2 formed carbonates in sedimentary rocks, leaving the outgassed N_2 to accumulate and become the most abundant species in the atmosphere. The pre-biological atmosphere thus seems to have been mildly reducing in nature.

Strongly reducing atmospheres of CH_4 and NH_3 must have been very short-lived, even if they existed at all. Photolysis of NH_3 (ultimate product N_2: cf. Section 9.2.2) reduces the lifetime to a few years at most for NH_3 mixing ratios between 1 and 100 p.p.m., while the lifetime against rainout is ~ 10 days. Methane is photochemically stable near the surface (because short-wavelength ultraviolet radiation is filtered out by H_2O vapour). However, the lifetime against oxidation by OH, initiated by the reaction

$$OH + CH_4 \rightarrow CH_3 + H_2O \qquad (9.18)$$

is only about 50 years. No known chemical processes generate either CH_4 or NH_3 in the atmosphere, and hydrocarbons (including CH_4) in geothermal emanations are ultimately of biological origin, according to carbon isotope measurements. Recent geological and geochemical data support the idea of a mildly reducing pre-biological atmosphere. The rock record begins 3.8 Gyr ago with highly metamorphosed sediments at Isua in West Greenland. These rocks were formed in the presence of H_2O and CO_2, but without abundant CH_4. Thus the starting point for the evolution of life and of the present atmosphere is likely to have been an atmosphere containing predominantly N_2, with some H_2O and CO_2. Minor constituents can include NO (formed by lightning acting on N_2, CO_2, and H_2O), HCl (from sea salt spray), trace amounts of volcanic H_2 and CO; and formaldehyde produced by photochemical processes.

Accumulation of free oxygen in the atmosphere led to a transition from reducing to oxidizing conditions. While the bulk of oxygen in the contemporary atmosphere is photosynthetic in origin (Section 1.5.1), some O_2 can be formed by inorganic photochemistry. Photolysis of water vapour

$$H_2O + h\nu \rightarrow OH + H \qquad (9.6)$$

$$OH + OH \rightarrow O + H_2O \qquad (9.19)$$

$$O + OH \rightarrow O_2 + H \qquad (9.20)$$

$$O + O \xrightarrow{\quad M \quad} O_2 \qquad (9.21)$$

and of carbon dioxide

$$CO_2 + h\nu \rightarrow CO + O \qquad (9.22)$$

$$O + O \xrightarrow{\quad M \quad} O_2 \qquad (9.21)$$

are the favoured routes, but two important limitations are placed on the amount of oxygen that can be produced. It has long been recognized that H_2O and CO_2 are photolysed by ultraviolet radiation in a spectral region that is absorbed by O_2 (say at $\lambda \leqslant 240$ nm for H_2O, and $\lambda \leqslant 230$ nm for CO_2). 'Shadowing' by the O_2 thus self-regulates photolysis at some concentration. Secondly, photolyses of H_2O vapour or CO_2 do not, on their own, constitute net sources of O_2. Reactions (9.6) and (9.19) to (9.21) have the effect of converting two H_2O molecules to one O_2 molecule and four H atoms. Only if atomic hydrogen is lost by exospheric escape is there a gain in O_2, because otherwise H_2O is re-formed. Addition of CO_2 photolysis [reaction (9.22)] to the scheme does not alter this conclusion, since the CO product interacts with OH

$$CO + OH \rightarrow CO_2 + H. \tag{9.23}$$

Water-vapour photolysis now follows a new route

$$3H_2O + 3hv \rightarrow 3OH + 3H \qquad 3 \times (9.6)$$

$$CO_2 + hv \rightarrow CO + O \tag{9.22}$$

$$CO + OH \rightarrow CO_2 + H \tag{9.23}$$

$$OH + OH \rightarrow H_2O + O \tag{9.20}$$

Net $\qquad 3(H_2O + hv_{H_2O}) + (CO_2 + hv_{CO_2}) \rightarrow 2O + 4H.$

but the outcome is still that for every O_2 molecule formed, four H atoms must be lost. Escape is thus the crucial event, and the rate is determined by the transport of all hydrogen species through lower levels of the atmosphere to the exosphere (Section 2.3.2). Loss of O_2, for example by reaction with crustal or oceanic Fe^{2+}, or with volcanic H_2, competes with production, and so further limits the amount of free O_2 that can build up without the help of photosynthesis.

Considerable difficulties arise in giving quantitative expression to the pre-biological formation of oxygen because of uncertainties in concentrations of precursor molecules (H_2O and CO_2), temperatures, and solar ultraviolet intensities. Carbon dioxide concentrations might have been much greater before the gas was converted to carbonate deposits, and water vapour levels would have been elevated had surface and atmospheric temperatures been higher than they are now. Observational data gathered through the International Ultraviolet Explorer (IUE) satellite have shown that young stars (the 'T-Tauri' stars), which resemble the Sun at the age of a few million years, emit 10^3 to 10^4 times as much *ultraviolet* radiation as the present Sun. If enhanced solar ultraviolet intensity was available during the pre-biological evolutionary period of our atmosphere, then the photolysis of H_2O and CO_2 are greatly enhanced, and become a significant source of O_2, especially if CO_2 concentrations are high. Photochemical models developed for interpretation of the modern troposphere and stratosphere (Sections 3.6 and 4.4.5) can be adapted for the palaeoatmosphere by incorporating appropriate source terms, temperature profiles, and boundary conditions. Recent results suggest that pre-biological O_2 at the surface would have been limited to about 2.5×10^{-14} of the present atmospheric level (PAL) had both [CO_2] and solar ultraviolet intensities been at their current values. With 100 times more CO_2, and 300 times more ultraviolet radiation from the young Sun, the surface

$[O_2]$ calculated is $\sim 5 \times 10^{-9}$ PAL. Unfortunately, the geologic record does not allow the wide range of possible oxygen concentrations to be defined much more closely. The simultaneous existence of oxidized iron and reduced uranium deposits in early rocks (> 2.2 Gyr old) requires $[O_2]$ to be more than 5×10^{-12} PAL, but less than 10^{-3} PAL, so that the range of values accommodated both by the model and by geochemistry seems to be roughly $5 \times 10^{-12} - 5 \times 10^{-9}$ PAL.

Pre-biological oxygen concentrations in the palaeoatmosphere are of importance in two ways connected with the emergence of life. Organic molecules are susceptible to thermal oxidation and photo-oxidation, and are unlikely to have accumulated in large quantities in an oxidizing atmosphere. Living organisms can develop mechanisms that protect against oxidative degradation but they are still photochemically sensitive to radiation at $\lambda \leqslant 290$ nm. Life as we know it depends on an ultraviolet screen provided by atmospheric oxygen and its photochemical derivative, ozone, because DNA and nucleic acids are readily destroyed (Section 1.4). Biological evolution therefore seems to have proceeded in parallel with the changes in our atmosphere from an oxygen-deficient to an oxygen-rich one.

Naturally occurring abiotic processes could have synthesized the organic molecules that were the precursors of terrestrial life. In the 1920s, the Russian biochemist A. I. Oparin and the British geneticist J. B. S. Haldane independently proposed that such synthesis could not take place in an oxidizing atmosphere, and they suggested that it required an atmosphere rich in methane, ammonia, hydrogen, and water. Stanley Miller and Harold Urey put these ideas on an experimental basis with their now-famous experiments in which they passed electric discharges through strongly reducing mixtures of CH_4, NH_3, H_2, and H_2O. Amongst the collection of products were found amino acids, essential to living systems. More recent research has shown that discharges through mixtures of CO_2, CO, N_2, and H_2O also result in all the gaseous precursors needed for the production of complex organic molecules. Thus it seems that organic compounds could have been synthesized in a palaeoatmosphere of the composition now thought most probable (CO_2, N_2, and H_2O) and it is not *necessary* to look for non-atmospheric sources, such as deposition of cometary material, that have been proposed from time to time. The alternative view finds its most extreme expression in the suggestions of Hoyle and Wickramasinghe, who propose that interstellar molecules accumulated within the heads of comets (cf. Sections 8.7 and 9.1.2). Chemical evolution occurring a few hundred metres below the cometary surface is seen as progressing as far as biopolymers and micro-organisms. Hoyle and Wickramasinghe even contend that some past and present epidemics were initiated by the viruses and bacteria falling to Earth in cometary dust, although these ideas have gained little acceptance so far.

The stages between organic molecules and 'life' cannot concern us here, however fascinating speculation about the steps may be. Suffice it to say that life on Earth extends *at least* as far back as the oldest sedimentary rocks. Fossils show the presence of abundant life from 3.5 Gyr BP (Before Present). Although the 'dawn of fossil records' has often been regarded as 550 Myr (5.5×10^8 yr) BP, with the emergence of the trilobites at the beginning of the *Cambrian* period, the large organisms that left conspicuous fossils were preceded by micro-organisms. Microscopic examination of rocks reveals much evidence of ancient microbial life. Bacterial cells initially changed the atmosphere from its oxygen content of $< 10^{-8}$ PAL towards much higher concentrations. Between 2 and 1.4 Gyr BP, a new type of cell seems to have appeared. The earlier cells lacked a nucleus, and are classified as *procaryotic*: bacteria fall into this category, and can operate anaerobically. Larger cells, that are almost certainly *eucaryotic*, or possessing a nucleus, are found in the fossil record for 1.4 Gyr BP. The importance of this finding for the interpretation of atmospheric evolution is that almost all eucaryotic cells require large quantities of oxygen to function. Cell division is preceded by a clustering and splitting of the chromosomes with the nucleus (*mitosis*), a process dependent on the protein actomyosin that cannot form in the absence of oxygen. As the atmospheric oxygen reached about 10^{-2} PAL, a revolution occurred, because eucaryotic cells, and then animal and plant life, emerged. Respiration and large-scale photosynthesis became of importance, enough free oxygen became available for the fibrous protein collagen to be formed, and the scene was set for the appearance of *metazoans*, or multi-celled species. About 550 million years ago the Cambrian period opened. According to earlier ideas, this period heralded an 'evolutionary explosion'. In many ways, the real significance of the Cambrian period is that the first animals with clear external skeletons are preserved as fossils whose identity has been recognized for centuries, while remains of earlier life-forms had not yet been discovered or understood. Metazoan fossils from the preceding 120 million years, the 'Ediacarian' period, are now known. Many of these are from species resembling jellyfish. Such organisms can absorb their oxygen through the external surfaces at concentrations of about seven percent of PAL. A reasonable estimate for when this level of oxygen was reached in the atmosphere can thus be set at about 670 Myr BP. The relatively impervious surface coverings of the Cambrian metazoans suggest that 120 million years later the oxygen concentration was approaching 10^{-1} PAL. Following the opening of the Cambrian, the complexity of life is known to have multiplied rapidly and the foundations for all modern phyla were laid. 'Advanced' life forms (i.e. non-microscopic) were found ashore by the Silurian age (420 Myr BP), and by the Early Devonian, only 30 Myr later, great forests had appeared. Soon afterwards, amphibian vertebrates ventured onto dry land.

According to one interpretation, due first to Berkner and Marshall, the evolution of O_2, and hence of 'protective' O_3, controlled the migration of life from the safety of stagnant pools and lakes, to the oceans, and then to dry land. In this scenario, a depth of say 10 m of *liquid* water will filter out much of the damaging ultraviolet radiation while allowing photosynthetic visible light to reach the organisms. Life in the oceans seems improbable at this stage, since organisms would be brought too near the surface by mechanical motions. When O_2 and O_3 had built up yet further, the ultraviolet zone of lethality would be restricted to a thin layer at the ocean surface, and life could spread to entire ocean areas, thus greatly enhancing photosynthetic activity. As the oxygen content of the atmosphere moved towards its present level, enough O_3 was available for no liquid water filter to be needed for protection, and life could finally be supported on dry land.

A major question surrounding the interpretation of Berkner and Marshall is whether the biological evolutionary events were linked causally to the atmospheric changes that undoubtedly occurred. If they were, then some kind of feedback mechanism (Gaia; see Section 1.6) may have been in operation, since the atmospheric evolution was certainly mediated by the biota. Resolution of this question will require further information: in the first place, it is necessary to know the time history of growth of O_2 and O_3 in the atmosphere. What we can do is to use atmospheric photochemical models to calculate the ozone concentrations that accompanied smaller O_2 levels in the early atmosphere. Figure 9.2 shows the results of one such calculation, in which the full chemistry of catalytic cycles (HO_x, NO_x, and ClO_x; see Section 4.4) was incorporated. These model calculations immediately reveal one interesting feature of the evolution of ozone concentrations in our atmosphere. At low

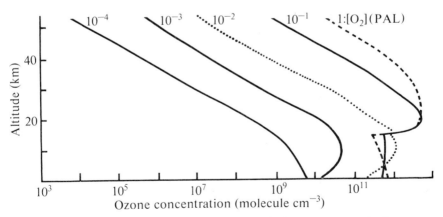

Fig. 9.2. Vertical distribution of ozone for different total atmospheric oxygen contents ranging from the present atmospheric level (PAL) to 10^{-4} PAL. The model used to obtain these results includes nitrogen, hydrogen, carbon, and chlorine chemistry; allowance is made for ozone loss at the planetary surface. [From Levine, J. S. *J. molec. Evol.* **18**, 161 (1982).]

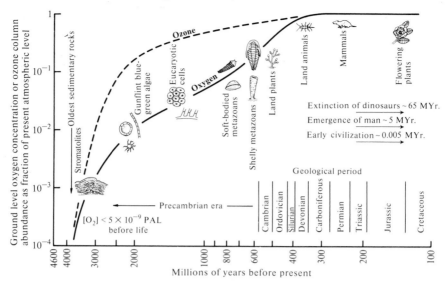

Fig. 9.3. Evolution of oxygen, ozone, and life on Earth. In the absence of life, surface oxygen concentrations are unlikely to have exceeded $\sim 5 \times 10^{-9}$ of the present value. The build-up of oxygen to its present level is largely a result of photosynthesis. Early organisms would have found high oxygen concentrations toxic, but eucaryotic (nucleated) cells require at least several per cent of the present level for their respiration. Soft-bodied metazoans could have survived at similar oxygen levels, but the reduced surface oxygen uptake area available once the species had developed shells must mean that the concentration was approaching one-tenth of its current value about 570 Myr ago. Considerations such as these are used in drawing up the oxygen growth curve. Ozone concentrations can be derived from a photochemical model. Life could not have become established on land until there was enough ozone to afford protection from solar ultraviolet radiation.

$[O_2]$, maximum ozone concentrations were found near the surface, but as oxygen concentrations increased, an ozone layer developed with its peak at successively higher altitudes.

Figure 9.3 summarizes the material presented so far in this section. Calculated ozone column abundances are shown for a growth in surface $[O_2]$ whose time evolution is based on the geochemical and fossil evidence.

From a biological standpoint, the characteristic of the ozone profiles that is important is the total column depth, since that is what determines how much ultraviolet radiation may leak through to the surface. Numerical values for the integrated areas under the curves of Fig. 9.2 are given in Table 9.2. The problem is to decide what flux of ultraviolet radiation is tolerable to life, and hence what column density of ozone furnishes an adequate screen. DNA is probably the most vulnerable and vital part of an evolving biological system. Absorption of radiation by DNA falls off at wavelengths longer than the broad maximum at $\lambda = 240$–280 nm, but most genetic damage may be caused by radiation at the longest absorbed wavelengths (say at $\lambda \sim 302$ nm). Damage would occur primarily at the few hours near mid-day, when the Sun is most nearly overhead, so that for all organisms with a generation time of less than a day, the maximum exposure is about four hours of high intensity light. Based on experiments with genetic damage to corn pollen, an arbitrary damaging dose at $\lambda = 302$ nm of 1.6×10^3 J m^{-2} may be specified. This translates to a flux of about 0.1 J m^{-2} s^{-1} at or below a wavelength of 302 nm. An adequate ultraviolet screen would be provided by an ozone column density of 7×10^{18} molecule cm^{-2} for a mid-day solar zenith angle of $30°$. Calculated ozone column densities are shown in Table 9.2 for several values of $[O_2]$: an effective ultraviolet shield is available from the atmosphere alone soon after oxygen passes the 10^{-1} PAL concentration. Also shown in the table are the fractional absorbances due to ozone at two wavelengths: $\lambda = 250$ nm corresponds to the maximum in DNA (and ozone) absorption, while $\lambda = 302$ nm is the estimate of the longest genetically active wavelength. The last two columns of the table show the thicknesses of liquid water needed to make the total ozone + water absorption equivalent to that of the 'standard' column (7×10^{18} molecule cm^{-2}) of ozone. Protective layers of considerable depth do seem necessary for $[O_2] < 10^{-1}$ PAL. Does the transition to $[O_2] > 10^{-1}$ PAL then really explain the appearance of life on dry land? Shelled organisms require dissolved oxygen that would be in equilibrium with $> 10^{-1}$ PAL in the atmosphere, so that the critical level of O_3 for biological protection would have already been passed when the organisms appeared abundantly in the Cambrian period (550 Myr BP). But life was probably not firmly established on dry land for a further

Table 9.2 Oxygen and ozone in the evolving atmosphere.

$[O_2]$ PAL	$[O_3]$ column[a] molecule cm^{-2}	Fractional absorption by ozone present at:		Water depth (in m) that brings total attenuation to 'standard' value[b]	
		$\lambda = 250$ nm	$\lambda = 302$ nm	$\lambda = 250$ nm	$\lambda = 302$ nm
10^{-4}	5.2×10^{15}	0.06	0.00	6.0	5.4
10^{-3}	7.0×10^{16}	0.54	0.03	5.2	5.3
10^{-2}	1.6×10^{18}	1.00	0.45	–	4.2
10^{-1}	5.9×10^{18}	1.00	0.89	–	0.8
1	9.7×10^{18}	1.00	0.97	–	–

[a] Calculated for chemistry including chlorine species. [Levine, J. S. *J. molec. Evol.* **18**, 161 (1982).]

[b] 'Standard' value is the screen provided by an ozone column of density 7×10^{18} molecule cm^{-2} [Ratner, M. I. and Walker, J. C. G. *J. atmos. Sci.* **29**, 803 (1972).]

170 Myr (late Silurian). Thus the possibility exists that the ozone screen may have been established before the Silurian period, and was not directly linked with the spread of life onto land. The connection between the emergence of life out of water and the development of the ozone shield remains a tantalizing one. Whether or not life moved onto dry land in response to the presence of sufficient ozone, it should be clear by now that there is, and always has been, an intimate relation between the existence of the biota and the composition of our atmosphere. The evolution of the one, ever since life appeared, has been the story of the evolution of the other.

9.5 Climates in the past

Evolution on dry land brought with it an increased rate of photosynthesis, and it seems possible that oxygen concentrations may have exceeded 1 PAL during the lush growth of vegetation during the carboniferous period (340 Myr BP). A possible consequence of this rapid photosynthesis is that atmospheric carbon dioxide could have been depleted if oxidation of organic matter did not regenerate CO_2 at a sufficient rate. Reduced radiation trapping would cause a drop in the Earth's temperature, and the lower temperatures would then lead to a decreased rate of photosynthesis and lowered atmospheric O_2 concentration. Cooling of the Earth from the depletion of CO_2 may well have been responsible for the Ice Ages of the Permian period (280 Myr BP). At the lower temperatures, CO_2 would be restored to a higher level, because of reduced rates of consumption in photosynthesis. Oxygen and CO_2 concentrations would thus swing in opposite senses, and may still be undergoing damped oscillations with a period of about 100 million years.

Changes of temperature, such as those described in the last paragraph, reflect one aspect of a changing *climate*. By 'climate', we understand the mean values and range of meteorological parameters such as temperature, pressure, humidity, precipitation, wind, and so on. A particular region, or even locality, can have its own climate, but the concept can also encompass wider areas, up to the global scale. Extrinsic factors that could alter the Earth's climate range from the well-established variations in solar luminosity and in our orbit round the Sun (probably responsible for the 10–50 000 year period 'Ice Ages' of the last two million years) to imaginative speculations such as alteration in atmospheric opacity resulting from dust clouds raised by impacting asteroids. Intrinsic factors include changes in surface albedo or in atmospheric reflectivity due to volcanic aerosol, variations in oceanic circulation, and alterations in atmospheric composition.

Although we called the glacial cold periods 'Ice Ages' in the last paragraphs, the term is often an exaggeration. In fact, one of the most interesting features of the Earth's climate is that it has been rather stable, at least since the emergence of life, some 3.5 Gyr ago. Ice Ages affect only those parts of the

Earth at latitude $> 45°$ north or south: only 30 per cent of the surface, some of which is, in any case, often partially frozen between glacial periods (as now). Both the geochemical record and the persistence of life itself indicate that the oceans can never have either frozen or boiled. Mean surface temperatures have probably never departed from the range of 5 to 50°C, and may have been highest at very early periods. But this result leads to a riddle! Standard stellar evolutionary models predict that 4.5 Gyr BP the Sun's luminosity was lower than it is today by 25 to 30 per cent. Such dimming has created the 'dim Sun paradox', because it translates into a decrease of Earth's effective temperature by 8 per cent, low enough to keep sea-water frozen for ~ 2 Gyr. Explanations to solve the problem include changes in albedo or increasing the greenhouse efficiency. Alterations in clouds, for example, could exert a negative feedback, stabilizing, effect, since lower temperatures would mean decreased cloud cover and reduced reflection away of solar radiation. Water vapour makes the largest contribution to the greenhouse effect in the contemporary atmosphere, but it is unlikely to be the agent of long-term temperature control. Its relatively high freezing and boiling points render its blanketing effect prone to unstable, *positive*, feedbacks by increasing ice and snow albedo at low temperatures (further reducing temperature), but increasing water vapour content at high temperatures (and yet further increasing the greenhouse effect). Whatever greenhouse gas or other mechanism kept the Earth warm, it must have been smoothly reduced to avoid exceeding the high temperature limit for life. Carbon dioxide seems the most likely greenhouse gas to have exerted thermostatic control of our climate. Negative feedback mechanisms can be identified for this gas. Non-biological control might include acceleration of the weathering of silicate minerals to carbonate deposits in response to increased temperatures. However, as we discussed in Section 1.6, present-day weathering is biologically determined, and the biota both sense and amplify temperature changes. This feedback regulation of climate is seen by its proponents as evidence in support of the Gaia hypothesis (Section 1.6). Models that include not only the CO_2 greenhouse effect, but also the consequential changes in cloud-albedo and water vapour trapping, suggest that $[CO_2]$ might have been 10^3 PAL at ~ 4.2 Gyr BP, 10^2 PAL at ~ 3 Gyr BP and 10 PAL at ~ 1.5 Gyr BP. Such concentrations seem reasonable if we accept that in Earth's pre-biological atmosphere CO_2 was a major component as it is on Venus or Mars today. Even 1000 times the present mixing ratio constitutes a partial pressure of CO_2 of only one-third of a (contemporary) atmosphere.

As we approach more closely (within a million years!) our own era, so the record of climate and atmospheric composition becomes richer and more detailed. One particularly fruitful source of information has proved to be the examination of the ice sheets covering Greenland and the Antarctic. A large fraction of the world's fresh water is contained in the Antarctic ice sheet, which is up to 5 km thick in parts. Virtually no melting of snow occurs even in the summer, so that each year a new layer of snow is added to the ice caps and compressed into solid ice. As the snow falls, it scavenges aerosols from the atmosphere, and these aerosols are trapped together with bubbles of air in the ice. Chemical or biological alteration to the trapped material is not expected, so that a core taken from the ice provides a stratigraphic record of the atmosphere with a resolution and stability greater than that available from any other sedimentary record.

In Antarctica, some places have so little snowfall that bore-holes penetrating 90 per cent of the ice sheet will generally cover time-spans of 20 000 to 200 000 years, and the oldest ice may be more than a million years old. One recent core from Greenland is 2035 m long and dates back 90 000 years. Dating of ice cores may be achieved in several ways. The most ancient samples have been dated by an analysis of ice sheet flow caused in response to horizontal stresses imposed by the overlying ice. Natural radioactive isotopes in the ice can be used to date absolutely ice samples up to about five half-lives old. For ^{14}C (half-life 5600 years), the oldest datable sample thus approaches 30 000 years of age. Younger samples can be very accurately dated by stratigraphic methods, in which ice layers are counted in the same way as tree rings. Regular seasonal cycles, and 'catastrophic' events such as volcanic eruptions, can be recognized.

Ice cores provide information both on climate and on atmospheric composition. One method for unlocking the climatic record involves measurement of stable isotope ratios such as D/H or $^{18}O/^{16}O$ in the ice. The method depends on the slightly lower vapour pressure of the heavier isotope species (e.g., about 1 per cent less for $H_2^{18}O$ compared with $H_2^{16}O$). Water vapour is in equilibrium with the oceans in the subtropical source regions of atmospheric humidity, but as the air masses move towards the polar regions they lose water irreversibly by precipitation over the pack ice and ice sheets. The precipitation removes *relatively* more of the heavier isotopic H_2O, thus depleting the remaining vapour yet more. During a climatically cool period, or, indeed, during winter, there is more rapid cooling at high latitudes than there is nearer the source regions. As a result, the depletion of D and ^{18}O in the water of polar ice cores shows long-term trends in parallel with changing climate, as well as seasonal oscillations.

Figure 9.4 shows the results obtained for oxygen isotope profiles in an Antarctic ice core (Byrd station). The change in depletion about 10 000 years ago of ^{18}O compared to the oceanic content of $H_2^{18}O$ clearly reflects the end of the last Ice Age. Mineral dusts showed dramatically increased concentrations in the middle of this Ice Age. Carbon dioxide concentrations, on the other hand, seem to have been 30 per cent lower during the period of glaciation than they were subsequently (Fig. 9.4). Both particle and carbon dioxide trends are believed to be *effects*, rather than *causes*, of the glaciation. A more vigorous circulation would result from the greater equator–pole temperature gradients, and Southern Hemisphere continents would be more arid and provide enhanced desert dust sources. Similarly, the rate of carbon dioxide exchange

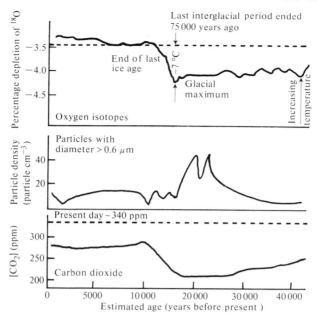

Fig. 9.4. Climatic record in deep ice cores. The core on which this figure is based is from Byrd Station in the Antarctic (80 °S, 120 °W), and is 2164 m deep. Ice drift results in an apparent slight rise in temperature since the end of the last ice age. The time axis is deliberately expanded in the first 10 000 yr. [From Peel, D. A. *New Scient.* **98**, 477 (1983).]

between ocean and atmosphere would be affected by lowered temperatures in such a way as to reduce the atmospheric CO_2 burden. Curiously, both the observed changes would themselves tend to cause cooling, because of increased albedo and reduced greenhouse trapping. Positive feedback effects could therefore have amplified both the glaciation and the rapid subsequent warming.

9.6 Climates of the future

The carbon dioxide concentrations shown for the last 10 000 years of Fig. 9.4 fall a long way short of the present-day values shown as the dotted line in the figure (and, in more detail, in Fig. 1.5). Detailed comparisons of calibration techniques using ice-core samples from Antarctica and Greenland show that in the period 800–2500 years before present, the CO_2 mixing ratio in the atmosphere was 2.6×10^{-4} (260 p.p.m.). In 1980, the mixing ratio exceeded 340 p.p.m. for the first time. The carbon dioxide increase has come about almost entirely since the Industrial Revolution, and mainly within the last 50 years, as a direct result of man's activities. Increased CO_2 emission is the primary cause, with burning of fossil fuels (98 per cent) and cement manufacture (2 per cent) as the major contributors. Deforestation in the tropics and changing agricultural practices reduce the efficiency of CO_2 recycling, thus providing a secondary aggravation of the CO_2 increase. Samples taken from individual tree rings[a] show a decreased specific activity of ^{14}C over the last 100–150 years. The result suggests dilution of ordinary atmospheric CO_2 (in which the ^{14}C content is continuously renewed) by fossil fuel releases, containing 'old' carbon whose ^{14}C has already decayed, rather than a reduced uptake of CO_2 by the biosphere.

From 1900 to 1973, the average annual rate of *increase* of CO_2 emission was roughly four per cent, but since the dramatic rise in fuel prices in 1973, the growth rate has halved, to about 2.3 per cent per annum. At present, 1.9×10^{13} kg of CO_2 are being released to the atmosphere each year, corresponding to roughly ten per cent of the rate of carbon assimilation by terrestrial plants. This release adds to the 2.6×10^{15} kg of CO_2 already in the atmosphere, but is initially partitioned between ocean and atmosphere in the ratio 0.47/0.53, so that atmospheric concentrations are increasing at the rate of 0.4 per cent per year ($\cong 1.3$ p.p.m. per year). *Eventually*, almost all of the added carbon dioxide will end up in the oceans, but removal from the atmosphere is limited by the slow mixing of the large volume of deep ocean water with the small volume surface layer, acting over a period of 500 years or more. Man is releasing carbon dioxide too fast for the oceans to cope, and it seems likely that in the next century the oceans will continue to take up less than half of the CO_2 added to the atmosphere each year.

Concern about increasing CO_2 levels is directed first towards possible consequences for our climate that would follow from increased radiation trapping. Progressively more sophisticated and complete models of the climate system have shown that mean surface temperatures rise by approximately 3°C for a doubling in $[CO_2]$. Positive feedbacks, such as decreased albedo because of shrinkage of the polar snow caps, can amplify small temperature changes. Warming in the Arctic would be three times larger than the average, and in the temperate zone of the Northern Hemisphere (where the majority of industrialized nations lie), the temperature rise for a carbon dioxide doubling should be 4–6°C. Warming trends in the Southern Hemisphere would be smaller because of the larger ocean area.

It is not only carbon dioxide levels that have been increasing in recent years. Methane is an important greenhouse gas, and atmospheric methane levels have been increasing at the rate of 1.7 per cent year since at least 1965.

---

[a] Tree rings afford another important piece of evidence about past climates, since the temperature during any one season can be estimated from the growth. Some trees (the Bristle Cone Pines of California) reach ages of > 1000 years. However, sub-fossil pieces of dead wood can be used to piece together a continuous series of rings extending nearly 10 millennia.

Greenland polar ice cores show that $[CH_4]$ is constant from depths of 250 m to 1950 m, corresponding to the period 1580 AD to about 25 000 BC! In 1580, methane levels began to rise from their mixing ratio of 0.7 p.p.m. at first slowly, but since 1918 much more rapidly, to reach their present mixing ratio of ~ 1.66 p.p.m. What has caused an accelerating increase in methane concentrations since the seventeenth century, after they had remained constant for 26 000 years? Man seems implicated because of the time-scale. Possibly some activity has reduced atmospheric $[OH]$, which is the primary sink for CH_4 (Section 5.3.2) through reaction (9.18). Alternatively, increasing $[CH_4]$ could result from increasing production of the gas. Atmospheric methane has largely biogenic sources, arising from anaerobic bacterial fermentation in swamp-lands, tropical rain forests, and the intestinal tracts of livestock and termites. The growth in rice paddy cultivation and in cattle farming, which are a response to population growth, could account for increased methane produc-tion. If this explanation is correct, then it seems probable that methane levels will continue to rise in the decades ahead. The increase in $[CH_4]$ already experienced is predicted to have increased global temperatures by 0.23°C, or 38 per cent of the effect estimated for the CO_2 increases since the industrial revolution to the present day. Increases in the future can only further elevate temperatures. However, there are additional *chemical* effects about which we can only speculate. Methane plays a central role in tropospheric chemistry (see Section 5.3). Increased concentrations of CH_4 may have caused con-sequential changes in the levels of other atmospheric constituents such as O_3, CO, and OH.

Methane may be involved in a positive-feedback process that will amplify temperature changes. Enormous quantities of methane are locked up as the clathrate hydrates such as $CH_4.6H_2O$, which are inclusion compounds of methane trapped in the interstices of ice crystals. Gas hydrates below the permafrost of polar regions may account for 2×10^{15} kg of carbon (three times the carbon content of the atmosphere), while estimates for the content of sea-floor sediments are even larger (up to 10^{17} kg). Phase diagrams for methane hydrate can be used to discover at what depths ($\equiv$ pressures) and temperatures the compound is only marginally stable and might therefore be released in a warmer climate. Arctic Ocean sediments seem the most likely to be labile and to be exposed to higher temperatures. Decomposition over a 100 yr period from a layer 40 m thick at ~ 300 m depth and halfway round the Arctic Ocean could release $\sim 8 \times 10^{12}$ kg yr^{-1} of CH_4, enough to produce a significant positive feedback effect on CO_2-induced warming.

Input of nitrogen into cropland is an additional way in which average world temperatures might be raised in the future. Commercial fertilizers and nitrogen-fixing leguminous crops (Sections 4.4.3 and 4.5.8) both have the potentiality of increasing emissions of N_2O, a greenhouse gas, from soils. The nitrogen fixed annually by combustion and in manufacturing fertilizers has now reached almost half of what plants produce naturally. Atmospheric

N_2O concentrations are now known to be increasing slowly (at ~ 0.2 per cent per annum for the period 1961–84). Once again, there is a coupling with chemistry, this time stratospheric, because N_2O is an important 'natural' source of the NO_x that controls ambient O_3 concentrations (Section 4.4.3).

Concern also arises over the release by man to the atmosphere of the chlorofluorocarbons (CFCs). As described in Section 4.5.4, the concentration of these synthetic compounds is building up in the troposphere (cf. Fig. 4.19). The compounds possess strong absorption bands in the infra-red region that, by chance, coincide with regions where CO_2 itself has relatively weak absorption. The CFCs thus have the potential of closing the atmospheric 'windows' through which radiation could escape to space, and the contributions of such compounds to greenhouse warming may consequently be much greater than the simple additive effect of the radiation trapped by the CFCs themselves. Chapter 4 concentrates on the impact that CFCs may have on depletion of ozone in the stratosphere, but it is evident that CFCs may have an additional adverse environmental impact by contributing to global warming. In general, the control strategies discussed in Section 4.5.6 are those needed to counter both threats, and reduced tropospheric lifetime is clearly one desirable quality in an 'alternative' halocarbon designed to replace the CFCs now in common use. However, it must be recognized that the compounds selected to reduce ozone depletion potentials (ODPs—Section 4.5.5) may not be ideal in terms of their infra-red absorption intensities and wavelengths, and due consideration must be given to these factors in assessing overall environmental impact.

Our list of possible perturbations to future climate is by no means exhaustive. And our discussion has considered only radiation trapping by the greenhouse effect, together with direct chemical intervention. There follow a multitude of feedbacks and couplings, such as the effects of the greenhouse gases as stratospheric cooling agents (Section 4.5.9) which produce consequential changes in chemical kinetics. Models are now beginning to treat climate and chemistry together in an attempt to simulate the 'real life' situation (Section 4.5.10).

At a simpler level, we have to ask whether or not we have yet experienced a global warming attributable to the increased concentrations of trapping gases. Theoretically, warmings of 0.5–$1.0°C$ should have arisen over the course of this century. However, 'noise' in the form of natural influences on climate make it extremely difficult to establish the existence of an enhanced greenhouse effect. The climatic influences include fluctuations of solar radiation intensity, the release of aerosols from major volcanic eruptions, and oceanic warmings and coolings. Maverick weather patterns towards the end of the 1980s and the beginning of the 1990s, including summer drought conditions in the USA, and hot summers and violent winter storms in Europe, have led to a popular conception that climate has already undergone a substantial change. Some ground-based observations give some support to this notion.

On average, 1989 was 0.23°C warmer than the reference period 1951–1980, and the decade of the 1980s was also significantly warmer than the reference period. Six of the 10 warmest years on record fell in the 1980s. For comparison, the warming over the preceding 134 years was less than 0.5°C. However, direct evidence that the global warming trend is attributable to an intensifying greenhouse effect is hard to discern in the temperature record. According to climatic models, enhanced greenhouse heating should lead to the most pronounced temperature increases in Arctic regions, but there is no consistent pattern in the experimental record. The situation is further confused by recent analysis of data from a network of satellites, launched in 1978, that carry microwave radiometers capable of measuring temperatures to 0.01°C. According to the satellite results, there is apparently a random pattern of temperature from year to year, and not a steady warming trend. These data suggest that the five years 1979–1983 were warmer than the period 1984–1989, even though 1987, followed by 1988, were the warmest years of all. Increases of global average temperatures by 0.5°C have been observed by the satellite network to occur in less than two weeks, thus casting severe doubt on the attribution of warmings of this magnitude to greenhouse trapping. The satellite measurements seem likely to be more accurate than the ground-based ones, since they are truly global, can be made remote from local sources of heating near population centres, and can be made in the middle troposphere.

In view of the conflict of views about whether greenhouse-related global warming has *already* occurred, it is evident that forecasts of the extent of global warming in the next century do not provide a rational basis for policy making. Furthermore, projections into the future about fossil fuel use, and hence of carbon dioxide release, are also fraught with uncertainty. Yet it is now that action needs to be taken if the untoward consequences of global warming are to be averted or alleviated. We must therefore consider what some of those consequences might be. At the present rate of growth, $[CO_2]$ will be double its pre-industrial value by the turn of the century. Even were the rate of increase of CO_2 release to fall by a factor of 10, the CO_2 mixing ratio would reach 370 p.p.m., possibly before the year 2000, and certainly by 2020. According to the 1990 report of the Intergovernmental Panel on Climate Change (IPCC), average global temperatures in 2020 will be between 1.3 and 2.5°C above pre-industrial levels. By 2070, the temperatures will exceed the pre-industrial values by 2.4 to 5.1°C, with a best estimate of 3.5°C. Temperatures will rise more over continental areas than over the oceans, and the Northern Hemisphere will warm more quickly than the Southern. As the CO_2 levels progress to a doubling of pre-industrial values, the southern parts of Europe and the United States would become truly tropical. The climate would be getting close to that which existed at the beginning of recorded history, some 4500 to 8000 years ago (the *Altithermal* period), when the world was definitely warmer.

A warmer Earth will have a smaller equator-to-pole temperature contrast, because the excess heating is concentrated in polar regions. Seasonal contrasts will become less pronounced. With less potential energy available in the system to convert to kinetic energy, the atmospheric heat engine will run more slowly. Large-scale circulation patterns will be influenced, and some regional climatic changes may be larger than the average. A warmer atmosphere and ocean will result in more evaporation and precipitation, but the important question is where the rain and snow will fall. Some indication of the expected precipitation patterns can be gleaned from reconstructions of the Altithermal period (see above), or from examination of the meteorology of particularly warm years in the immediate past. Mid- to low-latitude land masses seem likely to experience a wetter climate than at present, with the deserts of North Africa, north-west India, and the south-west USA becoming prairie-like. Conversely, northern Europe and most of the central parts of North America and the Soviet Union would become drier.

Warming in the polar regions could approach as much as 10°C by the middle of the next century. Formation of sea-ice will be reduced in both polar regions, and climatologists are tantalized by the possibility that Arctic Ocean ice might disappear and not return. Modification of the entire climate of the Arctic Basin would ensue, with profound ecological consequences on land and in the sea, and with possible release of much CH_4 from methane hydrate (cf. p. 411). The ice sheets of the Antarctic and Greenland (see Section 9.5) are more like geological formations than components of the hydrologic system. Since they rest on solid bedrock, their melting, or sliding into the ocean, would cause a rise in sea-level. Straightforward melting of the enormous masses of ice would probably take many thousands of years, but another mechanism could accelerate the process. Much of the West Antarctic ice sheet (and some of the East Antarctic ice sheet) rests on bedrock that is below sea-level. Warmer ocean water could work its way under the ice sheet, separating it from the bedrock and causing it to slide towards the ocean. Disintegration and melting of the ice would then be relatively rapid because of the more intimate contact with the water and because of the diminished thermal insulation. There is no sign that anything like this is happening now, and the time-scale involved for disintegration is probably more than 200 years. However, if the entire West Antarctic ice sheet were to disappear, a rise in sea-level of 5 to 7 m would result, with a serious impact on all the shorelines of the world. The best estimates given by the IPCC in 1990 for sea level rises are 0.2 m by 2020 and 0.4 m by 2070.

The socio-economic ramifications of climate change lie outside the scope of this book. Many practical issues will have to be faced, with some people being better off, and some faring worse, on a warmer Earth. Agriculture will be much affected, with mid-latitude farmers expecting about ten days longer growing season for every degree Centigrade rise in temperature. Large areas of Africa, the Middle East, India, and central China might cease to be water-deficient. On the other hand, the 'food basket' areas of North America and the Soviet Union would become much drier, and it would be harder to grow grain and other major food crops. A rise in sea-level could precipitate the migration of enormous populations from seaboard lowland regions, but it might not bother the people living inland, except indirectly. A higher level of CO_2 in the atmosphere and more rainfall in the subtropics would both enhance forest growth, and allow it to spread to places that are now too arid. Present dangerous trends in deforestation might therefore be arrested. Costs of space heating in winter would decrease, but air conditioning, where used, would become more expensive.

Inadvertent chemical modification by man of the stratosphere (Section 4.5) or troposphere (Section 5.5) seems likely to be dominated by the CO_2 release resulting from the burning of fossil fuels. The cause is an international activity, the effect is global in character, and the remedy will require the concerted action of the countries of the world. A major alteration of the global environment seems possible, or even probable, unless we decide to change the course of events. Man's greed for energy will either have to be satisfied by non-fossil sources or we shall have to prepare for the climatic change that is in store for us.

9.7 A doomed biosphere?

Man's interference with nature has been the theme of the last section. We should, however, recognize that nature herself may well make the environment hostile to life, even without our intervention. We saw in Section 9.5 that the Sun's luminosity is likely to have increased by 25 to 30 per cent over the past 4.5 Gyr, and there is no reason to suppose that the trend has ceased. Whether by active or passive control, changes in carbon dioxide concentration appear to have compensated for the increasing solar intensity in such a way as to keep the planetary temperature very nearly constant. But the capacity for control is now nearly exhausted, because $[CO_2]$ is approaching the lower limit tolerable for photosynthesis. If that limit is taken to be 150 p.p.m., then the CO_2 control of a climate favourable for life can continue for another 30–300 Myr. Some adaptation to lower CO_2 concentrations and to higher temperature is possible, but it would not buy much time. In human terms, the crisis is rather distant. Over the next centuries, our problem is to cope with, and then prevent, loss of temperature control because of *increases* in $[CO_2]$ (Section 9.6). There is some prospect that controlled nuclear fusion may be developed within the next century as a clean and almost inexhaustible supply of power. Our book therefore concludes by pointing out that the best hope for the protection of Earth's atmosphere may lie in harnessing, as a source of energy, the process out of which the Sun, planets, and atmospheres were born.

Bibliography

Section 9.1.1

Sources of atmospheric species and the evolution of atmospheres.

Planets and their atmospheres: origin and evolution. Lewis, J. S. and Prinn, R. G. (Academic Press, Orlando, 1984.)

The origin and evolution of planetary atmospheres. Henderson-Sellers, A. (Adam Hilger, Bristol, 1983.)

The elements: their origin, abundance, and distribution. Cox, P. A. (Oxford University Press, Oxford, 1989.)

Origin of the solar system. Cameron, A. G. W. *Ann. Rev. Ast. Astrophys.* **26**, 441 (1988).

Origin and evolution of planetary atmospheres: An introduction to the problem. Prinn, R. G. *Planet. Space Sci.* **30**, 741 (1982).

Formation of atmospheres during the accretion of the terrestrial planets. Ahrens, T. J., O'Keefe, J. D., and Lange, M. A. in *Origin and evolution of planetary and satellite atmospheres.* Atreya, S. K., Pollack, J. B., and Matthews, M. S., (eds.). (University of Arizona Press, Tucson, 1989.)

Origins of the atmospheres of the terrestrial planets. Cameron, A. G. W. *Icarus* **56**, 195 (1983).

Escape of atmospheres and loss of water. Hunten, D. M., Donahue, T. M., Walker, J. C. G., and Kasting, J. F. in *Origin and evolution of planetary and satellite atmospheres.* Atreya, S. K., Pollack, J. B., and Matthews, M. S. (eds.). (University of Arizona Press, Tucson, 1989.)

Venus: a contrast in evolution to Earth. Kaula, W. M. *Science* **247**, 1191 (1990).

Origin and evolution of the atmosphere of Venus. Donahue, T. M. and Pollack, J. B. in *Venus.* Hunten, D. M., Colin, L., Donahue, T. M., and Moroz, V. I. (University of Arizona Press, Tucson, 1983.)

The early environment and its evolution on Mars: implications for life. McKay, C. P., and Stoker, C. R. *Rev. Geophys.* **27**, 2 (1989).

Accretion of the planets

Dust in galaxies. Stein, W. A. and Soifer, B. T. *Annu. Rev. Astron. & Astrophys.* **21**, 177 (1983).

Formation of the terrestrial planets. Wetherill, G. W. *Annu. Rev. Astron. & Astrophys.* **18**, 77 (1980).

Formation of the Earth. Wetherill, G.W. *Annu. Rev. Earth & Planet. Sci.* **18**, 205 (1990).

Accumulation of planetesimals in the solar nebula. Nakagawa, Y., Hayashi, C., and Nakazawa, K. *Icarus* **54**, 361 (1983).

Origin and abundance of volatiles. Anders, E. and Owen, T. *Science* **198**, 435 (1977).

The origin of galaxies and clusters of galaxies. Peebles, P. J. E. *Science* **224**, 1385 (1984).

Section 9.1.2

Interstellar clouds and the chemical species in them.

Chemical abundances in molecular clouds. Irvine, W. M., Goldsmith, P. F., and Hjalmarson, A. In *Interstellar Processes.* Hollenbach, D. J. and Thronson, H. A. (eds.). (Reidel, Dordrecht, 1987.)

A new component of the interstellar matter: small grains and large aromatic molecules. Puget, J. L. and Léger, A. *Annu. Rev. Astron. Astrophys.* **27**, 161 (1989).

Detection of diffuse interstellar bands in the infra-red. Joblin, C., Maillard, J.P., d'Hendecourt, L., and Léger, A. *Nature* **346**, 729 (1990).

Molecular clouds, star formation and galactic structure. Scoville, N. and Young, J. S., *Scient. Am.* **250**, 30 (April 1984).

The structure and evolution of interstellar grains. Greenberg, J. M. *Scient. Am.* **250**, 96 (June 1984).

Giant molecular-cloud complexes in the galaxy. Blitz, L. *Scient. Am.* **246**, 72 (April 1982).

The spectra of interstellar molecules, Kroto, H. *Int. Rev. Phys. Chem.* **1**, 309 (1981).

Chemical conversions within interstellar clouds

The chemistry of interstellar gas and grains. Irvine, W. M. and Knacke, R. F. in *Origin and evolution of planetary and satellite atmospheres*. Atreya, S. K., Pollack, J. B., and Matthews, M. S. (eds.). (University of Arizona Press, Tucson, 1989.)

Chemistry in the interstellar medium. Special issue of *J. Chem. Soc. Faraday Trans.* **2**, Oct 1989.

Interstellar chemistry Duley, W. W. and Williams, D. A. (Academic Press, London, 1984.)

Interstellar chemistry: exotic molecules in space. Green, S. *Annu. Rev. phys. Chem.* **32**, 103 (1981).

The kinetic chemistry of dense interstellar clouds. Graedel, T. E., Langer, W. D., and Frerking, M. A. *Astrophys. J.* Suppl. **48**, 321 (1982).

The chemistry of interstellar molecules. Winneswisser, G. *Topics curr. Chem.* **99**, 39 (1981).

Interstellar ammonia. Ho, P. T. P. and Townes, C. H. *Annu. Rev. Astron. & Astrophys.* **21**, 239 (1983).

Ion-molecule reactions in the evolution of simple organic molecules in interstellar clouds and planetary atmospheres. Huntress, W. T., Jr. *Chem. Soc. Rev.* **6**, 295 (1977).

Laboratory studies of isotope exchange in ion-neutral reactions: interstellar implications. Smith, D. *Phil. Trans. R. Soc.* A **303**, 535 (1981).

Some H/D exchange reactions involved in the deuteration of interstellar molecules. Smith, D., Adams, N. G., and Alge, E. *Astrophys. J.* **263**, 123 (1981).

Quite complex compounds can be built up in interstellar chemistry, and the processes may at least mimic those responsible for the origin of life on Earth. The tholins are complex organic solids produced from cosmically abundant molecules.

Galactochemistry and the origin of life. Brown, R. D. *Chem. Br.* **15**, 570 (1979).

Molecular synthesis in interstellar clouds: recent laboratory studies of ionic reactions. Smith, D. and Adams, N. G. *Int. Rev. Phys. Chem.* **1**, 271 (1981).

Interstellar matter and chemical evolution. Peimbert, M., Serrano, A., and Torres-Peimbert, S. *Science* **224**, 35 (1984).

Tholins: organic chemistry of interstellar grains and gas. Sagan, C. and Khare, B. N. *Nature, Lond.* **277**, 102 (1979).

Extraterrestrial matter is certainly brought in to the Earth's atmosphere. An extreme view is that life, and even disease, is borne in from space.

Cometary delivery of organic molecules to the early Earth. Chyba, C.F., Thomas, P.J., Brookshaw, L., and Sagan, C. *Science* **249**, 366 (1990).

Martian gases in an antarctic meteorite. Bogard, D. D. and Johnson, P. *Science* **221**, 651 (1983).

Gas-rich meteorites: probe for particle environment and dynamical processes in the inner solar system. Goswami, J. N., Lal, D., and Wilkening, L. L. *Space Sci. Rev.* **37**, 111 (1984).

Interstellar matter in meteorites. Lewis, R. S. and Anders, E. *Scient. Am.* **249**, 54 (Aug 1983).

Evolution by bombardment. Prather, M. J. *Nature, Lond.* **308**, 604 (1984).

Extraterrestrial platinum group nuggets in deep-sea sediments. Brownlee, D. E., Bates, B. A., and Wheelock, M. M. *Nature, Lond.* **309**, 693 (1984).

Comets and the origin of life. (ed.) Ponnamperuma, C. (D. Reidel, Dordrecht, 1981.)

Does epidemic disease come from space? Hoyle, F. and Wickramasinghe, N. C. *New Scient.* **76**, 402 (1977).

Life from space—a history of panspermia. Kamminga, H. *Vistas in Astronomy* **26**, 67 (1982).

Impacts with cometary objects may also have led to periodic mass extinctions, for which there is geological evidence

Comet showers as a cause of mass extinctions. Hut, P., Alvarez, W., Elder, W. P., Hansen, T., Kauffman, E. G., Keller, G., Shoemaker, E. M., and Weissman, P. R. *Nature* **329**, 118 (1987).

The causes of mass extinction. Hallam, A., *Nature, Lond.* **308**, 686 (1984).

Section 9.2.1

Atmospheric compositions: key similarities and differences. Pepin, R. O. in *Origin and evolution of planetary and satellite atmospheres.* Atreya, S. K., Pollack, J. B., and Matthews, M. S. (eds.). (University of Arizona Press, Tucson, 1989.)

Terrestrial noble gases: constraints and implications on atmospheric evolution. Ozima, M. and Igarashi, G. in *Origin and evolution of planetary and satellite atmospheres.* Atreya, S. K., Pollack, J. B., and Matthews, M. S. (eds.). (University of Arizona Press, Tucson, 1989.)

Escape of atmospheres, ancient and modern. Hunten, D.M., *Icarus* **85**, 1 (1990).

Fractionation of noble gases by thermal escape from accreting planetesimals. Donahue, T. M. *Icarus* **66**, 195 (1986).

Mass fractionation of noble gases in diffusion-limited hydrodynamic hydrogen escape. Zahnle, K., Kasting, J. F., and Pollack, J. B. *Icarus* **84**, 502 (1990).

Noble gases in planetary atmospheres: implications for the origin and evolution of atmospheres. Pollack, J. B. and Black, D. C. *Icarus* **51**, 169 (1982).

Noble gases in the terrestrial planets. McElroy, M. B., and Prather, M. J. *Nature, Lond.* **293**, 535 (1981).

Noble gas geochemistry. Ozima, M. and Podosek, F. A. (Cambridge University Press, 1983.)

Helium on Venus: implications for uranium and thorium. Prather, M. J. and McElroy, M. B. *Science* **220**, 410 (1983).

Terrestrial inert gases: isotope tracer studies and clues to primordial components in the mantle. Lupton, J. E. *Annu. Rev. Earth & Planet. Sci.* **11**, 371 (1983).

Oldest reliable $^{40}Ar/^{39}Ar$ ages for terrestrial rocks: Barberton Mountain komatiites. Martinez, M. L., York, D., Hall, C. M., and Hanes, J. A. *Nature, Lond.* **307**, 352 (1984).

Circumstellar material in meteorites: noble gases, carbon, and nitrogen. Anders, E. in *Meteorites and the early solar system.* Kerridge, J. F., and Matthews, M. S. (eds.). (University of Arizona Press, Tucson, 1988.)

Local and exotic components of meteorites, and their origin. Anders, E. *Phil. Trans. Roy. Soc.* **A323**, 287 (1987).

Section 9.2.2

Many more references to Titan's atmosphere and its evolution are given in connection with Section 8.5.1.

Present state and chemical evolution of the atmospheres of Titan, Triton and Pluto. Lunine, J. I., Atreya, S. K., and Pollack, J. B. in *Origin and evolution of planetary and satellite atmospheres.* Atreya, S. K., Pollack, J. B., and Matthews, M. S. (eds.). (University of Arizona Press, Tucson, 1989.)

How primitive are the gases in Titan's atmosphere? Owen, T. *Adv. Space. Res.* **7**, 51 (1987).

Clathrate and ammonia hydrate at high pressure: application to the origin of methane on Titan. Lunine, J. I. and Stevenson, D. J. *Icarus* **70**, 61 (1987).

The composition and origin of Titan's atmosphere. Owen, T. *Planet. Space Sci.* **30**, 833 (1982).

Shock waves produced by impacting bodies have been suggested as a further means of initiating chemical change in Titan's atmosphere

High temperature shock formation of N_2 and organics on primordial Titan. McKay, C. P., Scattergood, T. W., Pollack, J. B., Borucki, W. J., and van Ghysegham, H. T. *Nature* **332**, 520 (1988).

Estimated impact shock production of N_2 and organic compounds on early Titan. Jones, T. D. and Lewis, J. S. *Icarus* **72**, 381 (1987).

Chemistry and evolution of Titan's atmosphere. Strobel, D. F. *Planet. Space Sci.* **30**, 839 (1982).

Section 9.3

Deuterium in the solar system. Geiss, J. and Reeves, H. *Astron. Astrophys.* **93**, 189 (1981).

Deuterium in the outer solar system: evidence for two distinct reservoirs. Owen, T., Lutz, B. L., and de Bergh, C. *Nature* **320**, 244 (1986).

Mechanisms and observations for isotope fractionation in planetary atmospheres. Kaye, J. A. *Rev. Geophys.* **25**, 1609 (1987).

Kinetic isotopic fractionation and the origin of HDO and CH_3D in the solar system. Yung, Y. L., Friedl, R. R., Pinto, J. P., Bayes, K. D., and Wen, J. -S. *Icarus* **74**, 121 (1988).

Escape of hydrogen from Venus. McElroy, M. B., Prather, M. J., and Rodriguez, J. M. *Science* **215**, 1614 (1982).

Venus was wet: A measurement of the ratio of deuterium to hydrogen. Donahue, T. M., Hoffman, J. H., Hodges, R. R., Jr., and Watson, A. J. *Science* **216**, 630 (1982).

Identification of deuterium ions in the ionosphere of Venus. Hartle, R. and Taylor, H. A., Jr. *Geophys. Res. Lett.* **10**, 965 (1983).

Loss of oxygen from Venus. McElroy, M. B., Prather, M. J., and Rodriguez, J. M. *Geophys. Res. Lett.* **9**, 649 (1982).

Water on Mars. Carr, M. H. *Nature* **326**, 30 (1987).

HDO in the Martian atmosphere: implications for the abundance of crustal water. Yung, Y. L., Wen, J. -S., Pinto, J. P., Allen, M., Pierce, K. K., and Paulson, S. *Icarus* **76**, 146 (1988).

Deuterium on Mars: the abundance of HDO and the value of D/H. Owen, T., Maillard, J. P., de Bergh, C., and Lutz, B. *Science* **240**, 1767 (1988).

Photochemistry and evolution of Mars' atmosphere: A Viking perspective. McElroy, M. B., Kong, T. Y., and Yung, Y. L. *J. geophys. Res.* **82**, 4379 (1977).

Monodeuterated methane in the outer solar system. III Its abundance on Titan. de Bergh, C., Lutz, B. L., Owen, T., and Chauville, J. *Astrophys. J.* **329**, 951 (1988).

D to H ratio and the origin and evolution of Titan's atmosphere. Pinto, J. P., Lunine, J. I., Kim, S. -J., and Yung, Y. L. *Nature* **319**, 388 (1986).

Monodeuterated methane in the outer solar system. IV Its detection and abundance on Neptune. de Bergh, C., Lutz, B. L., Owen, T. and Maillard, J. P. *Astrophys. J.* **355**, 661 (1990).

Cosmogonical implications of elemental and isotopic abundances in atmospheres of the giant planets. Gautier, D. and Owen, T. *Nature, Lond.* **304**, 691 (1983).

The D/H ratio in water from Halley. Eberhardt, P., Dolder, U., Schulte, W., Krankowsky, D., Lämmerzahl, P., Hoffman, J. H., Hodges, R. R., Berthelier, J. J., and Illiano, J. M. *Astron. Astrophys.* **187**, 435 (1987).

Section 9.4

Introductory reviews on the evolution of the Earth's atmosphere.

Origin and evolution of the atmosphere. Wayne, R. P. *Chem. Brit.* **24**, 225 (1988).

The photochemistry of the early atmosphere. Levine, J. S. in *The photochemistry of atmospheres.* Levine, J. S. (ed.). (Academic Press, Orlando, 1985.)

The chemical evolution of the atmosphere and oceans. Holland, H. D. (Princeton University Press, Princeton, New Jersey, 1982.)

Evolution of the atmosphere. Walker, J. C. G. (Macmillan, London, 1978.)

The earliest atmosphere of the earth. Walker, J. C. G. *Precambr. Res.* **17**, 147 (1982).

The photochemistry of the paleoatmosphere. Levine, J. S. *J. molec. Evol.* **18**, 161 (1982).

Solar intensities may have varied considerably over geological time scales. The latest evidence suggests that the radiation in the ultraviolet may have been more intense than it is now.

The evolution of solar ultraviolet luminosity. Zahnle, K. J. and Walker, J. C. G. *Rev. Geophys. & Space Phys.* **20**, 280 (1982).

UV radiation from the young Sun and oxygen and ozone levels in the pre-biological palaeoatmosphere. Canuto, V. M., Levine, J. S., Augustsson, T. R., and Imhoff, C. L. *Nature, Lond.* **296**, 816 (1982).

The young Sun and the early Earth, its atmosphere and photochemistry. Canuto, V. M., Levine, J. S., Augustsson, T. R., Imhoff, C. L., and Giampapa, M. S. *Nature, Lond.* **305**, 281 (1983).

Formation of oxygen and ozone, and fixation of nitrogen, in the absence of life.

Photochemistry of CO and H_2O: analysis of laboratory experiments and applications to the prebiotic Earth's atmosphere. Wen, J. -S., Pinto, J. P., and Yung, Y. L. *J. geophys. Res.* **94**, 14957 (1989).

Photochemical reactions of water and CO in Earth's primitive atmosphere. Bar-Nun, A. and Chang, S. *J. geophys. Res.* **88**, 6662 (1983).

Oxygen and ozone in the early Earth's atmosphere. Canuto, V. M., Levine, J. S., Augustsson, T. R., and Imhoff, C. L., *Precambr. Res.* **20**, 109 (1983).

Prebiotic atmospheric oxygen levels. Carver, J. H. *Nature, Lond.* **292**, 136 (1981).

Limits on oxygen concentration in the prebiological atmosphere and the rate of abiotic fixation of nitrogen. Kasting, J. F. and Walker J. C. G. *J. geophys. Res.* **86**, 1147 (1981).

Rates of fixation by lightning of carbon and nitrogen in possible primitive atmospheres. Chameides, W. L. and Walker, J. C. G. *Orig. Life* **11**, 291 (1981).

Fixation of nitrogen in the prebiotic atmosphere. Yung, Y. L. and McElroy, M. B. *Science* **203**, 1002 (1979).

The influence of the biosphere on the development of the atmosphere. The first article discusses the evolution of life on Earth and its relationship to atmospheric oxygen concentrations. This, together with the other papers, also gives a guide to current thinking about the origins of life and the earliest living organisms.

The biosphere. Cloud, P., *Scient. Am.* **249**, 132 (Sept 1983).

The origin and early evolution of life on Earth. Oró, J., Miller, S.L., and Lazcano, A. *Annu. Rev. Earth & Planet. Sci.* **18**, 317 (1990).

The first organisms. Cairns-Smith, A. G. *Scient. Am.* **252** (3), 74 (June 1985).

Primordial organic chemistry. Ponnamperuma, C. *Chem. Br.* **15**, 560 (1979).

The early fossil record. Brasier, M. D. *Chem. Br.* **15**, 588 (1979).

The oldest eucaryotic cells. Vidal, G. *Scient. Am.* **250**, 32 (February 1984).

Designing the first organism. Andrew, S. P. S. *Chem. Br.* **15**, 580 (1979).

Organisms of the first kind. Cairns-Smith, G. *Chem. Br.* **15**, 576 (1979).

The evolution of Earth's atmosphere after life was established. Several of these papers seek a connection between the protection afforded by the ozone shield and the stages of evolutionary development of living organisms. The first paper is the prototype of such studies.

On the origin and rise of oxygen concentration in the Earth's atmosphere. Berkner, L. V. and Marshall, L. C. *J. atmos. Sci.* **22**, 225 (1965).

The evolutionary role of atmospheric ozone. Blake, A. J. and Carver, J. H. *J. atmos. Sci.* **34**, 720 (1977).

Evolution of oxygen and ozone in Earth's atmosphere. Kasting, J. F. and Donahue, T. M. in *Life in the universe* (ed. Billingham, J.) pp. 149–62. (MIT Press, Cambridge, Mass., 1981.)

Possible variation of ozone in the troposphere during the course of geologic time. Chameides, W. L. and Walker, J. C. G. *Am. J. Sci.* **275**, 737 (1975).

Atmospheric constraints on the evolution of metabolism. Walker, J. C. G. *Orig. Life* **10**, 93 (1980).

Atmospheric ozone and the history of life. Ratner, M. I. and Walker, J. C. G. *J. atmos. Sci.* **29**, 803 (1972).

Atmospheres and evolution. Margulis, L. and Lovelock, J. E., in *Life in the universe* (ed. Billingham, J.), pp. 79–100. (MIT Press, Cambridge, Mass, 1981.)

Section 9.5

> *An introduction to palaeoclimatology. Changes in the solar luminosity or in the Earth's orbit may have had an influence on climate.*

The first book is a popular account of evolution and the climate

Ice time: climate, science, and life on Earth. Levenson, T. (Harper and Row, New York, 1989.)

Isotopes and climate. Bowen, R. (Elsevier Applied Science, London, 1990.)

Understanding climate change. Berger, A., Dickinson, R. E., Kidson, J. W. (eds.). *Geophys. Monographs*, Vol. 52 (1989).

Climate variability: past, present, and future. Mitchell, J. M. *Climate Change* **16**, 231 (1990).

The environmental record in glaciers and ice sheets. Oeschger, H. and Langway, C. C. Jr. (eds.). (Wiley, New York, 1989.)

Fossils and climate. Brenchley, P. J. (ed.). (Wiley, Chichester, 1984.)

Climate and history. Wigley, T. M. L., Ingram, M. J., and Farmer, G. (eds.). (Cambridge University Press, Cambridge, 1981.)

Climate variability on the scale of decades to centuries. Stockton, C. W. *Climate Change* **16**, 173 (1990).

Aspects of climate variability in the Pacific and the Western Americas. Peterson, D. H. (ed.). *Geophys. Monographs*, Vol. 55 (1990).

Climates throughout geologic time. Frakes, L. A. (Elsevier, Amsterdam, 1979.)

Climates: present, past and furture. Lamb, H. H. (2 vols) (Methuen, London, 1977.)

> *This article describes the transformation of our climate from that of a greenhouse to that of an icehouse, and provides a summary of the kind of evidence used to infer climates of the past.*

Fifty million years ago. McGowran, B. *Amer. Sci.* **78**, 30 (1990).

The ancient Sun. Pepin, R. O., Eddy, J. A., and Merrill, R. B. (eds.). (Pergamon Press, Oxford, 1981.)

Solar-terrestrial influences on weather and climate. Gregory, J. *Nature, Lond.* **299**, 401 (1982).

The Earth's orbit and the ice ages. Covey, C. *Scient. Am.* **250**, 42 (Feb. 1984).

The variable sun. Foukal, P. V. *Scient. Am.* **262** (2), 26 (Feb. 1990).

What drives glacial cycles. Broecker, W. S. and Denton, G. H. *Sci. Amer.* **262** (1), 42 (Jan. 1990).

Climate evolution on the terrestrial planets. Kasting, J. F., and Toon, O. B. in *Origin and evolution of planetary and satellite atmospheres.* Atreya, S. K., Pollack, J. B., and Matthews, M. S. (eds.). (University of Arizona Press, Tucson, 1989.)

The recent climate record: what it can and cannot tell us. Karl, T. R., Tarpley, J. D., Quayle, R. G., Diaz, H. F., Robinson, D. A., and Bradley, R. S. *Rev. Geophys.* **27**, 405 (1989).

Recent climatic change: a regional approach. Gregory, S. (ed.). (Bellhaven Press, London, 1988.)

How climate evolved on the terrestrial planets. Kasting, J. F., Toon, O. B., and Pollack, J. B. *Sci. Amer.* **252** (2), 46 (Feb. 1988).

Volcanoes can certainly have an effect on climate. The last reference shows how an historical record of a stratospheric dust cloud can be associated with evidence of volcanic eruptions.

Volcanoes and the climate. Toon, O. B. and Pollack, J. B. *Natural History* **86**, 8 (1977).
Atmospheric aerosols and climate. Toon, O. B. and Pollack, J. B. *Am. Scient.* **68**, 268 (1980).
The faint young sun-climate paradox: volcanic influences. Schatten K. H. and Endal. A. S. *Geophys Res. Lett.* **9**, 1309 (1982).
Volcanic CO_2 and solar forcing of Northern and Southern Hemisphere surface air temperatures. Gilliland, R. L. and Schneider, S. H. *Nature, Lond.* **310**, 38 (1984).
Mystery cloud of AD 536. Stothers, R. B. *Nature, Lond.* **307**, 344 (1984).

The use of ice cores to provide data on ancient atmospheric gas concentrations.

Antarctic ice: the frozen time capsule. Peel, D. A. *New Scient.* **98**, 477 (1983).
Carbon isotope data in core V19–30 confirm reduced carbon dioxide concentration in ice age atmosphere. Shackleton, N. J., Hall, M. A., Line, J., and Chang Shuxi *Nature, Lond.* **306**, 319 (1983).
Comparison of CO_2 measurements by two laboratories on air from bubbles in polar ice. Barnola, J. M., Raynaud, D., Neftel, A., and Oeschger, H. *Nature, Lond.* **303**, 410 (1983).
Methane: the record in polar ice cores. Craig. H. and Chou, C. C. *Geophys. Res. Lett.* **9**, 1221 (1982).

Tree ring dating and the extraction of a climatic record.

Climate from tree rings. Hughes, M. K., Kelly, P. M., Pilcher, J. R., and LaMarche, V. C., Jr. (eds.) (Cambridge University Press, Cambridge, 1982.)
Frost rings in trees as records of major volcanic eruptions. LaMarche, V. C., Jr. and Hirschboeck, K. M. *Nature, Lond.* **307**, 121 (1984).
Reconstruction of precipitation history in North American corn belt using tree rings. Blasing, T. J. and Duvick, D. *Nature, Lond.* **307**, 143 (1984).
July-August temperatures at Edinburgh between 1721 and 1975 from tree-ring density and width data. Hughes, M. K., Schweingruber, F. H., Cartwright, D., and Kelly, P. M. *Nature, Lond.* **308**, 341 (1984).
Trace elements in tree rings: evidence of recent and historical air pollution. Baes, C. F., III, and McLaughlin, S. B. *Science* **224**, 494 (1984).

This reference is given to summarize the use of radiocarbon dating, since the method is often employed in cross-calibrating tree-ring and ice-core records.

Radiocarbon dating. Seuss, H. E. *Endeavour* (new series) **4**, 113 (1980).

Climates in the past, and their possible regulation by passive or active feedback mechanisms.

Feedbacks between weathering and atmospheric CO_2 over the last 100 million years. Volks, T. *Am. J. Sci.* **287**, 763 (1987).
The carbonate–silicate geochemical cycle and its effect on atmospheric carbon dioxide

over the past 100 million years. Berner, R. A., Lasaga, A. C. and Garrels, R. M. *Am. J. Sci.* **283**, 641 (1983).

Effects of increased CO_2 concentrations on surface temperature of the early Earth. Kuhn, W. R. and Kasting, J. F. *Nature, Lond.* **301**, 53 (1983).

Cloud Feedback: a stabilizing effect for the early Earth. Rossow, W. G., Henderson-Sellers, A., and Weinreich, S. K. *Science* **217**, 1245 (1982).

A negative feedback mechanism for the long-term stabilization of Earth's surface temperatures. Walker, J. C. G., Hays, P. B., and Kasting, J. F. *J. geophys Res.* **86**, 9776 (1981).

The regulation of carbon dioxide and climate: Gaia or geochemistry? Lovelock, J. E. and Watson, A. J. *Planet. Space Sci.* **30**, 785 (1982).

Polar glaciation and the genesis of the ice ages. Hunt, B. G., *Nature, Lond.* **308**, 48 (1984).

Snow cover and atmospheric variability. Walsh, J. E. *Am. Scient.* **72**, 50 (1984).

Carbon dioxide, other greenhouse gases, and the supposed threat to world climate in the future: introductory reviews.
The following books provide popular accounts of the greenhouse effect

Winds of change: living in the global greenhouse. Gribbin, J. and Kelly, M. (Hodder and Stoughton, London, 1989.)

Hothouse Earth: the greenhouse effect and Gaia. Gribbin, J. (Bantam Press, London, 1990.)

The greenhouse effect. Boyle, S. and Ardill, J. (Hodder and Stoughton, Sevenoaks, 1989.)

Turning up the heat; our perilous future in the global greenhouse. Pearce, F. (The Bodley Head, London, 1989.)

The great climate debate. White, R.M. *Scient. Am.* **263**, 18 (July 1990).

The greenhouse effect and climate change. Mitchell, J. F. B. *Rev. Geophys.* **27**, 115 (1989).

The greenhouse effect, climatic change, and ecosystems. Bolin, B. [SCOPE **29**] (Wiley, Chichester, 1986.)

Climate modelling. Schneider, S. H. *Sci. Amer.* **256** (5), 72 (May 1987).

The changing climate. Schneider, S. H. *Sci. Amer.* **261** (3), 38 (Sept. 1989).

Global climatic change. Houghton, R. A. and Woodwell, G. M. *Sci. Amer.* **260** (4), 18 (April 1989).

Sun and dust versus greenhouse gases: an assessment of their relative roles in global climate change. Hansen J.E. and Lacis, A.A. *Nature* **346**, 713 (1990).

Carbon dioxide increase in the atmosphere and oceans and possible effects on climate. Chen, C. -T. and Drake, E. T. *Annu. Rev. Earth Planet. Sci.* **14**, 201 (1986).

Observational constraints on the global atmospheric CO_2 budget. Tans, P. P., Fung, I. Y., and Takahasi, T. *Science* **247**, 1431 (1990).

Carbon dioxide: friend or foe. An inquiry into the climatic and agricultural consequences of the rapidly rising CO_2 content of the Earth's atmosphere. Idso, S. B. (IBR Press, Tempe, Arizona, 1982.)

Changing climate. Report of the Carbon Dioxide Assessment Committee of the National Research Council Board on Atmospheric Sciences and Climate. (National Academy Press, Washington, DC, 1983.)

Our threatened climate. Bach, W. (D. Reidel Co., Dordrecht, 1983.)

Carbon dioxide and world climate. Revelle, R. *Scient. Am.* **247**, 35 (Aug. 1982).

Carbon dioxide and climate change. Manabe, S. *Adv. Geophys.* **25**, 39 (1983).

The climatic effect of CO_2: a different view. Ellsaesser, H. W. *Atmos. Environ.* **18**, 431 (1984).

> *This book is a major work discussing the atmosphere, sources, and measurements of CO_2 and other greenhouse gases, and climatic effects.*

Carbon dioxide review: 1982. Clark, W. C. (ed.) (Oxford University Press, 1982.)

> *Although deforestation may reduce the uptake of CO_2 from the atmosphere, increased emission rates seem largely responsible for elevated atmospheric concentrations.*

Global deforestation: contribution to atmospheric carbon dioxide. Woodwell, G. M., Hobbie, J. E., Houghton, R. A., Melillo, J. M., Moore, B., Peterson, B. J., and Shaver, G. R. *Science* **222**, 1081 (1983).

Response of the global climate to changes in atmospheric chemical composition due to fossil fuel burning. Hameed, S., Cess, R. D., and Hogan, J. S. *J. geophys. Res.* **85**, 7537 (1980).

The effects of pollutants on global climate. Mitchell, J. F. B. *The Meteorological Magazine* **113**, 1 (1984).

Man's emission of carbon dioxide into the atmosphere. Hirschler, M. M. *Atmos. Environ.* **15**, 719 (1981).

> *These publications indicate the considerations and the types of model employed in predicting the climatic effect of increased atmospheric carbon dioxide concentrations.*

A climate modelling primer. Henderson-Sellars, A. and McGoffin, K. (John Wiley, Chichester, 1987.)

Managing atmospheric CO_2. Harvey, L. D. D. *Climatic Change* **15**, 343 (1989).

Feedback mechanisms in the climate system affecting future levels of carbon dioxide. Kellogg, W. W. *J. geophys. Res.* **88**, 1263 (1983).

A new model for the role of the oceans in determining atmospheric P_{CO_2}. Sarmiento, J. L. and Toggweiler, J. R. *Nature, Lond.* **308**, 621 (1984).

CO_2 induced climate change and spectral variations in the outgoing radiation. Charlock, T. P. *Tellus* **36B**, 139 (1984).

Changes in atmospheric CO_2: influence of the marine biota at high latitude. Knox, F. and McElroy, M. B. *J. geophys. Res.* **89**, 4629 (1984).

Marine biological controls of atmospheric CO_2 and climate. McElroy, M. B. *Nature, Lond.* **302**, 328 (1983).

Transient climate response to increasing atmospheric carbon dioxide. Bryan, K., Komro, F. G., Manabe, S., and Spelman, M. J. *Science* **215**, 56 (1982).

Radiative effects of changing atmospheric water vapour. Doherty, G. M. and Newell, R. E. *Tellus* **36B**, 149 (1984).

Coupled effects of atmospheric N_2O and O_3 on the Earth's climate. Wang, W. -C. and Sze, N. D. *Nature, Lond.* **286**, 589 (1980).

> *One climatic effect that could exert a positive feedback influence is the melting of polar ice. Antarctic ice loss can lead also to an increase in ocean depths.*

Variability of Antarctic sea ice and changes in carbon dioxide. Zwally, H. J., Parkinson, C. L., and Comiso, J. C. *Science* **220**, 1005 (1983).

Summer ice and carbon dioxide. Kukla, G. and Gavin, J. *Science* **214**, 497 (1981).

> *Carbon dioxide is not the only greenhouse gas known to be increasing in concentration. Water vapour, CH_4, N_2O, and other trace gases can trap radiation, so that chemical processes in the atmosphere have a direct influence on climate. The Bibliographies for Sections 1.5 and 5.4 provide further references to papers describing the changing atmospheric burden of trace gases.*

Photochemistry, composition, and climate. Kuhn, W. R. in *The photochemistry of atmospheres*. Levine, J. S., (ed.). (Academic Press, Orlando, 1985.)

The role of atmospheric chemistry in climate change. Wuebbles, D. J., Grant, K. E., Connell, P. S., and Penner, J. E. *APCA J.* **39**, 22 (1989).

Atmospheric trace gases and global climate: a seasonal model study. Wang, W. -C., Molnar, G., Ko, M. K. W., Goldenberg, S., and Sze, N. D. *Tellus* **42B**, 149 (1990).

A comparison of the contribution of various gases to the greenhouse effect. Rohde, H. *Science* **248**, 1217 (1990).

Relative contributions of greenhouse gas emissions to global warming. Lashof, D. A. and Ahuja, D. R. *Nature* **344**, 529 (1990).

Climate–chemical interactions and effects of changing atmospheric trace gases. Ramanathan, V., Callis, L., Cess, R., Hansen, J., Isaksen, I., Huhn, W., Lacis, A., Luther, F., Mahlman, J., Reck, R., and Schlesinger, M. *Rev. Geophys.* **25**, 1441 (1987).

Trace gases and other potential perturbations to global climate. Wang, W.-C., Wuebbles, D. J., Washington, W. M., Isaacs, R. G., and Molnar, G. *Rev. Geophys.* **24**, 110 (1986).

The changing atmosphere. Graedel, T. E., and Crutzen, P. J. *Sci. Amer.* **261** (3), 28 (Sept. 1989).

The changing atmosphere. Rowland, F. S., and Isaksen, I. S. A. (eds.). (Wiley, Chichester, 1988.)

> *On the detection of temperature changes that may have occurred already as a result of enhanced greenhouse heating.*

Precise monitoring of global temperature trends from satellites. Spencer, R. W. and Christy. J. R. *Science* **247**, 1558 (1990).

Global warming trends. Jones, P.D. and Wigley, T.M.L. *Scient. Am.* **263**, 66 (Aug. 1990).

> *The articles to which references are given here explore not only the influences of man on climate, but also some of the social and economic consequences of climate change.*

Uncertainties of estimates of climatic change: a review. Dickinson, R. E. *Climatic Change* **15**, 5 (1989).

Scientific perspectives on the greenhouse problem. Nierenberg, W. A., Jastrow, R., and Seitz, F. (Marshall Institute, Washington, DC, 1989.)

Assessing the social implications of climate fluctuations: a guide to climate impact studies. Riesbaume, W. E. (UNEP, Nairobi, 1989.)

Greenhouse gas emissions: environmental consequences and policy responses. Oppenheimer, M. (ed.). *Climatic Change*, Special Issue (Vol 15, Nos 1–2, 1989).

Carbon dioxide and other greenhouse gases: climatic and associated impacts. Fantechi, R. and Ghazi, A. (Kluwer Academic Publishers, Dordrecht, 1989.)

Ozone depletion, greenhouse gases, and climate change. (National Academy Press, Washington, 1989.)

Halocarbon ozone depletion and global warming potentials. Chapter 4, Volume 1 of Scientific assessment of stratospheric ozone: 1989. World Meteorological Organization, Global ozone research and monitoring project: report no. 20. (WMO, Geneva, 1990.)

Model calculations of the relative effects of CFCs and their replacements on global warming. Fisher, D. A., Hales, C. H., Wang, W. -C., Ko, M. K. W., and Sze, N. D. *Nature* **344**, 513 (1990).

Preliminary measurements of the greenhouse warming implications of halocarbon substitutes for CFC-11 and CFC-12. Montague, D. C. and Perrine, R. L. *Atmos. Env.* **24A**, 1331 (1990).

Influences of mankind on climate. Kellogg, W. W. *Annu. Rev. Earth & Planet. Sci.* **7**, 63 (1979).

Global climate change and US agriculture. Adams, R. M., Rosenzweig, C., Peart, R. M., Ritchie, J. T., McCarl, B. A., Glyer, J. D., Curry, R. B., Jones, J. W., Boote, K. J., and Allen, L. H., Jr. *Nature* **345**, 219 (1990).

Climate change and society: consequences of increasing atmospheric carbon dioxide. Kellogg, W. W. and Schware, R. (Westview Press, Boulder, Colorado, 1981.)

Society, science and climate change. Kellogg, W. W. and Schware, R. *Foreign Affairs* **60**, 1076 (1982).

The changing climate. Responses of the natural flora and fauna. Ford, M. J. (George Allen & Unwin, London, 1982.)

At the time of writing, the most authoritative and strongly supported statement on climate change that has ever been made by the international scientific community is the final report of Working Group 1 of the Intergovernmental Panel on Climate Change (IPCC). The IPCC is sponsored jointly by the World Meteorological Organization and the United Nations Environment Programme, and several hundred scientists from 25 countries participated in the preparation and review of the scientific data. Issues confronted with full rigour include global warming, greenhouse gases, the greenhouse effect, sea level changes, forcing of climate, and the history of the Earth's changing climate.

Climate change: the IPCC scientific assessment. Houghton, J.T., Jenkins, G.J. and Ephraums, J.J. (eds.) (Cambridge University Press, Cambridge, 1990.)

Section 9.7

Carbon dioxide may not be able to act as a thermostatic regulator for more than a limited period in the future.

Life span of the biosphere. Lovelock, J. E. and Whitfield, M. *Nature, Lond.* **296**, 561 (1982).

Index